AF789934

TEACHER'S EDITION

ACTIVE™ PHYSICAL SCIENCE

An Inquiry Approach to Physical Science

TEACHER'S EDITION

ACTIVE PHYSICAL SCIENCE™

An Inquiry Approach to Physical Science

Dr. Arthur Eisenkraft

Volume 2 Teacher's Edition

IT's ABOUT TIME®
HERFF JONES EDUCATION DIVISION

HERFF JONES EDUCATION DIVISION
84 Business Park Drive, Armonk, NY 10504 Phone (914) 273-2233
Fax (914) 273-2227 Toll Free (888) 698-TIME (8463) www.its-about-time.com

It's About Time, Founder
Laurie Kreindler

Creative/Art Director
John Nordland

Design/Production
Burmar Technical Corporation
Murray Fuchs
Robin Hoffmann
Jennifer Von Holstein

Technical Art
Burmar Technical Corporation
Jennifer Von Holstein
Dennis Falcon

Illustrations
Tomas Bunk

Project Managers
Barbara Zahm
John Nordland

Project Editor
Ruta Demery

Physics Reviewers
John Hubrisz
Dwight Neuenschwander
John Roeder
Patty Rourke

Photo Research
Jennifer Von Holstein

Safety Reviewer
Ed Robeck

Active Physical Science™
©Copyright 2005: It's About Time, Herff Jones Education Division

All student activities in this textbook have been designed to be as safe as possible, and have been reviewed by professionals specifically for that purpose. As well, appropriate warnings concerning potential safety hazards are included where applicable to particular activities. However, responsibility for safety remains with the student, the classroom teacher, the school principals, and the school board.

It's About Time® and *Active Physics*® are registered trademarks of It's About Time. Registered names and trademarks, etc., used in this publication, even without specific indication thereof, are not to be considered unprotected by law.

Active Physics® is copyrighted by It's About Time, Herff Jones Education Division. *Active Physics*® was developed originally by the American Association of Physics Teachers with financial support from the National Science Foundation (grant no. MDR-9150111). Herff Jones has acquired all rights to *Active Physics*® and is solely responsible for all text and curriculum material related to *Active Physics*®.

All rights reserved. No part of this publication may be reproduced, stored in a retrieval system, or transmitted, in any form or by any means, electronic, mechanical, photocopying, recording, or otherwise, without the prior written permission of the copyright owner.

Care has been taken to trace the ownership of copyright material contained in this publication. The publisher will gladly receive any information that will rectify any reference or credit line in subsequent editions.

Printed and bound in the United States of America

ISBN #1-58591-322-7 3 Volume Set ISBN #1-58591-320-0

1 2 3 4 5 VH 08 07 06 05 04

This project was supported, in part, by the
National Science Foundation
Opinions expressed are those of the authors and not necessarily those of the National Science Foundation

© It's About Time

Project Director, Active Physics

Arthur Eisenkraft has taught high school physics for over 28 years and is currently the Distinguished Professor of Science Education and a Senior Research Fellow at the University of Massachusetts, Boston. Dr. Eisenkraft is the author of numerous science and educational publications. He holds U.S. Patent #4447141 for a Laser Vision Testing System (which tests visual acuity for spatial frequency).

Dr. Eisenkraft has been recognized with numerous awards including: Presidential Award for Excellence in Science Teaching, 1986 from President Reagan; American Association of Physics Teachers (AAPT) Excellence in Pre-College Teaching Award, 1999; AAPT Distinguished Service Citation for "excellent contributions to the teaching of physics", 1989; Science Teacher of the Year, Disney American Teacher Awards in their American Teacher Awards program, 1991; Honorary Doctor of Science degree from Rensselaer Polytechnic Institute, 1993; Tandy Technology Scholar Award 2000.

In 1999 Dr. Eisenkraft was elected to a 3-year cycle as the President-Elect, President and Retiring President of the National Science Teachers Association (NSTA), the largest science teacher organization in the world. In 2003, he was elected a fellow of the American Association for the Advancement of Science (AAAS).

Dr. Eisenkraft has been involved with a number of projects and chaired many competition programs, including: the Toshiba/NSTA ExploraVisions Awards (1991 to the present); the Toyota TAPESTRY Grants (1990 to the present); the Duracell/NSTA Scholarship Competitions (1984 to 2000). He was a columnist and on the Advisory Board of *Quantum* (a science and math student magazine that was published by NSTA as a joint venture between the United States and Russia; 1989 to 2001). In 1993, he served as Executive Director for the XXIV International Physics Olympiad after being Academic Director for the United States Team for six years. He has served on a number of committees of the National Academy of Sciences including the content committee that helped write the National Science Education Standards.

Dr. Eisenkraft has appeared on *The Today Show*, *National Public Radio*, *Public Television*, *The Disney Channel* and numerous radio shows. He serves as an advisor to the ESPN Sports Figures Video Productions.

He is a frequent presenter and keynote speaker at National Conventions. He has published over 100 articles and presented over 200 papers and workshops. He has been featured in articles in *The New York Times*, *Education Week*, *Physics Today*, *Scientific American*, *The American Journal of Physics* and *The Physics Teacher*.

© It's About Time

TABLE OF CONTENTS

© It's About Time

© It's About Time

Acknowledgements

Active Physical Science

was developed by a team of leading physicists, university educators and classroom teachers with financial support from the National Science Foundation.

NSF Program Officer

Gerhard Salinger
Instructional Materials Development (IMD)

Principal Investigators

Bernard V. Khoury
American Association of Physics Teachers

Dwight Edward Neuenschwander
American Institute of Physics

Project Director

Arthur Eisenkraft
University of Massachusetts

Primary and Contributing Authors

Richard Berg
University of Maryland
College Park, MD

Howard Brody
University of Pennsylvania
Philadelphia, PA

Chris Chiaverina
New Trier Township High School
Crystal Lake, IL

Ron DeFronzo
Eastbay Ed. Collaborative
Attleboro, MA

Ruta Demery
Blue Ink Editing
Stayner, ON

Carl Duzen
Lower Merion High School
Havertown, PA

Jon L. Harkness
Active Physics Regional Coordinator
Wausau, WI

Ruth Howes
Ball State University
Muncie, IN

Douglas A. Johnson
Madison West High School
Madison, WI

Ernest Kuehl
Lawrence High School
Cedarhurst, NY

Robert L. Lehrman
Bayside, NY

Salvatore Levy
Roslyn High School
Roslyn, NY

Tom Liao
SUNY Stony Brook
Stony Brook, NY

Charles Payne
Ball State University
Muncie, IN

Mary Quinlan
Radnor High School
Radnor, PA

Harry Rheam
Eastern Senior High School
Atco, NJ

Bob Ritter
University of Alberta
Edmonton, AB, CA

John Roeder
The Calhoun School
New York, NY

John J. Rusch
University of Wisconsin, Superior
Superior, WI

Patty Rourke
Potomac School
McLean, VA

Ceanne Tzimopoulos
Omega Publishing
Medford, MA

Larry Weathers
The Bromfield School
Harvard, MA

David Wright
Tidewater Comm. College
Virginia Beach, VA

Consultants

Peter Brancazio
Brooklyn College of CUNY
Brooklyn, NY

Robert Capen
Canyon del Oro High School
Tucson, AZ

Carole Escobar

Earl Graf
SUNY Stony Brook
Stony Brook, NY

Jack Hehn
American Association of Physics Teachers
College Park, MD

Donald F. Kirwan
Louisiana State University
Baton Rouge, LA

Gayle Kirwan
Louisiana State University
Baton Rouge, LA

James La Porte
Virginia Tech
Blacksburg, VA

Charles Misner
University of Maryland
College Park, MD

Robert F. Neff
Suffern, NY

Ingrid Novodvorsky
Mountain View High School
Tucson, AZ

John Robson
University of Arizona
Tucson, AZ

Mark Sanders
Virginia Tech
Blacksburg, VA

Brian Schwartz
Brooklyn College of CUNY
New York, NY

Bruce Seiger
Wellesley High School
Newburyport, MA

Clifford Swartz
SUNY Stony Brook
Setauket, NY

© It's About Time

Barbara Tinker
The Concord Consortium
Concord, MA

Robert E. Tinker
The Concord Consortium
Concord, MA

Joyce Weiskopf
Herndon, VA

Donna Willis
American Association of Physics Teachers
College Park, MD

Safety Reviewer

Gregory Puskar
University of West Virginia
Morgantown, WV

Equity Reviewer

Leo Edwards
Fayetteville State University
Fayetteville, NC

Physics at Work

Alex Straus
writer
New York, NY

Mekea Hurwitz
photographer

Physics InfoMall

Brian Adrian
Bethany College
Lindsborg, KS

First Printing Reviewer

John L. Hubisz
North Carolina State University
Raleigh, NC

Unit Reviewers

Robert Adams
Polytech High School
Woodside, DE

George A. Amann
F.D. Roosevelt High School
Rhinebeck, NY

Patrick Callahan
Catasaugua High School
Center Valley, PA

Beverly Cannon
Science and Engineering Magnet High School
Dallas, TX

Barbara Chauvin

Elizabeth Chesick
The Baldwin School
Haverford, PA

Chris Chiaverina
New Trier Township High School
Crystal Lake, IL

Andria Erzberger
Palo Alto Senior High School
Los Altos Hills, CA

Elizabeth Farrell Ramseyer
Niles West High School
Skokie, IL

Mary Gromko
President of Council of State Science Supervisors
Denver, CO

Thomas Guetzloff

Jon L. Harkness
Active Physics Regional Coordinator
Wausau, WI

Dawn Harman
Moon Valley High School
Phoenix, AZ

James Hill
Piner High School
Sonoma, CA

Bob Kearney

Claudia Khourey-Bowers
McKinley Senior High School

Steve Kliewer
Bullard High School
Fresno, CA

Ernest Kuehl
Roslyn High School
Cedarhurst, NY

Jane Nelson
University High School
Orlando, FL

Mary Quinlan
Radnor High School
Radnor, PA

John Roeder
The Calhoun School
New York, NY

Patty Rourke
Potomac School
McLean, VA

Gerhard Salinger
Fairfax, VA

Irene Slater
La Pietra School for Girls

Pilot Test Teachers

John Agosta

Donald Campbell
Portage Central High School
Portage, MI

John Carlson
Norwalk Community Technical College
Norwalk, CT

Veanna Crawford
Alamo Heights High School
New Braunfels, TX

Janie Edmonds
West Milford High School
Randolph, NJ

© It's About Time

Eddie Edwards
Amarillo Area Center for Advanced Learning
Amarillo, TX

Arthur Eisenkraft
Fox Lane High School
Bedford, NY

Tom Ford

Bill Franklin

Roger Goerke
St. Paul, MN

Tom Gordon
Greenwich High School
Greenwich, CT

Ariel Hepp

John Herrman
College of Steubenville
Steubenville, OH

Linda Hodges

Ernest Kuehl
Lawrence High School
Cedarhurst, NY

Fran Leary
Troy High School
Schenectady, NY

Harold Lefcourt

Cherie Lehman
West Lafayette High School
West Lafayette, IN

Kathy Malone
Shady Side Academy
Pittsburgh, PA

Bill Metzler
Westlake High School
Thornwood, NY

Elizabeth Farrell Ramseyer
Niles West High School
Skokie, IL

Daniel Repogle
Central Noble High School
Albion, IN

Evelyn Restivo
Maypearl High School
Maypearl, TX

Doug Rich
Fox Lane High School
Bedford, NY

John Roeder
The Calhoun School
New York, NY

Tom Senior
New Trier Township High School
Highland Park, IL

John Thayer
District of Columbia Public Schools
Silver Spring, MD

Carol-Ann Tripp
Providence Country Day
East Providence, RI

Yvette Van Hise
High Tech High School
Freehold, NJ

Jan Waarvick

Sandra Walton
Dubuque Senior High School
Dubuque, IA

Larry Wood
Fox Lane High School
Bedford, NY

Field Test Coordinator

Marilyn Decker
Northeastern University
Acton, MA

Field Test Workshop Staff

John Carlson

Marilyn Decker

Arthur Eisenkraft

Douglas Johnson

John Koser

Ernest Kuehl

Mary Quinlan

Elizabeth Farrell Ramseyer

John Roeder

Field Test Evaluators

Susan Baker-Cohen

Susan Cloutier

George Hein

Judith Kelley

all from Lesley College, Cambridge, MA

Field Test Teachers and Schools

Rob Adams
Polytech High School
Woodside, DE

Benjamin Allen
Falls Church High School
Falls Church, VA

Robert Applebaum
New Trier High School
Winnetka, IL

Joe Arnett
Plano Sr. High School
Plano, TX

Bix Baker
GFW High School
Winthrop, MN

Debra Beightol
Fremont High School
Fremont, NE

Patrick Callahan
Catasaugua High School
Catasaugua, PA

George Coker
Bowling Green High School
Bowling Green, KY

Janice Costabile
South Brunswick High School
Monmouth Junction, NJ

Stanley Crum
Homestead High School
Fort Wayne, IN

Russel Davison
Brandon High School
Brandon, FL

Christine K. Deyo
Rochester Adams High School
Rochester Hills, MI

Jim Doller
Fox Lane High School
Bedford, NY

Jessica Downing
Esparto High School
Esparto, CA

Douglas Fackelman
Brighton High School
Brighton, CO

© It's About Time

Rick Forrest
Rochester High School
Rochester Hills, MI

Mark Freeman
Blacksburg High School
Blacksburg, VA

Jonathan Gillis
Enloe High School
Raleigh, NC

Karen Gruner
Holton Arms School
Bethesda, MD

Larry Harrison
DuPont Manual High School
Louisville, KY

Alan Haught
Weaver High School
Hartford, CT

Steven Iona
Horizon High School
Thornton, CO

Phil Jowell
Oak Ridge High School
Conroe, TX

Deborah Knight
Windsor Forest High School
Savannah, GA

Thomas Kobilarcik
Marist High School
Chicago, IL

Sheila Kolb
Plano Senior High School
Plano, TX

Todd Lindsay
Park Hill High School
Kansas City, MO

Malinda Mann
South Putnam High School
Greencastle, IN

Steve Martin
Maricopa High School
Maricopa, AZ

Nancy McGrory
North Quincy High School
N. Quincy, MA

David Morton
Mountain Valley High School
Rumford, ME

© It's About Time

Charles Muller
Highland Park High School
Highland Park, NJ

Fred Muller
Mercy High School
Burlingame, CA

Vivian O'Brien
Plymouth Regional High School
Plymouth, NH

Robin Parkinson
Northridge High School
Layton, UT

Donald Perry
Newport High School
Bellevue, WA

Francis Poodry
Lincoln High School
Philadelphia, PA

John Potts
Custer County District
High School
Miles City, MT

Doug Rich
Fox Lane High School
Bedford, NY

John Roeder
The Calhoun School
New York, NY

Consuelo Rogers
Maryknoll Schools
Honolulu, HI

Lee Rossmaessler
Mott Middle College High School
Flint, MI

John Rowe
Hughes Alternative Center
Cincinnati, OH

Rebecca Bonner Sanders
South Brunswick High School
Monmouth Junction, NJ

David Schilpp
Narbonne High School
Harbor City, CA

Eric Shackelford
Notre Dame High School
Sherman Oaks, CA

Robert Sorensen
Springville-Griffith Institute
and Central School
Springville, NY

Teresa Stalions
Crittenden County High School
Marion, KY

Roberta Tanner
Loveland High School
Loveland, CO

Anthony Umelo
Anacostia Sr. High School
Washington, D.C.

Judy Vondruska
Mitchell High School
Mitchell, SD

Deborah Waldron
Yorktown High School
Arlington, VA

Ken Wester
The Mississippi School for Mathematics
and Science
Columbus, MS

Susan Willis
Conroe High School
Conroe, TX

Meeting *Active Physical Science* for the First Time

Welcome! A Five-Minute Introduction

Active Physical Science is a different species of physical science course. In *Active Physical Science* students are introduced to physics and chemistry concepts on a need-to-know basis as they explore issues that deal with sports, transportation, the home, roller coasters, atomic structure, special effects, educational games, and chemistry shows. Each chapter is independent of any other chapter. You can begin the physical science course with any chapter.

As an example, let's begin with **Chapter 3, The Track and Field Championship**. On Day One, students are introduced to the **Chapter Challenge**. Your school has been invited to send a team to the Penn Relays, the country's largest track and field event. Some schools hire professionals to help improve the performances of their athletes. However, your school cannot afford to hire any professionals. Can the physical science class help prepare the track and field team for the Penn Relays?

© It's About Time

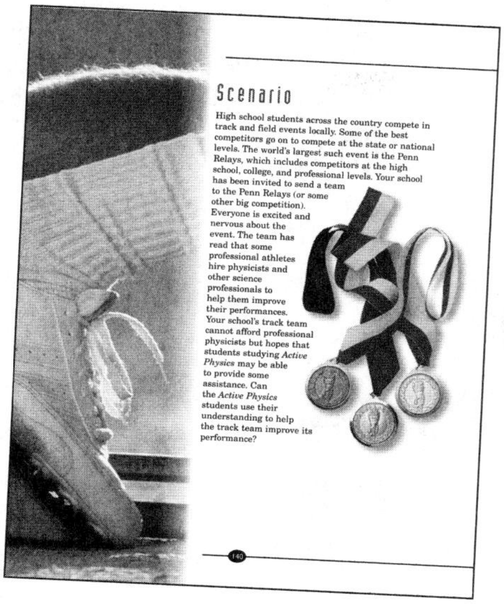

Scenario

High school students across the country compete in track and field events locally. Some of the best competitors go on to compete at the state or national levels. The world's largest such event is the Penn Relays, which includes competitors at the high school, college, and professional levels. Your school has been invited to send a team to the Penn Relays (or some other big competition). Everyone is excited and nervous about the event. The team has read that some professional athletes hire physicists and other science professionals to help them improve their performances. Your school's track team cannot afford professional physicists but hopes that students studying *Active Physics* may be able to provide some assistance. Can the *Active Physics* students use their understanding to help the track team improve its performance?

140

The **Chapter Challenge** has three parts:

1. Help the students on the track and field team compare themselves with their competitors.
2. Provide a description of the physics principles as they relate to track events.
3. Provide specific techniques to improve the performance of the athletes.

How can students get started? How can students complete such a challenge without the requisite physical science knowledge? Before the chapter activities begin, a discussion takes place about the criteria for success. The class discusses what is expected in an excellent response to the challenge. How will they be graded? For instance, should the track and field manual contain physics principles, charts, past records, relevant equations, definitions, and specific techniques? Once the students decide what should be included, they will then have to decide on how each element will be assessed. Should the physics principles be given more weight than past records? Students will have a sense of what will be required from them even before they begin. This grading system will be revisited before work on the **Chapter Challenge** is finalized.

Day Two begins the first of ten activities. Each successive day begins with another activity. *Active Physical Science* is an activity-based curriculum. **Let's look at Activity 5: Acceleration.**

Each activity begins with a **What Do You Think?** feature. This particular activity begins by mentioning, "Accelerating out of the starting blocks is important if a runner is going to win a race." The students are then given a couple of questions to think about. One asks the students what determines who wins a race if all runners have the same top speed?

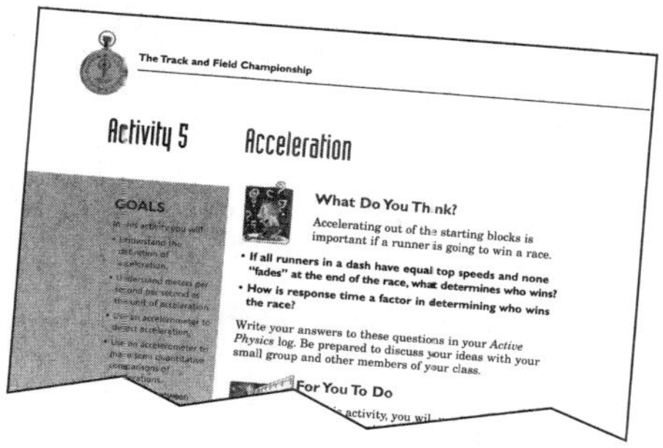

The Track and Field Championship

Activity 5 Acceleration

What Do You Think?

Accelerating out of the starting blocks is important if a runner is going to win a race.

- If all runners in a dash have equal top speeds and none "fades" at the end of the race, what determines who wins?
- How is response time a factor in determining who wins the race?

Write your answers to these questions in your *Active Physics* log. Be prepared to discuss your ideas with your small group and other members of your class.

GOALS

For You To Do

These **What Do You Think?** questions are intended to find out what students already know about acceleration—to get into the "acceleration" part of their brains. Formally, we say that these questions are to elicit the students' prior understanding and is part of the constructivist approach. Students write a response for one minute and discuss for another two minutes. But we don't reach closure. The questions open the conversation.

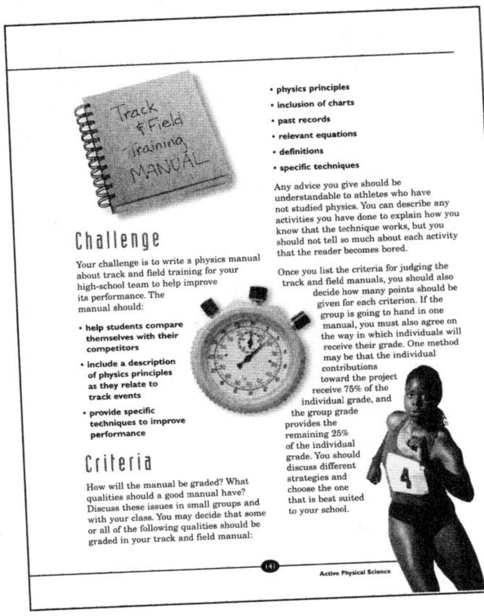

Challenge

Your challenge is to write a physics manual about track and field training for your high-school team to help improve its performance. The manual should:

- **help students compare themselves with their competitors**
- **include a description of physics principles as they relate to track events**
- **provide specific techniques to improve performance**

Criteria

How will the manual be graded? What qualities should a good manual have? Discuss these issues in small groups and with your class. You may decide that some or all of the following qualities should be graded in your track and field manual:

- **physics principles**
- **inclusion of charts**
- **past records**
- **relevant equations**
- **definitions**
- **specific techniques**

Any advice you give should be understandable to athletes who have not studied physics. You can describe any activities you have done to explain how you know that the technique works, but you should not tell so much about each activity that the reader becomes bored.

Once you list the criteria for judging the track and field manuals, you should also decide how many points should be given for each criterion. If the group is going to hand in one manual, you must also agree on the way in which individuals will receive their grade. One method may be that the individual contributions toward the project receive 75% of the individual grade, and the group grade provides the remaining 25% of the individual grade. You should discuss different strategies and choose the one that is best suited to your school.

141 Active Physical Science

© It's About Time

Students then begin the **For You To Do** (or **Investigate**) part of the activity.

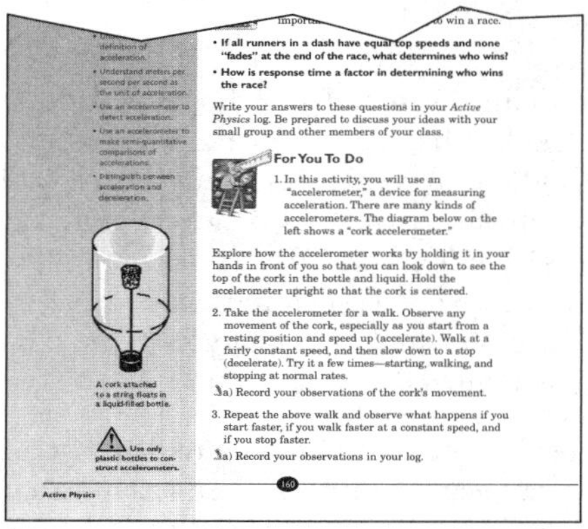

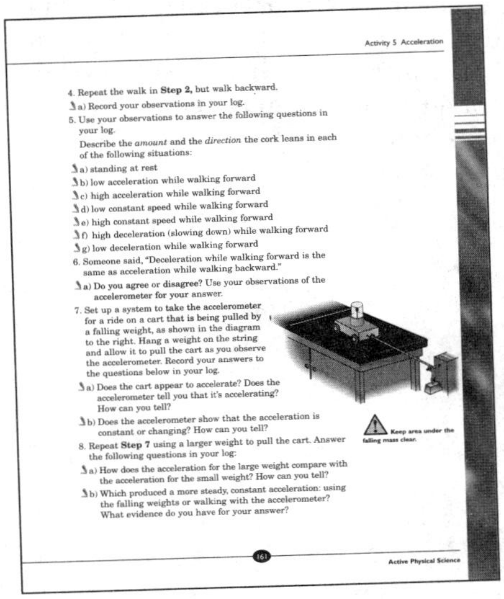

In this activity students investigate acceleration. First they explore how an accelerometer that they have constructed works. They stand still, walk forwards, walk backwards, and change speed to determine how the accelerometer will "read" their movements. Then they use the accelerometer to record the acceleration of a cart attached to a falling weight.

The **Physics Talk** (or **Chem Talk**) focuses the students' attention on the principles of physical science introduced in the activity. In this activity students are presented with an equation that relates acceleration to speed and time. It also provides a sample problem for the students.

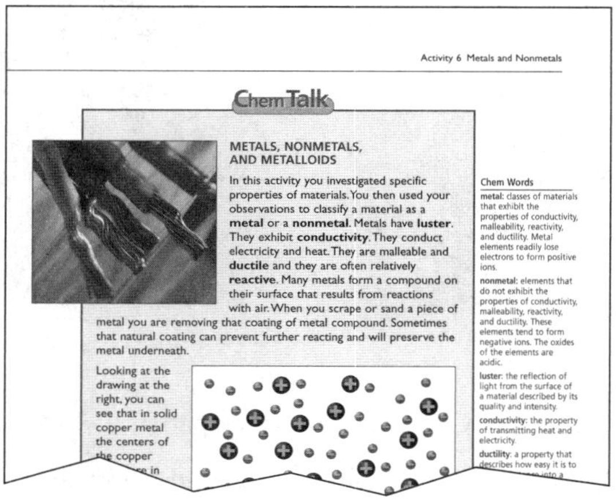

A **For You to Read** explains the difference between scalar and vector quantities.

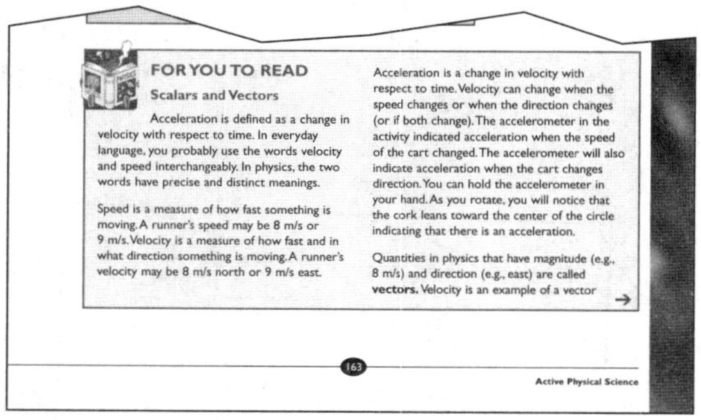

© It's About Time

A **Reflecting on the Activity and the Challenge** relates the activity to the larger challenge of creating a training manual for the track and field team. "Since a race always begins with a speed of zero and ends with runners in motion, acceleration is part of every race." Students have been given another piece of the jigsaw puzzle. They can use acceleration at the starting blocks as part of the advice they provide the track and field team.

This chapter also has activities that help students understand how measurement is important in track and field, how to analyze trends from a graph, ways to increase top speed, and the components of projectile motion.

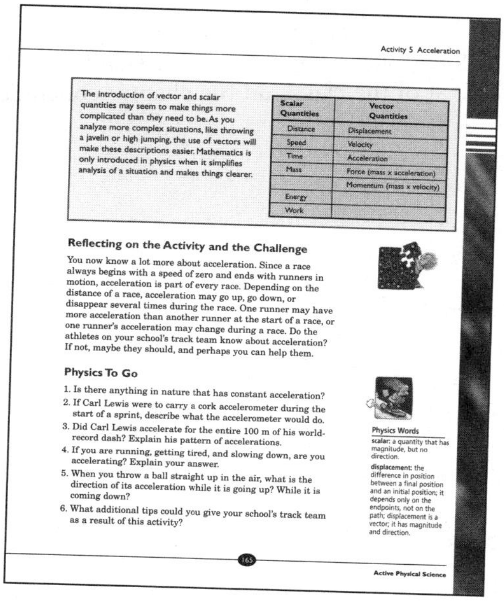

Activity 5 Acceleration

The introduction of vector and scalar quantities may seem to make things more complicated than they need to be. As you analyze more complex situations, like throwing a javelin or high jumping, the use of vectors will make these descriptions easier. Mathematics is only introduced in physics when it simplifies analysis of a situation and makes things clearer.

Scalar Quantities	Vector Quantities
Distance	Displacement
Speed	Velocity
Time	Acceleration
Mass	Force (mass x acceleration)
	Momentum (mass x velocity)
Energy	
Work	

Reflecting on the Activity and the Challenge

You now know a lot more about acceleration. Since a race always begins with a speed of zero and ends with runners in motion, acceleration is part of every race. Depending on the distance of a race, acceleration may go up, go down, or disappear several times during the race. One runner may have more acceleration than another runner at the start of a race, or one runner's acceleration may change during a race. Do the athletes on your school's track team know about acceleration? If not, maybe they should, and perhaps you can help them.

Physics To Go

1. Is there anything in nature that has constant acceleration?
2. If Carl Lewis were to carry a cork accelerometer during the start of a sprint, describe what the accelerometer would do.
3. Did Carl Lewis accelerate for the entire 100 m of his world-record dash? Explain his pattern of accelerations.
4. If you are running, getting tired, and slowing down, are you accelerating? Explain your answer.
5. When you throw a ball straight up in the air, what is the direction of its acceleration while it is going up? While it is coming down?
6. What additional tips could you give your school's track team as a result of this activity?

Physics Words
scalar: a quantity that has magnitude, but no direction.
displacement: the difference in position between a final position and an initial position; it depends only on the endpoints, not on the path; displacement is a vector; it has magnitude and direction.

165 Active Physical Science

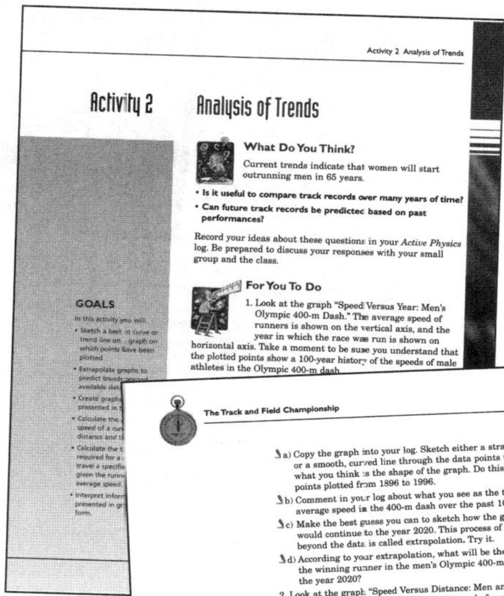

Activity 2 Analysis of Trends

Activity 2 Analysis of Trends

What Do You Think?

Current trends indicate that women will start outrunning men in 65 years.

- Is it useful to compare track records over many years of time?
- Can future track records be predicted based on past performances?

Record your ideas about these questions in your *Active Physics* log. Be prepared to discuss your responses with your small group and the class.

For You To Do

1. Look at the graph "Speed Versus Year: Men's Olympic 400-m Dash." The average speed of runners is shown on the vertical axis, and the year in which the race was run is shown on horizontal axis. Take a moment to be sure you understand that the plotted points show a 100-year history of the speeds of male athletes in the Olympic 400-m dash.

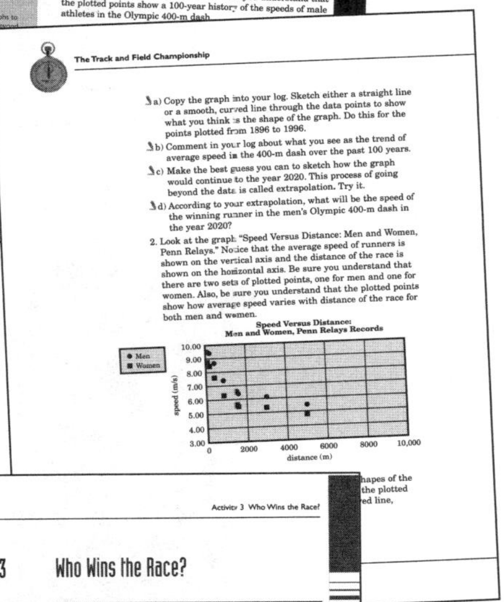

The Track and Field Championship

a) Copy the graph into your log. Sketch either a straight line or a smooth, curved line through the data points to show what you think is the shape of the graph. Do this for the points plotted from 1896 to 1996.

b) Comment in your log about what you see as the trend of average speed in the 400-m dash over the past 100 years.

c) Make the best guess you can to sketch how the graph would continue to the year 2020. This process of going beyond the data is called extrapolation. Try it.

d) According to your extrapolation, what will be the speed of the winning runner in the men's Olympic 400-m dash in the year 2020?

2. Look at the graph "Speed Versus Distance: Men and Women, Penn Relays." Notice that the average speed of runners is shown on the vertical axis and the distance of the race is shown on the horizontal axis. Be sure you understand that there are two sets of plotted points, one for men and one for women. Also, be sure you understand that the plotted points show how average speed varies with distance of the race for both men and women.

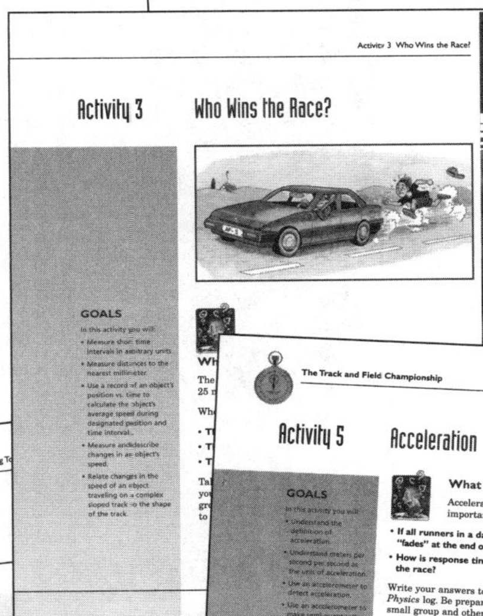

Activity 3 Who Wins the Race?

Activity 3 Who Wins the Race?

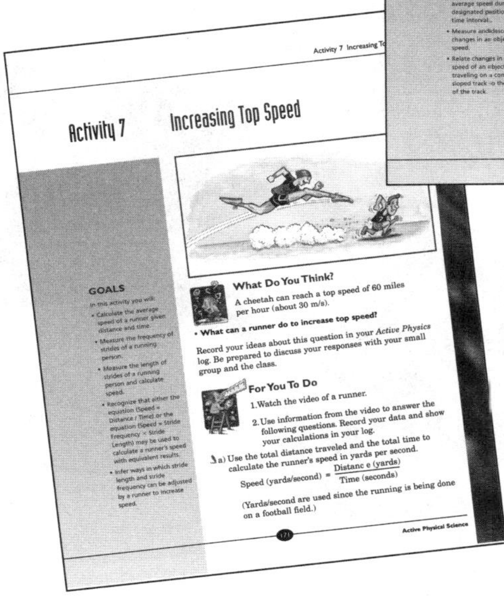

Activity 7 Increasing Top Speed

Activity 7 Increasing Top Speed

What Do You Think?

A cheetah can reach a top speed of 60 miles per hour (about 30 m/s).

- What can a runner do to increase top speed?

Record your ideas about this question in your *Active Physics* log. Be prepared to discuss your responses with your small group and the class.

For You To Do

1. Watch the video of a runner.
2. Use information from the video to answer the following questions. Record your data and show your calculations in your log.

a) Use the total distance traveled and the total time to calculate the runner's speed in yards per second.

Speed (yards/second) = Distance (yards) / Time (seconds)

(Yards/second are used since the running is being done on a football field.)

171 Active Physical Science

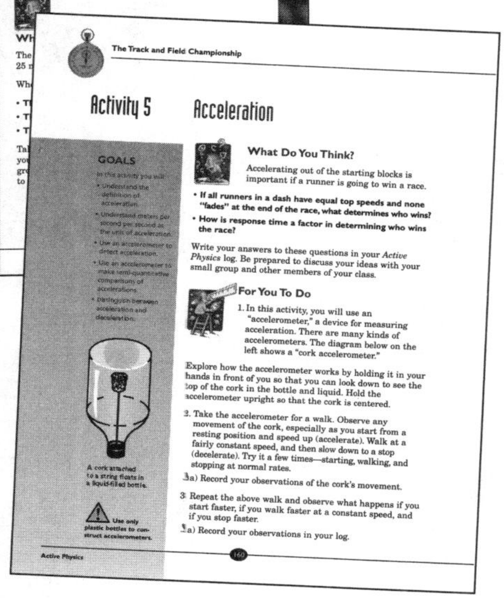

The Track and Field Championship

Activity 5 Acceleration

What Do You Think?

Accelerating out of the starting blocks is important if a runner is going to win a race.

- If all runners in a dash have equal top speeds and none "fades" at the end of the race, what determines who wins?
- How is response time a factor in determining who wins the race?

Write your answers to these questions in your *Active Physics* log. Be prepared to discuss your ideas with your small group and other members of your class.

For You To Do

1. In this activity, you will use an "accelerometer," a device for measuring acceleration. There are many kinds of accelerometers. The diagram below on the left shows a "cork accelerometer."

Explore how the accelerometer works by holding it in your hands in front of you so that you can look down to see the top of the cork in the bottle and liquid. Hold the accelerometer upright so that the cork is centered.

2. Take the accelerometer for a walk. Observe any movement of the cork, especially as you start from a resting position and speed up (accelerate). Walk at a fairly constant speed, and then slow down to a stop (decelerate). Try it a few times—starting, walking, and stopping at normal rates.

a) Record your observations of the cork's movement.

3. Repeat the above walk and observe what happens if you start faster, if you walk faster at a constant speed, and if you stop faster.

a) Record your observations in your log.

A cork attached to a string floats in a liquid-filled bottle.

Use only plastic bottles to construct accelerometers.

Active Physics 160

The activity concludes with a **Physics to Go** (or **Chemistry to Go**) homework (or in-class) assignment. Here students are asked about the specifics of the activity. In this activity students are required to explain how acceleration may affect an athlete's performance in a race. They are also given problems to solve based on the sample problem provided in the **Physics Talk** section.

© It's About Time

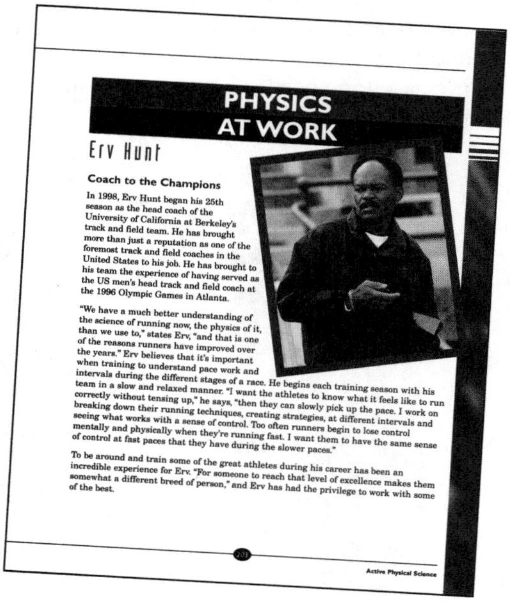

PHYSICS AT WORK

Erv Hunt

Coach to the Champions

In 1998, Erv Hunt began his 25th season as the head coach of the University of California at Berkeley's track and field team. He has brought more than just a reputation as one of the foremost track and field coaches in the United States to his job. He has brought to his team the experience of having served as the US men's head track and field coach at the 1996 Olympic Games in Atlanta.

"We have a much better understanding of the science of running now, the physics of it, than we use to," states Erv, "and that is one of the reasons runners have improved over the years." Erv believes that it's important when training to understand pace work and intervals during the different stages of a race. He begins each training season with his team in a slow and relaxed manner. "I want the athletes to know what it feels like to run correctly without tensing up," he says, "then they can slowly pick up the pace. I work on breaking down their running techniques, creating strategies, at different intervals and seeing what works with a sense of control. Too often runners begin to lose control mentally and physically when they're running fast. I want them to have the same sense of control at fast paces that they have during the slower paces."

To be around and train some of the great athletes during his career has been an incredible experience for Erv. "For someone to reach that level of excellence makes them somewhat a different breed of person," and Erv has had the privilege to work with some of the best.

203

Active Physical Science

The chapter concludes with a **Physics at Work** profile where students are introduced to someone whose job is related to the **Chapter Challenge**. In this chapter Erv Hunt, a former coach of the track and field team at the University of California explains the importance of the physics principles involved in the "pace work and intervals during the different stages of a race."

With the results of all of the activities before them, student teams now complete the **Chapter Challenge**. They put the jigsaw-puzzle pieces of distance, time, speed, acceleration, trajectory motion, and potential and kinetic energy together to create a manual for the track and field team. The teams share their work with the rest of the class and the chapter concludes.

4

Thrills and Chills

Scenario

You are excited and scared as you sit back into the seat. You pull the safety restraints into place. The next thing you know, you are beginning a slow but steady ascent into the sky. Then, with a sudden jolt, you reach the top. This is where the thrill or nightmare begins. You hurtle down the track at ever-increasing speeds. You are flung against one side of your seat as you scream around a curve. You shriek as you hang upside down, fortunately, firmly secured to your seat. All the time, your stomach has no idea where it is or you are. Finally, you come to rest where you began. What a ride! Want to go again?

Roller coasters have been enjoyed for many years. However, the roller coaster that may appeal to you, may not appeal to your parents or other friends and relatives.

206

© It's About Time

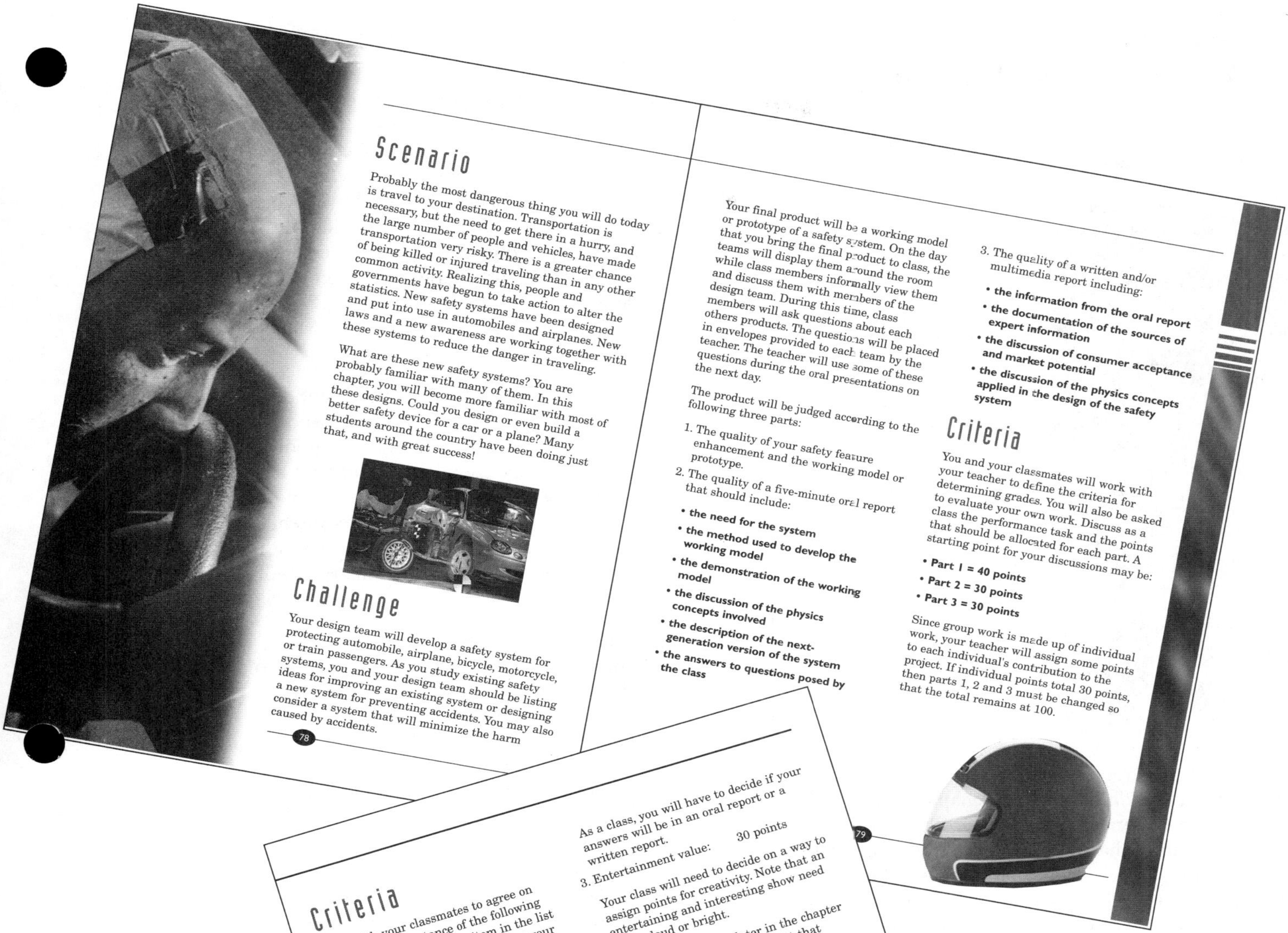

Scenario

Probably the most dangerous thing you will do today is travel to your destination. Transportation is necessary, but the need to get there in a hurry, and the large number of people and vehicles, have made transportation very risky. There is a greater chance of being killed or injured traveling than in any other common activity. Realizing this, people and governments have begun to take action to alter the statistics. New safety systems have been designed and put into use in automobiles and airplanes. New laws and a new awareness are working together with these systems to reduce the danger in traveling.

What are these new safety systems? You are probably familiar with many of them. In this chapter, you will become more familiar with most of these designs. Could you design or even build a better safety device for a car or a plane? Many students around the country have been doing just that, and with great success!

Challenge

Your design team will develop a safety system for protecting automobile, airplane, bicycle, motorcycle, or train passengers. As you study existing safety systems, you and your design team should be listing ideas for improving an existing system or designing a new system for preventing accidents. You may also consider a system that will minimize the harm caused by accidents.

78

Your final product will be a working model or prototype of a safety system. On the day that you bring the final product to class, the teams will display them around the room while class members informally view them and discuss them with members of the design team. During this time, class members will ask questions about each others products. The questions will be placed in envelopes provided to each team by the teacher. The teacher will use some of these questions during the oral presentations on the next day.

The product will be judged according to the following three parts:

1. The quality of your safety feature enhancement and the working model or prototype.
2. The quality of a five-minute oral report that should include:
   - **the need for the system**
   - **the method used to develop the working model**
   - **the demonstration of the working model**
   - **the discussion of the physics concepts involved**
   - **the description of the next-generation version of the system**
   - **the answers to questions posed by the class**
3. The quality of a written and/or multimedia report including:
   - **the information from the oral report**
   - **the documentation of the sources of expert information**
   - **the discussion of consumer acceptance and market potential**
   - **the discussion of the physics concepts applied in the design of the safety system**

Criteria

You and your classmates will work with your teacher to define the criteria for determining grades. You will also be asked to evaluate your own work. Discuss as a class the performance task and the points that should be allocated for each part. A starting point for your discussions may be:

- **Part 1 = 40 points**
- **Part 2 = 30 points**
- **Part 3 = 30 points**

Since group work is made up of individual work, your teacher will assign some points to each individual's contribution to the project. If individual points total 30 points, then parts 1, 2 and 3 must be changed so that the total remains at 100.

79

Criteria

Work with your classmates to agree on the relative importance of the following assessment criteria. Each item in the list has a point value given after it, but your class must decide what kind of grading system to use.

1. The variety and number of physics concepts used to produce the light and sound effects: 30 points
   - four or more concepts: 30 points
   - three concepts: 25 points
   - two concepts: 20 points
   - one concept: 10 points
2. Your understanding of the physics concepts: 40 points

   Following your production, you will be asked to:
   - a) Name the physics concepts that you used. 10 points
   - b) Explain each concept. 10 points
   - c) Give an example of something that each concept explains or an example of how each concept is used. 10 points
   - d) Explain why each concept is important. 10 points

   As a class, you will have to decide if your answers will be in an oral report or a written report.
3. Entertainment value: 30 points

   Your class will need to decide on a way to assign points for creativity. Note that an entertaining and interesting show need not be loud or bright.

You will have a chance later in the chapter to again discuss these criteria. At that time, you may have more information on the concepts and how you might produce your show. You may want to then propose changes in the criteria and the point values.

325

© It's About Time

Features of Active Physical Science

1. Scenario

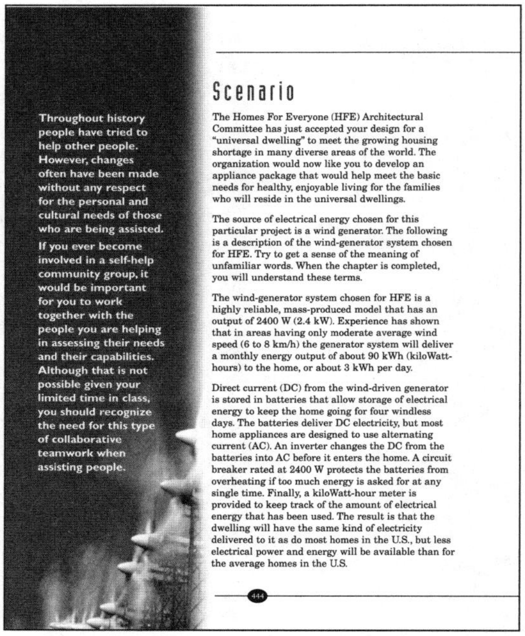
Throughout history people have tried to help other people. However, changes often have been made without any respect for the personal and cultural needs of those who are being assisted.

If you ever become involved in a self-help community group, it would be important for you to work together with the people you are helping in assessing their needs and their capabilities. Although that is not possible given your limited time in class, you should recognize the need for this type of collaborative teamwork when assisting people.

Scenario

The Homes For Everyone (HFE) Architectural Committee has just accepted your design for a "universal dwelling" to meet the growing housing shortage in many diverse areas of the world. The organization would now like you to develop an appliance package that would help meet the basic needs for healthy, enjoyable living for the families who will reside in the universal dwellings.

The source of electrical energy chosen for this particular project is a wind generator. The following is a description of the wind-generator system chosen for HFE. Try to get a sense of the meaning of unfamiliar words. When the chapter is completed, you will understand these terms.

The wind-generator system chosen for HFE is a highly reliable, mass-produced model that has an output of 2400 W (2.4 kW). Experience has shown that in areas having only moderate average wind speed (6 to 8 km/h) the generator system will deliver a monthly energy output of about 90 kWh (kiloWatt-hours) to the home, or about 3 kWh per day.

Direct current (DC) from the wind-driven generator is stored in batteries that allow storage of electrical energy to keep the home going for four windless days. The batteries deliver DC electricity, but most home appliances are designed to use alternating current (AC). An inverter changes the DC from the batteries into AC before it enters the home. A circuit breaker rated at 2400 W protects the batteries from overheating if too much energy is asked for at any single time. Finally, a kiloWatt-hour meter is provided to keep track of the amount of electrical energy that has been used. The result is that the dwelling will have the same kind of electricity delivered to it as do most homes in the U.S., but less electrical power and energy will be available than for the average homes in the U.S.

444

Each *Active Physical Science* chapter opens with an engaging **Scenario**. Students from diverse backgrounds and localities have been interviewed in order to find situations which are not only realistic but meaningful to the high school population. The scenarios (only a paragraph or two in length) set the stage for the **Chapter Challenge** which immediately follows. Many teachers choose to read the scenario aloud to the class as a way of introducing the new chapter.

2. Challenge

The **Chapter Challenge** is the heart and soul of *Active Physical Science*. It provides a purpose for all of the work that will follow. The challenges provide the rationale for learning. One of the common complaints teachers hear from students is, "Why am I learning this?" In *Active Physical Science*, no students raise this criticism. Similarly, no teacher has to answer, "Because one day it will be useful to you." The complaint is avoided because on Day One of the chapter students are presented with a challenge that, in essence, becomes their job for the next few weeks.

In *Safety*, Chapter 2, students are challenged to design and build an improved safety device for an automobile. The study of momentum, forces and Newton's Laws will be integral to their understanding of the required features in a safety device.

In *Electricity for Everyone*, Chapter 7, students must create an appliance package that can be used in developing nations. The appliance package is limited by the wind generator available to the households. Students must also supply a rationale for how each suggested appliance will enhance the well-being of the family using it. This requires students to be able to differentiate between power and energy. It also provides a basis for students to reflect on quality of life issues in parts of the globe that they learn about in their social studies classes.

The beauty of the challenges lies in the variety of tasks and opportunities for students of different talents and skills to excel. Students who express themselves artistically will have an opportunity to shine in some challenges, while the student who can design and build may be the group leader in another challenge. Some challenges have a major component devoted to writing while others require oral or visual presentations. All challenges require the demonstration of solid physics understanding.

The challenges are not contrived situations for high school students. Professional engineers also design and build improved safety devices. Medical writers and illustrators design posters and pamphlets. The challenge in Chapter 4, *Thrills and Chills*, requires students to adapt a roller coaster design for a select group of riders. The group can consist of young children, daredevils, or physically challenged people. This challenge has been successfully completed by 9th grade students, 12th grade physics students, and by mechanical engineers. The expectations may be different for each of these audiences, but the challenge is consistent.

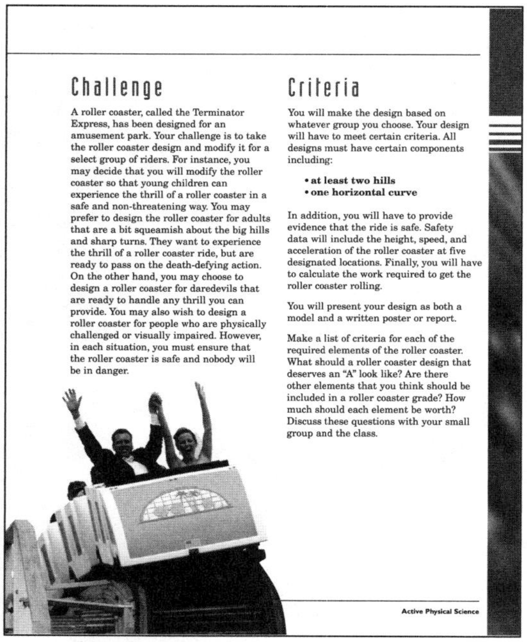
Challenge

A roller coaster, called the Terminator Express, has been designed for an amusement park. Your challenge is to take the roller coaster design and modify it for a select group of riders. For instance, you may decide that you will modify the roller coaster so that young children can experience the thrill of a roller coaster in a safe and non-threatening way. You may prefer to design the roller coaster for adults that are a bit squeamish about the big hills and sharp turns. They want to experience the thrill of a roller coaster ride, but are ready to pass on the death-defying action. On the other hand, you may choose to design a roller coaster for daredevils that are ready to handle any thrill you can provide. You may also wish to design a roller coaster for people who are physically challenged or visually impaired. However, in each situation, you must ensure that the roller coaster is safe and nobody will be in danger.

Criteria

You will make the design based on whatever group you choose. Your design will have to meet certain criteria. All designs must have certain components including:

- **at least two hills**
- **one horizontal curve**

In addition, you will have to provide evidence that the ride is safe. Safety data will include the height, speed, and acceleration of the roller coaster at five designated locations. Finally, you will have to calculate the work required to get the roller coaster rolling.

You will present your design as both a model and a written poster or report.

Make a list of criteria for each of the required elements of the roller coaster. What should a roller coaster design that deserves an "A" look like? Are there other elements that you think should be included in a roller coaster grade? How much should each element be worth? Discuss these questions with your small group and the class.

Active Physical Science

© It's About Time

3. Criteria

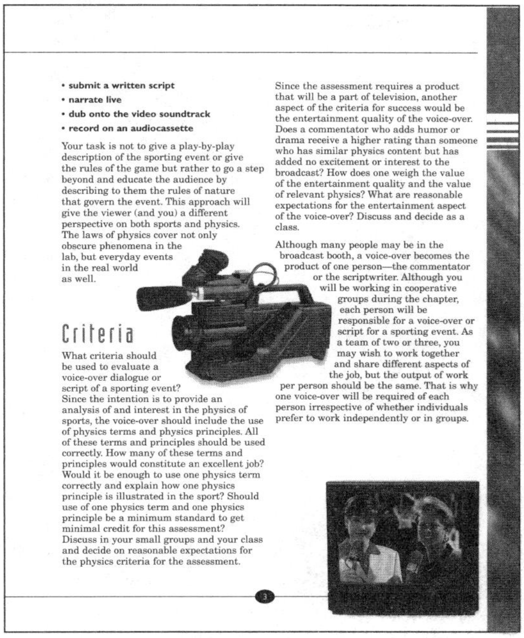
- **submit a written script**
- **narrate live**
- **dub onto the video soundtrack**
- **record on an audiocassette**

Your task is not to give a play-by-play description of the sporting event or give the rules of the game but rather to go a step beyond and educate the audience by describing to them the rules of nature that govern the event. This approach will give the viewer (and you) a different perspective on both sports and physics. The laws of physics cover not only obscure phenomena in the lab, but everyday events in the real world as well.

Criteria

What criteria should be used to evaluate a voice-over dialogue or script of a sporting event? Since the intention is to provide an analysis of and interest in the physics of sports, the voice-over should include the use of physics terms and physics principles. All of these terms and principles should be used correctly. How many of these terms and principles would constitute an excellent job? Would it be enough to use one physics term correctly and explain how one physics principle is illustrated in the sport? Should use of one physics term and one physics principle be a minimum standard to get minimal credit for this assessment? Discuss in your small groups and your class and decide on reasonable expectations for the physics criteria for the assessment.

Since the assessment requires a product that will be a part of television, another aspect of the criteria for success would be the entertainment quality of the voice-over. Does a commentator who adds humor or drama receive a higher rating than someone who has similar physics content but has added no excitement or interest to the broadcast? How does one weigh the value of the entertainment quality and the value of relevant physics? What are reasonable expectations for the entertainment aspect of the voice-over? Discuss and decide as a class.

Although many people may be in the broadcast booth, a voice-over becomes the product of one person—the commentator or the scriptwriter. Although you will be working in cooperative groups during the chapter, each person will be responsible for a voice-over or script for a sporting event. As a team of two or three, you may wish to work together and share different aspects of the job, but the output of work per person should be the same. That is why one voice-over will be required of each person irrespective of whether individuals prefer to work independently or in groups.

3

In creating *Active Physical Science*, we had thought that the generation of the challenge was good enough. Upon reflection, we soon realized that criteria for success must also be included. When students agree to the matrix by which they will be measured, the research has shown that the students will perform better and achieve more. It makes sense. In the simplest situation of cleaning a lab room, the teacher may simply state, "Please clean up the lab." The results are often a minimal cleanup. If the teacher begins by asking, "What does a clean lab room look like?" and students and teacher jointly list the attributes of a clean lab room (i.e., no paper on the floor, all beakers put away, all materials on the back of the lab tables, all power supplies unplugged and all water removed), the students respond differently and the cleanup is better. When students are asked to include physical science principles in an explanation, the students should know whether the expectation is for three physical science principles or five.

The discussion of grading criteria and the creation of a grading rubric is a crucial ingredient for student success. *Active Physical Science* requires a class discussion, after the introduction of the challenge, about the grading criteria. How much is required? What does an "A" presentation look like? Should creativity be weighed more than delivery? The criteria can be visited again at the end of the chapter, but at this point it provides a clarity to the challenge and the expectation level that the students should set for themselves.

© It's About Time

4. What Do You Think?

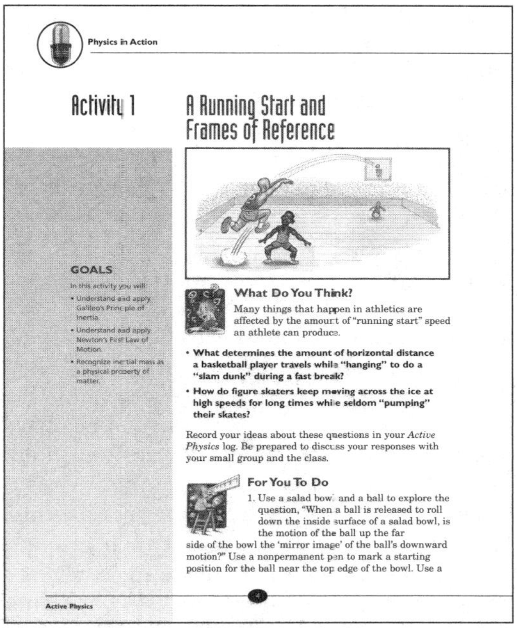
Physics in Action

Activity 1

A Running Start and Frames of Reference

GOALS

In this activity you will:
- Understand and apply Galileo's Principle of Inertia.
- Understand and apply Newton's First Law of Motion.
- Recognize inertial mass as a physical property of matter.

What Do You Think?

Many things that happen in athletics are affected by the amount of "running start" speed an athlete can produce.

- **What determines the amount of horizontal distance a basketball player travels while "hanging" to do a "slam dunk" during a fast break?**
- **How do figure skaters keep moving across the ice at high speeds for long times while seldom "pumping" their skates?**

Record your ideas about these questions in your *Active Physics* log. Be prepared to discuss your responses with your small group and the class.

For You To Do

1. Use a salad bowl and a ball to explore the question, "When a ball is released to roll down the inside surface of a salad bowl, is the motion of the ball up the far side of the bowl the 'mirror image' of the ball's downward motion?" Use a nonpermanent pen to mark a starting position for the ball near the top edge of the bowl. Use a

Active Physics

4

During the past few years much has been written about a constructivist approach to learning. Videos of Harvard graduates, in caps and gowns, show that the students are not able to explain correctly why it is colder in the winter than it is in the summer. These students have previously answered these questions correctly in 4th grade, in middle school, and then again in high school. How else would they have gotten into Harvard? We believe that they never internalized the logic and understanding of the seasons. One reason for this problem is that they were never confronted by what they did believe, and were never adequately shown why they should give up that belief system. Certainly, it is worth writing down a "book's perfect answer" on a test to secure a good grade, but to actually believe requires a more thorough examination of competing explanations.

The best way to ascertain a student's prior understanding is through extensive interviewing. Much of the research literature in this area includes the results of these interviews. In a classroom, this one-on-one dialogue is rarely possible. The **What Do You Think?** questions introduce each activity in a way in which to elicit prior understandings. It gives students an opportunity to verbalize what they think about speed, or energy, or elements and compounds, before they embark on an activity. The brief discussion of the range of answers brings the student a little closer in touch with that part of his/her brain which understands speed, energy, or elements and compounds. The **What Do You Think?** question is not intended to produce a correct answer or a discussion of the features of the questions. It is not intended to bring closure. The activity which follows will provide that discussion as experimental results are analyzed. The **What Do You Think?** question should take no more than a few minutes of class time. It is the lead into the physical science investigation. Students should be strongly encouraged to write their responses to the questions in their logs, to ensure that they have in fact addressed their prior conceptions. After students have discussed their responses in their small groups, activate a class discussion. Ask students to volunteer other students' answers which they found interesting. This may encourage students to exchange ideas without the fear of personally giving a "wrong" answer.

5. For You To Do (or Investigate)

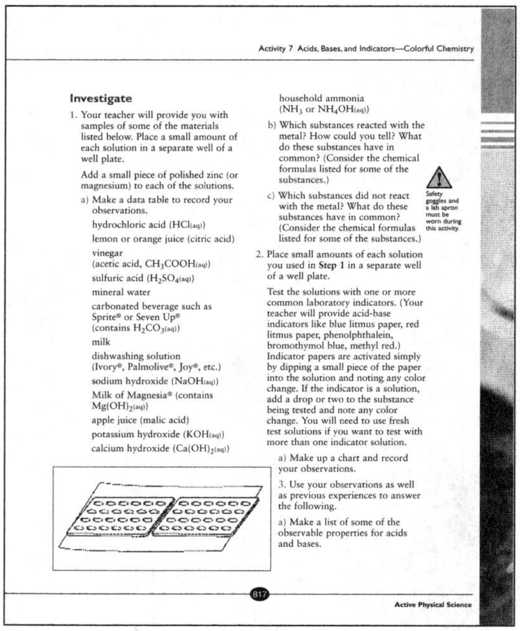

Activity 7 Acids, Bases, and Indicators—Colorful Chemistry

Investigate

1. Your teacher will provide you with samples of some of the materials listed below. Place a small amount of each solution in a separate well of a well plate.

   Add a small piece of polished zinc (or magnesium) to each of the solutions.

   a) Make a data table to record your observations.

   hydrochloric acid ($HCl_{(aq)}$)
   lemon or orange juice (citric acid)
   vinegar (acetic acid, $CH_3COOH_{(aq)}$)
   sulfuric acid ($H_2SO_{4(aq)}$)
   mineral water
   carbonated beverage such as Sprite® or Seven Up® (contains $H_2CO_{3(aq)}$)
   milk
   dishwashing solution (Ivory®, Palmolive®, Joy®, etc.)
   sodium hydroxide ($NaOH_{(aq)}$)
   Milk of Magnesia® (contains $Mg(OH)_{2(aq)}$)
   apple juice (malic acid)
   potassium hydroxide ($KOH_{(aq)}$)
   calcium hydroxide ($Ca(OH)_{2(aq)}$)
   household ammonia (NH_3 or $NH_4OH_{(aq)}$)

   b) Which substances reacted with the metal? How could you tell? What do these substances have in common? (Consider the chemical formulas listed for some of the substances.)

   c) Which substances did not react with the metal? What do these substances have in common? (Consider the chemical formulas listed for some of the substances.)

Safety goggles and a lab apron must be worn during this activity.

2. Place small amounts of each solution you used in **Step 1** in a separate well of a well plate.

   Test the solutions with one or more common laboratory indicators. (Your teacher will provide acid-base indicators like blue litmus paper, red litmus paper, phenolphthalein, bromothymol blue, methyl red.) Indicator papers are activated simply by dipping a small piece of the paper into the solution and noting any color change. If the indicator is a solution, add a drop or two to the substance being tested and note any color change. You will need to use fresh test solutions if you want to test with more than one indicator solution.

   a) Make up a chart and record your observations.

3. Use your observations as well as previous experiences to answer the following.

   a) Make a list of some of the observable properties for acids and bases.

817 Active Physical Science

Active Physical Science is a hands-on, minds-on curriculum. Students *do* physical science; they do not *read* about doing physical science. Each activity has instructions for each part of the investigation. The pencil icons in the Student Edition are provided to remind students that data, hypotheses, or conclusions should be recorded in their log or laboratory manual.

Activities are the opportunity for students to garner the knowledge that they will need to complete the **Chapter Challenge**. Students will understand the physical science principle involved because they have investigated it. In *Active Physical Science*, if a student is asked, "How do you know?" the response is, "Because I did an experiment!"

Recognizing that many students know how to read, but do not like reading, Background information is provided within the context of the activity. Students have demonstrated that they will read when the information is required for them to continue withtheir exploration.

Occasionally, the activity will require the entire class to participate in a large, single demonstration simultaneously. The teacher, on other occasions, may decide that a specific activity is best done as a demonstration. This would be appropriate if there is limited equipment for that one activity, or the facilities are not available. Viewing demonstrations on an ongoing basis, however, is not what these 12 chapters are about.

There are specific **For You To Do** (or **Investigate**) activities where computer spreadsheets, force transducers, or specific electronic equipment is required. Most of these activities have 'low-tech' alternatives provided in the Teacher's Edition. In the initial teaching of *Active Physical Science*, the low-tech alternative may be the only reasonable approach. As the course becomes a staple of the school offerings, it is hoped that funds can be set aside to improve the students' access to equipment.

© It's About Time

Most of the **For You To Do** (or **Investigate**) activities require between one and two class periods. With the present trend toward block scheduling, there are so many time structures that it is difficult to predict how *Active Physical Science* will best fit with your schedule. The other impact on time is the achievement and preparation level of the students. In a given activity, students may be required to complete a graph of their data. This is considered one small part of the activity. If the students have never been exposed to graphing, this could require a two-period lesson to teach the rudiments of graphing with suitable practice in interpretation. *Active Physical Science* is accessible to all students. The teacher is in the best position to make accommodations in time reflecting the needs of the students.

6. Physics Talk and Chem Talk

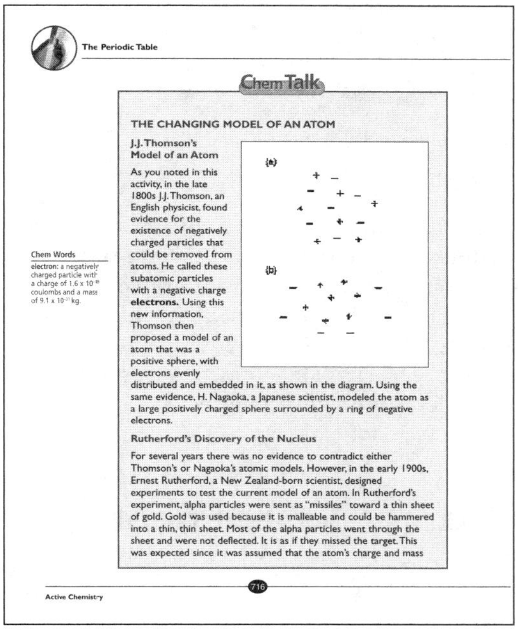
The Periodic Table

Chem Talk

THE CHANGING MODEL OF AN ATOM

J.J. Thomson's Model of an Atom

As you noted in this activity, in the late 1800s J.J. Thomson, an English physicist, found evidence for the existence of negatively charged particles that could be removed from atoms. He called these subatomic particles with a negative charge **electrons.** Using this new information, Thomson then proposed a model of an atom that was a positive sphere, with electrons evenly distributed and embedded in it, as shown in the diagram. Using the same evidence, H. Nagaoka, a Japanese scientist, modeled the atom as a large positively charged sphere surrounded by a ring of negative electrons.

Chem Words

electron: a negatively charged particle with a charge of 1.6×10^{-19} coulombs and a mass of 9.1×10^{-31} kg.

Rutherford's Discovery of the Nucleus

For several years there was no evidence to contradict either Thomson's or Nagaoka's atomic models. However, in the early 1900s, Ernest Rutherford, a New Zealand-born scientist, designed experiments to test the current model of an atom. In Rutherford's experiment, alpha particles were sent as "missiles" toward a thin sheet of gold. Gold was used because it is malleable and could be hammered into a thin, thin sheet. Most of the alpha particles went through the sheet and were not deflected. It is as if they missed the target. This was expected since it was assumed that the atom's charge and mass

716

Active Chemistry

Equations and formulas are often the simplest, most straightforward, most concise, and clearest way of expressing physical science principles. *Active Physical Science* limits the mathematics to the 9th grade curriculum. Students who have shied away from studying physics because of the mathematics prerequisites find that they are welcomed into *Active Physical Science*. **Physics** and **Chem Talk** is a means by which specific attention can be given to the mathematical equations. It also provides an opportunity to illustrate a problem solution or to derive a complex equation. For some students, there is a need to guide them through the algebraic manipulation which shows the equivalence of $F = ma$ and $a = F/m$. Where appropriate, this manipulation is explicitly shown. Finally, sample problems required for the **Chapter Challenge** will also be in **Physics Talk**.

7. For You To Read

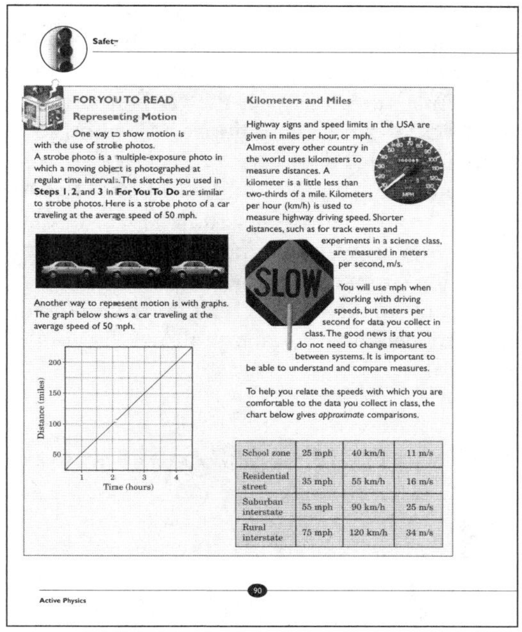
Safety

FOR YOU TO READ

Representing Motion

One way to show motion is with the use of strobe photos. A strobe photo is a multiple-exposure photo in which a moving object is photographed at regular time intervals. The sketches you used in **Steps 1, 2,** and **3** in **For You To Do** are similar to strobe photos. Here is a strobe photo of a car traveling at the average speed of 50 mph.

Another way to represent motion is with graphs. The graph below shows a car traveling at the average speed of 50 mph.

Kilometers and Miles

Highway signs and speed limits in the USA are given in miles per hour, or mph. Almost every other country in the world uses kilometers to measure distances. A kilometer is a little less than two-thirds of a mile. Kilometers per hour (km/h) is used to measure highway driving speed. Shorter distances, such as for track events and experiments in a science class, are measured in meters per second, m/s.

You will use mph when working with driving speeds, but meters per second for data you collect in class. The good news is that you do not need to change measures between systems. It is important to be able to understand and compare measures.

To help you relate the speeds with which you are comfortable to the data you collect in class, the chart below gives *approximate* comparisons.

School zone	25 mph	40 km/h	11 m/s
Residential street	35 mph	55 km/h	16 m/s
Suburban interstate	55 mph	90 km/h	25 m/s
Rural interstate	75 mph	120 km/h	34 m/s

90

Active Physics

The **For You To Read** inserts provide students with some reading at the 9th grade level. This section may be used to tie together concepts from the present activity or a set of activities. It may also be used to provide a glimpse into the history of the physics principle being investigated. Finally, **For You To Read** may provide background information which will help clarify the meaning of the physics principle investigated in **For You To Do** (or **Investigate**).

8. Reflecting on the Activity and the Challenge

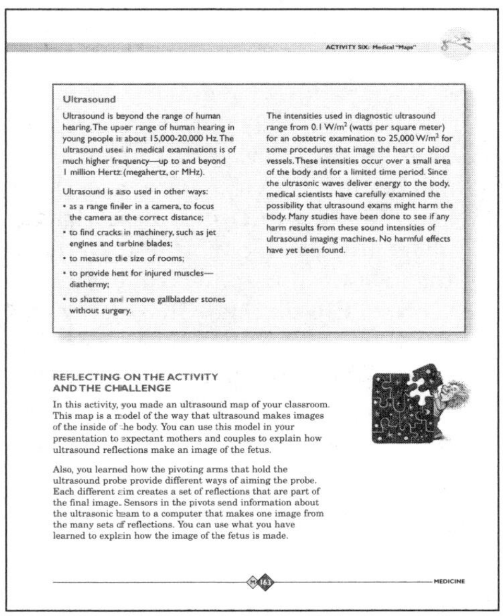
ACTIVITY SIX: Medical "Maps"

Ultrasound

Ultrasound is beyond the range of human hearing. The upper range of human hearing in young people is about 15,000-20,000 Hz. The ultrasound used in medical examinations is of much higher frequency—up to and beyond 1 million Hertz (megahertz, or MHz).

Ultrasound is also used in other ways:

- as a range finder in a camera, to focus the camera at the correct distance;
- to find cracks in machinery, such as jet engines and turbine blades;
- to measure the size of rooms;
- to provide heat for injured muscles—diathermy;
- to shatter and remove gallbladder stones without surgery.

The intensities used in diagnostic ultrasound range from 0.1 W/m^2 (watts per square meter) for an obstetric examination to 25,000 W/m^2 for some procedures that image the heart or blood vessels. These intensities occur over a small area of the body and for a limited time period. Since the ultrasonic waves deliver energy to the body, medical scientists have carefully examined the possibility that ultrasound exams might harm the body. Many studies have been done to see if any harm results from these sound intensities of ultrasound imaging machines. No harmful effects have yet been found.

REFLECTING ON THE ACTIVITY AND THE CHALLENGE

In this activity, you made an ultrasound map of your classroom. This map is a model of the way that ultrasound makes images of the inside of the body. You can use this model in your presentation to expectant mothers and couples to explain how ultrasound reflections make an image of the fetus.

Also, you learned how the pivoting arms that hold the ultrasound probe provide different ways of aiming the probe. Each different aim creates a set of reflections that are part of the final image. Sensors in the pivots send information about the ultrasonic beam to a computer that makes one image from the many sets of reflections. You can use what you have learned to explain how the image of the fetus is made.

M 163

MEDICINE

© It's About Time

At the close of each activity, the student is often so involved with the completion of the single experiment that the larger context of the investigation is lost. **Reflecting on the Activity and the Challenge** is the opportunity for students to place the new insights and information into the context of the chapter and the **Chapter Challenge**. If the **Chapter Challenge** is considered a completed picture, each activity is a jigsaw piece. By completing enough of the **For You To Do** (or **Investigate**) activities, the students will be able to fit the jigsaw pieces together and complete the challenge. This summary section ensures that the students do not forget about the larger context and continue their personal momentum toward completion of the challenge.

9. Physics and Chemistry To Go

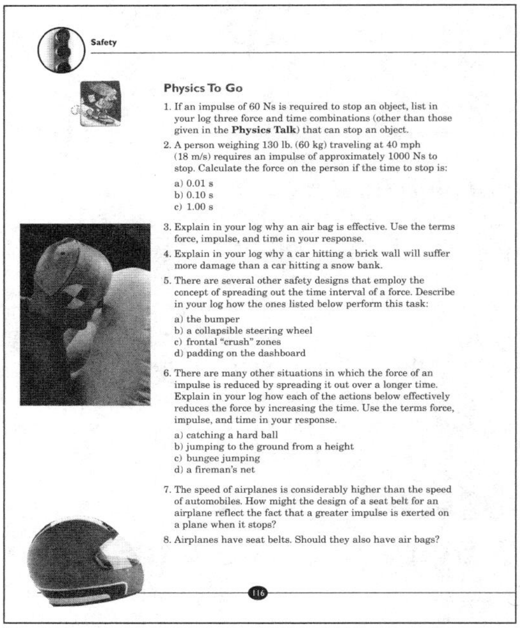
Safety

Physics To Go

1. If an impulse of 60 Ns is required to stop an object, list in your log three force and time combinations (other than those given in the **Physics Talk**) that can stop an object.
2. A person weighing 130 lb. (60 kg) traveling at 40 mph (18 m/s) requires an impulse of approximately 1000 Ns to stop. Calculate the force on the person if the time to stop is:
   a) 0.01 s
   b) 0.10 s
   c) 1.00 s
3. Explain in your log why an air bag is effective. Use the terms force, impulse, and time in your response.
4. Explain in your log why a car hitting a brick wall will suffer more damage than a car hitting a snow bank.
5. There are several other safety designs that employ the concept of spreading out the time interval of a force. Describe in your log how the ones listed below perform this task:
   a) the bumper
   b) a collapsible steering wheel
   c) frontal "crush" zones
   d) padding on the dashboard
6. There are many other situations in which the force of an impulse is reduced by spreading it out over a longer time. Explain in your log how each of the actions below effectively reduces the force by increasing the time. Use the terms force, impulse, and time in your response.
   a) catching a hard ball
   b) jumping to the ground from a height
   c) bungee jumping
   d) a fireman's net
7. The speed of airplanes is considerably higher than the speed of automobiles. How might the design of a seat belt for an airplane reflect the fact that a greater impulse is exerted on a plane when it stops?
8. Airplanes have seat belts. Should they also have air bags?

116

These sections provide additional questions and problems that can be completed outside of class. Some of the problems are applications of the principles involved in the preceding activity. Others are replication of the work in the **For You To Do** (or **Investigate**) activity. Still others provide an opportunity to transfer the results of the investigation to the context of the **Chapter Challenge**. **Physics** and **Chemistry To Go** provide a means by which students can be working on the larger **Chapter Challenge** in smaller chunks during the chapter.

10. Stretching Exercises (or Inquiring Further)

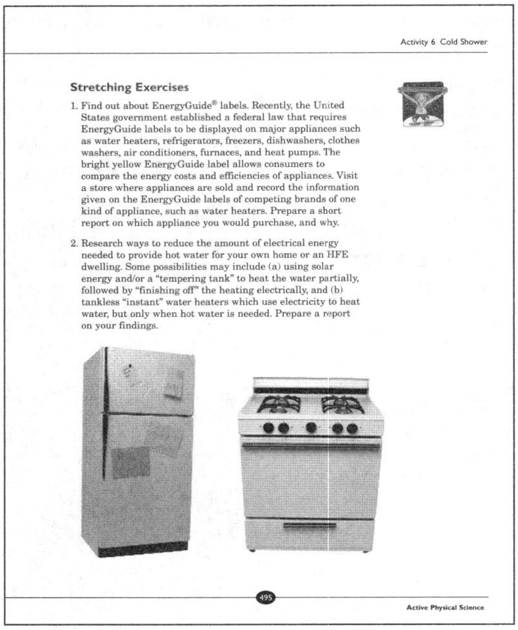
Activity 6 Cold Shower

Stretching Exercises

1. Find out about EnergyGuide® labels. Recently, the United States government established a federal law that requires EnergyGuide labels to be displayed on major appliances such as water heaters, refrigerators, freezers, dishwashers, clothes washers, air conditioners, furnaces, and heat pumps. The bright yellow EnergyGuide label allows consumers to compare the energy costs and efficiencies of appliances. Visit a store where appliances are sold and record the information given on the EnergyGuide labels of competing brands of one kind of appliance, such as water heaters. Prepare a short report on which appliance you would purchase, and why.
2. Research ways to reduce the amount of electrical energy needed to provide hot water for your own home or an HFE dwelling. Some possibilities may include (a) using solar energy and/or a "tempering tank" to heat the water partially, followed by "finishing off" the heating electrically, and (b) tankless "instant" water heaters which use electricity to heat water, but only when hot water is needed. Prepare a report on your findings.

495 Active Physical Science

Some students express additional interest in a specific topic or an extension to a topic. The **Stretching Exercises** (or **Inquiring Further**) provide an avenue in which to pursue that interest. **Stretching Exercises** (or **Inquiring Further**) often require additional readings or interviews. They may be given for extra credit to students who wish to attempt a more in-depth problem or a tougher exercise.

11. Chapter Assessment

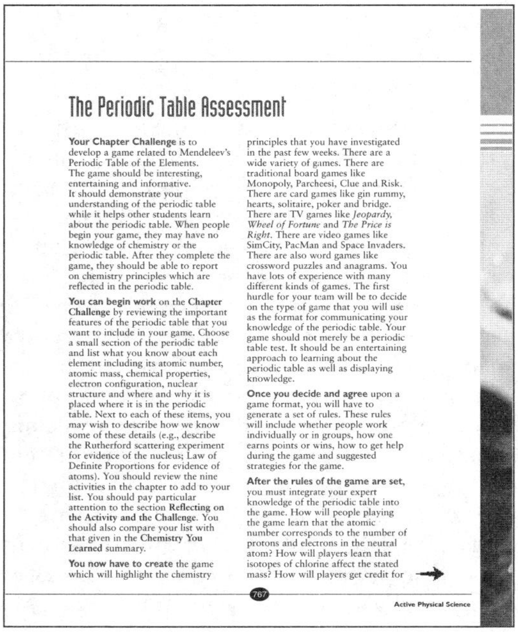
The Periodic Table Assessment

Your Chapter Challenge is to develop a game related to Mendeleev's Periodic Table of the Elements. The game should be interesting, entertaining and informative. It should demonstrate your understanding of the periodic table while it helps other students learn about the periodic table. When people begin your game, they may have no knowledge of chemistry or the periodic table. After they complete the game, they should be able to report on chemistry principles which are reflected in the periodic table.

You can begin work on the **Chapter Challenge** by reviewing the important features of the periodic table that you want to include in your game. Choose a small section of the periodic table and list what you know about each element including its atomic number, atomic mass, chemical properties, electron configuration, nuclear structure and where and why it is placed where it is in the periodic table. Next to each of these items, you may wish to describe how we know some of these details (e.g., describe the Rutherford scattering experiment for evidence of the nucleus; Law of Definite Proportions for evidence of atoms). You should review the nine activities in the chapter to add to your list. You should pay particular attention to the section **Reflecting on the Activity and the Challenge**. You should also compare your list with that given in the **Chemistry You Learned** summary.

You now have to create the game which will highlight the chemistry principles that you have investigated in the past few weeks. There are a wide variety of games. There are traditional board games like Monopoly, Parcheesi, Clue and Risk. There are card games like gin rummy, hearts, solitaire, poker and bridge. There are TV games like *Jeopardy, Wheel of Fortune* and *The Price is Right*. There are video games like SimCity, PacMan and Space Invaders. There are also word games like crossword puzzles and anagrams. You have lots of experience with many different kinds of games. The first hurdle for your team will be to decide on the type of game that you will use as the format for communicating your knowledge of the periodic table. Your game should not merely be a periodic table test. It should be an entertaining approach to learning about the periodic table as well as displaying knowledge.

Once you decide and agree upon a game format, you will have to generate a set of rules. These rules will include whether people work individually or in groups, how one earns points or wins, how to get help during the game and suggested strategies for the game.

After the rules of the game are set, you must integrate your expert knowledge of the periodic table into the game. How will people playing the game learn that the atomic number corresponds to the number of protons and electrons in the neutral atom? How will players learn that isotopes of chlorine affect the stated mass? How will players get credit for

767 Active Physical Science

© It's About Time

The **Chapter Assessment** is the return to the **Chapter Challenge and Criteria**. The students are ready to complete the challenge. They are able to view the challenge with a clarity that has emerged from the completion of the **For You To Do** (or **Investigate**) activities of the chapter. Students are able to review the chapter as they discuss the synthesis of the information into the required context of the challenge. The students should have some class time to work together to complete the challenge and to present their project. In many physical science courses, all students are expected to converge on the same solution. In *Active Physical Science*, each group is expected to have a unique solution. All solutions must have correct physical science, but there is ample room for creativity on the students' part. This is one of the features that captures the imagination of students who have often previously chosen not to enroll in physical science classes.

12. Physics and Chemistry You Learned

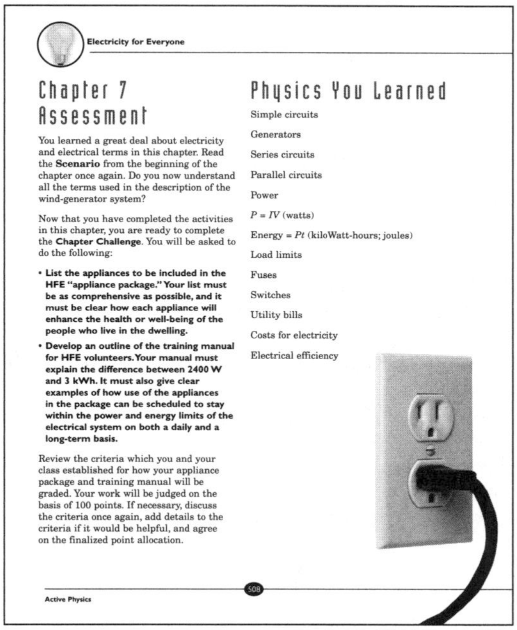
Electricity for Everyone

Chapter 7 Assessment

You learned a great deal about electricity and electrical terms in this chapter. Read the **Scenario** from the beginning of the chapter once again. Do you now understand all the terms used in the description of the wind-generator system?

Now that you have completed the activities in this chapter, you are ready to complete the **Chapter Challenge**. You will be asked to do the following:

- **List the appliances to be included in the HFE "appliance package." Your list must be as comprehensive as possible, and it must be clear how each appliance will enhance the health or well-being of the people who live in the dwelling.**
- **Develop an outline of the training manual for HFE volunteers. Your manual must explain the difference between 2400 W and 3 kWh. It must also give clear examples of how use of the appliances in the package can be scheduled to stay within the power and energy limits of the electrical system on both a daily and a long-term basis.**

Review the criteria which you and your class established for how your appliance package and training manual will be graded. Your work will be judged on the basis of 100 points. If necessary, discuss the criteria once again, add details to the criteria if it would be helpful, and agree on the finalized point allocation.

Physics You Learned

Simple circuits

Generators

Series circuits

Parallel circuits

Power

$P = IV$ (watts)

Energy = Pt (kiloWatt-hours; joules)

Load limits

Fuses

Switches

Utility bills

Costs for electricity

Electrical efficiency

508

Active Physics

This small section at the end of the chapter provides a list of physical science concepts and equations which were studied in the context of the **For You To Do** (or **Investigate**) activities. It provides students with a sense of accomplishment and serves as a quick review of all that was learned during the preceding weeks.

13. Physics and Chemistry At Work

PHYSICS AT WORK

Loretta Wright

GRANTING WISHES

How would you feel if you could make people's dreams come true? And how would you feel if you knew you could make a major difference in this world? Loretta Wright does that every day. She is a project officer for the Annenberg/CPB-Math and Science Project which distributes over six million dollars a year in grants. The project's goals are to improve math and science education throughout the country. As project officer, Loretta determines whether various grants will be funded.

"There are many factors we consider in making a decision on who should receive a grant. First, it must be established that the project under review fits our mission and has the potential to reach a large number of people," explains Loretta. "The project must have a reasonable budget, goals and activities that we feel are appropriate and achievable and, of course, experienced personnel."

One exciting example of an Annenberg/CPB-funded project that Loretta is very proud of is the Tennessee Valley Project. "This is a rural telecommunication project in which teachers and students used the Internet and Web resources along with resources in their communities as tools to learn more about science. For example, students who were learning about water resources went into their community and tested water samples. They then posted their results on the Web and contacted students and scientists all around the country to discuss those results, compare them to the water quality in other communities, and find solutions if water quality was a problem."

"Being a project officer for Annenberg is a tremendous privilege for me. As a youth I never dreamed I would one day work in the corporate world." Loretta graduated from Risley High School in Brunswick, Georgia and went to Fort Valley State College in ...a. She later received a master's degree in biology from Atlanta Univ... educator in several public school systems for ... eacher and administrator had some ... foundations and ...

Cool Chemistry Show

Chemistry at Work

Marc Pollack

President, Flix FX

"Everyone talks about 'Movie Magic,'" says Marc Pollack, president of the prestigious Hollywood special effects company Flix FX. "So I guess that makes me a magician." But Pollack is clearly more comedian than magician. The 'magic' he creates for movies like *Blackhawk Down*, *Men In Black* and *Cast Away*, in addition to scores of television commercials, museum installments and Las Vegas casinos, is the product not of mysterious hocus-pocus but rather fundamental principles of science. "One of the most important aspects of our work," he continues, "is to push the limits of how chemicals are designed to be used." Among other things, Pollack and his crew at Flix FX use vacuum-forming thermo-plastics to make tin-based silicon molds for everything from prehistoric creatures to futuristic robots. Through a combination of trial and error experimentation and traditional research science, they've perfected the process. "Silicon is what we call an R.T.V.," Pollack explains. "That stands for room temperature vulcanization. So depending on the type and amount of catalyst we use, the mold will cure at different rates and with slightly differing properties." By manipulating the ratio of silicon to catalysts they can make strong, realistic molds in the most efficient way possible. "Increasing the amount of catalyst will speed up the curing process but too much catalyst will shorten the life of the mold," he says. "Every job is different so determining that balance is one of our many challenges."

Pollack, who is now a master in the art of using chemicals like silicon, polypropylene, urethane and urethane elastimers, is not a chemist by trade. He actually graduated from film school at SUNY Purchase in the hopes of becoming "the next Steven Spielberg." Then, through a twist of fate, he became a special effects nut and eventually founded Flix FX in 1990. "Now," Pollack says, "Spielberg may one day come to me."

Special effects — Pollack creates both mechanical and physical — is an industry in a constant state of transformation. "The industry is always trying new stuff and that's exciting," Pollack says. "For instance, someone just developed a great water-based breakaway glass for stunts called Smash Glass. It's similar to fiberglass without the dangerous elements associated with that material and can be made to break into either large chunks or tiny little pieces. I can't wait to get my hands on it and break it over someone's head. That's part of my job these days and I love it."

832

Active Chemistry

These sections highlight an individual whose work or hobby is illustrative of the **Chapter Challenge. Physics** and **Chemistry At Work** speaks to the authenticity of the **Chapter Challenges.** The profiles illustrate how knowledge of physical science is important and valuable in different walks of life. The choice of profiles span the ethnic, racial, and gender diversity that we find in our nation.

© It's About Time

Active Physical Science and the National Science Education Standards

Active Physical Science was designed and developed to provide teachers with instructional strategies that model the following from the Standards:

Guide and Facilitate Learning

- Focus and support inquiries while interacting with students.
- Orchestrate discourse among students about scientific ideas.
- Challenge students to accept and share responsibility for their own learning.
- Recognize and respond to student diversity; encourage all to participate fully in science learning.
- Encourage and model the skills of scientific inquiry as well as the curiosity openness to new ideas and data and skepticism that characterize science.

Engage in ongoing assessment of their teaching and student learning

- Use multiple methods and systematically gather data about student understanding and ability.
- Analyze assessment data to guide teaching.
- Guide students in self-assessment.

Design and manage learning environments that provide students with time, space and resources needed for learning science

- Structure the time available so students are able to engage in extended investigations.
- Create a setting for student work that is flexible and supportive of science inquiry.
- Make available tools, materials, media, and technological resources accessible to students.
- Identify and use resources outside of school.

Develop communities of science learners that reflect the intellectual rigor of scientific attitudes and social values conducive to science learning

- Display and demand respect for diverse ideas, skills, and experiences of students.
- Enable students to have significant voice in decisions about content and context of work and require students to take responsibility for the learning of all members of the community.
- Nurture collaboration among students.
- Structure and facilitate ongoing formal and informal discussion based on shared understanding of rules.
- Model and emphasize the skills, attitudes and values of scientific inquiry.

Assessment Standards

- Features claimed to be measured are actually measured.
- Students have adequate opportunity to demonstrate their achievement and understanding.
- Assessment tasks are authentic and developmentally appropriate, set in familiar context, and engaging to students with different interests and experiences.
- Assesses student understanding as well as knowledge.
- Improve classroom practice and plan curricula.
- Develop self-directed learners.

© It's About Time

Active Physical Science Addresses Key NSES Recommendations

Scenario-Driven

There are 12 chapters in *Active Physical Science*. Each chapter begins with an engaging **Scenario** or project assignment that challenges the students and sets the stage for the learning activities and chapter assessments to follow. Chapter contents and activities are selectively aimed at providing the students with the knowledge and skills needed to address the introductory challenge, thus providing a natural content filter in the "less is more" curriculum.

Flexibly Formatted

Units are designed to stand alone, so teachers have the flexibility of changing the sequence of presentation of the units, omitting the entire unit, or not finishing all of the chapters within a unit. Although intended to serve as a full-year physics course, the units of *Active Physical Science* could be adapted to spread across a four-year period in an integrated high school curriculum.

Multiple Exposure Curriculum

The thematic nature of the course requires students to continually revisit fundamental physics principles throughout the year, extending and deepening their understanding of these principles as they apply them in new contexts. This repeated exposure fosters the retention and transferability of learning, and promotes the development of critical thinking skills.

Constructivist Approach

Students are continually asked to explore how they think about certain situations. As they investigate new situations, they are challenged to either explain observed phenomena using an existing paradigm or to develop a more consistent one. This approach can be helpful in including situations to abandon previously held notions in favor of the more powerful ideas and explanations offered by scientists.

Authentic Assessment

For the culmination of each chapter, students are required to demonstrate the usefulness of their newly acquired knowledge by adequately meeting the challenge posed in the chapter introduction. Students are then evaluated on the degree to which they accomplish this performance task. The curriculum also includes other methods and instruments for authentic assessments as well as non-traditional procedures for evaluating and rewarding desirable behaviors and skills.

© It's About Time

Cooperative Grouping Strategies

Use of cooperative groups is integral to the course as students work together in small groups to acquire the knowledge and information needed to address the series of challenges presented through the chapter scenarios. Ample teacher guidance is provided to assure that effective strategies are used in group formation, function, and evaluation.

Math Skills Development/Graphing Calculators and Computer Spreadsheets

The presentation and use of math in *Active Physical Science* varies substantially from traditional high school physics courses. Math, primarily algebraic expressions, equations, and graphs is approached as a way of representing ideas symbolically. Students begin to recognize the usefulness of math as an aid in exploring and understanding the world around them. Finally, since many of the students in the target audience are insecure about their math backgrounds, the course engages and provides instruction for the use of graphing calculators and computer spreadsheets to provide math assistance.

Minimal Reading Required

Because it is assumed that the target audience reads only what is absolutely necessary, the entire course is activity-driven. Reading passages are presented mainly within the context of the activities, and are written at the ninth grade level.

Use of Educational Technologies

Videos which capture students' attention explore a variety of the *Active Physical Science* topics. Opportunities are also provided for students to produce

their own videos in order to record and analyze events. Computer software programs make use of various interfacing devices.

Problem Solving

For the curriculum to be both meaningful and relevant to the target population, problem-solving related to technological applications and related issues is an essential component of the course. Problem solving ranges from simple numerical solutions where one result is expected, to more involved decision-making situations where multiple alternatives must be compared.

Challenging Learning Extensions

Throughout the text, a variety of **Stretching Exercises** are provided for more motivated students. These extensions range from more challenging design tasks, to enrichment readings, to intriguing and unusual problems. Many of the extensions take advantage of the frequent opportunities the curriculum provides for oral and written expression of student ideas.

© It's About Time

Cooperative Learning

Benefits of Cooperative Learning

Cooperative learning requires you to organize and structure a lesson so that students work with other students to jointly accomplish a task. Group learning is an essential part of balanced methodology. It should be blended with whole-class instruction and individual study to meet a variety of learning styles and expectations as well as maintain a high level of student involvement.

Cooperative learning has been thoroughly researched and agreement has been reached on a number of results. Cooperative learning:

- promotes trust and risk-taking
- elevates self-esteem
- encourages acceptance of individual differences
- develops social skills
- permits a combination of a wide range of backgrounds and abilities
- provides an inviting atmosphere
- promotes a sense of community
- develops group and individual responsibility
- reduces the time on a task
- results in better attendance
- produces a positive effect on student achievement
- develops key employability skills

As with any learning approach, some students will benefit more than others from cooperative learning. Therefore, you may question as to what extent you should use cooperative learning strategies. It is important to involve the student in helping decide which type of learning approaches they prefer, and to what extent each is used in the classroom. When students have a say in their learning, they will accept to a greater extent any method which you choose to use.

© It's About Time

Phases of Cooperative Learning Lessons

Organizational Pre-lesson Decisions

What academic and social objectives will be emphasized? In other words, what content and skills are to be learned and what interaction skills are to be emphasized or practiced?

What will be the group size? Or, what is the most appropriate group size to facilitate the achievement of the academic and social objectives? This will depend on the amount of individual involvement expected (small groups promote more individual involvement), the task (diverse thinking is promoted by larger groups), nature of the task or materials available and the time available (shorter time demands smaller groupings to promote involvement).

Who will make up the different groups? Teacher-selected groups usually have the best mix, but this can only happen after the teacher gets to know his/her students well enough to know who works well together. Heterogeneous groupings are most successful in that all can learn through active participation. The duration of the groups' existence may have some bearing on deciding the membership of groups.

How should the room be arranged? Practicing routines where students move into their groups quickly and quietly is an important aspect. Having students face-to-face is important. The teacher should still be able to move freely among the groups.

What Materials and/or Rewards Might be Prepared in Advance?

Setting the Lesson

Structure for Positive Interdependence: When students feel they need one another, they are more likely to work together—goal interdependence becomes important. Class interdependence can be promoted by setting class goals which all teams must achieve in order for class success.

Explanation of the Academic Task: Clear explanations and sometimes the use of models can help the students. An explanation of the relevance of the activity is importance. Checks for clear understanding can be done either before the groups form or after, but they are necessary for delimiting frustrations.

Explanation of Criteria for Success: Groups should know how their level of success will be determined.

Structure for Individual Accountability: The use of individual follow-up activities for tasks or social skills will provide for individual accountability.

Specification of Desired Social Behaviors: Definition and explanations of the importance of values of social skills will promote student practice and achievement of the different skills.

Monitoring/Intervening During Group Work

Through monitoring students' behaviors, intervention can be used more appropriately. Students can be involved in the monitoring by being "a team observer," but only when the students have a very clear understanding of the behavior being monitored.

Interventions to increase chances for success in completing the task or activity and for the teaching of collaborative skills should be used as necessary—they should not be interruptions. This means that the facilitating teacher should be moving among the groups as much as possible. During interventions, the problem should be turned back to the students as often as possible, taking care not to frustrate them.

Evaluating the Content and Process of Cooperative Group Work

Assessment of the achievement of content objectives should be completed by both the teacher and the students. Students can go back to their groups after an assignment to review the aspects in which they experienced difficulties.

When assessing the accomplishment of social objectives, two aspects are important: how well things proceeded and where/how improvements might be attempted. Student involvement in this evaluation is a very basic aspect of successful cooperative learning programs.

Organizing and Monitoring Groups

An optimum size of group for most activities appears to be four; however, for some tasks, two may be more efficient. Heterogeneous groups organized by the teacher are usually the most successful. The teacher will need to decide what factors should be considered in forming the heterogeneous groups. Factors which can be considered are: academic achievement, cultural background, language proficiency, sex, age, learning style, and even personality type.

Level of academic achievement is probably the simplest and initially the best way to form groups. Sort the students on the basis of marks on a particular task or on previous year's achievement. Then choose a student from each quartile to form a group. Once formed, groups should be flexible. Continually monitor groups for compatibility and make adjustments as required.

Students should develop an appreciation that it is a privilege to belong to a group. Remove from group work any student who is a poor participant or one who is repeatedly absent. These individuals can then be assigned the same tasks to be completed in the same time line as a group. You may also wish to place a ten percent reduction on all group work that is completed individually.

The chart on the next page presents some possible group structures and their functions.

What Does Cooperative Learning Look Like?

During a cooperative learning situation, students should be assigned a variety of roles related to the particular task at hand. Following is a list of possible roles that students may be given. It is important that students are given the opportunity of assuming a number of different roles over the course of a semester.

Leader:

Assigns roles for the group. Gets the group started and keeps the group on task.

Organizer:

Helps focus discussion and ensures that all members of the group contribute to the discussion. The organizer ensures that all of the equipment has been gathered and that the group completes all parts of the activity.

Recorder:

Provides written procedures when required, diagrams where appropriate and records data. The recorder must work closely with the organizer to ensure that all group members contribute.

Researcher:

Seeks written and electronic information to support the findings of the group. In addition, where appropriate, the researcher will develop and test prototypes. The researcher will also exchange information gathered among different groups.

Encourager:

Encourages all group members to participate. Values contributions and supports involvement.

Checker:

Checks that the group has answered all the questions and the group members agree upon and understand the answers.

Diverger:

Seeks alternative explanations and approaches. The task of the diverger is to keep the discussion open. "Are other explanations possible?"

© It's About Time

Some Possible Group Structures and Their Functions*

Structure		Brief Description	Academic and Social Functions
Team Building	Round-robin	Each student in turn shares something with his/her teammates.	Expressing ideas and opinions, creating stories. Equal participation, getting acquainted with each other.
Class Building	Corners	Each student moves to a group in a corner or location as determined by the teacher through specified alternatives. Students discuss within groups, then listen to and paraphrase ideas from other groups.	Seeing alternative hypotheses, values, and problem solving approaches. Knowing and respecting differing points of view.
Mastery	Numbered heads together	The teacher asks a question, students consult within their groups to make sure that each member knows the answer. Then one student answers for the group in response to the number called out by the teacher.	Review, checking for knowledge comprehension, analysis, and divergent thinking. Tutoring.
	Color coded co-op cards	Students memorize facts using a flash card game or an adaption. The game is structured so that there is a maximum probability for success at each step, moving from short to long-term memory. Scoring is based on improvement.	Memorizing facts. Helping, praising.
	Pairs check	Students work in pairs within groups of four. Within pairs students alternate — one solves a problem while the other coaches. After every problem or so, the pair checks to see if they have the same answer as the other pair.	Practicing skills. Helping, praising.
Concept Development	Three-step interview	Students interview each other in pairs, first one way, then the other. Each student shares information learned during interviews with the group.	Sharing personal information such as hypotheses, views on an issue, or conclusions from a unit. Participation, involvement.
	Think-pair-share	Students think to themselves on a topic provided by the teacher; they pair up with another student to discuss it; and then share their thoughts with the class.	Generating and revising hypotheses, inductive and deductive reasoning, and application. Participation and involvement.
	Team word-webbing	Students write simultaneously on a piece of paper, drawing main concepts, supporting elements, and bridges representing the relation of concepts/ideas.	Analysis of concept into components, understanding multiple relations among ideas, and differentiating concepts. Role-taking.
Multifunctional	Roundtable	Each student in turn writes one answer as a paper and a pencil are passed around the group. With simultaneous roundtables, more than one pencil and paper are used.	Assessing print knowledge, practicing skills, recalling information, and creating designs. Team building, participation of all.
	Partners	Students work in pairs to create or master content. They consult with partners from other teams. Then they share their products or understandings with the other partner pair in their team.	Mastery and presentation of new material, concept development. Presentation and communication skills.
	Jigsaw	Each student from each team becomes an "expert" on one topic area by working with members from other teams assigned to the same topic area. On returning to their own teams, each one teaches the other members of the group and students are assessed on all aspects of the topic.	Acquisition and presentation of new material review and informed debate. Independence, status equalization.

* Adapted from Spencer Kagan (1990), "*The Structural Approach to Cooperative Learning*," Educational Leadership, December 1989/January 1990.

© It's About Time

Active Listener:

Repeats or paraphrases what has been said by the different members of the group.

Idea Giver:

Contributes ideas, information, and opinions.

Materials Manager:

Collects and distributes all necessary material for the group.

Observer:

Completes checklists for the group.

Questioner:

Seeks information, opinions, explanations, and justifications from other members of the group.

Reader:

Reads any textual material to the group.

Reporter:

Prepares and/or makes a report on behalf of the group.

Summarizer:

Summarizes the work, conclusions, or results of the group so that they can be presented coherently.

Timekeeper:

Keeps the group members focused on the task and keeps time.

Safety Manager:

Responsible for ensuring that safety measures are being followed, and the equipment is clean prior to and at the end of the activity.

Group Assessment

Assessment should not end with a group mark. Students and their parents have a right to expect marks to reflect the students' individual contributions to the task. It is impossible for you as the instructor to continuously monitor and record the contribution of each individual student. Therefore, you will need to rely on the students in the group to assign individual marks as merited.

There are a number of ways that this can be accomplished. The group mark can be multiplied by the number of students in the group, and then the total mark can be divided among the students, as shown in the graphics that follow.

© It's About Time

Activity:______________________

Group Mark: 8/10

Number in Group: 4

Total Marks: 32/40

Distribution of Marks

Student's Name	Mark	Signature
Ahmed	8/10	____________________
Jasmin	8/10	____________________
Mike	7/10	____________________
Tabitha	9/10	____________________

Another way to share group marks is to assign a factor to each student. The mark factors must total the number of students in the group. The group mark is then multiplied by this factor to arrive at each student's individual mark which best represents their contribution to the task, as shown below.

Activity:______________________

Group Mark: 8/10

Number in Group: 4

Mark Factors and Individual Marks

Student's Name	Mark Factor	Individual Mark	Signature
Ahmed	1.0	8/10	____________________
Jasmin	1.0	8/10	____________________
Mike	0.9	7.2/10	____________________
Tabitha	1.1	8.8/10	____________________
Total Mark Factor	4		

In any case, students must sign to show that they are in agreement with the way the individual marks were assigned.

You may also wish to provide students with an **Assessment Rubric** similar to the one shown which they can use to assess the manner in which the group worked together.

© It's About Time

Assessment Rubric for Group Work: Individual Assessment of the Group

Individual's name: ______________________________

Names of group members: ______________________________

Name of activity: ______________________________

Circle the appropriate number: #1 is excellent, #2 is good, #3 is average, and #4 is poor.

1. The group worked cooperatively. Everyone assumed a role and carried it out.	1	2	3	4
2. Everyone contributed to the discussion. Everyone's opinion was valued.	1	2	3	4
3. Everyone assumed the roles assigned to them.	1	2	3	4
4. The group was organized. Materials were gathered, distributed, and collected.	1	2	3	4
5. Problems were addressed as a group.	1	2	3	4
6. All parts of the task were completed within the time assigned.	1	2	3	4
Comments:				

If you were to repeat the activity, what things would you change?

© It's About Time

Chapter 5

LET US ENTERTAIN YOU

Chapter 5- Let Us Entertain You

National Science Education Standards

Chapter Challenge

The use of sound and light in the entertainment industry provides the scenario for this chapter. Students are challenged to design a sound and light show that demonstrates the physics principles they learned, yet is low budget. They are limited to using only sounds that come from human voices or homemade instruments and light from conventional household lamps.

Chapter Summary

To gain understanding of science principles necessary to meet this challenge, students work collaboratively on activities to learn about wave motion, sound waves, light rays, and how mirrors and lenses change the direction of light rays and result in formation of images. They learn to use the iterative process of engineering design, refining designs based on the physics they learn. These experiences engage students in the following content from the National Science Education Standards.

Content Standards

Unifying Concepts

- Evidence, models and explanations
- Constancy, change, and measurement

Science as Inquiry

- Identify questions and concepts that guide scientific investigations
- Use technology and mathematics to improve investigations
- Design and conduct scientific investigations

Science and Technology

- Abilities of technological design
- Identify a problem or design an opportunity
- Propose designs and choose between alternate solutions

Physical Science

- Structure and properties of matter
- Motions and forces
- Interactions of energy and matter

© It's About Time

Key Physics Concepts and Skills

Activity Summaries	Physics Principles
Activity 1: Making Waves Students begin the chapter by making waves with a Slinky®, observing pulses, periodic, standing, and compressional waves. From these observations, then measurements, they establish the relationships among wavelength, frequency, and speed of the wave.	• Wave motion • Periodic, standing, and compressional waves • Wavelength – Frequency, speed
Activity 2: Sounds in Strings To connect waves to sound, students observe the vibration of a plucked string and compare how vibration and pitch vary when the tension of the string changes. They then explore the affect on vibrations and pitch when the length of the string is changed. Reading explains the physics concepts in the observed phenomena.	• Sound waves • Wave motion • Tension and pitch • Frequency
Activity 3: Sounds from Vibrating Air Drinking straws and test tubes filled with water are used to model instruments that use columns of vibrating air to produce sounds. The relationship of pitch to length of the column of air provides another look at frequency and wavelength, helping students understand how sound is produced by compressional and standing waves.	• Sound waves • Wave motion • Compressional waves • Frequency and wavelength
Activity 4: Reflected Light In this activity, students begin looking at how light can be incorporated into the chapter challenge. They explore the result of changing the angle at which light rays are aimed at a mirror and learn to predict and control where images will be visible.	• Light rays • Reflection of light • Real images—focus, focal length
Activity 5: Curved Mirrors Shining a light beam on concave and convex mirrors increases student understanding of the variables that are involved in creating an image. They apply what they have learned to predict the path of a light beam reflected off a mirror.	• Reflection of light • Real images—focus, focal length • Controlling variables
Activity 6: Refraction of Light In this activity, a block of gelatin allows students to explore what happens when light goes from air into another substance. They observe and measure the angle of incidence and the angle of refraction as they learn about Snell's Law and how to mathematically predict where the beam of light can be observed.	• Refraction of light • Snell's Law
Activity 7: Effect of Lenses on Light Shining a light through different lenses enables students to observe how focal length and the size of the image changes as the light source moves closer to, then farther away from a lens. They then consider how the variables in this phenomenon can enhance their sound and light show for the chapter challenge.	• Refraction of light • Lenses and image formation • Focus, focal length
Activity 8: Color This final activity adds to the study of light with observations of shadows. By carefully tracing the light ray and noting the areas without any light and the areas of gray light, students begin to learn about diffusion of light. They extend their investigations to include the effect of shining different colored lights on objects.	• Light and shadows • White light • Color addition

© It's About Time

GETTING STARTED WITH EQUIPMENT NEEDED TO CONDUCT THE ACTIVITIES.

Items needed—not supplied in Material Kits

Preparing the equipment needed for each activity in this chapter is an important procedure. There are some items, however, needed for the chapter that are not supplied in the It's About Time material kitpackage. Many of these items may already be in your school and would be an unnecessary expense to duplicate. Please read carefully the list of items to the right which are not found in the supplied kits and locate them before beginning activities.

© It's About Time

Items needed—not supplied by It's About Time:

- **Copies of Blackline Master of Sine Waves**
- **Copies of Blackline Master of Wavelength Tables**
- **Copies Of Blackline Master Pitch vs. Tube Length**
- **Ripple Tank**
- **Pencil**
- **Water, about 300 cc in a container**
- **Overhead Projector**

© It's About Time

Equipment List For Chapter 5

PART	ITEM	QTY	ACTIVITY	TO SERVE
22-2000-C1	40-W Bulb and Socket, Ceramic w/Switch	1	5, 7, 8	group
AC-2279-C1	Acrylic Block	1	6	group
AS-6105-C1	Acetate Sheet, 8 cm Square	1	7	group
BS-1608-C1	AA Battery	1	4	group
BS-5903-C1	Bulb, Clear 100 Watt	1	7	group
BU-0004-C1	Bulb, Set/4, Red	1	8	group
BU-0003-C1	Bulb, Set/4, Blue	1	8	group
BU-0002-C1	Bulb, Set/4, Green	1	8	group
BU-0001-C1	Bulb, Set/4, White	1	8	group
CS-3246-C1	C-Clamp	1	2	group
CS-6111-C1	Cardboard Sheet, 14x14 Square	1	4, 5, 6	group
CS-5011-C1	Unlined Index Cards	100	1, 5, 6	group
FF-7704-C1	File Folder	6	8	group
FL-0030-C1	Fishing Line Spool	1	2	group
GS-5016-C1	Goggles, Safety	1	2	group
LH-6534-S-C1	Laser Pointer	1	4, 5, 6	group
LS-0015-C1	1.5 Volt Miniature Light Bulb	1	4	group
LS-1001-C1	Lens, Convex Focal Point 10 cm	1	7	group
LS-4119-C1	Stand for Clip-on Lamp	3	8	group
LS-6600-C1	Light Source for Optical Bench Apparatus	1	4	group
LS-9000-C1	Lamp, Clip-On w/Shade	3	8	group
MH-2621-C1	Mass Hanger	1	2	group
MS-1006-C1	Convex/Concave Mirror Surface	1	5	group
MS-1425-C1	Marker, Felt-Tip	1	7	group
MS-2114-C1	Mirror, Plane, 3x4x1/2	2	4	group
	Makeup Mirror, Convex/Concave Surface	1	4	group
OH-1404-C1	Optical Bench Apparatus	1	4	group
PH-3070-C1	Pulley on Mount	1	2	group
PM-0521-C1	Protractor	1	4, 6	group
PR-0001-C1	Right Angle Prism	1	4, 5, 6	group
RH-6131-C1	Rod, Glass	1	4, 5, 6	group
RS-2826-C1	Ruler, Metric	1	3, 4, 5, 6, 7	group
SH-3639-C1	Heavy Duty Slinky	1	1	group
SM-1676-C1	Stick, Meter, 100 cm, Hardwood	1	1, 2, 5, 7	group
SS-1281-C1	Steel Scissors	1	1, 3, 7, 8	group
SS-6113-C1	Drinking Straws	10	3	group
SS-6133-C1	Support for Mirrors	2	4	group
SS-6135-C1	Screen, White	4	8	group
SS-7778-C1	Stopwatch	1	1	group
TS-0337-C1	Test Tube	5	3	group
TS-2662-C1	Tape, Masking 3/4x60 yds.	1	1, 5	group
WH-0500-C1	Hooked Mass Weight, 500 Grams	2	2	group
WH-9012-C1	Slotted Weight Set	1	2	group
WS-6136-C1	Wood Block to Raise Light Source	3	6	group
ZZ-0066-C1	66 qt. Clear Plastic Storage Container	1		group
	THINGS NEEDED NOT SUPPLIED			
	Ripple Tank	1	1	class
	Pencil	1	2	group
	Key or other small metal object	1	2	group
	Water, about 300 cc in a container	1	3	group
	Various Wind Instruments	1	3	class
	Sound probe and software	1	3	class
	Small amount of play sand	1	4	group
	Oscilloscope	1	4	class
	Cosmetics Mirror	1	6	group
	Overhead Projector	1	8	class
	File cards one with an arrow shaped opening	1	8	group

© It's About Time

ITEMS NEEDED – NOT SUPPLIED BY IT'S ABOUT TIME				
PART	ITEM	QTY	ACTIVITY	TO SERVE
	Copy of Blackline Master of Sine Waves	1	1	student
	Copy of Blackline Master of Wavelength Tables	1	1	student
	Copy of Blackline Master Pitch vs. Tube Length	1	3	student
	Ripple Tank	1	1	class
	Pencil		2	
	Water, about 300 cc in a container		3	
	Overhead Projector		8	class

Organizer for Materials Available in Teacher's Edition

Activity in Student Text	Additional Material	Alternative / Optional Activities
ACTIVITY 1: Making Waves, p. 326	Amplitude and Wavelength Measurements, p. 33 Sine Waves for Wave Model, p. 34	Activity 1 A: Exploring Water Waves p. 32
ACTIVITY 2: Sounds in Strings, p. 340		
ACTIVITY 3: Sounds from Vibrating Air, p. 346	Pitch vs. Tube Length, p. 55	Activity 3 A: Wind Instruments and Resonant Frequencies, p. 54
ACTIVITY 4: Reflected Light, p. 354	Angles of Incidence and Reflection, p. 69	Activity 4 A: Low-Tech Reflection and Mirror Symmetry, p. 68
ACTIVITY 5: Curved Mirrors, p. 362	Bulb and Image Distances, p. 84 Graph Paper, p. 85	Activity 5 A: Low-Tech Reflection: Curved Mirror, p. 83
ACTIVITY 6: Refraction of Light, p. 371	Sines of Angles of Incidence and Reflection, p. 95	
ACTIVITY 7: Effects of Lenses on Light, p. 377	Determining Distances of Objects, and Images and Appearances, p. 108	
ACTIVITY 8: Color, p. 386		

© It's About Time

Scenario

Most entertainment today comes from the communication of sound and light signals. You look forward to television shows, movies, and rock concerts. The sound signals that entertain you come from voices or musical instruments. Light signals make the images you see on TV or in the movies. Specially designed light patterns add to the effect of a rock concert.

Challenge

You have been made part of a committee to design a two- to four-minute sound and light show to entertain other students your age. But unlike the professionals, you have neither the funds nor the technology available to rock stars or MTV™ productions. All the sounds you use must come from musical instruments or sound makers that you build yourself, or from human voices. Some of these sounds may be prerecorded and then played back during your show. If your teacher has a laser and is willing to allow you to use it, you may do so. All other light must come from conventional household lamps.

324

© It's About Time

Chapter and Challenge Overview

Chapter 5 challenges groups of students to produce a two–to four–minute light show. In addition, each student must write a report that lists the physics concepts used in the show, gives an example of each, and explains why each concept is important. The shows are evaluated for creativity, and the reports are evaluated for the understanding of the concepts.

Sound and light are the basis of much of the entertainment industry. Beyond fireworks and light shows, sound is transmitted by radio, sound and light are transmitted by television, and the videotape and CD store the information for bringing us the sound and light on demand. A sound and light show is an excellent opportunity for your students to show what they have learned and to have some fun at the same time.

Students begin the unit in **Activity 1** by making pulses and standing waves on the Slinky®. They determine the relationship among speed, frequency, and wavelength. Also, they build a model to help understand wave motion. Next, students begin the study of musical instruments. In **Activity 2** they investigate the sounds made by a vibrating string. They observe the effect of changes in string length and tension on the pitch the string produces. Students continue the study of musical instruments by investigating the resonance of pipes in **Activity 3**. They observe how the pitch produced depends on the length of the pipe and whether or not one end is closed. Through reading, they then relate these observations to the patterns of standing waves in a tube.

Next, the students begin the investigation of mirrors. In **Activity 4**, they investigate reflection from a plane mirror. They measure the angles of incidence and refraction for reflected light beams, locate the reflected image, and observe the reversals in reflections of letters of the alphabet. Finally, they investigate multiple reflections. **Activity 5** introduces curved mirrors. In this activity, students investigate concave and convex mirrors. Students begin by shining parallel light beams at curved mirrors to find the focal length and the location of the focus. With a light bulb, students look for real and virtual images and investigate image distance vs. object distance.

To prepare for the study of lenses, students investigate refraction in **Activity 6** by shining a beam of light into a rectangular acrylic block. The students measure the angles of incidence and refraction. They also observe total internal reflection and measure the critical angle. Having studied refraction, students then explore convex lenses in **Activity 7**. They observe the real image made by a convex lens and how its size and position change as the object distance changes. To prepare for the light show, the students project the image of a slide onto a wall. In **Activity 8** the students investigate additive color mixing. They observe shadows of objects illuminated by combinations of colored lights. For each such combination they make a drawing to record the pattern of the shadow colors.

© It's About Time

Criteria

Work with your classmates to agree on the relative importance of the following assessment criteria. Each item in the list has a point value given after it, but your class must decide what kind of grading system to use.

1. The variety and number of physics concepts used to produce the light and sound effects: 30 points

   four or more concepts: 30 points

   three concepts: 25 points

   two concepts: 20 points

   one concept: 10 points

2. Your understanding of the physics concepts: 40 points

   Following your production, you will be asked to:

   a) Name the physics concepts that you used. 10 points
   b) Explain each concept. 10 points
   c) Give an example of something that each concept explains or an example of how each concept is used. 10 points
   d) Explain why each concept is important. 10 points

   As a class, you will have to decide if your answers will be in an oral report or a written report.

3. Entertainment value: 30 points

   Your class will need to decide on a way to assign points for creativity. Note that an entertaining and interesting show need not be loud or bright.

You will have a chance later in the chapter to again discuss these criteria. At that time, you may have more information on the concepts and how you might produce your show. You may want to then propose changes in the criteria and the point values.

325

© It's About Time

Assessment Rubric: Sound and Light Show

Meets the standard of excellence. **5**	• Four or more physics concepts are appropriately used to produce the sound and light show. • All sounds produced have excellent quality and come from more than one musical instrument or sound maker designed and constructed by the student. • Additional research, beyond basic concepts presented in the chapter, is evident. • Physics concepts are integrated in a creative manner into the show. • The show is entertaining and interesting.
Approaches the standard of excellence. **4**	• Three physics concepts are appropriately used to produce the sound and light show. • All sounds produced have good quality and come from more than one musical instrument or sound maker designed and constructed by the student. • Physics concepts are integrated in a creative manner into the show. • The show is entertaining.
Meets an acceptable standard. **3**	• Two physics concepts are appropriately used to produce the sound and light show. • Sounds produced have a reasonable quality and come from a musical instrument or sound maker designed and constructed by the student. • The presentation shows creativity. • The show has some entertainment value.
Below acceptable standard and requires remedial help. **2**	• One physics concept is used to produce the sound and light show. • An attempt has been made to design and construct a musical instrument. Quality of sound produced is extremely poor. • The presentation lacks creativity. • The show has little entertainment value.
Basic level that requires remedial help or demonstrates a lack of effort. **1**	• No physics concepts are used in the production of the sound and light show. • No attempt has been made to design and construct a musical instrument. • The presentation lacks creativity. • The show has no entertainment value.

© It's About Time

For use with *Let Us Entertain You*, Chapter 5

Assessment Rubric: Written Report

Meets the standard of excellence. **5**	• Four physics concepts are named and clearly explained. • The importance of each concept is explained. • An example is provided for each concept. • Scientific vocabulary and symbols for units are used consistently and precisely. • Sentence structure is consistently controlled. Spelling, punctuation, and grammar are consistently used in an effective manner. • Where appropriate, data is organized into tables or presented by graphs or diagrams.
Approaches the standard of excellence. **4**	• Three physics concepts are named and clearly explained. • The importance of each concept is explained. • An example is provided for each concept. • Scientific vocabulary and symbols for units are used appropriately. • Sentence structure is consistently controlled. Spelling, punctuation, and grammar are consistently used in an effective manner. • Some data is organized into tables or presented by graphs or diagrams.
Meets an acceptable standard. **3**	• Two physics concepts are named and clearly explained. • The importance of each concept is explained. • An example is provided for each concept. • Use of scientific vocabulary and symbols for units is evident. • Sentence structure is generally controlled. Spelling, punctuation, and grammar do not impede the meaning. • Very limited presentation of data by diagrams, tables, or graphs.
Below acceptable standard and requires remedial help. **2**	• One physics concept is named and explained. • An attempt is made to explain the importance of the concept. • Limited use of scientific vocabulary and symbols for units. Usage is not always consistent or precise. • Sentence structure is poorly controlled. Spelling, punctuation, and grammar impedes the meaning. • No presentation of data by diagrams, tables, or graphs.
Basic level that requires remedial help or demonstrates a lack of effort. **1**	• No physics concepts are named and explained or physics concepts named are explained incorrectly. • No attention to use of scientific vocabulary or scientific symbols for units of measurement. • Sentence structure is poorly controlled. Spelling, punctuation, and grammar impedes the meaning.

© It's About Time

For use with *Let Us Entertain You*, Chapter 5

What is in the Physics InfoMall for Chapter 5?

While teaching a physics course, it is often nice to have a source of information other than only the textbook used in the course. One source of information for not just physics, but how to teach physics, is the Physics InfoMall CD-ROM. This duo-platform CD-ROM is a huge database containing 19 textbooks, thousands of articles from physics journals, thousands of homework-style problems (many with solutions available to you), numerous demonstration and laboratory ideas, catalogs, and books covering related topics. In addition to this, you can also find historical information, and more.

The organization of the database is similar to a shopping mall; there are various stores that contain related types of information. The Textbook Trove has textbooks, the Book Basement has related material, the Article & Abstracts Attic has thousands of journal articles, etc. You can browse the stores, just looking to see if there is something interesting you have not seen before, or you can ask for specific information. The database is not interactive; it is a collection of text and graphics, with hyperlinks and a powerful search engine. The power of the search engine in nearly as important as the volume of material on the CD-ROM.

If you are already familiar with using the search engine, you may wish to skip this paragraph and the next. The search engine for the InfoMall enables the user to locate almost any word or passage included in the many articles, textbooks, etc. If you use a Macintosh (remember—the CD-ROM works on both PC and Mac), you can search the entire database (that is, all stores) at once. This discussion was prepared using a Mac. If you use a PC, there will be minor differences but the database is identical. The same information can be found on either platform. There are two categories for searches: Simple and Compound. Simple searches look for a single word or phrase. Compound searches can locate multiple words or phrases that may not occur together. The Compound search also provides search options that are not available in the Simple search. Since a Compound search can always be done for a single word, it provides the same function as a Simple search. All searches in this discussion will be done using the Compound search.

The search option can be found under the "Functions" menu. When you select "Compound Search", you will get a window that has all the options you need for just about any search you are likely to want. Among these choices are database selection (which store or stores you want searched), category (perhaps only certain articles), and search words. You can enter any words or words you want the engine to search for. These words can be combined with the logical operators AND, OR, and NOT. That is, you can find passages that contain, for example, "phone" AND "Bell" but NOT "monopoly". In addition, you can use an asterisk * as a wild character. That is, search word "fun*" will search for any word beginning with "fun", such as "fun", "funny", "funding", "fundamental", etc. Wild characters allow great freedom in searching, but as this example shows, you need to be careful. And finally, if your search is too broad, you will get "Too Many Hits" and you will have to restrict your search parameters to something that will not be found so many times on the CD-ROM. Such restriction techniques include searching in fewer stores, using fewer wild characters, or searching for less common words followed by a search of only the search hits. With a little practice, you should have very little trouble finding the information you desire.

When using the Physics InfoMall CD-ROM, it is easy to get in the "search" mode, where every time you want something, you engage the search engine. Don't forget that that InfoMall is also great for simply browsing. While searching for items, it is good to keep an open mind. When looking up an article, for example, there might be another article in the same journal that you can use later. You can locate it later by browsing, or place a Bookmark or Note there, so you can quickly come back to it when convenient. (Notes and Bookmarks are not used in this discussion, but you may find them useful. Consult the User's Guides for more information. Incidentally, the User's Guides are on the CD-ROM in the Utility Closet.)

© It's About Time

ACTIVITY 1
Making Waves

Background Information

Waves are a collective phenomena. The neighboring parts of the wave medium—the matter the wave moves through—interact and propogate a disturbance from one place to another. For spring waves, each part of the spring influences the neighboring parts through the force of the stretched spring. A disturbance in one part of the spring creates forces on neighboring parts, and the disturbance moves. The disturbance is a stretching of the spring from its equilibrium position, which is a straight spring. In **Activity 1**, the students disturb this equilibrium by whipping one end of spring back and forth to make a pulse, as shown in the drawing on page 327 of the Student Edition.

In the pulse, the spring coils are stretched. This stretching exerts forces on these coils and pulls them back towards the equilibrium position. The stretching also puts forces on the nearby parts of the spring. These forces are reaction forces. When part of the spring is pulled one way, these reaction forces pull adjacent parts the other way. That is how the pulse moves.

Notice that the spring does not have to be motionless to be in the equilibrium position. Look again at the diagram of the standing wave on page 328 of the Student Edition. For all standing waves, the whole spring goes through the equilibrium position twice in each wave cycle. As suggested above, the spring is moving at this time, so all of the energy in the spring is kinetic as the spring moves through the equilibrium position.

The most important wave properties are amplitude, frequency, wavelength, and speed. The first example discussed in teacher notes for **WDYT** is how a tsunami grows. For this, amplitude depends on wave speed. When two waves pass through each other, the two amplitudes simply add at every point, as shown. This is the principle of superposition.

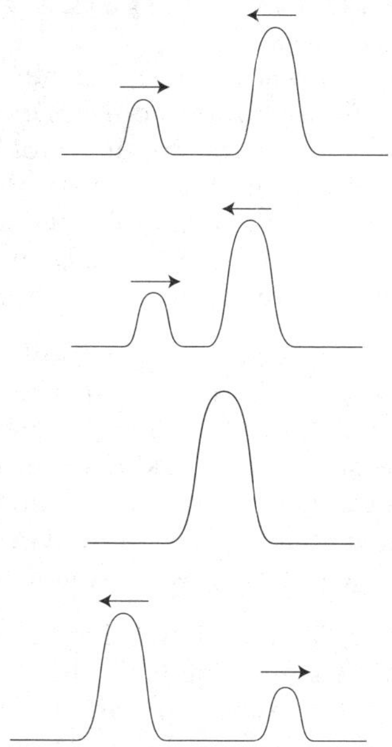

If you took a snapshot of the wave, with a meter stick right above the crests, you could measure the wavelength (the distance between adjacent crests). If you made a movie of the motion, with a clock in the frame, you could measure the frequency of the waves by timing one complete cycle. The period is simply the inverse of the frequency. Notice how the units invert as well. One hertz is a cycle per second. It is the inverse of the period, which is measured in seconds (cycles are dimensionless).

Suppose you are watching water waves washing over a rock. If you watch a particular crest for a time of one wave period, it moves one wavelength. The frequency gives the number of wave periods in one second. Thus the speed is simply the wavelength times the frequency.

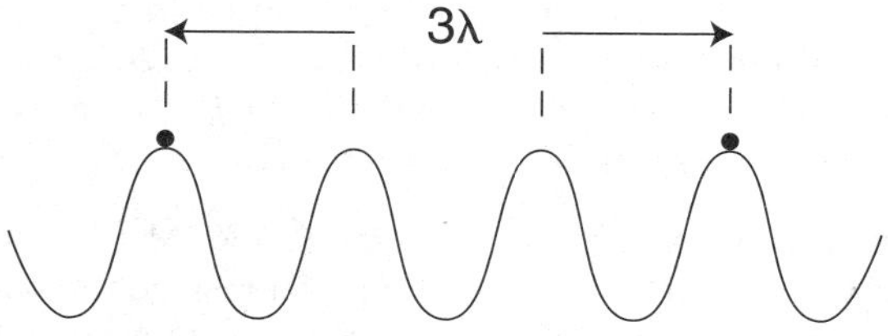

In 1s, a point on this wave moves 3λ.
Wave speed = λf

A wave moves through a medium. The spring is the medium for Slinky waves, water is the medium for water waves, and air is the medium for sound waves. But there is one wave that does not require a medium. That is the electromagnetic wave.

© It's About Time

Electromagnetic waves move freely through space, which is an excellent vacuum. The Sun gives off light and radio waves. Various distant objects in the universe give off every kind of electromagnetic radiation, and all these waves travel through the emptiness of space to reach the Earth.

In a standing wave, the spring vibrates but no waves are seen to move along its length. Standing waves occur only at certain frequencies, which correspond to simple patterns of motion. In the lowest-frequency standing wave on a spring, the spring bows out in the middle, with the ends, of course, fixed in place. If the frequency is doubled, there is a node, a point of no motion, in the middle of the spring. (See the drawings on page 328.)

In a standing wave, a small amount of oscillation with just the right frequency can excite a huge oscillation in the vibrating system, a phenomena called resonance. Moving the frequency away from a resonant frequency produces disorganized motion with a very small amplitude.

Active-ating the Physics InfoMall

As you might imagine, there is plenty of information on "waves" on the InfoMall. Let's take a hint from the **What Do You Think?** question, and search for tsunamis. Since this word may be spelled poorly (perhaps by various authors on the InfoMall?), we can save some time and trouble by searching for "tsunam*". Note that by doing this, some spelling errors are avoided, plus we get the benefit of a search for all forms of the word (how many words begin with these six letters?). While we might not expect much from such a search, we are greatly rewarded for our efforts.

The first hit is from The Fascination of Physics, Chapter 14, Making Waves, found in the Textbook Trove. This has great information, graphics, and even the necessary wave vocabulary. All this is live text, meaning you can copy and paste your findings right into your own handouts. This includes the graphics! Much of the information you may want related to **Activity 1** is right in this single search hit. There are nice drawings of standing waves, discussion of wave speed, interference, and much more.

Other hits include a rather inexpensive experiment with a "tsunami": "Tsunamis, sealing wax, and string:" Citation: F. L. Curzon B. Neilson Physics Department, University of British Columbia, Vancouver, British Columbia V6T 1W5, Canada *Am. J. Phys.*, Vol. 44, No. 11, November 1976 Pages 1073 - 1076: A ground glass screen on the side of a wave tank has been used to record the profile left by the peak of a tsunami propagating along a wave tank with a sloping base. Using traditional sealing wax-and-string technology and old-fashioned but cheap data processing, we have carried out a $12 verification of tsunami growth. The features of the experiment are presented in enough detail to enable anyone with the necessary $20 (allowing for inflation) to investigate interesting anomalies which personal circumstances prevented us from pursuing.

You are encouraged to try this search for yourself. These two items were not the only hits from the search. Of course, you might want other types of information. Suppose you want something on "wave speed." Search with these keywords, and you find that our earlier hit is again first. Look a little further down the list and you can find some information on teaching about waves in *Teaching Physics: A Guide For The Non-Specialist*, found in the Book Basement: "Teaching of the subject of waves usually starts with an exploration of wave behaviour on springs. There are two types of springs commonly in use for this, a tightly coiled fairly massive spring, about 2 m in length, and the Slinky. The latter is very suitable for demonstration but not for use by pupils as it easily ties itself in knots. Pupils can perform set exercises with these springs to explore the factors determining wave speed, what happens to waves that cross and what happens to a wave on reflection. Essentially this is an exercise in observation which pupils enjoy but find difficult to draw the correct conclusions." You should look this up to find what difficulties students have with this topic.

Of course, waves are discussed in nearly all physics textbooks. This means that many of the words you may want to search for occur so many times that you may get too many search hits to handle effectively. This is where browsing may help (of course, you can always limit searches to single stores, for example). Simply go to the InfoMall Entrance and click on the Textbook Trove. Select any text you like. Note that there are different levels

© It's About Time

of texts, from conceptual to those that use calculus. Choose the one that best suits your students. Then simply look over the table of contents of the book you select, find the appropriate chapter, and read away! You can easily find the same information that is contained in **Physics Talk**, along with more examples of how to use the equations.

For activities, you may want to search the Demo & Lab Shop. Or use a rather crude search: "wave demo*". This produces a few hits, and some are good, but you can probably (and should try to) find better searches.

As an example of other ways you may wish to use the InfoMall, see **Physics To Go Step 3**. Students are asked to consider using a photograph to take measurements. Search the InfoMall for "measurement*" AND "photograph*". You will find several references that discuss the benefits of doing exactly that — using photographs to make measurements. Check these out.

Planning for the Activity

Time Requirements:

two class periods

Materials Needed

For the class:

- ripple tank

For each group:

- masking tape, 3/4" × 60 yds
- meter stick, 100 cm, hardwood
- Slinky®, heavy duty
- steel scissors
- stopwatch
- unlined index cards

NOTES

© It's About Time

NOTES

© It's About Time

Activity Overview

To help understand basic wave properties, students make pulses on a Slinky and measure the speed of the pulse for different wave amplitudes. They create standing waves on the Slinky and determine the relationship among speed, frequency, and wavelength. Students make both transverse and compressional waves on the Slinky and contrast the motion of the two kinds of waves. Finally, they build a model to help relate basic wave properties to observations of wave motion.

Student Objectives

Students will:

- Observe the motion of a pulse.
- Measure the speed of a wave.
- Observe standing waves.
- Investigate the relationship among wave speed, wavelength, and frequency.
- Make a model of wave motion.

Answers for the Teacher Only

What Do You Think?

A tsunami begins when an earthquake shifts the ocean bottom and creates a small step in the height of the water surface. This step is usually only a few tens of centimeters high. This small disturbance in the surface spreads out in circular waves from the point above the earthquake, but the amplitude remains small as long as the water depth does not change.

However, when the wave reaches more shallow water, the wave speed drops, and the back of the wave overruns the front, increasing the amplitude just as if the waves were piling up on top of each other. This is exactly the same effect, but on a much larger scale, that creates breaking waves at ocean beaches.

Let Us Entertain You

Activity 1 Making Waves

GOALS

In this activity you will:

- Observe the motion of a pulse.
- Measure the speed of a wave.
- Observe standing waves.
- Investigate the relationship among wave speed, wavelength, and frequency.
- Make a model of wave motion.

What Do You Think?

One of the largest tsunamis (tidal waves) grew from about 0.7 m high in the open ocean to 35 m high when it reached shore.

- **How does water move to make a wave?**
- **How does a wave travel?**

Record your ideas about these questions in your *Active Physics* log. Be prepared to discuss your responses with your small group and with your class.

Active Physics 326

© It's About Time

For You To Do

1. In an area free of obstacles, stretch out a Slinky® so the turns are a few centimeters apart. Mark the positions of the end of the Slinky by sticking pieces of tape on the floor. Measure the distance between the pieces of tape.
   a) Record the distance between the pieces of tape in your log.
2. With the Slinky stretched out to the tape, grab the spring near one end, as shown in the drawing, and pull sideways 20 cm and back. To move it correctly, move your wrist as if snapping a whip. Observe what happens. You have made a transverse pulse.
   a) In what direction does the spring move as the pulse goes by?
   b) A dictionary definition of *transverse* is: "Situated or lying across." Why is *transverse* a good name for the wave you observed?

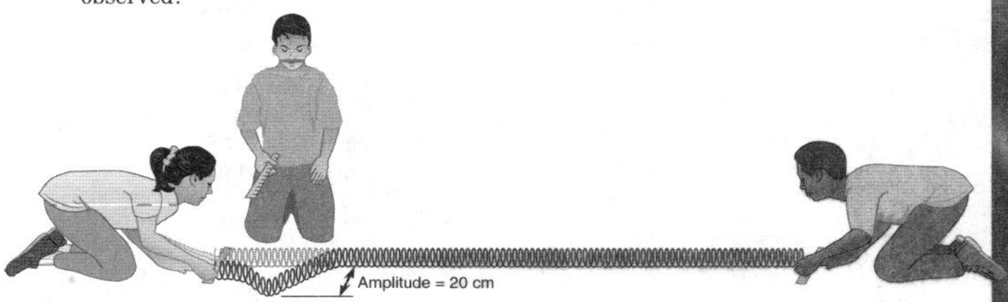

   c) Measure and record the amplitude of the wave. The distance you disturbed the spring is called the amplitude. The amplitude tells how much the spring is displaced.
3. After you have experimented with making pulses, measure the speed of the pulse. You will need to measure the time it takes the pulse to go the length of the spring. Take several measurements and then average the values.
   a) Record your data in the second and third rows of a table like the one on the following page.

ANSWERS

For You To Do

1. Answers will vary.
2. a) Sideways.
   b) The motion of the Slinky itself is at right angles to the motion of the pulse. Like a bridge built across a river, the Slinky motion is "across" the Slinky itself.
   c) Answers will vary.
3. The students' data for the speed of the Slinky waves will depend on the Slinky material (metal or plastic) and on how often they stretch the spring. Sample data is shown below.

3. a) Sample data

Length of Slinky = 2.5 m

time (s)	average time (s)	speed (m/s)
.47		
.44	.47	5.3
.5		

Also shown is data for a standing wave. Here the wavelength is twice the length of the Slinky.

wavelength (m)	frequency (cycle/s)	wavelength x frequency (m/s)
5	1	5
2.5	2	5
1.7	3	5

© It's About Time

ANSWERS

For You To Do
(continued)

4. a) Answers will vary.
   b) The speed is independent of the pulse size.
5. a) The motion will probably be disorganized.
6. Now the motion is organized. The spring looks like a sine wave.
7. a) Many wavelengths are possible.

Amplitude	Time for pulse to travel from one end to the other	Average time	Speed = $\frac{\text{length of spring}}{\text{average time}}$

4. Measure the speed of the pulses for two other amplitudes, one larger and one smaller than the value used in **Step 3**.
   a) Record the results in the table in your log.
   b) How does the speed of the pulse depend on the amplitude?
5. Now make waves! Swing one end back and forth over and over again along the floor. The result is called a periodic wave.
   a) Describe the appearance of the periodic wave you created.
6. To make these waves look very simple, change the way you swing the end until you see large waves that do not move along the spring. You will also see points where the spring does not move at all. These waves are called standing waves.

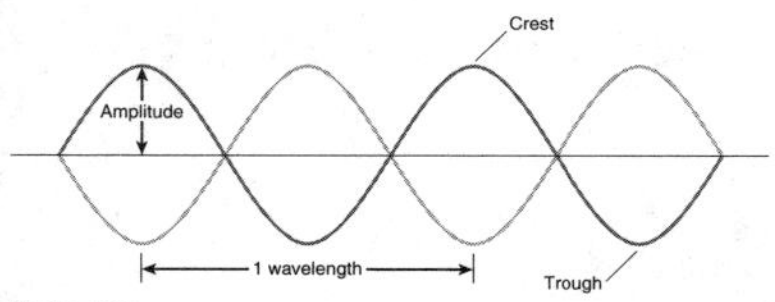

Standing wave

7. The distance from one crest (peak) of a wave to the next is called the wavelength. Notice that you can find the wavelength by looking at the points where the spring does not move. The wavelength is twice the distance between these points. Measure the wavelength of your standing wave.
   a) Record the wavelength of your standing wave in your log.
8. You can also measure the wave frequency. The frequency is the number of times the wave moves up and down each second. Measure the frequency of your standing wave. [Hint: Watch the hands of the person shaking the spring. Time a certain number of back-and-forth motions. The frequency is the number of back-and-forth motions of the hand in one second.]

© It's About Time

a) Record the wave frequency in your log. The unit of frequency is the hertz (Hz).

9. Make several different standing waves by changing the wave frequency. Try to make each standing wave shown in the drawings (at right). Measure the wavelength. Measure the frequency.

a) Record both in a table like the one below.

Wavelength (m/cycle)	Frequency (cycles/s or Hz)	Speed (m/s) wavelength × frequency

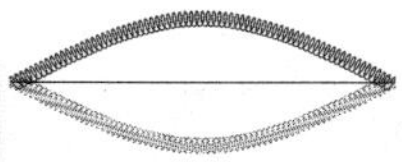

Wavelength = twice Slinky length

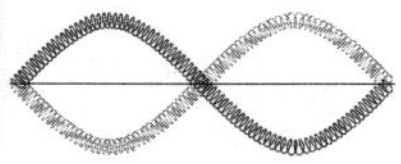

Wavelength = Slinky length

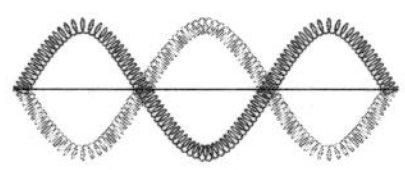

Wavelength = 2/3 Slinky length

b) For each wave, calculate the product of the wavelength and the frequency. Compare these values with the average speed of the pulse that you found in **Steps 3** and **4** above.

10. All the waves you have made so far are transverse waves. A different kind of wave is the compressional (or longitudinal) wave. Have the members of your group stretch out the Slinky between the pieces of tape and hold the ends firmly. To make a compressional wave, squeeze part of the spring and let it go. Measure the speed of the compressional wave and compare it with the speed of the transverse wave.

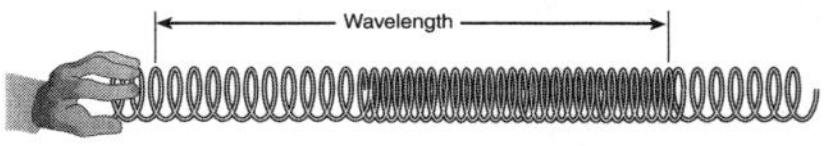

a) Record your results in a table partly like the one after **Step 3**.

b) In what direction does the Slinky move as the wave goes by?

ANSWERS

For You To Do

(continued)

8. a) The frequency measured will depend on how the spring is shaken.

9. a) Wavelength and frequency are inversely proportional. If one doubles the other is cut in half.

   b) The speeds probably agree to within ten or twenty percent.

10. a) Student data

   b) Looking along the string it is alternately stretched and compressed. There is no sideways motion. Back and forth along the direction the Slinky is stretched.

© It's About Time

ANSWERS

For You To Do
(continued)

10. c) The wave consists of compressions and extensions of the spring. The wave runs lengthwise along the spring.

11.-12. Student activity,

13. If you stand in water and watch how the water moves as a wave passes by, you would see the same up-and-down motion.

14. a) As in **Step 13**, you will see a part of the wave going up and down.

c) A dictionary definition of *compressional* is: "*a*. The act or process of compressing. *b*. The state of being compressed." A dictionary definition of *longitudinal* is: "Placed or running lengthwise." Explain why *compressional* or *longitudinal wave* is a suitable name for this type of wave.

Physics Words

periodic wave: a repetitive series of pulses; a wave train in which the particles of the medium undergo periodic motion (after a set amount of time the medium returns to its starting point and begins to repeat its motion).

crest: the highest point of displacement of a wave.

trough: the lowest point on a wave.

amplitude: the maximum displacement of a particle as a wave passes; the height of a wave crest; it is related to a wave's energy.

11. To help you understand waves better, construct a wave viewer by cutting a slit in a file card and labeling it as shown.

12. Make a drawing of a transverse wave on a strip of adding machine tape. Place this strip under the wave viewer so you can see one part of the wave through the slit.

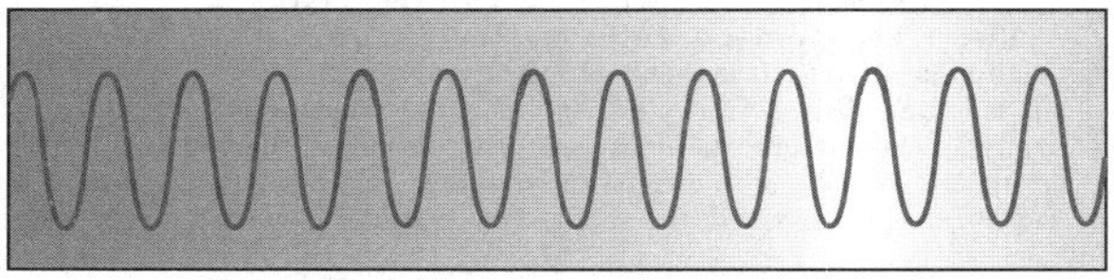

13. With the slit over the tape, pull the tape so that the wave moves. You will see a part of the wave (through the slit) going up and down.

14. Draw waves with different wavelengths on other pieces of adding machine tape. Put these under the slit and pull the adding machine tape at the same speed.

a) Describe what you see.

© It's About Time

FOR YOU TO READ

Wave Vocabulary

In this activity, you were able to send energy from one end of the Slinky to the other. You used chemical energy in your muscles to create mechanical energy in your arms that you then imparted to the Slinky. The Slinky had energy. A card at the other end of the Slinky would have moved once the wave arrived there. The ability to move the card is an indication that energy is present. The total energy is transferred but it is always conserved.

Of course, you could have used that same mechanical energy in your arm to throw a ball across the room. That would also have transferred the energy from one side of the room to the other. It would have also moved the card.

There is a difference between the Slinky transferring the energy as a wave and the ball transferring the energy. The Slinky wave transferred the energy, but the Slinky basically stayed in the same place. If the part of the Slinky close to one end were painted red, the red part of the Slinky would not move across the room. The Slinky wave moves, but the parts of the Slinky remain in the same place as the wave passes by. A wave can be defined as a transfer of energy with no net transfer of mass.

Leonardo da Vinci stated that "the wave flees the place of creation, while the water does not." The water moves up and down, but the wave moves out from its center.

In discussing waves, a common vocabulary helps to communicate effectively. You observed waves in the lab activity. We will summarize some of the observations here and you can become more familiar with the terminology.

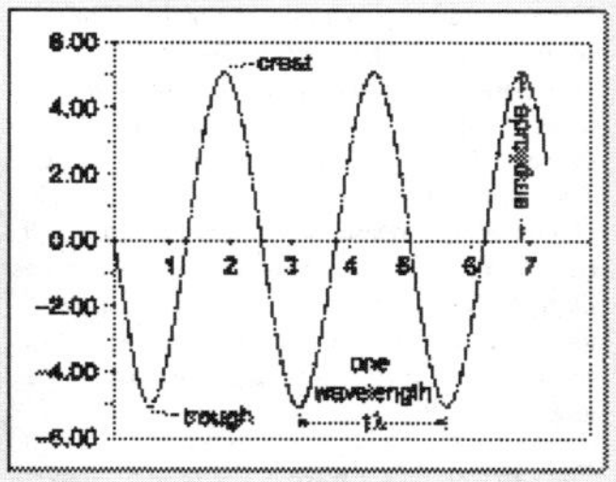

A **periodic wave** is a repetitive series of pulses. In the periodic wave shown in the diagram above, the highest point is called the **crest.** The lowest point is called the **trough.** The maximum disturbance, the **amplitude,** is 5.00 cm. Notice that this is the height of the crest or the height of the trough. It is *not* the distance from the crest to the trough.

© It's About Time

The **wavelength** of a periodic wave is the distance between two consecutive points in phase. The distance between two crests is one wavelength or 1 λ. (The Greek letter lambda is used to signify wavelength.) The wavelength of the wave in the diagram is 2.5 cm.

The amplitude of a periodic wave is the maximum disturbance. A large amplitude corresponds to a large energy. In sound, the large amplitude is a loud sound. In light, the large amplitude is a bright light. In Slinkies, the large amplitude is a large disturbance.

The wavelength of the wave in the diagram is 2.5 cm. It is the distance between two crests or the distance between two troughs.

The **frequency** is the number of vibrations occurring per unit time. A frequency of 10 waves per second may also be referred to as 10 vibrations per second, 10 cycles per second, 10 per second, 10 s^{-1}, 10 Hz (hertz). The human ear can hear very low sounds (20 Hz) or very high sounds (20,000 Hz). You can't tell the frequency by examining the wave in the diagram. The "snapshot" of the wave is at an instant of time. To find the frequency, you have to know how many crests pass by a point in a given time.

The **period**, T, of a wave is the time it takes to complete one cycle. It is the time required for one crest to pass a given point. The period and the frequency are related to one another. If three waves pass a point every second, the frequency is three waves per second. The period would be the time for one wave to pass the point, which equals $\frac{1}{3}$ s. If 10 waves pass a point every second, the frequency is 10 waves per second. The period would be the time for one wave to pass the point, which equals $\frac{1}{10}$ second. Mathematically, this relationship can be represented as:

$$T = \frac{1}{f} \text{ or } f = \frac{1}{T}$$

Points in a periodic wave can be "in phase" if they have the same displacement and are moving in the same direction. All crests of the wave shown below are "in phase."

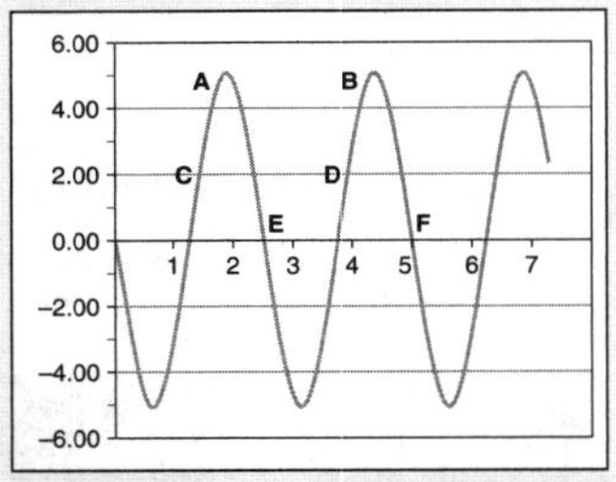

In the wave shown, the following pairs of points are in phase A and B, C and D, E and F.

A **node** is a spot on a standing wave where the medium is motionless. There are places along the medium that do not move as the standing wave moves up and down. The locations of these nodes do not change as the standing wave vibrates. A **transverse wave** is a wave in which the motion of the medium is perpendicular to the motion of the wave. A **longitudinal wave** is a wave in which the motion of the medium is parallel to the direction of the motion of the wave.

© It's About Time

PHYSICS TALK

Calculating the Speed of Waves

You can find the speed of a wave by measuring the distance the crest moves during a certain change in time.

$$\text{speed} = \frac{\text{change in distance}}{\text{change in time}}$$

In mathematical language:

$$v = \frac{\varnothing d}{\varnothing t}$$

where v = speed

d = distance

t = time

Suppose the distance the crest moves is 2 m in 0.2 s. The speed can be calculated as follows:

$$v = \frac{\varnothing d}{\varnothing t} = \frac{2\text{ m}}{0.2\text{ s}} = 10\text{ m/s}$$

The distance from one crest of a wave to the next is the wavelength. The number of crests that go by in one second is the frequency. Imagine you saw five crests go by in one second. You measure the wavelength to be 2 m. The frequency is 5 crests/second, so the speed is $(5 \times 2) = 10$ m/s. Thus, the speed can also be found by multiplying the wavelength and the frequency.

$$\text{speed} = \text{frequency} \times \text{wavelength}$$

In mathematical language:

$$v = f\lambda$$

where v = speed

f = frequency

λ = wavelength

→

Physics Words

wavelength: the distance between two identical points in consecutive cycles of a wave.

frequency: the number of waves produced per unit time; the frequency is the reciprocal of the amount of time it takes for a single wavelength to pass a point.

period: the time required to complete one cycle of a wave.

node: a point on a standing wave where the medium is motionless.

transverse pulse or wave: a pulse or wave in which the motion of the medium is perpendicular to the motion of the wave.

longitudinal pulse or wave: a pulse or wave in which the motion of the medium is parallel to the direction of the motion of the wave.

© It's About Time

Standing waves happen anywhere that the length of the Slinky and the wavelength have a particular mathematical relationship. The length of the Slinky must equal $\frac{1}{2}$ wavelength, 1 wavelength, $1\frac{1}{2}$ wavelengths, 2 wavelengths, etc. Mathematically, this can be stated as:

$$L = \frac{n\lambda}{2}$$

where L is the length of the Slinky,
λ is the wavelength
n is a number (1, 2, 3...)

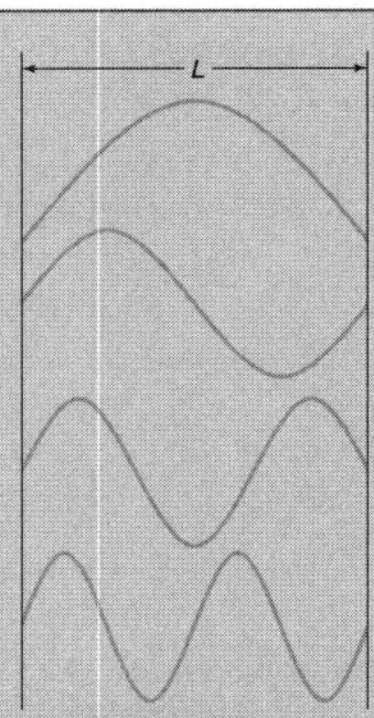

Sample Problem I

You and your partner sit on the floor and stretch out a Slinky to a length of 3.5 m. You shake the Slinky so that it forms one loop between the two of you. Your partner times 10 vibrations and finds that it takes 24.0 s for the Slinky to make these vibrations.

a) How much of a wave have you generated and what is the wavelength of this wave?

Strategy: Draw a sketch of the wave you have made and you will notice that it looks like one-half of a total wave. It is! This is the maximum wavelength that you can produce on this length of Slinky. You can use the equation that shows the relationship between the length of the Slinky and the wavelength.

Givens:

$L = 3.5$ m

$n = 1$

© It's About Time

Solution:

$$L = \frac{n\lambda}{2}$$

Rearrange the equation to solve for λ.

$$\lambda = \frac{2L}{n}$$

$$= \frac{2\,(3.5\text{ m})}{1}$$

$$= 7.0\text{ m}$$

b) What is the period of vibration of the wave?

Strategy: The period is the amount of time for one vibration. You have the amount of time for 10 vibrations.

Solution:

$$T = \frac{\text{time for 10 vibrations}}{10} = \frac{24.0\text{ s}}{10} = 2.4\text{ s}$$

c) Calculate the wave frequency.

Strategy: The frequency represents the number of vibrations per second. It is the reciprocal of the period.

Given:

$$T = 2.4\text{ s}$$

Solution:

$$f = \frac{\text{number of vibrations}}{\text{time}} \text{ or } f = \frac{1}{T}$$

$$= \frac{1}{2.4\text{ s}}$$

$$= 0.42\text{ vibrations per second}$$

$$= 0.42\text{ s}^{-1}\text{ or }0.42\text{ Hz}$$

→

© It's About Time

d) Determine the speed of the wave you have generated on the Slinky.

Strategy: The speed of the wave may be found by multiplying the frequency times the wavelength.

Givens:

$f = 0.42\ \text{Hz}$

$\lambda = 7.0\ \text{m}$

Solution:

$$v = f\lambda$$
$$= 0.42\ \text{Hz} \times 7.0\ \text{m}$$
$$= 29\ \text{m/s}$$

Remember that Hz may also be written as 1/s so the unit of speed is m/s.

Sample Problem 2

You stretch out a Slinky to a length of 4.0 m, and your partner generates a pulse that takes 1.2 s to go from one end of the Slinky to the other. What is the speed of the wave on the Slinky?

Strategy: Use your kinematics equation to determine the speed.

Givens:

$d = 4.0\ \text{m}$

$t = 1.2\ \text{s}$

Solution:

$$v = \frac{d}{t}$$
$$= \frac{4.0\ \text{m}}{1.2\ \text{s}}$$
$$= 3.3\ \text{m/s}$$

© It's About Time

Reflecting on the Activity and the Challenge

Slinky waves are easy to observe. You have created transverse and compressional Slinky waves and have measured their speed, wavelength, and frequency. For the **Chapter Challenge**, you may want to create musical instruments. You will receive more guidance in doing this in the next activities. Your instruments will probably not be made of Slinkies. You may, however, use strings that behave just like Slinkies. When you have to explain how your instrument works, you can relate its production of sound in terms of the Slinky waves that you observed in this activity.

Physics To Go

1. a) Four characteristics of waves are amplitude, wavelength, frequency, and speed. For each characteristic, tell how you measured it when you worked with the Slinky.
   b) For each characteristic, give the units you used in your measurement.
   c) Which wave characteristics are related to each other? Tell how they are related.
2. a) Suppose you shake a long Slinky slowly back and forth. Then you shake it rapidly. Describe how the waves change when you shake the Slinky more rapidly.
   b) What wave properties change?
   c) What wave properties do not change?
3. Suppose you took a photograph of a wave on a Slinky. How can you measure wavelength by looking at the photograph?
4. Suppose you mount a video camera on a tripod and aim the camera at one point on a Slinky. You also place a clock next to the Slinky, so the video camera records the time. When you look at the video of a wave going by on the Slinky, how could you measure the frequency?
5. a) What are the units of wavelength?
   b) What are the units of frequency?
   c) What are the units of speed?
   d) Tell how you find the wave speed from the frequency and the wavelength.

ANSWERS

Physics To Go

1. a) *amplitude*: measure the distance of a point on the Slinky from its rest position.

   wavelength: make a standing wave; measure the distance between the points of the Slinky that do not move; twice this distance is the wavelength.

   frequency: measure how many times you shook the Slinky in ten seconds and divide by ten.

   speed: make a pulse; measure distance traveled and time elapsed; divide time into distance to find the speed.

   b) amplitude and wavelength: meters
   frequency: 1/s
   speed: m/s

   c) Speed is frequency times wavelength.

ANSWERS

Physics To Go
(continued)

2. a) When you shake the Slinky more rapidly, you can see more crests and the crests are closer together. If you are making standing waves, the more rapidly you shake the Slinky, the more complete waves you will see.

   b) The wavelength and frequency change.

   c) The speed does not change.

3. If there is a meter stick in the photograph, use its image to measure the distance between one crest and the next.

4. Measure the time for a point on the Slinky to go through one complete cycle. You may have to measure for ten cycles and divide the time by ten.

5. a) m

   b) 1/s

   c) m/s

   d) speed = wavelength 3 frequency

© It's About Time

ANSWERS

Physics To Go
(continued)

5. e) m/s = m × 1/s

6. a) A standing wave is a repeating back-and-forth motion that does not move from one place to another. For transverse waves on a Slinky, the parts of the spring move from side-to-side, but nothing moves along the spring (nothing moves like the pulse did).

   b) See diagram on page 328.

   c) See diagram on page 328. At places where the wave amplitude is zero, there is no wave motion. At other places, the wave amplitude increases, reaches a maximum, then decreases, goes through zero, changes sign, reaches a maximum, goes back through zero, etc.

   d) Find the length of one complete wave of the spring. Or, figure out what fraction of a wave you see on the spring and from that compute the wavelength.

7. a) In a compressional wave, the back-and-forth motion is in the same direction that the disturbance moves (or in the opposite direction). In a transverse wave, the back-and-forth motion is perpendicular to the direction the disturbance moves.

   b) In transverse waves, the Slinky moves back-and-forth perpendicular to its length. In compressional waves, the Slinky moves back-and-forth along its length.

8. a) You shook the spring at a higher frequency.

   b) You shook the spring at a lower frequency.

e) Using your answer to **Part (d)**, show how the units of speed are related to the units of wavelength and frequency.

6. a) What is a standing wave?
   b) Draw a standing wave.
   c) Add labels to your drawing to show how the Slinky moves.
   d) Tell how to find the wavelength by observing a standing wave.

7. a) Explain the difference between transverse waves and compressional waves.
   b) Slinky waves can be either transverse or compressional. Describe how the Slinky moves in each case.

8. a) When you made standing waves, how did you shake the spring (change the frequency) to make the wavelength shorter?
   b) When you made standing waves, how did you shake the spring (change the frequency) to make the wavelength longer?

9. Use the wave viewer and adding machine tape to investigate what happens if the speed of the wave increases. Pull the tape at different speeds and report your results.

10. A Slinky is stretched out to 5.0 m in length between you and your partner. By shaking the Slinky at different frequencies, you are able to produce waves with one loop, two loops, three loops, four loops, and even five loops.
    a) What are the wavelengths of each of the wave patterns you have produced?
    b) How will the frequencies of the wave patterns be related to each other?

11. A tightrope walker stands in the middle of a high wire that is stretched 10 m between the two platforms at the ends of the wire. He is bouncing up and down, creating a standing wave with a single loop and a period of 2.0 s.

ANSWERS

Physics To Go *(continued)*

9. If the speed of the wave seen with the viewer increases, the frequency increases but the wavelength remains the same. If these were water waves and you photographed them, the wavelength would remain the same no matter what the speed of the wave (you could measure the wavelength from a still photo). But if you made a video, you would see more crests go by if the wave speed increased. That is consistent with the relationship speed = wavelength × frequency.

10. The number of loops and the wavelength are related by the following equation:

$$\frac{n\lambda}{2} = L$$

then $\lambda = \dfrac{2L}{n}$

© It's About Time

Activity 1 Making Waves

a) What is the wavelength of the wave he is producing?
b) What is the frequency of this wave?
c) What is the speed of the wave?

12. A clothesline is stretched 9 m between two trees. Clothes are hung on the line as shown in the diagram. When a particular standing wave is created in the line, the clothes remain stationary.

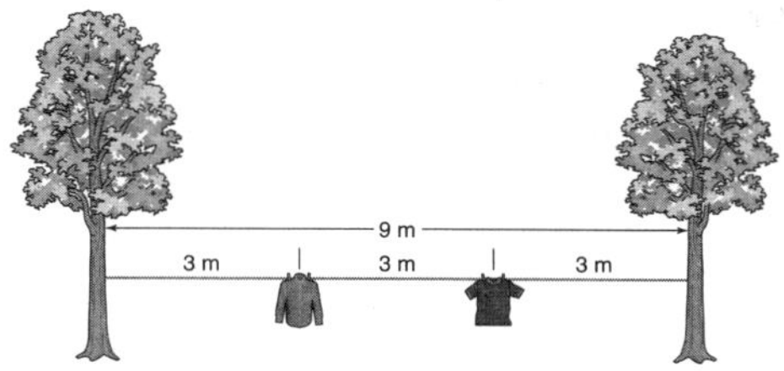

a) What is the term for the positions occupied by the clothes?
b) What is the wavelength of this standing wave?
c) What additional wavelengths could exist in the line such that the clothes remain stationary?

13. During the Slinky lab, your partner generates a wave pulse that takes 2.64 s to go back and forth along the Slinky. The Slinky stretches 4.5 m along the floor. What is the speed of the wave pulse on the Slinky?

14. A drum corps can be heard practicing at a distance of 1.6 km from the field. What is the time delay between the sound the drummer hears ($d = 0$ m) and the sound heard by an individual 1.6 km away? (Assume the speed of sound in air to be 340.0 m/s.)

339

ANSWERS

Physics To Go *(continued)*

a) for one loop: $\lambda = \frac{2(5.0\text{m})}{1}$

$= 10\text{m}$

two loops: $\lambda = 5.0\text{m}$

three loops: $\lambda = 3.3\text{m}$

four loops: $\lambda = 52.5\text{m}$

five loops: $\lambda = 2.0\text{m}$

b) As the wavelength decreases, the frequency at which you will need to shake the Slinky increases.

11. First find the wavelength, then the frequency, and finally put them together to find the wave speed.

ANSWERS

Physics To Go
(continued)

a) $\lambda = \frac{2L}{n}$

$= \frac{2(10\text{m})}{1}$

$= 20\text{m}$

b) $f = \frac{1}{T}$

$= \frac{1}{2.0\ \text{s}}$

$= 0.5\ \text{s}^{-1}$ of 0.5 Hz

c) $v = f\lambda$

$= 0.5\ \text{Hz} \times 20\text{m}$

$= 10$ m/s

12.a) The clothes are located at the nodes.

b) $\lambda = 6$ m

c) Wavelengths in which nodes will occur at the same locations will allow the clothes to remain stationary. These would include (but not limited to): l = 0.75 m, 2 m and 3 m.

13. You can solve this problem using the basic kinematics equation:

$d = vt$

so $v = \frac{d}{t}$

$= \frac{4.5\ \text{m}}{2.64\ \text{s}}$

$= 1.7$ m/s

14. $v = d/t$

$t = d/v$

$= (1600\ \text{m}) / (340\ \text{m/s})$

$= 4.7$ s

© It's About Time

Activity 1 A

Exploring Water Waves

FOR YOU TO DO

If you have ripple tanks, set them up and let students explore ripples and also water waves with a single frequency and wavelength.

© It's About Time

For use with *Let Us Entertain You*, Chapter 5, Activity 1: Making Waves

Amplitude and Wavelength Measurements

Amplitude	Time for pulse to travel from one end to another	Average time	Speed= $\frac{\text{length of spring}}{\text{average time}}$

Wavelength (m/cycles)	Frequency (cycles/s or Hz)	Speed (m/s) wavelength x frequency

© It's About Time

For use with *Let Us Entertain You*, Chapter 5, Activity 1: Making Waves

Sine Waves for Wave Model

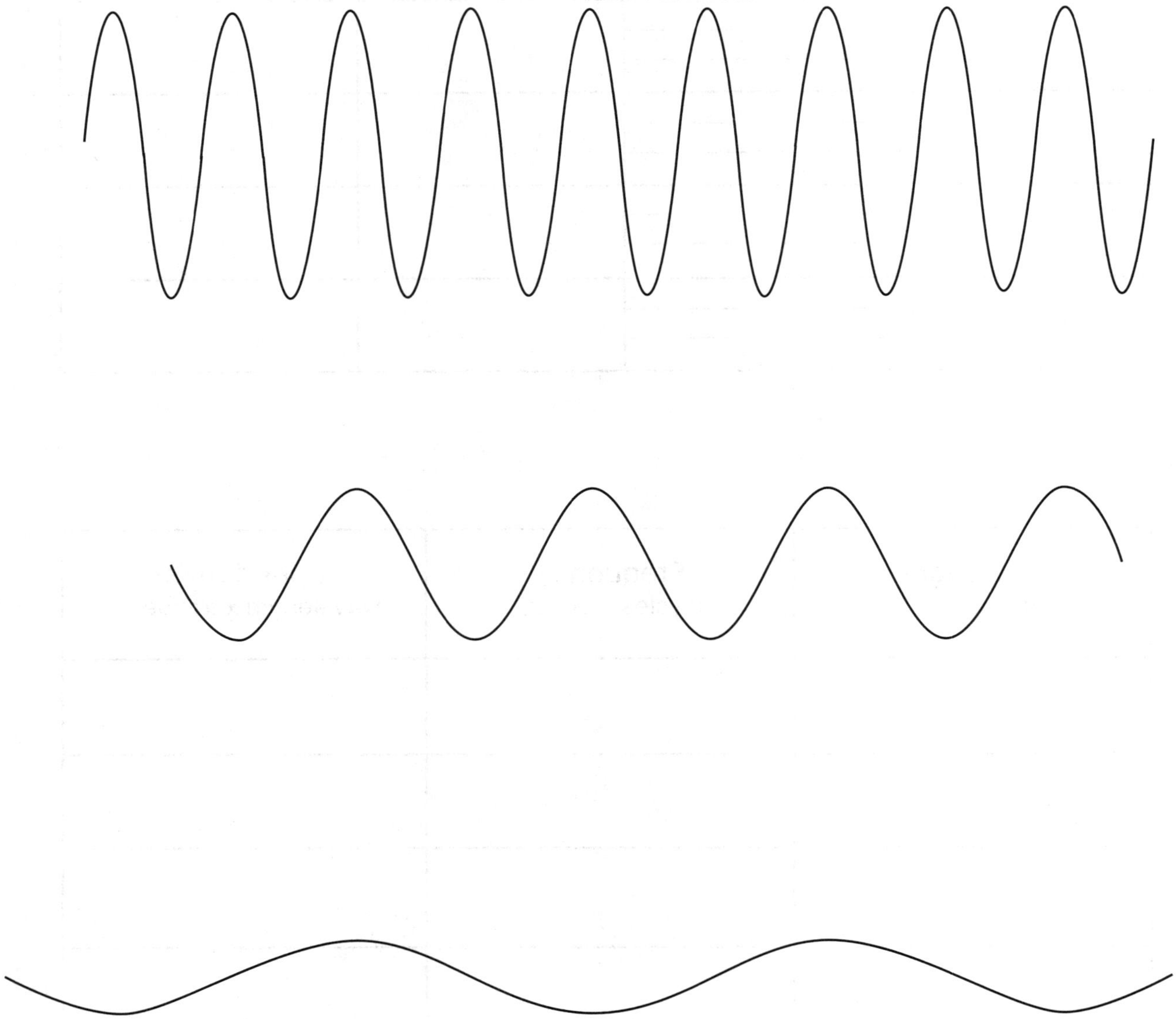

© It's About Time

For use with *Let Us Entertain You*, Chapter 5, Activity 1: Making Waves

NOTES

© It's About Time

ACTIVITY 2
Sounds in Strings

Background Information

A tensioned string is a vibrating system. When you pull back the string and release it, the string tension accelerates the string back towards its equilibrium position, as the drawing shows.

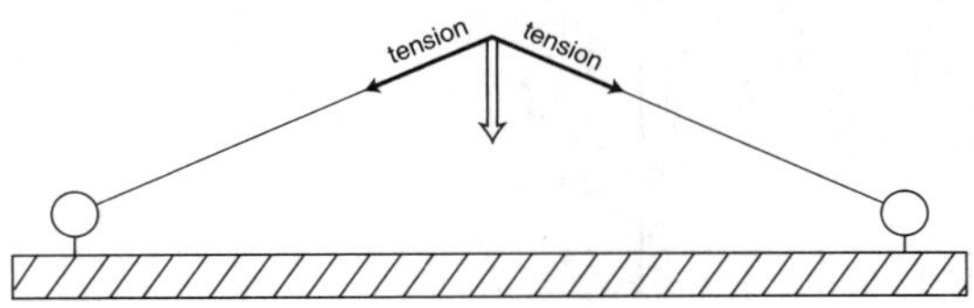

The variables that affect the frequency of a vibrating string are the string tension, the string length, and the string's mass per unit length. When the string is pulled back, the restoring force is equal to the sum of the two components of the tension that are perpendicular to the undisturbed string. The larger the tension, the larger the restoring force, the more rapidly the string will accelerate back towards the equilibrium position. The higher the tension, the higher the frequency, and, when the sound is heard, the higher the pitch.

Suppose you plucked a long and a short string. The drawing shows what happens.

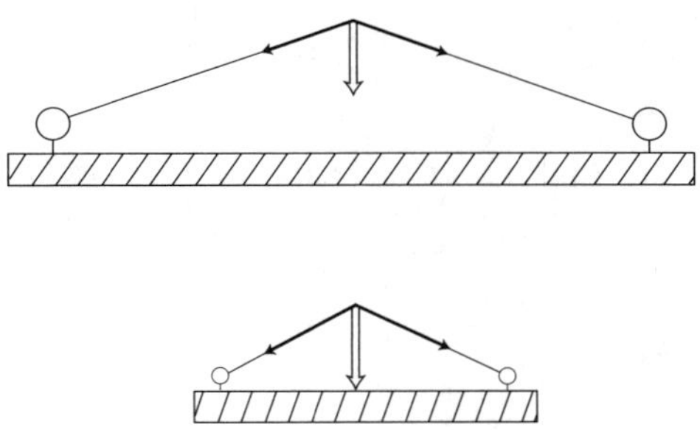

The component of the tension that is restoring the string to the equilibrium position is much higher for the shorter string. For the same amplitude, the shorter string experiences a larger restoring force. Consequently, its frequency will be correspondingly higher, as will the pitch it produces.

The third variable is the mass per unit length of the string (although students do not investigate it in this activity). Take a look at piano strings or guitar strings. The bass strings of a piano are much more heavily wound than the midrange or treble strings. In a nylon-string guitar, the base strings are wound with metal wire, and the lower the string, the heavier the wire. The larger the mass per unit length of the string, the less it responds to the tension that pulls it back towards the equilibrium position. The larger the mass per unit length, the lower the frequency.

Standing waves form on the string in the same way as on the spring.

When the string in a musical instrument vibrates, it moves in a combination, or superposition, of many standing wave patterns. A guitarist can suppress certain of these standing waves to emphasize the remaining ones by making "chimes." The technique is to simultaneously pluck the string and lightly tap it at the right spot for the particular chime. If the guitarist taps the string right in the middle, that suppresses the fundamental and all other odd harmonics with large amplitudes in the center of the string (the fundamental is the first harmonic).

NOTE: In this activity students find the wavelength of the fundamental from the length of the string. Since the activity does not introduce harmonics, the wavelength is called "the wavelength of the sound."

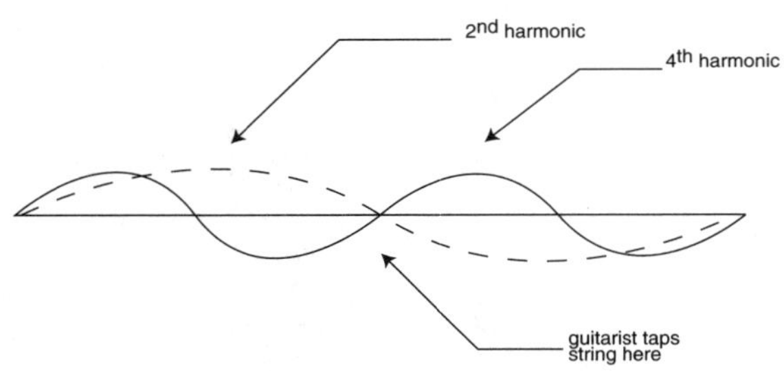

Active-ating the Physics InfoMall

Here we get a great chance to really test the search engine. Sounds in strings? Stringed instruments? What should we search for? After all, are they "string" or "stringed" instruments? Why bother with deciding? Search for "string* instrument*". That's right—two wild characters in one search term. Is the InfoMall up to this task? You bet! Try it.

© It's About Time

One of the hits is from the Sources of Musical Sounds chapter of *Modern College Physics*, found in the Textbook Trove. There is wonderful discussion as well as some nice graphics. One of the sections in this chapter is on stringed instruments. A discussion of wind instruments is also in this chapter (it might be useful later).

Stretching Exercises, Step 2, mentions a frequency meter. Search for key words "frequency meter" and you will find journal articles on using a frequency meter, plus a reference to the Vernier catalog on how to order the frequency meter software.

Also in the **Stretching Exercises**, students are asked to create a graph. You may wish to search for "student difficult*" AND "graph*". This produces several wonderful hits, including "Student difficulties with graphical representations of negative values of velocity," in *The Physics Teacher*, vol. 27, issue 4 on the InfoMall. Also found in that search is "Student difficulties in connecting graphs and physics: Example from kinematics," in the *Physics American Journal*, vol. 55, issue 6. Don't let these titles make you think that graphs are a bad thing! It is good to be aware that students do not always understand graphs, a powerful tool in the study of physics. Note that many of the hits you will find from this search involve the use of computers. Perhaps you may wish to have your students create their graphs on computers.

Planning for the Activity

Time Requirements:

one and a half class periods

Materials Needed

For each group:

- c-clamp
- fishing line spool
- goggles, safety
- hooked mass weight, 500 gm
- key or other small metal object
- mass hanger
- meter stick, 100 cm, hardwood
- pencil
- pulley on mount
- slotted weight set

© It's About Time

Advance Preparation and Setup

Set up the experiment in advance to find out how long the string must be. Use nylon monofilament rated at least 15 lb. test. Cut the string in advance, and be sure the string is long enough to drape over the pulley and tie to the mass hanger.

Teaching Notes

You may have to help students use the drawing for **FYTD, Step 5**, of the string fretted in different places. Encourage them to order the lengths they pluck from longest to shortest and look for a pattern in the pitches they observe. To facilitate ordering the pitches, encourage the students to compare the strings in pairs. After recording each result, they can readily order the pitches and corresponding string lengths.

Encourage the students to work quietly so they can observe the pitch of the vibrating string.

Students should wear safety goggles while working with the tensioned string.

Students may relate amplitude and frequency, so they would interpret louder sounds as lower in pitch. This prior conception could hinder students from recognizing that amplitude and frequency are independent variables. To help, set up a frequency generator, amplifier, and speaker. Then vary the frequency and amplitude to demonstrate that all possible combinations of pitch and volume can exist.

Activity Overview

Students begin the study of musical instruments by investigating two variables—tension and length—that determine the sounds made by a vibrating string. They hang a weight on the end of the string to set the tension. With the tension fixed, they vary the length of the string and observe the pitch produced. Then with the length fixed, students vary the tension and again observe the pitch. They make general statements about the effect of changes in string length and tension.

Student Objectives

Students will:

- Observe the effect of string length and tension upon pitch produced.
- Control the variables of tension and length.
- Summarize experimental results.
- Calculate wavelength of a standing wave.
- Organize data in a table.

Activity 2 Sounds in Strings

GOALS

In this activity you will:

- Observe the effect of string length and tension upon pitch produced.
- Control the variables of tension and length.
- Summarize experimental results.
- Calculate wavelength of a standing wave.
- Organize data in a table.

What Do You Think?

When the ancient Greeks made stringed musical instruments, they discovered that cutting the length of the string by half or two-thirds produced other pleasing sounds.

- **How do guitarists or violinists today make different sounds?**

Record your ideas about this question in your *Active Physics* log. Be prepared to discuss your responses with your small group and with your class.

For You To Do

1. Carefully mount a pulley over one end of a table. Securely clamp one end of a string to the other end of the table.
2. Tie the other end of the string around a mass hanger. Lay the string over the pulley. Place a pencil under the

ANSWERS FOR THE TEACHER ONLY

What Do You Think?

Both guitarists and violinists make different sounds in a similar way. The player pushes the string against the neck of the instrument at different positions to play different notes. One significant difference is that the guitar has frets, thin metal strips that run across the neck right under the strings. The guitarist pushes the string down against the fret, whereas the violinist must find the correct position to push the string down to make a particular note.

In addition, each string is tuned to a different note, so each string has a different range. Guitarists and violinists tune their instruments by adjusting the tension in the strings. Guitarists turn knobs on the tuning machine, a set of gears that increase torque through mechanical advantage. The violinist turns pegs, which have larger knobs than those on the guitar but which are connected directly to the strings.

ANSWERS

For You To Do

1. Be sure the masses hang beyond the pulley but all are up off the floor.
2. If the pencil moves when the sting is plucked, tape or clamp the pencil in place.

© It's About Time

string near the clamp, so the string can vibrate without hitting the table, as shown in the drawing.

3. Hang one 500-g mass on the mass hanger. Pluck the string, listen to the sound, and observe the string vibrate.
   a) Record your observations in your log in a table similar to the following:

Make sure the area under the hanging mass is clear (no feet, legs). Also monitor the string for fraying.

Length of vibrating string	Load on mass hanger	Pitch (high, medium, low)

4. Use a key or some other small metal object. Press this object down on the string right in the middle, to hold the string firmly against the table. Pluck each half of the string.
   a) Record the result in your table.
5. To change the string length, press down with the key at the different places shown in the diagrams on the next page. Pluck each part of the string.
   a) Record the results in your table.

ANSWERS

For You To Do *(continued)*

3. a) If the strings are the same material and length, the sounds will have the same pitch.
4. a) The frequency will be higher.
5. a) The shorter the length, the higher the pitch.

© It's About Time

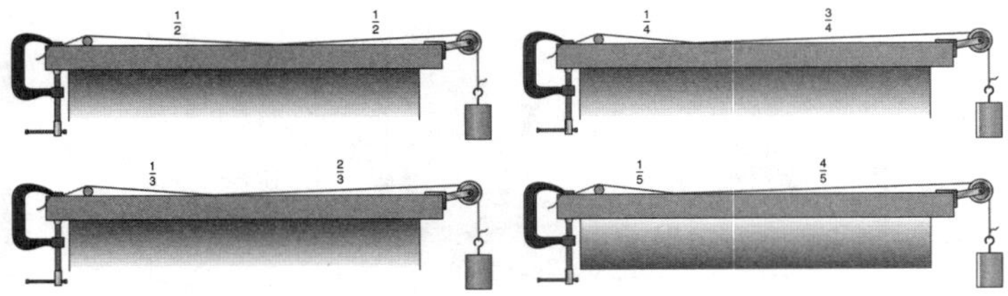

6. When you pluck the string, it does not move at the ends. Look at the drawing under **Step 9** of the **For You To Do** section in **Activity 1**. Measure the length of your string, and find the wavelength of the vibration for each string length.
   a) Record the wavelength in your table.
   b) Look over the data in your table. Make a general statement about what happens to the pitch you hear as you change the length of the string.

Make sure the string is capable of holding 2 kg.

7. Remove the key, so the string is its original length. Pluck the string. To investigate the effect of tightening the string, add a second 500-g mass to the mass hanger. Pluck the string again, observe the vibration, and listen to the pitch of the sound.
   a) Make up a table to record your data in your log.
   b) Add a description of the pitch of the sound to your table. Continue adding weights and observing the sound until the total mass is 2000 g.
   c) Look over your data. As the mass increases, the string becomes tighter, and its tension increases. Make a general statement about what happens to the pitch you hear as you change the tension on the string.

Answers

For You To Do *(continued)*

6. a) Assume the tone produced is the fundamental which is shown in the top drawing on p. 329. The wavelength is twice the length of the part of the string that is vibrating.

   b) As the string is made shorter, the pitch it makes goes up.

7. a) See the table on 341. Here the variables are the load on the mass hanger and pitch.

   b) The results are qualitative. (Have the students compare the pitch produced by each load.)

   c) As the mass on the mass hanger goes up, the pitch the string makes goes up.

© It's About Time

Activity 2 Sounds in Strings

FOR YOU TO READ

Changing the Pitch

Sound comes from vibration. You observed the vibration of the string as it produced sound. You investigated two of the variables that affect the sound of a vibrating string.

When you pushed the vibrating string down against the table, the length of the string that was vibrating became shorter. Shortening the string increased the **pitch** (resulted in a higher pitch). In the same way, a guitarist or violinist pushes the string against the instrument to shorten the length that vibrates and increases the pitch.

When you hung weights on the end of the string, that increased the pitch too. These weights tightened the string, so they created more tension in it. As the string tension increased, the pitch of the sound also increased. In tuning a guitar or violin, the performer changes the string tension by turning a peg attached to one end of a string. As the peg pulls the string tighter, the pitch goes up.

Combining these two results into one expression, you can say that increasing the tension or decreasing the length of the string will increase the pitch.

The string producing the pitch is actually setting up a standing wave between its endpoints. The length of the string determines the wavelength of this standing wave. Twice the distance between the endpoints is the wavelength of the sound. The pitch that you hear is related to the frequency of the wave. The higher the pitch, the higher the frequency. The speed of the wave is equal to its frequency multiplied by its wavelength.

$$v = f\lambda$$

where v = speed

f = frequency

λ = wavelength

If the speed of a wave is constant, a decrease in the wavelength will result in an increase in the frequency or a higher pitch. A shortened string produces a higher pitch.

Reflecting on the Activity and the Challenge

Part of the **Chapter Challenge** is to create a sound show. In this activity you investigated the relationship of pitch to length of the string and tension of the string: the shorter the string, the higher the pitch; the greater the tension, the higher the pitch. You also learned that the string is setting up a standing wave between its two ends, just like the standing wave that you created in the Slinky in **Activity 1**. That's the physics of stringed instruments! If you wanted to create a stringed or multi-string instrument for your show, you would now know how to adjust the length and tension to produce the notes you want. If you were to make such a stringed instrument, you could explain how you change the pitch by referring to the results of this activity.

Physics Words

pitch: the quality of a sound dependent primarily on the frequency of the sound waves produced by its source.

© It's About Time

ANSWERS

Physics To Go

1. a) By changing the weight hung on the end of the string (or, in an instrument, by turning the knobs on the tuning machine).

   b) Increasing the tension increases the pitch of the sound.

2. a) By pressing the string down against the tabletop.

   b) Decreasing the length of the string increases the pitch of the sound.

3. The right combination of decreasing the length and decreasing the tension (or increasing both) can keep the pitch the same.

4. In general, the pitch would change. (See answer to **Question 3.**)

5. a) For the guitar, by pressing the string down on the neck of the guitar; for the piano, by striking a different string.

   b) For the guitar, by changing the string tension; for the piano, also by changing the string tension.

6. a) To change the tension of the strings.

   b) To make it easier to press the string against the neck at the right place, so the note has the right pitch.

   c) No.

   d) Because the player must judge where to press the string down so each note has the correct pitch.

7. a)-c) Student activity.

Physics To Go

1. a) Explain how you can change the tension of a vibrating string.
   b) Tell how changing the tension changes the pitch.
2. a) Explain how you can change the length of a vibrating string.
   b) Tell how changing the length changes the sound produced by the string.
3. How would you change both the tension and the length and keep the pitch the same?
4. Suppose you changed both the length and the tension of the string at the same time. What would happen to the sound?
5. a) For the guitar and the piano, tell how a performer plays different notes.
   b) For the guitar and the piano, tell how a performer (or tuner) changes the pitch of the strings to tune the instrument.
6. a) Look at a guitar. Find the tuners (at the end of the neck). Why does a guitar need tuners?
   b) What is the purpose of the frets on a guitar?
   c) Does a violin or a cello have frets?
   d) Why do a violinist and a cellist require more accuracy in playing than a guitarist?
7. a) Using what you have learned in this activity, design a simple two-stringed instrument.
   b) Include references to wavelength, frequency, pitch, and standing waves in your description.
   c) Use the vocabulary of wavelength, frequency, and standing waves from **Activity 1** to describe how the instrument works.

© It's About Time

Stretching Exercises

1. Set up the vibrating string as you did in the preceding **For You To Do**. This time, you will measure the frequency of the sound. Set up a frequency meter on your computer. Pick up the sound with a microphone. Investigate how changing the length of the string changes the frequency of the sound. Create a graph to describe the relationship.

2. Set up the vibrating string, computer, and microphone as you did in **Stretching Exercise 1**. This time, investigate how changing the string tension changes the frequency of the sound. Create a graph to describe the relationship.

3. Design an investigation to find how the diameter (thickness) of the string or the type of material the string is made of affects the pitch you hear. Submit your design to your teacher for approval before proceeding to carry out your experiments.

ANSWERS

Stretching Exercises

See **Program Resources** for packages of computer sound probe and software.

© It's About Time

ACTIVITY 3
Sounds from Vibrating Air

Background Information

The phenomena of generating sound by blowing into a tube is an example of resonance. In resonance, a system that has a natural frequency is excited by an oscillation of the same frequency. A familiar example is pushing a child on a swing. The swing is a pendulum, with a single natural frequency. When the child is pushed so that the push is in the same direction as the child's velocity, energy feeds into the system rapidly and the child swings higher and higher (the system here is the child on the swing). If the push is opposed to the child's velocity, energy drains out of the system rapidly.

Sound, too, can be pictured as a regular series of pushes and pulls. The sound wave is compressional, so the wave is a pattern of regular pressure changes.

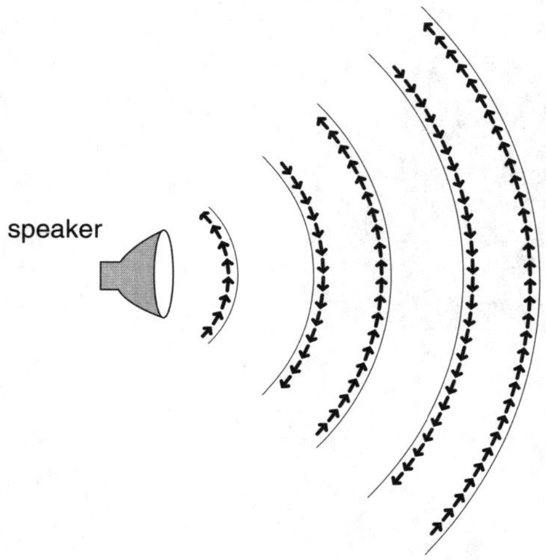

One important difference between blowing into a tube and pushing a child on a swing is that the push is made at just the right time to get the child swinging with large amplitude. The push has a single frequency. By contast, the sound of blowing into a tube is noise—a random spectrum of unrelated frequencies. But when that spectrum contains a frequency that makes the tube resonate, then the sound level for that frequency in the tube builds up and up. Listening to a sea shell is the same effect.

Now imagine a tube, with length "*L*," that is closed at one end. A sound wave reaches the open end of the tube. The wave moves to the end of the tube, reflects, and returns to the open end, which requires a time of $2L/v$, as shown.

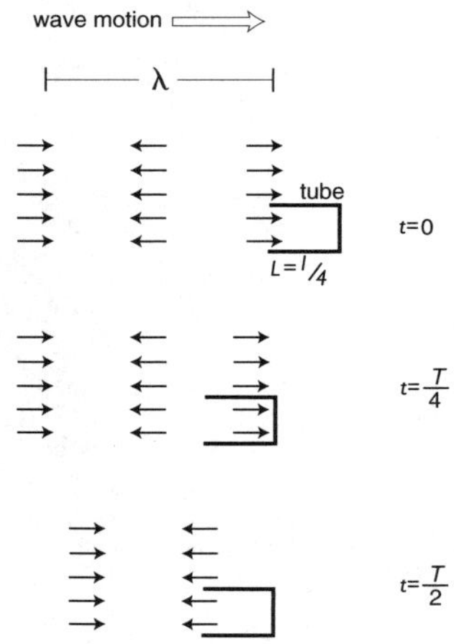

When the wave returns to the open end, the wave will be pushing the air at this end out of the tube. For resonance to occur, the sound wave exciting the tube must have gone through a half a cycle since it entered the tube, which requires a time of half the period *T*.

$2L/v = T/2 = 1/(2f)$

Substituting $v = \lambda \times f$ gives

$L = \lambda/4$

This equation expresses the same result as the drawings on page 348 in the Student Edition.

Active-ating the Physics InfoMall

As we found in the previous activity, there is a passage on wind instruments in the Sources of Musical Sounds chapter of Modern College Physics, found in the Textbook Trove.

Of course, you may want to perform a new search for this activity. One idea is to search for "flute". The very first hit is "Acoustics of the flute," from *Physics Today*, vol. 21, issue 11, 1968. The second hit is "Observations on the acoustical characteristics of the English flute," from the *American Journal of Physics*, vol. 27, issue 1, 1959. The is also a hit for "The physics of the organ flue pipe," in the *American Journal of Physics*, vol. 21, issue 5, 1953.

Also in this search is a chapter from *Household Physics*, from the Textbook Trove. There is an entire chapter on Musical Sounds and Musical Instruments. This chapter includes absolutely wonderful discussions of the topics we have in this activity as well as the previous activities. The graphics are good, too. There are discussions of vibrating air columns (with pipes open or closed at

© It's About Time

one end), pitch, the human voice and vocal cords, and more.

If you want more problems than you find in **Physics To Go**, simply visit the Problems Place. All six sources in the Problems Place have sections on Sound or Waves. This includes *Schaum's 3000 Solved Problems in Physics*, so don't hesitate to use this store for problems or examples for your students.

Planning for the Activity

Time Requirements:

one class period

Materials Needed

For the class:

- sound probe and software
- various wind instruments

For each group:

- drinking straws (10)
- ruler, metric
- steel scissors
- test tubes (5)
- water, about 300 cc, in a container

For each student:

- 1 copy of the Blackline Master, Pitch vs. Tube Length

Advance Preparation and Setup

Make one copy of the Blackline Master for each student. Be sure that the each group has enough water to partly fill the test tubes as explained in the student book.

Teaching Notes

Encourage the students to make direct comparisons by blowing into first one tube and then another. By making systematic comparisons of pairs of tube, they can order the tubes according to pitch.

Keep the student groups as well separated as possible.

Students are reasonably successful in associating vibration with sound when they can directly observe the vibrating object, such as a guitar string. A vibrating column of air is much more difficult to conceptualize, as is the idea that vibrations can move out from the column towards a listener. If you have both a guitar and a recorder in the room, have students play a note on each. First have the class trace the vibration from the guitar string to the students' ears. Then do the same with the recorder.

© It's About Time

Activity Overview

Students continue the study of musical instruments by investigating the resonance of pipes. They blow into different lengths of drinking straw and observe the tone produced and also observe how the tone changes if they cover one end of the straw. They perform a similar experiment with a test tube partially filled with water. Through reading they then relate these observations to the patterns of standing waves in a tube.

Student Objectives

Students will:

- Identify resonance in different kinds of tubes.
- Observe how resonance pitch changes with length of tube.
- Observe the effect of closing one end of the tube.
- Summarize experimental results.
- Relate pitch observations to drawings of standing waves.
- Organize observations to find a pattern.

Activity 3 Sounds from Vibrating Air

GOALS

In this activity you will:

- Identify resonance in different kinds of tubes.
- Observe how resonance pitch changes with length of tube.
- Observe the effect of closing one end of the tube.
- Summarize experimental results.
- Relate pitch observations to drawings of standing waves.
- Organize observations to find a pattern.

What Do You Think?

The longest organ pipes are about 11 m long. A flute, about 0.5 m long, makes musical sound in the same way.

- **How do a flute and organ pipes produce sound?**

Record your ideas about this question in your *Active Physics* log. Be prepared to discuss your responses with your small group and with your class.

ANSWERS FOR THE TEACHER ONLY

What Do You Think?

A flute and an organ produce sound from a resonating air column. The flutist's lips direct a stream of air into the mouthpiece to set up the vibration of the air inside the instrument. The strength, direction of the air stream, and length of the air column determine the harmonic content of the vibration and thereby the octave of the musical note. In the organ, compressed air is blown into each tube to create the resonance.

Each pipe in the organ has a different length and produces a different note, with longer pipes producing lower notes. The flutist produces different notes by pressing keys to expose openings in the instrument that, in effect, change its length.

© It's About Time

For You To Do

1. Carefully cut a drinking straw in half. Cut one of the halves into two quarters. Cut one of the quarters into two eighths. Pass one part of the straw out to one member of your group.

2. Gently blow into the top of the piece of straw.
   a) Describe what you hear.
   b) Listen as the members of your group blow into their straw pieces one at a time. Describe what you hear.
   c) Write a general statement about how changing the length of the straw changes the pitch you hear.

3. Now cover the bottom of your straw piece and blow into it again. Uncover the bottom and blow again.
   a) Compare the sound the straw makes when the bottom is covered and then uncovered.
   b) Listen as the members of your group blow into their straw pieces, with the bottom covered and then uncovered. Write a general statement about how changing the length of the straw changes the pitch you hear when one end is covered.

4. Obtain a set of four test tubes. Leave one empty. Fill the next halfway with water. Fill the next three-quarters of the way. Fill the last one seven-eighths of the way.

5. Give each test tube to one member of your group. Blow across your test tube.
   a) Describe what you hear.
   b) Listen as the members of your group blow, one at a time, across their test tubes. Record what you hear.
   c) What pattern do you find in your observations?
   d) Compare the results of blowing across the straws with blowing across the test tubes. How are the results consistent?

Make sure the outsides of the tubes are dry.

ANSWERS

For You To Do

1. Student activity.

2. a)-b) A faint tone is produced.
   c) The shorter the straw, the higher the pitch.

3. a) When the bottom of the straw is covered, the pitch is much lower than when both ends are open.
   b) When one end of the straw is covered, the shorter the straw, the higher the pitch.

4. Student activity.

5. a) A faint tone is produced.
   b)-c) The shorter the air space in the tube, the higher the pitch.
   d) The shorter the length, the higher the pitch.

© It's About Time

PHYSICS TALK

Vibrating Columns of Air

The sound you heard when you blew into the straw and test tube was

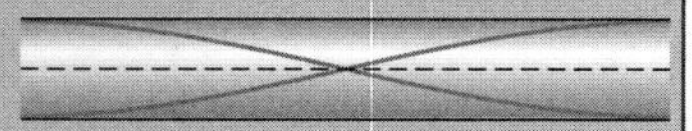

Tube is open at both ends.
1/2 wavelength fits in straw.

produced by a standing wave. If both ends of the straw are open, the air at both ends moves back and forth. The above drawing shows the movement of the air as a standing wave.

When you covered the other end of the straw, you prevented the

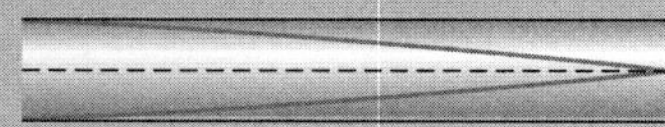

Tube is closed at one end.
1/4 wavelength fits in straw.

air from moving at the covered end. This drawing shows the movement of the air as a standing wave.

The velocity of a wave is equal to the frequency multiplied by the wavelength. Therefore,

$$\text{frequency} = \frac{\text{wave speed}}{\text{wavelength}}$$

Using mathematical symbols,

$$f = \frac{v}{\lambda}$$

As the wavelength increases, the frequency decreases. The wavelength in the open straw is half the wavelength in the straw closed at one end. This equation predicts that the frequency of the standing wave in the open straw is twice the frequency of the standing wave in the straw closed at one end.

Physics Words

diffraction: the ability of a sound wave to spread out as it emerges from an opening or moves beyond an obstruction.

© It's About Time

FOR YOU TO READ

Compressing Air to Make Sound

Sound is a compression wave. The molecules of air bunch up or spread apart as the sound wave passes by.

At the end where the tube is closed, the air cannot go back and forth, because its motion is blocked by the end of the tube. That's why the wave's amplitude goes to zero at the closed end. At the open end, the amplitude is as large as it can possibly be. This back-and-forth motion of air at the open end makes a sound wave that moves from the tube to your ear.

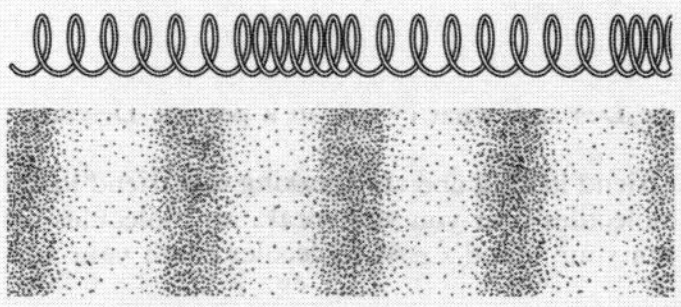

In the compressional Slinky wave, the coils of the Slinky bunched up in a similar fashion when the Slinky wave passed by.

Wave Diffraction

As the sound wave leaves the test tube in this activity, it spreads out. In the same way, when you speak to a friend, the sound waves leave your mouth and spread out. You can speak to a group of friends because the sound leaves your mouth and moves out to the front and to the sides.

This ability of the sound wave to spread out as it emerges from an opening is called **diffraction**. The smaller the opening, the more spreading of the sound. The spreading of the wave as it emerges from two holes can be shown with a diagram.

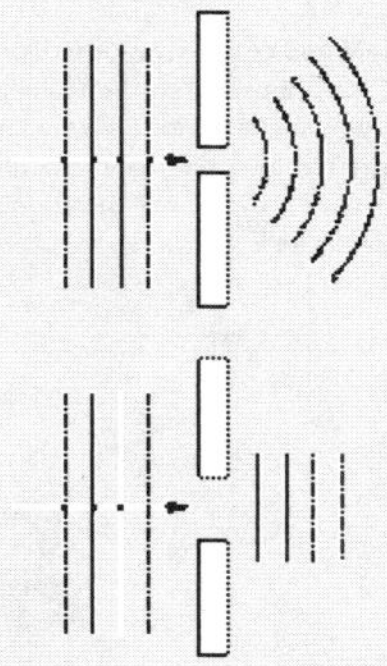

The wave on the top is going through a small opening (in comparison to its wavelength) and diffracts a great deal. The wave on the bottom is going through a large opening (in comparison to its wavelength) and shows little diffraction.

Cheerleaders use a megaphone to limit the diffraction. With a megaphone, the mouth opening becomes larger. The sound wave spreads out less, and the cheering crowd in front of the cheerleader hears a louder sound.

→

© It's About Time

A new musical instrument that you can make uses a straw instead of a test tube.

Take a straw and cut the ends to form a V as show below.

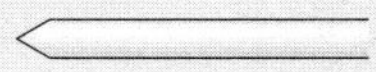

Flatten the V end of the straw and blow this "trumpet." You can shorten the trumpet, decrease the wavelength of the standing wave, and increase the frequency of the sound. Try making a sound. As you emit the sound, use scissors to cut ends off the straw. Listen to the different tones.

You can probably make a trombone by inserting one straw within another.

The sound diffracts from the small opening. You can add a horn to one end of the straw and limit the diffraction. The effect will be that the sound appears louder because it doesn't spread out. You can make a horn out of a piece of paper, as shown.

You may want to adapt this idea of a trumpet and a megaphone and make diffraction a part of your light and sound show.

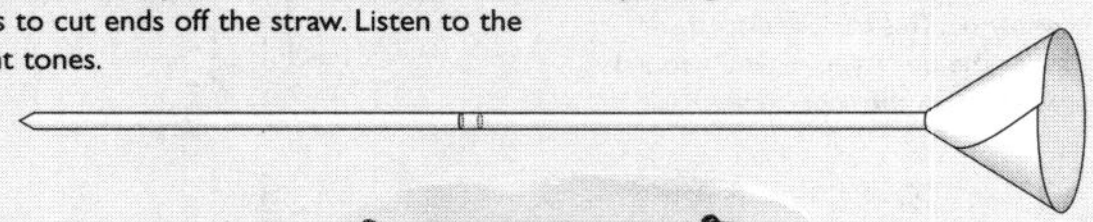

© It's About Time

Reflecting on the Activity and the Challenge

In this activity you have observed the sounds produced by different kinds of pipes. If the pipe is cut to a shorter length, the pitch of the sound increases. Also, when the pipe is open at both ends, the pitch is much higher than if the pipe were open at only one end. You have seen how simple drawings of standing waves in these tubes help you find the wavelength of the sound. If the tube is closed at one end, the air has zero displacement at that end. If the tube is open at one end, the air has maximum displacement there.

For your sound show, you may decide to create some "wind" instruments using test tubes or straws, or other materials approved by your teacher. When it comes time to explain how these work, you can refer to this activity to get the physics right.

Physics To Go

1. a) You can produce a sound by plucking a string or by blowing into a pipe. How are these two ways of producing sound similar?
   b) How are these two ways different?

2. a) For each piece of straw your group used, make a full-sized drawing to show the standing wave inside. Show both the straw closed at one end and open at both ends.
   b) Next to each drawing of the standing waves, make a drawing, at the same scale, of one full wavelength. You may need to tape together several pieces of paper for this drawing.
   c) Frequency times the wavelength is the wave speed. The speed is the same for all frequencies. From your answer to **Part (b)**, what can you predict about the frequencies of the standing waves in the straw pieces?
   d) How well do your predictions from **Part (c)** agree with your observations in this activity?

Answers

Physics To Go

1. a) The string is vibrating, and the air in the pipe is also vibrating.
   b) The string vibration is transverse, and the air vibration is compressional.

2. a)-b) See drawings on page 348.
   c) If you double the length of the straw, you cut the frequency it produces in half. When both ends are open, the frequency will be twice as high as when one end is closed.

© It's About Time

d) There should be good agreement.

3. a) 11 m

b) See drawings on page 348.

c) 4 × 11 m = 44 m

d) The speed of sound is independent of the wavelength. But the product of wavelength and frequency equals speed, so this product is a constant. Thus, a longer wavelength means a lower frequency.

4. a) $\lambda = 4 \times 3$ m = 12 m

b) $f = v/\lambda$ = 340/12 Hz = 28 Hz

c) 2 × 3 m = 6 m

d) f = 340/6 Hz = 57 Hz

5. The frequency is three times higher.

6. a) Waves spreading into a region behind an obstruction is called diffraction.

b) Students draw a diagram similar to the one in the For You to Read section.

3. a) What is the length, in meters, of the longest organ pipe?
   b) Assume this pipe is closed at one end. Draw the standing wave pattern.
   c) For this pipe, how long is the wavelength of this standing wave?
   d) Why does a long wavelength indicate that the frequency will be low? Give a reason for your answer.

4. a) Suppose you are listening to the sound of an organ pipe that is closed at one end. The pipe is 3 m long. What is the wavelength of the sound in the pipe?
   b) The speed of sound in air is about 340 m/s. What is the frequency of the sound wave?
   c) Now suppose you are listening to the sound of an organ pipe that is open at both ends. As before, the pipe is 3 m long. What is the wavelength of the sound in the pipe?
   d) What is the frequency of the sound wave?

5. Suppose you listen to the sound of an organ pipe that is closed at one end. This pipe is only 1 m long. How does its frequency compare with the frequency you found in **Question 4, Part (b)**?

6. Waves can spread into a region behind an obstruction.
   a) What is this wave phenomenon called?
   b) Draw a diagram to illustrate this phenomenon.

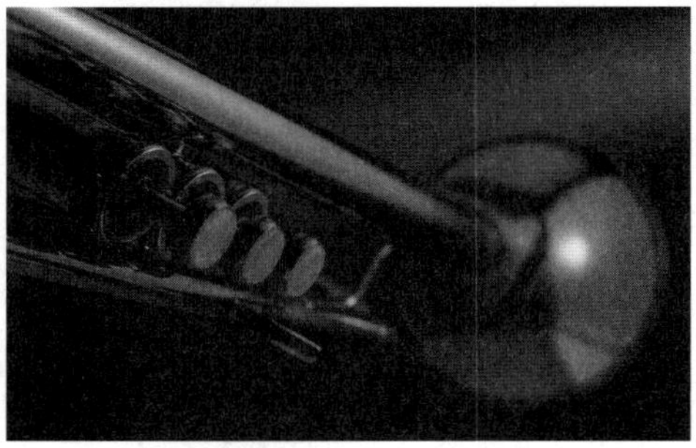

© It's About Time

Activity 3 Sounds from Vibrating Air

Stretching Exercises

1. If you have a good musical ear, add water to eight test tubes to make a scale. Play a simple piece for the class.
2. Obtain a 2- to 3- meter-long piece of a 7- to 10-centimeter-diameter plastic pipe, like that used to filter water in small swimming pools. In an area free of obstructions, twirl the pipe overhead. What can you say about how the sound is formed? Place some small bits of paper on a stool. Twirl the pipe and keep one end right over the stool. What happens to the paper? What does that tell you about the air flowing through the pipe? Try to play a simple tune by changing the speed of the pipe as you twirl it.
3. Carefully cut new straw pieces, as you did in **For You To Do, Step 1**. This time, you will measure the frequency of the sound. Set up a frequency meter on your computer. Place the microphone near an open end of the straw.
   As before, each person blows into only one piece of straw. Make the sound and record the frequency. Now cover the end of the straw and predict what frequency you will measure. Make the measurement and compare it with your prediction. Repeat the measurements for all of the lengths of straw. Record your results, and tell what patterns you find.

ANSWERS

Stretching Exercises

Try the lengths for a major scale:

11 cm
9.8 cm
8.8 cm
8.3 cm
7.4 cm
6.6 cm
5.9 cm
5.5 cm

Movement of the air in the pipe produced a standing wave. If paper bits are sucked into the pipe, that shows the direction of the air flow.

© It's About Time

Activity 3 A

Wind Instruments and Resonant Frequencies

FOR YOU TO DO

1. Wind Instruments: Have students bring wind instruments to class. Discuss whether each instrument is open at one end or both ends. Explain how the performer plays different notes. Also discuss how to tune the instrument.

2. Measuring Resonant Frequencies of Tubes: With sound probe and software, measure the resonant frequency of tubes of various lengths. Make the measurements for tubes open at both ends and closed at one end. Compare your results with the predictions of the standing wave diagrams.

© It's About Time

For use with *Let Us Entertain You*, Chapter 5, Activity 3: Sounds from Vibrating Air

Pitch vs. Tube Length

Length of test tube or straw	End closed or open	Length of air column	Pitch

© It's About Time

For use with *Let Us Entertain You*, Chapter 5, Activity 3: Sounds from Vibrating Air

ACTIVITY 4
Reflected Light

Background Information

When light strikes an ordinary object, the light is reflected in all directions. This reflected light can be represented by light rays that fan out from each point on the object.

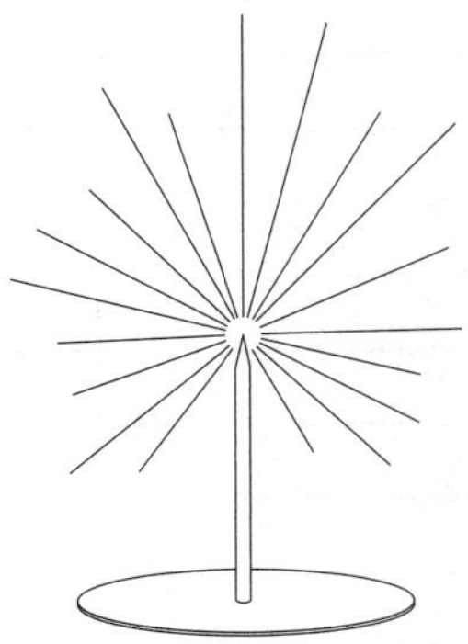

When the object is placed before a mirror, the light rays that fan out from the object can be extended to the mirror. As shown in the drawing on page 356 of the Student Edition, the angles of incidence and reflection for these rays are equal. These angles are straightforward to measure, but there is a minor complication to interpreting the results. Most lab mirrors are rear-surface mirrors. That is, the reflecting coating is on the rear of the mirror, so the light must first pass through glass before being reflected, and then must pass through glass again before again entering the air. The light is refracted at each air-glass interface, as shown in the

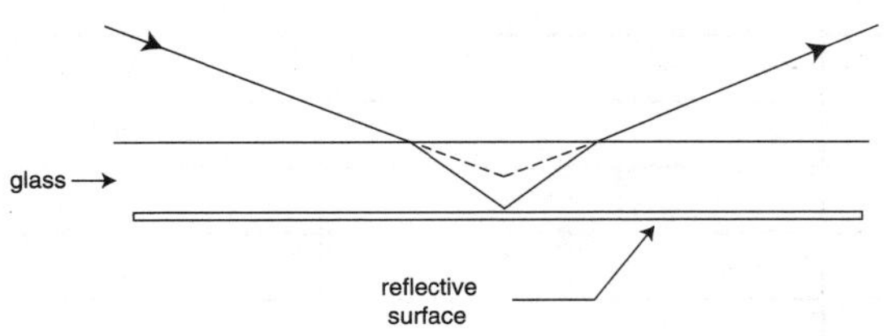

drawing. The net effect is that the light acts as if it had been reflected from a surface located about in the middle of the lab mirror (where the dotted lines meet in the drawing). Notice that the angle of incidence is still equal to the angle of reflection, by symmetry. However, when the reflected ray is extended behind the mirror to locate the image, the extension will be shifted away from the point where the incident ray hits the mirror. The size of the shift increases with the angle of incidence (and the shift is zero for a light beam coming in along the normal). The result is that the image is slightly blurred.

Extending the paths of the reflected rays back behind the mirror produces the location of the image.

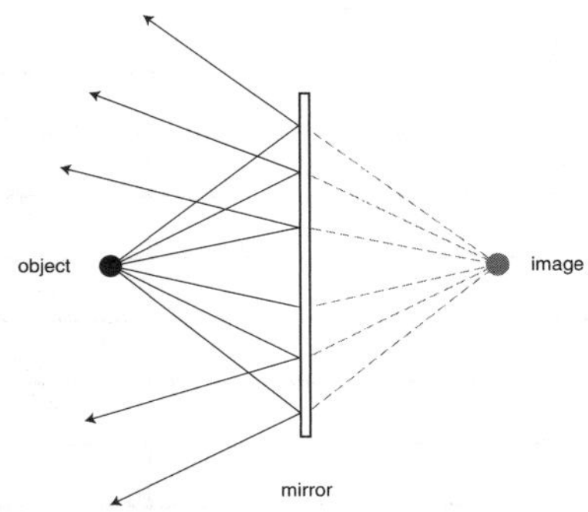

Note that the image is as far behind the mirror as the object is in front. Also, the image is located on the line that goes through the object and is perpendicular to the mirror surface. If you walk towards a mirror, your reflection seems to walk towards you. If you reach out to shake hands with your reflection, a reflected hand reaches towards you . . . but it's a left hand. What happened? Move your hand to the left, and the reflection moves to the left. Move your hand to the right, and the reflection moves to the right. But move your hand towards the mirror, and the reflection moves towards you (in the opposite direction). This is the source of the inversion of your hand. What is inverted in the mirror is not right to left, and not up to down, but rather towards the mirror and away from the mirror. The reflection of an asymmetrical letter like a "R" shows what happens. Note that the reflection of the left hand is a right hand.

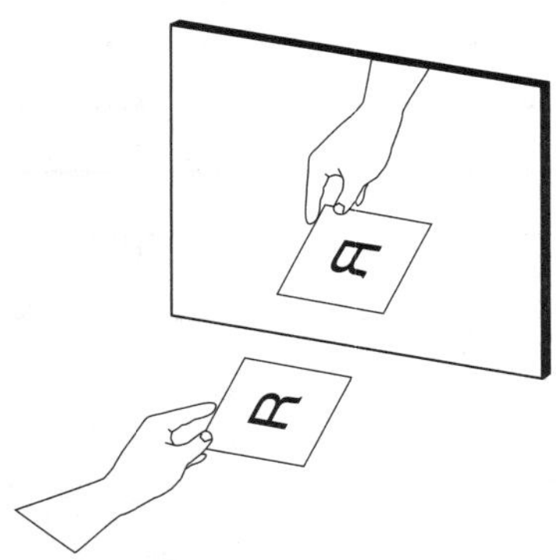

© It's About Time

Active-ating the Physics InfoMall

This is the first of several activities that involve light and optics. A good strategy is to find what problems students may have with this area of physics. One way to find out is to search for "student difficult*" AND "optics". You may not want to limit these terms to "the same paragraph" for the first attempt. Indeed, there are several nice articles found in this search. (Note that most textbooks do not directly address the issue of student difficulties or misconceptions. Articles, being more recent and flexible in most cases, do treat these topics.)

The first hit on the list is "An investigation of student understanding of the real image formed by a converging lens or concave mirror," in the *American Journal of Physics*, vol. 55, issue 2, 1987. Note that this article is actually on the CD-ROM for your use. But, since this activity is concerned with plane mirrors, we may wish to keep this article for later. Just a little lower on the list is "Student difficulties in understanding image formation by a plane mirror," from *The Physics Teacher*, vol. 24, issue 8, 1986. This is something you should read.

A related search is "misconcept*" and "optics". This provides a larger number of hits, but includes both articles above. One new hit is "Misinterpretation of theories of light," from the *American Journal of Physics.*

Suppose the question about the mirror on the moon gets you curious. Search for "mirror" AND "moon" to find some information.

What if you want to find additional activities or demonstrations for your class? You can search, or go straight to the Demo & Lab Shop. You can choose any of the books you find there; by chance, the first book found while preparing this document was Physics Lab Experiments and Computer Aids, which has an activity similar to the *Active Physical Science* activity, with graphics that might be helpful. In addition, there are many more ideas you may wish to try yourself.

In **For You To Do Step 11**, students are asked to place two mirrors at right angles. To see what might happen, search for "mirror*" AND "right angle*", but restrict the search to " "within paragraph." Not only will your question be answered, but we find several articles that might be useful for other things. For example, "Optics: It's Elementary," from Teachers Treasures, has some great ideas. "Some reflections on plane mirrors and images," from *The Physics Teacher*, vol. 29, issue 7, 1991, is also worth finding. Repeat this search for some great sources.

© It's About Time

Planning for the Activity

Time Requirements:

one and a half class periods

Materials Needed

For the class:

- oscilloscope

For each group:

- AA battery
- cardboard sheet, 14" × 14" square
- glass rod
- laser pointer or ray bar
- miniature light bulb, 1.5 vVolt
- optical bench apparatus
- light source for optical bench apparatus
- plane mirrors, 3 × 4 × 1/2 (2)
- makeup mirror, convex/concave surface
- protractor
- right angle prism
- ruler, metric
- small amount of play sand
- supports for mirrors (2)

For each student:

- 1 copy of the Blackline Master: Angles of Incidence and Angles of Reflection

Advance Preparation and Setup

Make a copy, for each student, of the Blackline Master Angles of Incidence and Angles of Reflection. You can support the mirrors on blocks of wood or tape them to boxes. Also, you can place small mirrors in slit rubber stoppers. You may also want to insert the glass rod in a rubber stopper and hold the stopper in a test tube clamp, mounted on a ring stand. Be sure that each group has two mirrors that can meet edge-to-edge to make the multiple reflections most effectively. Mirrors in frames cannot meet in this way. Test each light source in advance. Be sure that all mirrors are clean. Darken the room as much as possible.

Teaching Notes

When the students measure the angle of incidence vs. the angle of reflection, encourage them to take data throughout the range of the angle of incidence (zero to 90 degrees). Every fifteen degrees is a good interval. Encourage the students to sight carefully as they place the dots on the paper to mark the position of the beam. Point out that they must be directly over the beam when they place their dots. You might mention the effect of parallax—an object seems to shift position when seen from different places. To help students visualize the position of the image, have them hold a meter stick or ruler perpendicular to the mirror, as shown.

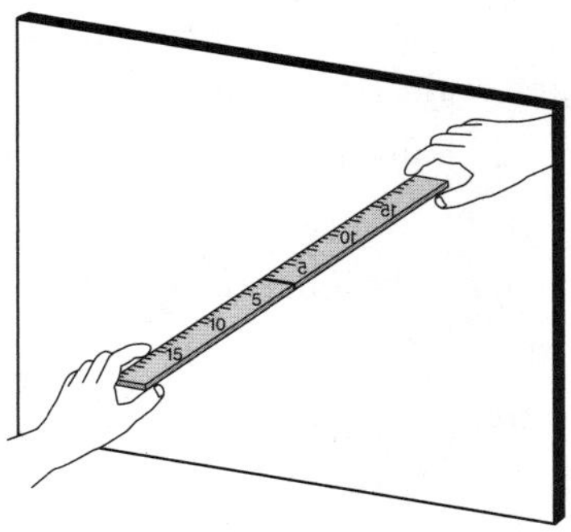

If the students use lasers as the light source, each group must keep its laser's beam under control at all times. Have each group set up a backstop behind its mirrors to block the beam. A large piece of cardboard makes a good backstop.

An easy way to attach the glass rod to the laser is to drill a hole through a flat rubber eraser slightly smaller than the barrel of the pen laser. Push the pen laser through the hole until the end of the laser just comes through. Then tape the glass rod to the rummer eraser so that it is positioned in the laser beam.

SAFETY PRECAUTION: Warn the students not to look into the laser beam. Also emphasize that the laser should only be turned on when it is resting on the tabletop.

Many secondary students have not yet understood the path of light from a light source to an object to their eye. Some may still believe that vision is something that emanates from the eyes, so the reflection of light from the object is not important. It may help to darken the room and illuminate an object with a flashlight. Have the students trace the path of light (or even walk along the path) from the flashlight to their eyes. Ask about the fact that many students, in different locations in the classroom, can see the object, and what that must mean in terms of the reflected light (that it's reflected in many directions).

There should be good agreement (within a few degrees) between the values of the angles of incidence and angles of refraction, especially if the mirrors are made of polished metal. Remember that if the students are using plastic or glass mirrors, these are usually rear-surfaced. In effect, the reflection happens in the middle of the mirror, but the angle of reflection is still equal to the angle of incidence. The main effect will be to shift the extension of the beam used to locate the image.

© It's About Time

NOTES

© It's About Time

Activity Overview

Students begin the study of mirrors by investigating reflection from a plane mirror. To find the direction of a reflected light ray, they measure the angles of incidence and refraction. Then they locate the reflected image and observe the reversals in reflections of letters of the alphabet. Finally, students investigate multiple reflections from two mirrors and the resulting symmetrical patterns.

Student Objectives

Students will:

- Identify the normal of a mirror.
- Measure angles of incidence and reflection.
- Observe the relationship between the angle of incidence and the angle of reflection.
- Observe changes in the reflections of letters.
- Identify patterns in multiple reflections.

Activity 4 Reflected Light

GOALS

In this activity you will:

- Identify the normal of a mirror.
- Measure angles of incidence and reflection.
- Observe the relationship between the angle of incidence and the angle of reflection.
- Observe changes in the reflections of letters.
- Identify patterns in multiple reflections.

What Do You Think?

Astronauts placed a mirror on the Moon in 1969 so that a light beam sent from Earth could be reflected back to Earth. By timing the return of the beam, scientists found the distance between the Earth and the Moon. They measured this distance to within 30 cm.

- **How are you able to see yourself in a mirror?**
- **If you want to see more of yourself, what can you do?**

Record your ideas about these questions in your *Active Physics* log. Be prepared to discuss your responses with your small group and with your class.

For You To Do

1. Place a piece of paper on your desk. Carefully aim the laser pointer, or the light from a ray box, so the light beam moves horizontally, as shown on the opposite page.

ANSWERS FOR THE TEACHER ONLY

What Do You Think?

You can see yourself in a mirror by observing light reflected from you that the mirror reflects back into your eyes. The mirror creates a virtual image of you the same distance behind the mirror that you are in front of it. But you can see only that part of your body that faces the mirror, since light from the back of your head moves off in another direction and does not hit the mirror. To see the back of your head requires a second mirror, which is why hair stylists hold up a second mirror to show their clients their new haircut. If you covered the walls of a room with mirrors, you could see many different reflections of the back of your head.

ANSWERS

For You To Do

1. Student activity.

© It's About Time

2. Place a glass rod in the light beam so that the beam spreads up and down. Shine the beam on the piece of paper to be sure the beam passes through the glass rod.

3. Carefully stand the plane mirror on your desk in the middle of the piece of paper. Draw a line on the paper along the

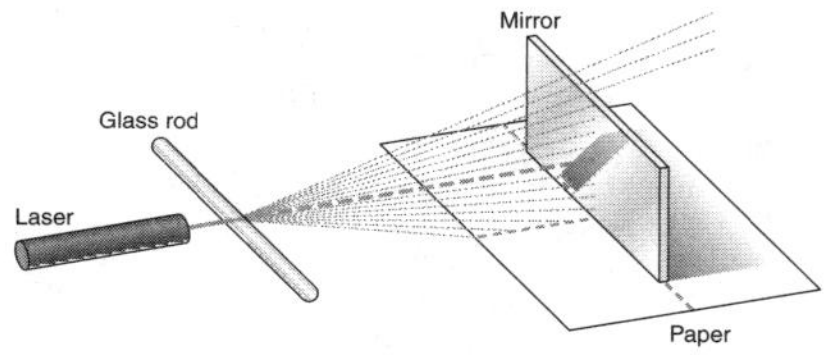

front edge of the mirror. Now remove the mirror and draw a dotted line perpendicular to the first line, as shown. This dotted line is called the **normal**.

4. Aim the light source so the beam approaches the mirror along the normal. Be sure the glass rod is in place to spread out the beam.
   a) What happens to the light after it hits the mirror?

5. Make the light hit the mirror at a different angle.
   a) What happens now?
   b) On the paper, mark three or more dots under the beam to show the direction of the beam as it travels to the mirror. The line you traced shows the incident ray. Also make dots to show the light going away from the mirror. This line shows the reflected ray. Label this pair of rays to show they go together.

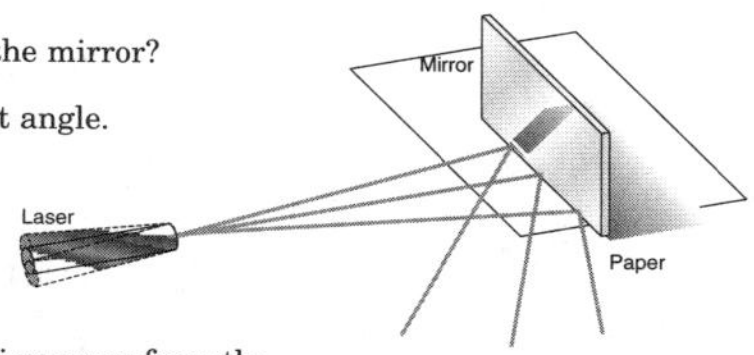

6. Turn the light source so it starts from the same point but strikes the mirror at different angles. For each angle, mark dots on the paper to show the direction of the incident and reflected rays. Also, label each pair of rays.

Do not use mirrors with chipped edges. Make sure the ends of the glass rod are polished.

Never look directly at a laser beam or shine a laser beam into someone's eyes. Always work above the plane of the beam and beware of reflections from shiny surfaces.

Physics Words
normal: at right angles or perpendicular to.

ANSWERS

For You To Do *(continued)*

2. Find some way to support the glass rod, such as with the optional stopper, test-tube holder and ringstand (or the rod can be taped to a stack of books).

3. Student activity. Support the mirror, using the mirror support in the kit.

4. a) It reflects back along the normal.

5. a) The beam reflects but not along the normal.
   b) Encourage students to look straight down to avoid parallax.

6. Student activity.

© It's About Time

7. Most lab mirrors have the reflecting surface on the back. In addition, the light bends as it enters and leaves the glass part of the mirror. In your drawing, the rays may not meet at the mirror surface. Extend the rays until they do meet.
   - a) Measure these angles for one pair of your rays.

8. Turn off the light source and remove the paper. Look at one pair of rays. The diagram shows a top view of the mirror, the normal, and an incident and reflected ray. Notice the angle of incidence and the angle of reflection in the drawing. Using a protractor, measure these angles for one pair of rays.
   - a) Record your data in a table.
   - b) Measure and record the angles of incidence and reflection for all of your pairs of rays.

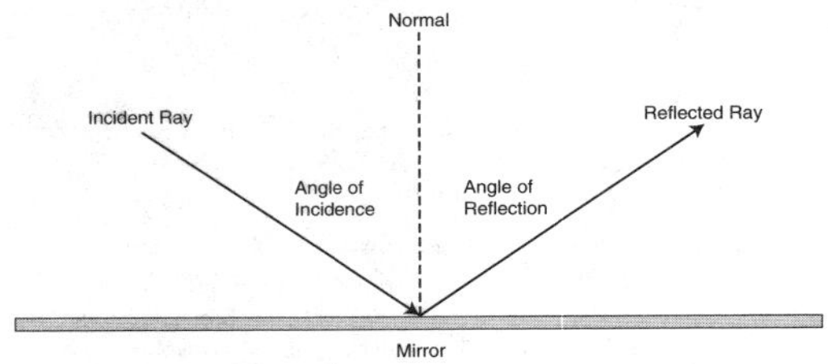

   - c) What is the relationship between the angles of incidence and reflection?
   - d) Look at the reflected rays in your drawing. Extend each ray back behind the mirror. What do you notice when you have extended all the rays? The position where the rays meet is the location of the *image* of the light source. All of the light rays leave one point in front of the mirror. The reflected rays all seem to emerge from one point behind the mirror. Wherever you observed the reflection, you would see the source at this point behind the mirror.
   - e) Tape a copy of your diagram in your log.

9. Hold the light source, or any object, near the mirror and look at the reflection. Now hold the object far away and again look at the reflection.

ANSWERS

For You To Do *(continued)*

7. See drawing in **Backround Information**. The angles should be equal.

8. a)-c) The angle of incidence is (approximately) equal to the angle of reflection.

   d) The extended rays intersect in the same place (or nearly so).

   e) See drawing of extended rays in **Background Information**.

© It's About Time

a) How is the position of the reflection related to the position of the object?

10. Set up a mirror on another piece of paper, and draw the normal on the paper. Write your name in block capital letters along the normal (a line perpendicular to the mirror). Observe the reflection of your name in the mirror.
    a) How can you explain the reflection you see?
    b) Which letters in the reflection are closest to the mirror? Which are farthest away?
    c) In your log, make a sketch of your name and its reflection.

11. Carefully stand up two mirrors so they meet at a right angle. Be sure they touch each other, as shown in the drawing.

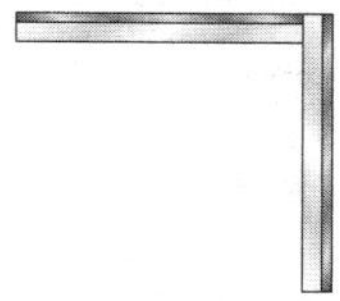

12. Place an object in front of the mirrors.
    a) How many images do you see?
    b) Slowly change the angle between the mirrors. Make a general statement about how the number of images you see changes as the angle between the mirrors changes.

Physics Words

angle of incidence: the angle a ray of light makes with the normal to the surface at the point of incidence.

angle of reflection: the angle a reflected ray makes with the normal to the surface at the point of reflection.

ray: the path followed by a very thin beam of light.

FOR YOU TO READ

Images in a Plane Mirror

An object like the tip of a nose reflects light in all directions. That is why everybody in a room can see the tip of a nose. Light reflects off a mirror in such a way that the **angle of incidence** is equal to the **angle of reflection**. You can look at the light leaving the tip of a nose and hitting a mirror to see how an image is produced and where it is located. Each **ray** of light leaves the nose at a different angle. Once it hits the mirror, the angle of incidence must equal the angle of reflection. There are now a set of rays diverging from the mirror. If you assume that the light always travels in straight lines, you can extend these rays behind the mirror and find where they "seem" to emerge from. That is the location of the image.

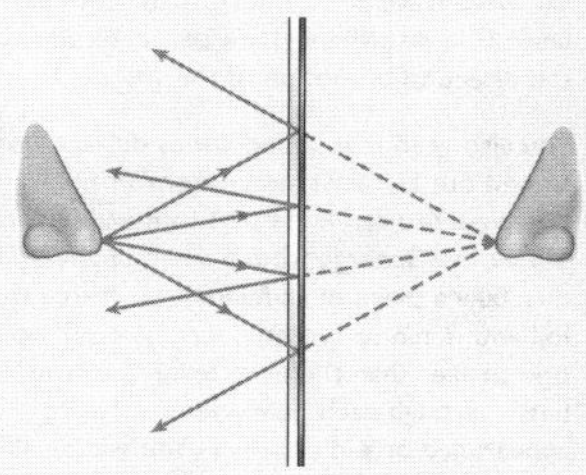

The mirror does such a good job of reflecting that it looks as if there is a tip of a nose (and all other parts of the face) behind the mirror. If you measure the distance of the image behind the mirror, you will find that it is equal to the distance of the nose (object) in front of the mirror. This can also be proved using geometry. →

ANSWERS

For You To Do *(continued)*

9. a) The position of the reflection is behind the mirror. It is the same distance behind the mirror that the object was in front of the mirror.

10. a)-b) The letters that are closest to the mirror make reflections that look closest to the mirror. The letters that are farthest from the mirror make reflections that are farthest from the mirror. Moreover, within a letter, the part of letter closer to the mirror makes a reflection closer to the mirror than the rest of the letter.

    c) Student diagrams will differ.

11. Student activity.

12. a) Three (one is behind each mirror, as before; the third image is behind the corner where the mirrors meet).

    b) As the angle between the mirrors reduces, the number of reflections increases.

© It's About Time

Diffraction of Light

As you begin to study the reflection of light rays, it is worthwhile to recognize that light is a wave and has properties similar to sound waves.

In studying sound waves, you learned that sound waves are compressional or longitudinal. The disturbance is parallel to the direction of motion of the wave. In sound waves, the compression of the air is left and right as the wave travels to the right. You saw a similar compressional wave using the compressed Slinky.

Light waves are transverse waves. They are similar to the transverse waves of the Slinky. In a transverse wave, the disturbance is perpendicular to the direction of the wave. In the Slinky, the disturbance was up and down as the wave traveled to the right. In light, the fields (the disturbance) are perpendicular to the direction of motion of the waves.

You also read that sound waves diffract—they spread out as they emerge from small openings. You can find out if light waves spread out as they emerge from a small opening. Try this: Take a piece of aluminum foil. Pierce the foil with a pin to create a succession of holes, one smaller than the next. Shine the laser beam through each hole and observe its appearance on a distant wall. You will be able to observe the diffraction of light.

Sample Problem

Light is incident upon the surface of a mirror at an angle of 40°.

a) Sketch the reflection of the ray.

Strategy: The angles of incidence and reflection are always measured from the normal. The Law of Reflection states that the angle of incidence is equal to the angle of reflection. Since the angle of incidence is equal to 40°, the angle of reflection is also 40°.

Given:

$\theta_i = 40°$

Solution:

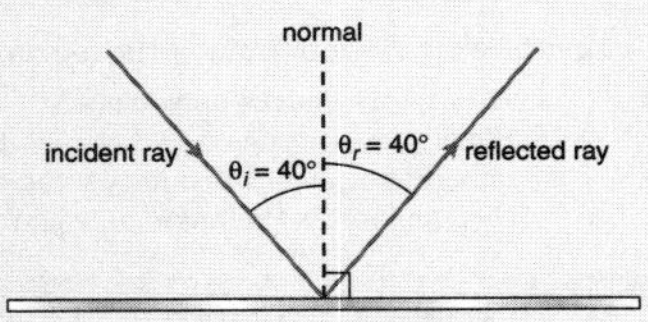

b) At what angle, as measured from the surface of the mirror, did the beam strike the mirror?

Strategy: The angle of incidence is measured from the normal. The question is asking for the complementary angle.

Solution:

$\theta_i = \theta_r = 40°$

$90° - 40° = 50°$

The angle between the light beam and the mirror is 50°.

© It's About Time

Reflecting on the Activity and the Challenge

In this activity you aimed light rays at mirrors and observed the reflections. From the experiment you discovered that the angle of incidence is equal to the angle of reflection. Therefore, you can now predict the path of a reflected light beam. You also experimented with reflections from two mirrors. When you observed the reflection in two mirrors, you found many images that made interesting patterns.

This activity has given you experience with many interesting effects that you can use in your sound and light show. For instance, you may want to show the audience a reflection in one mirror or two mirrors placed at angles. You can probably create a kaleidoscope. You will also be able to explain the physics concept you use in terms of reflected light.

Physics To Go

1. How is the way light reflects from a mirror similar to the way a tennis ball bounces off a wall?
2. a) What is the normal to a plane mirror?
   b) When a light beam reflects from a plane mirror, how do you measure the angle of incidence?
   c) How do you measure the angle of reflection?
   d) What is the relationship between the angle of incidence and the angle of reflection?
3. Make a top-view drawing to show the relationships among the normal, the angle of incidence, and the angle of reflection.
4. a) Suppose you are experimenting with a mirror mounted vertically on a table, like the one you used in this activity. Make a top-view drawing, with a heavy line to represent the mirror and a dotted line to represent the normal.
   b) Show light beams that make angles of incidence of 0°, 30°, 45°, and 60° to the normal.
   c) For each of the above beams, draw the reflected ray. Add a label if necessary to show where the rays are.

ANSWERS

Physics To Go

1. For the tennis ball and for light, the angle of incidence equals the angle of reflection (except for the effect of the spin of the ball).
2. a) The normal is a line perpendicular to the mirror surface.
   b) The angle of incidence is the angle between the incident beam and the normal. You measure this angle by tracing the path of the incident beam and then drawing the normal where the beam hits the mirror. Use a protractor to measure the angle.
   c) The angle of reflection is the angle between the reflected beam and the normal. See answer to **Part (b)**.
   d) They are equal.
3. See the drawing on page 356.
4. a) See the drawing on page 356.
   b)-c) See below.

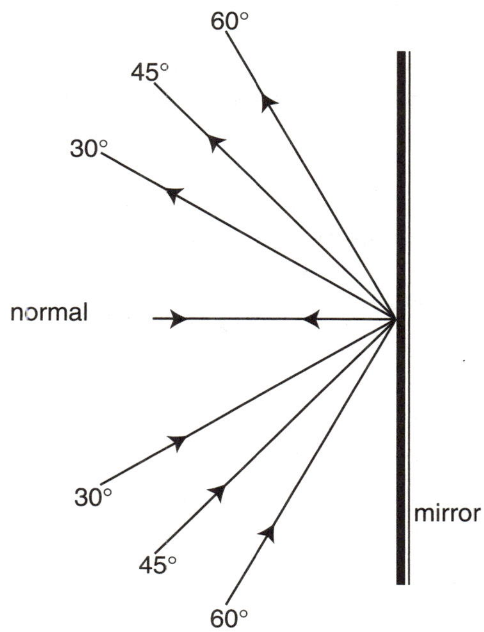

© It's About Time

ANSWERS

Physics To Go

(continued)

5. a) Student activity.

   b) The reflection moves towards the mirror and towards you.

   c) The reflection moves away from the mirror and away from you.

   d) The reflection is the same distance behind the mirror as the object is in front of the mirror. If you move your hand away from the mirror, the distance to the mirror increases, so the reflection is further from the mirror, too.

6. a) The letters A, I, T, M, O, W, U, and V.

   b) MOM, TOT, WOW.

   c) K, N, and S.

   d) See drawing below.

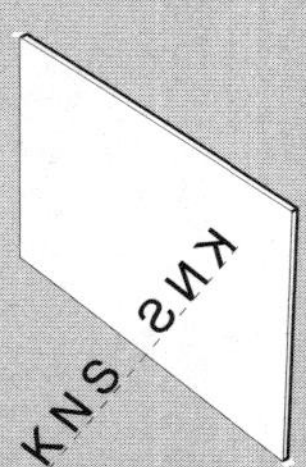

7. It is usually seen in a rear-view mirror in which it would read normally.

8. Placement of the original light ray is random. But once the ray has struck the mirror, angles of incidence to the normal and angles of reflection from the normal must be carefully measured.

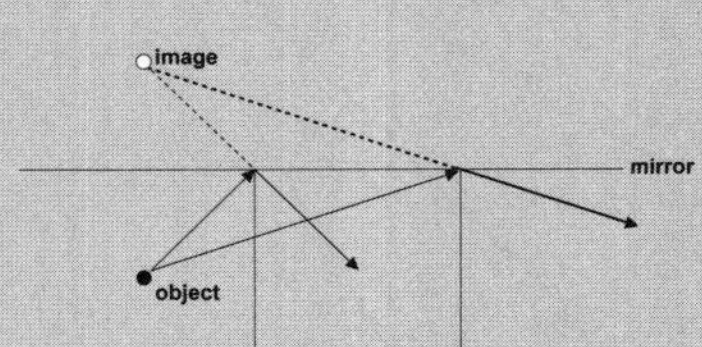

Let Us Entertain You

5. a) Stand in front of a mirror.
   b) Move your hand toward the mirror. Which way does the reflection move?
   c) Move your hand away from the mirror. Which way does the reflection move?
   d) Use what you learned about the position of the mirror image to explain your answers to **Parts (b)** and **(c)**.

6. Suppose you wrote the whole alphabet along the normal to a mirror in the way you wrote your name in **Step 10** of **For You To Do**.
   a) Which letters would look just like their reflections?
   b) Write three words that would look just like their reflections.
   c) Write three letters that would look different from their reflections.
   d) Draw the reflection of each letter you gave in **Part (c)**.

7. Why is the word *Ambulance* written in an unusual way on the front of an ambulance?

8. Use a ruler and protractor and a ray diagram to locate the image of an object placed in front of a plane mirror. Be careful! You must measure as carefully as you can to obtain the most accurate answer.

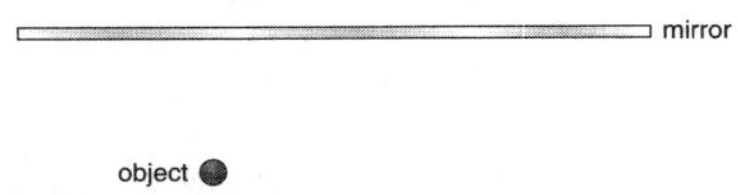

9. Locate the image of the lamp shown in the diagram.

360

Active Physics

ANSWERS

Physics To Go *(continued)*

9.

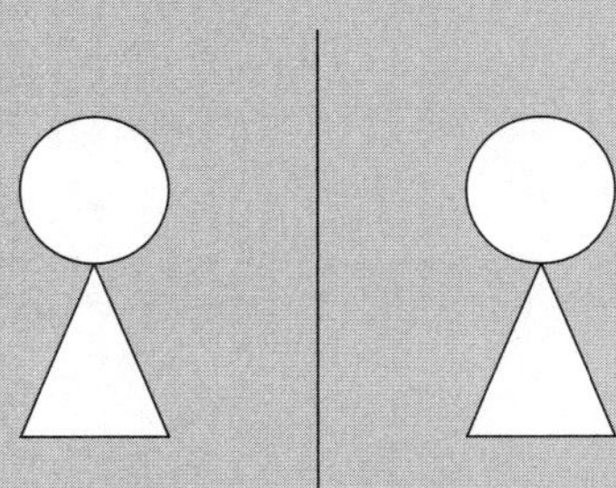

© It's About Time

10. After reflecting off mirrors A, B, and C, which target will the ray of light hit?

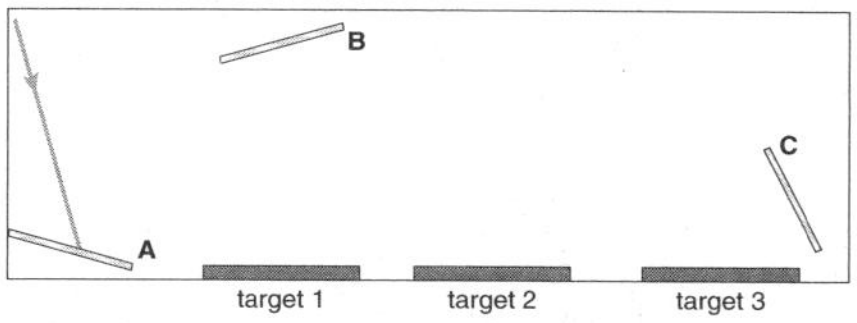

Stretching Exercises

1. Carefully tape together three small mirrors to make a corner reflector. Shine a flashlight down into the corner. Where does the reflected beam go?
2. Build a kaleidoscope by *carefully* inserting two mirrors inside a paper towel holder. You can also use three identical mirrors. Do not force the mirrors into the tube. Tape the edges of the mirrors together, with the mirrored surfaces inside. Describe what you see through your kaleidoscope.
3. Carefully tape together one edge of two mirrors so they can move like a hinge, with the mirrored surfaces facing each other. Place a small object between the mirrors. Investigate how the number of images you see depends on the angle between the mirrors. You will need a protractor to measure this angle. Plot a graph of the results. What mathematical relationship can you find between the angle and the number of images?

Answers

Stretching Exercises

1. The beam reflects back along the direction it entered the corner mirror.
2. Student activity.
3. Number of images = (360/angle); however when the formula dictates a whole number of images, some of the images overlap and can only be seen as the angle between the mirrors is further decreased.

Answers

Physics To Go *(continued)*

10. Target 3

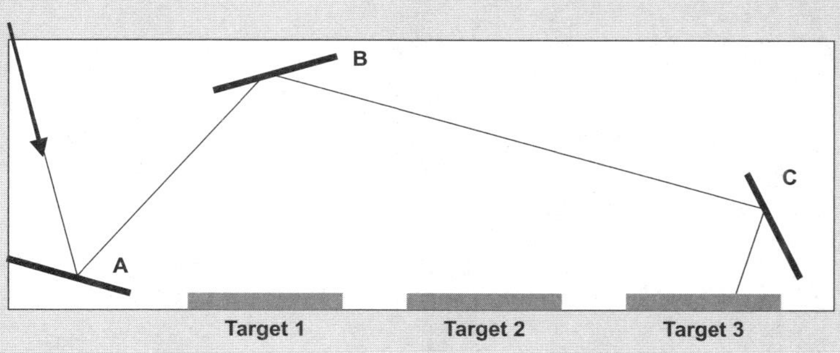

© It's About Time

Activity 4 A

Low-Tech Reflection and Mirror Symmetry

FOR YOU TO DO

1. Low-Tech Reflection: You can do the Angle of Incidence/Angle of Reflection experiment with a flashlight. The results are less accurate, of course, than they would be with a laser beam, but students will still get the idea. You might give the students string and have them stretch it along the normal, from the flashlight to the mirror, and from the mirror to the reflection. Then they can measure the angles between the strings.

2. Mirror Symmetry: Give each group a simple design. Have them make a drawing to predict how the reflection of the design will look. Then they hold the design up to a mirror to observe the reflection. Arrows on the design point to the location where the mirror will be. When the students have made and discussed their predictions, give them the mirrors.

© It's About Time

For use with *Let Us Entertain You*, Chapter 5, Activity 4: Reflected Light

Table of Angles of Incidence	Angles of Reflection

For use with Let Us Entertain You, Chapter 5, Activity 4: Reflected Light

Table of Angles of Incidence	Angles of Reflection

For use with *Let Us Entertain You*, Chapter 5, Activity 4: Reflected Light

© It's About Time

ACTIVITY 5
Curved Mirrors

Background Information

A good starting point for thinking about the curved mirror is the plane mirror. The plane mirror makes an erect, virtual image.

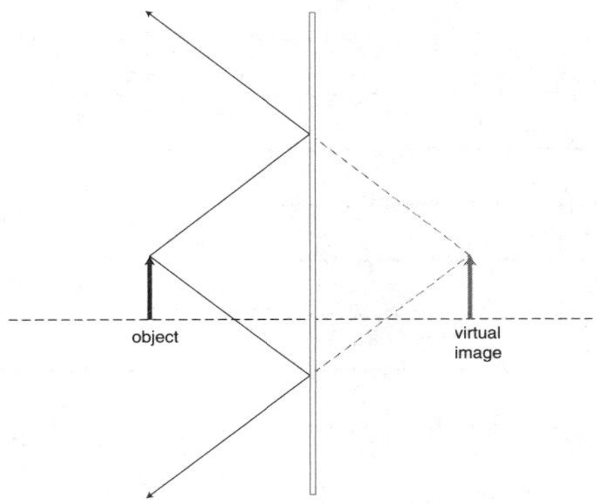

Now imagine the mirror is bowed out to make a convex mirror. How does the image change? The light rays show that the image is still behind the mirror, so the image is still virtual.

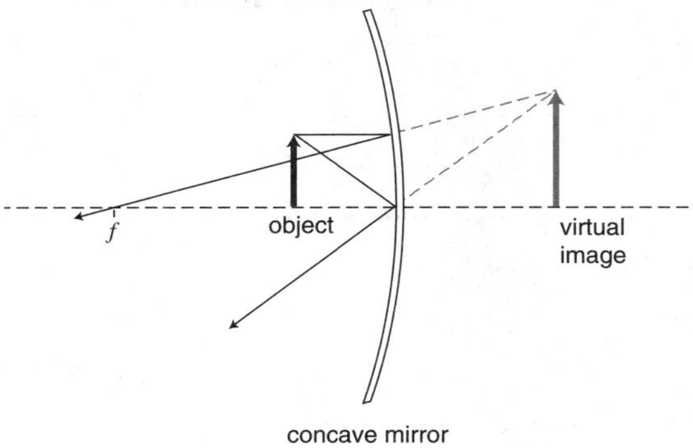

concave mirror

Moreover, the image is reduced, since parallel rays are reflected out away from the mirror. Also, the image is closer to the mirror than is the object. The focus is labeled. The more strongly the mirror is curved, the shorter the focal length and the more the virtual image is reduced compared to the object.

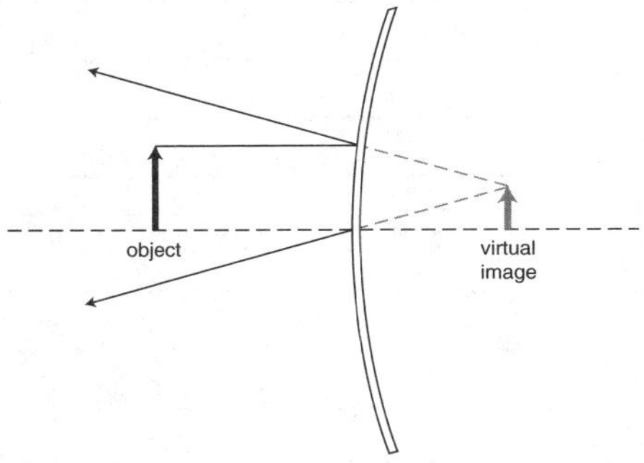

convex mirror

Now imagine bowing the plane mirror inward, to make a concave mirror.

Since parallel rays now bend in towards the mirror axis, objects now appear magnified when the object is close to the mirror. Only when the object is within a focal length of the mirror is the object virtual. For such object positions, the image is further from the mirror than is the object. If the object is outside the focal length, the reflected rays cross, so a real image is formed, as shown in the drawings in **Physics Talk** on page 366 in the Student Edition. This image can be seen on a card. It can, of course, also be seen by looking at the mirror, rather than on a card, and this is the basis of the famous "floating coin" illusion made by two parabolic mirrors fastened together. The concave mirror is exactly analogous to the convex lens, which is a magnifier when the object is up close and which produces a real image otherwise. The lens equation also describes the location of the image and object for mirrors:

$1/f = 1/D_o + 1/D_i$

Image formation in curved mirrors, including the mathematical description of the equation above, are exactly analagous to image formation in lenses, the subject of **Activity** 7.

Active-ating the Physics InfoMall

This activity is all about images formed by curved mirrors. Research has been conducted to see how students understand this phenomenon, as reported in "An investigation of student understanding of the real image formed by a converging lens or concave mirror," from the *American Journal of Physics*, vol. 55, issue 2, 1987, mentioned above. This article may prepare you for problems that might arise with your students.

© It's About Time

The first search to perform here might be "curved mirror*". This does produce a few good hits. But if you want to find more, or better, sources of information, you may want to be more specific: search for concave or convex mirrors.

One item you may find while searching is "Physics Review Text" from the Regent's Physics Review in the Problems Place. In Opt. Unit 4 – Geometric Optics you can find a wonderful set of instructions for drawing ray diagrams. Again, these can be copied right into your own documents. As the InfoMall goes, the graphics in this section are great! You may also want to look in the Optical Elements section of *Physics Including Human Applications*, found in the Textbook Trove. Of course, these are just a fraction of the things you can find on the InfoMall.

Step 11 of **For You To Do** uses an equation. If you want to see what the InfoMall has for this, go back to the InfoMall Entrance (the colorful screen with the directory map). In the lower right-hand corner you will see a button with equations on it. Clicking on this takes you to the equations dictionary. You will be only a few clicks away from this equation, along with examples.

Planning for the Activity

Time Requirements:

two class periods

Materials Needed

For each group:

- 40-W bulb and socket, ceramic w/switch
- cardboard sheet, 14" × 14" square (for a backstop to absorb a laser beam)
- convex/concave mirror surface
- glass rod
- laser pointer
- masking tape, 3/4" × 60 yds.
- paper, plain
- right angle prism
- ruler, metric
- unlined index cards (100)
- stick, meter, 100 cm hardwood

For each student:

- 1 copy of Blackline Master: Object and Image Distances
- 1 copy of Blackline Master: Graph Paper
- Second copy of Blackline Master: Graph Paper (if **PTG #5** is assigned)

Advance Preparation and Setup

Make copies of the Object and Image Distances and Graph Paper, for each student. If you assign **Question 5** from **Physics To Go**, make an additional copy of the Graph Paper for each student. If possible, find a large-diameter mailing tube to make the forms for the mirrors used in the ray-tracing. Cut a section about 5 cm high. Give each group a complete section and also one cut in half (to be the form for the concave lens). If it is necessary to use a flashlight as a light source, you will need a very large-diameter form and a large piece of mylar to trace the parallel rays, because the flashlight beam will spread out so much. When the students investigate real images, use a large, well-shaped concave mirror so the image will be bright and clear. A cosmetics mirror with a diameter of about 15 cm works well. If your students use lasers, collect large pieces of cardboard to use as backstops.

Teaching Notes

If you have neither a laser nor a light box, and you must use a flashlight for a light source, you can still encourage the students to get the general idea of how parallel light beams are reflected from curved mirrors (towards the mirror axis or away, as discussed in **Background Information**). The students can still see images made by a light bulb in the second half of the activity. When the students observe images, give them the best mirrors you have available.

Once the students have observed the images, you may want to remind them about **Steps 1** to **7**, when they observed reflections of individual light beams. Encourage them to use their observations to predict which mirrors make real images.

You may want to set up a light bulb in the front of the room where it is visible to all groups. This bulb can serve as the distant object, to minimize how the students must move about the lab. If your students use lasers, emphasize the importance of keeping their laser beam within their work area.

© It's About Time

If students use a laser, have each group prop up a large piece of cardboard to absorb the beam. If the students use a 110-V bulb as a light source, caution them to handle the bulb carefully and to leave the bulb in its socket at all times. Be sure there are no exposed connections to the socket. If the socket is ceramic and has screw-terminal posts underneath, screw the socket to a piece of wood so the terminals are inaccessible.

Many students believe that the virtual image is located inside the mirror. Many others believe that the image of an object moves if the observer moves. To help students, you can have them set the mirror on a table and place a small object nearby. Have them observe the reflection of the object from different positions and report what they see. You might also have them repeat this observation with a plane mirror.

This data was taken with a cosmetics mirror.

D_o (m)	D_i (m)
26.5	0.37
5.13	0.45
2.74	0.52
1.37	0.52
0.51	1.30

In the above data, note how in going from 5.13 m to 26.5 m, more than a factor of five in object distance, the image moved in less than fifteen percent. The focal length of the mirror is very close to .37 m. Also notice how the image distance increases rapidly as the object approaches the focus of the mirror.

© It's About Time

NOTES

© It's About Time

Activity Overview

Having worked with plane mirrors in **Activity 4**, students now investigate curved mirrors. As in **Activity 4**, students begin by shining parallel light beams at the curved mirrors to record the pattern of reflection. From these patterns students find the focal length and the location of the focus. With a small light bulb, students look for real and virtual images. For the concave mirror, students draw a graph of the image distance vs the object distance. To use a quantitative model of reflection, students examine an equation that relates image distance, object distance, and focal length.

Student Objectives

Students will:

- Identify the focus and focal length of a curved mirror.
- Observe virtual images in a convex mirror.
- Observe real and virtual images in a concave mirror.
- Measure and graph image distance vs object distance for a convex mirror.
- Summarize observations in a sentence.

Answers for the Teacher Only

What Do You Think?

The reflection in a flat mirror has the same proportions as the object. To explain this observation, recall that the reflected image is the same distance behind the mirror as the object is in front, and on a line from the object perpendicular to the mirror. A displacement of the object produces an equal displacement of the image, so the proportions are preserved. But in a curved mirror, this proportionality is lost.

The ray diagrams on pages 363 and 364 show how curved mirrors reflect a parallel beam of light. Imagine standing close to these mirrors and looking at the reflection of the distant object. Since the reflection is seen along the extension of the reflected light rays, the concave mirror makes the object look larger (in the direction of the curvature) and the convex mirror makes the object look smaller.

Activity 5 Curved Mirrors

GOALS

In this activity you will:

- Identify the focus and focal length of a curved mirror.
- Observe virtual images in a convex mirror.
- Observe real and virtual images in a concave mirror.
- Measure and graph image distance versus object distance for a convex mirror.
- Summarize observations in a sentence.

What Do You Think?

The curved mirror of the Palomar telescope is five meters across. Mirrors with varying curvatures are used in amusement parks as fun-house mirrors. Store mirrors and car side-view mirrors are also curved.

- **How is what you see in curved mirrors different from what you see in ordinary flat mirrors?**

Record your ideas about this question in your *Active Physics* log. Be prepared to discuss your responses with your small group and with your class.

For You To Do

1. Carefully aim a laser pointer, or the light from a ray box, so the light beam moves horizontally, as you did in the previous activity. Place a glass rod in the light beam so that the beam spreads up and down.

Answers

For You To Do

1. Support the glass rod in the same way, as in **Activity 4**.

© It's About Time

2. Place a convex mirror in the light beam, as shown in the diagram.

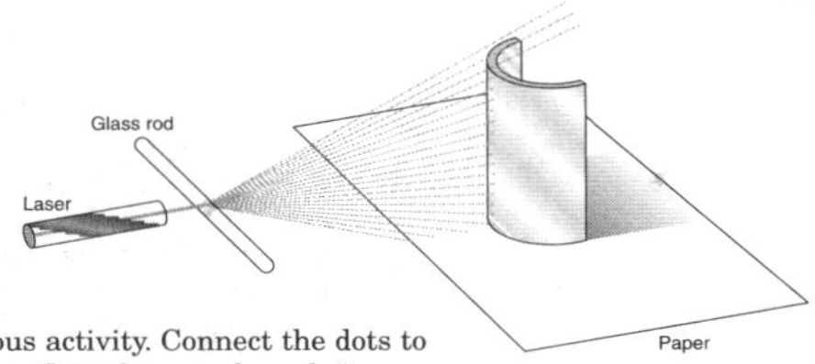

3. Shine a beam directly at the center of the mirror. This is the incident beam. Show its path by placing three or more dots on the paper, as you did in the previous activity. Connect the dots to make a straight line. Find the reflected ray and mark its path in a similar way. Label the two lines so you will know they go together.

Never look directly at a laser beam or shine a laser beam into someone's eyes. Always work above the plane of the beam and beware of reflections from shiny surfaces.

4. You will move the light source sideways to make a series of parallel beams. To make sure the incident beams are parallel, line up each one with the dots you made to show the incoming beam in **Step 3**. Mark the path of the incoming ray with three dots.

5. Each parallel beam makes a reflected beam. Show the path of each of these reflected rays. Label each incident and reflected beam so you will know that they go together.
   a) Write a sentence to tell what happens to the parallel beams after they are reflected.
   b) Make a drawing in your *Active Physics* log to record the path of the light.

Physics Words

focus: the place at which light rays converge or from which they appear to diverge after refraction or reflection; also called focal point.

focal length: the distance between the center of a lens and either focal point.

6. Remove the mirror. With a ruler, extend each reflected ray backwards to the part of the paper that was behind the mirror.
   a) You probably noticed that all the lines converge in a single point. The place where the extended rays meet is called the **focus** of the mirror. The distance from this point to the mirror is called the **focal length**. Measure and record this focal length.

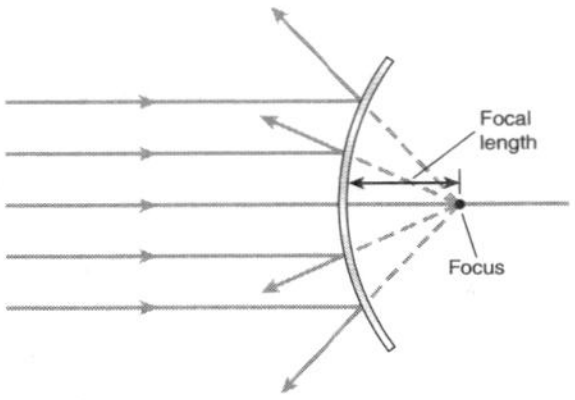

7. Place the concave side of the mirror in the light beam. To help you remember the name *concave*, think of the

ANSWERS

For You To Do *(continued)*

2. Pull the mylar tight to make a smooth curved surface.

3. Student activity.

4. It may be easier to pull the paper sideways to move the mirror instead of the light source.

5. a) After reflecting from the convex mirror, the parallel beams spread apart.

   b) See the drawing on page 363.

6. a) The rays may not meet at the same place, but the students should get the idea.

© It's About Time

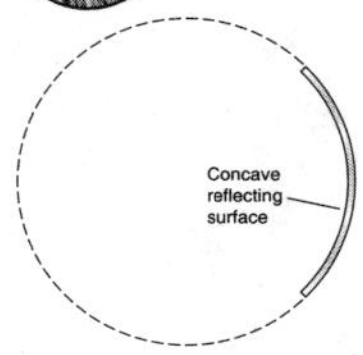

concave mirror as "caving in." Repeat **Steps 3** through **5** for this mirror.

a) Write a sentence to tell what happens to the parallel beams after they are reflected from the concave mirror.

b) Make a drawing in your *Active Physics* log to record the path of the light. The place where the beams cross is called the focus. The distance from the focus to the mirror is the focal length.

c) Measure and record the focal length.

d) How do concave and convex mirrors reflect light differently? Record your answer in your log.

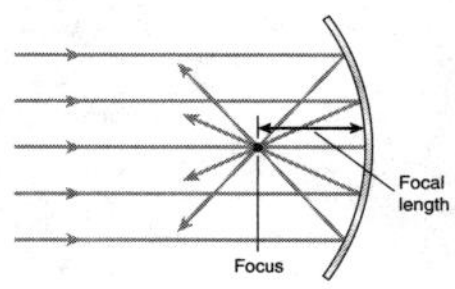

8. Use the concave mirror. Use a 40-W light bulb or a candle as a light source, which will be called the "object." Carefully mount your mirror so it is at the same height as the light source. Place a light bulb about a meter away from the mirror. Put the bulb slightly off the center line, as shown, so that an index card will not block the light from hitting the mirror.

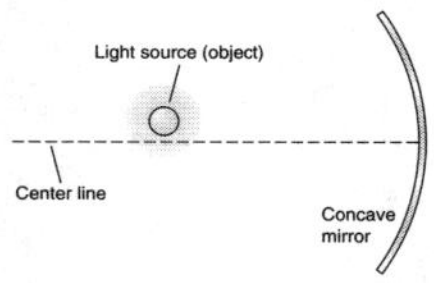

9. Try to find the image of the object on an index card. Move the card back and forth until the image is sharp. The image you found is called a *real* image because you are able to project it on a card.

a) Record the distance of the bulb from the mirror and of the image on the file card from the mirror. Put your results in the first line of a table like the one below.

Distance of bulb from mirror	Distance of image from mirror

Answers

For You To Do *(continued)*

7. a) After reflecting from the concave mirror, the parallel beams come together.

b) See the drawing on page 364.

c) If the students used the same form for both the convex and concave mirror the focal lengths should be approximately the same.

d) A convex mirror spreads out a light beam. A concave mirror brings a light beam to a focus.

8.-9. a) Students may need help in finding the image.

© It's About Time

10. Carefully move the mirror closer to the object. Find the sharp image, as before, by moving the index card back and forth.
   - a) Record the image and object distances in your table.
   - b) Repeat the measurement for at least six object locations.
   - c) Draw a graph of the image distance (y-axis) versus the object distance (x-axis).
   - d) Write a sentence that describes the relationship between the image distance and the object distance.

11. A mathematical relation that describes concave mirrors is

$$\frac{1}{f} = \frac{1}{D_o} + \frac{1}{D_i}$$

where

f is the focal length of that particular mirror

D_o is the object distance

D_i is the image distance

You have measured D_o and D_i. Calculate $\frac{1}{D_o}$ and $\frac{1}{D_i}$. Find their sum for each pair of data.

   - a) Record your calculations in your log.
   - b) Are your sums approximately equal? If so, you have mathematically found the value of $\frac{1}{f}$ for the mirror you used.

12. A convex mirror cannot form a real image that can be projected onto a screen. It can form an image behind the mirror, like a plane mirror.
   - a) Record in your log descriptions of the image in a convex mirror when the mirror is held close and when the mirror is held far from the object.

ANSWERS

For You To Do *(continued)*

10. a) See sample data.
    b) See sample data.
    c) See graph in the margin of page 366.
    d) As the object distance decreases, the image distance increases (and vice versa).

11. a)-b) The sums should be equal to within 10% to 15%.

12. a) The students should see a reduced image in the convex mirror at any distance. If the object is very close, the image looks similar to the image in the plane mirror.

Sample Data

D_o (m)	D_i (m)
26.5	0.37
5.13	0.45
2.74	0.52
1.37	0.52
0.51	1.30

© It's About Time

Physics Words

real image: an image that will project on a screen or on the film of a camera; the rays of light actually pass through the image.

PHYSICS TALK

Making Real Images

To find how a concave mirror makes a **real image**, you can view a few rays of light. Each ray of light obeys the relation you found for plane mirrors (angle of incidence = angle of reflection). In this case, you choose two easily drawn rays.

Look at the drawing. It shows rays coming into a concave mirror from a point on a light bulb. One ray comes in parallel to the dotted line, which is the axis of the mirror. This ray reflects through the focus. The other ray hits the center of the mirror. This ray reflects and makes the same angle with the mirror axis going out as it did coming in. Where these rays meet is the image of the top of the light bulb.

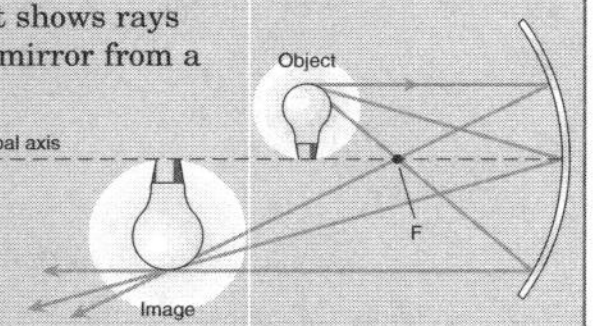

The next drawing shows the same mirror, but with the object much further from the mirror. Notice how the image in this second drawing is much smaller and much closer to the focus.

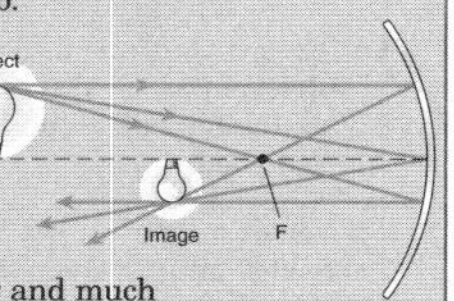

As you have seen, the position of the object and image are described by the equation below.

$$\frac{1}{f} = \frac{1}{D_o} + \frac{1}{D_i}$$

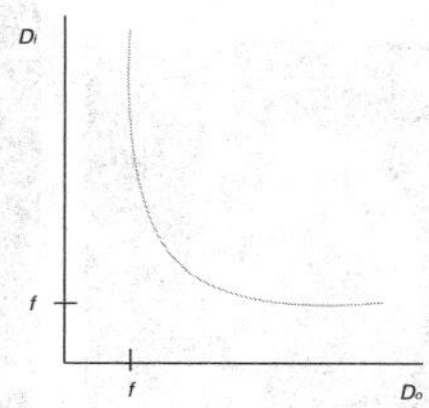

Look at the graph of this equation at left. Notice that as the object distance decreases, the image distance becomes very large. As the object distance increases, the image distance moves towards the focal length (*f*). Also notice that neither the object distance nor the image distance can be less than the focal length.

© It's About Time

Reflecting on the Activity and the Challenge

You have observed how rays of light are reflected by a curved mirror. You have seen that a concave mirror can make an upside-down real image (an image on a screen). You have also seen that the image and object distances are described by a simple mathematical relationship. In addition, you have seen that there is no real image in a convex mirror, and the image is always smaller than the object.

You may want to use a curved mirror in your sound and light show. You may want to project an image on a screen or produce a reflection that the audience can see in the mirror. What you have learned will help you explain how these images are made.

Since the image changes with distance, you may try to find a way to have a moving object so that the image will automatically move and change size. A ball suspended by a string in front of a mirror may produce an interesting effect. You may also wish to combine convex and concave mirrors so that some parts of the object are larger and others are smaller. Convex and concave mirrors could be shaped to make some kind of fun-house mirror.

Remember that your light show will be judged partly on creativity and partly on the application of physics principles. This activity has provided you with some useful principles that can help with both criteria.

Physics To Go

1. a) Make a drawing of parallel laser beams aimed at a convex mirror.
   b) Draw lines to show how the beams reflect from the mirror.
2. a) Make a drawing of parallel laser beams aimed at a concave mirror.
   b) Draw lines to show how the beams reflect from the mirror.
3. a) Look at the back of a spoon. What do you see?
   b) Look at the inside of a spoon. What do you see?

ANSWERS

Physics To Go

1. a)-b) See drawing on page 363.

2. a)-b) See drawing on page 364.

3. a) You see a right-side-up image of yourself and most of the room you are in. If you hold your finger very close to the spoon, you see a right-side-up, reduced image.

   b) You see an upside-down image of yourself. If you hold your finger very close to the spoon, you see a right-side-up, enlarged image.

© It's About Time

ANSWERS

Physics To Go
(continued)

4\. a) It should be concave, so it will magnify.

b. One side is concave and the other is convex. The concave side provides magnification. The convex side provides a wider view.

c. This curved mirror is convex. It provides a wide-angle view, as does the outside of the spoon, so you can see cars out to the side, in what is called the blind spot. But the image is smaller, so cars appear further behind than they actually are.

d. A dentist uses a curved mirror to get a magnified view of your teeth.

5\. a)

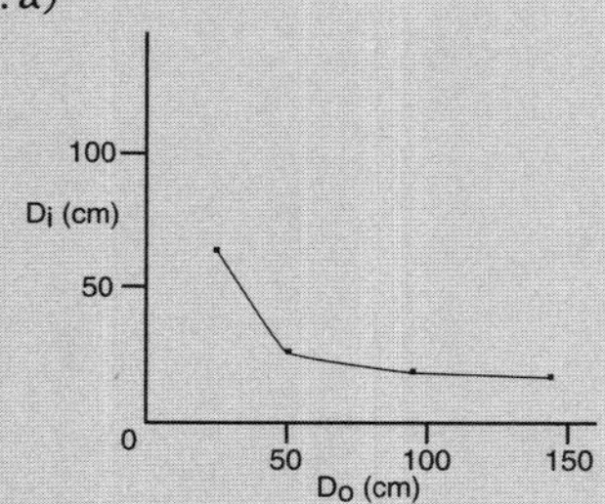

4\. a) If you were designing a shaving mirror, would you make it concave or convex? Explain your answer.

b) Why do some makeup mirrors have two sides? What do the different sides do? How does each side produce its own special view?

c) How does a curved side mirror on a car produce a useful view? How can this view sometimes be dangerous?

d) Why does a dentist use a curved mirror?

D_i (cm)	D_o (cm)
549	15
56	25
20	50
18	91
14	142

5\. a) A student found the real image of a light bulb in a concave mirror. The student moved the light bulb to different positions. At each position, the student measured the position of the image and the light bulb. The results are shown in the table on the left. Draw a graph of this data.

b) Make a general statement to summarize how the image distance changes as the object distance changes.

c) If the object were twice as far away as the greatest object distance in the data, estimate where the image would be.

d) If the object were only half as far from the mirror as the smallest object distance in the data, estimate what would happen to the image.

6\. A ball is hung on a string in front of a flat mirror. The ball swings toward the mirror and back. How would the image of the ball in the mirror change as the ball swings back and forth?

7\. a) A ball is hung on a string in front of a concave mirror. The ball swings toward the mirror and back. How would the image of the ball in the mirror change as the ball swings back and forth?

b) How could you use this swinging ball in your light show?

8\. Outdoors at night, you use a large concave mirror to make an image on a card of distant auto headlights. You make the image on a card. What happens to the image as the car gradually comes closer?

b) As the object distance decreases, the image distance increases.

c) The greatest object distance (142 cm) is already so large that the image (at 14 cm) is very close to the focus. Doubling the object distance would bring the image even closer to the focus, but the change might be impossible for the students to detect.

d) The smallest object distance is 15 cm, and the focal length of the mirror is about 14 cm (see **Part (c)** above). If the smallest object distance were halved, the object would be within one focal length of the mirror, so there would be no real image.

6\. As the ball swings towards the mirror, the image swings towards the mirror, too.

7\. a) The answer depends on whether or not the ball swings through the focus of the mirror. If it does, then while swinging in, the image goes from upside-down to right-side-up and magnified. At the focus, the image fills the mirror.

b) You could have several balls swinging in concave mirrors with different focal lengths. You could ask a member of the audience to move his or her head like the ball and then report what he or she sees.

8\. The image gradually moves away from the mirror and becomes larger.

© It's About Time

9. The diagram shows a light ray *R* parallel to the principal axis of a spherical concave (converging) mirror. Point *F* is the focal point and *C* is the center of curvature. Draw the reflected light ray.

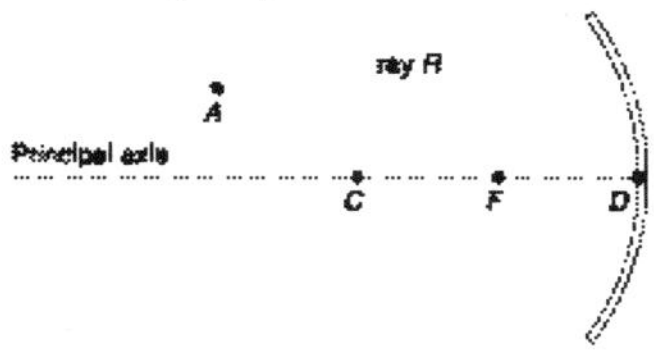

10. The diagram shows a curved mirror surface and a light bulb and its image. In relation to the focal point of the mirror, where is the light bulb (object) most likely located?

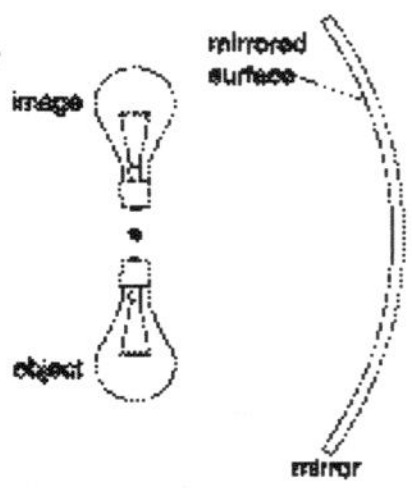

11. A candle is located beyond the center of curvature, *C*, of a concave spherical mirror having a principal focus, *F*, as shown in the diagram. Sketch the image of the candle.

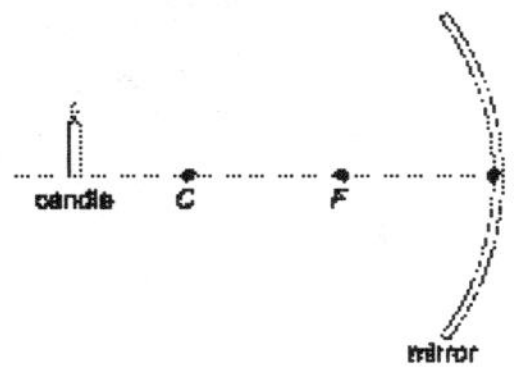

ANSWERS

Physics To Go
(continued)

9.

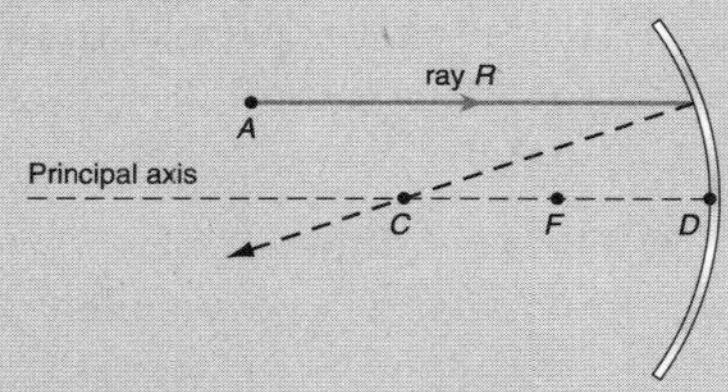

10. The object is most likely located at the center of curvature.

11.

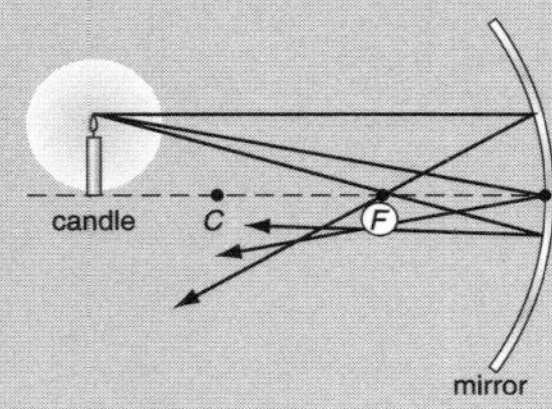

© It's About Time

ANSWERS

Physics To Go

(continued)

12.a) The ray that is parallel to the principal axis will pass through *F*.

b) The size of the image produced will get smaller.

12. The diagram shows four rays of light from object *AB* incident on a spherical mirror with a focal length of 0.04 m. Point *F* is the principal focus of the mirror, point *C* is the center of curvature, and point *O* is located on the principal axis.

a) Which ray of light will pass through *F* after it is reflected from the mirror?

b) As object *AB* is moved from its position toward the left, what will happen to the size of the image produced?

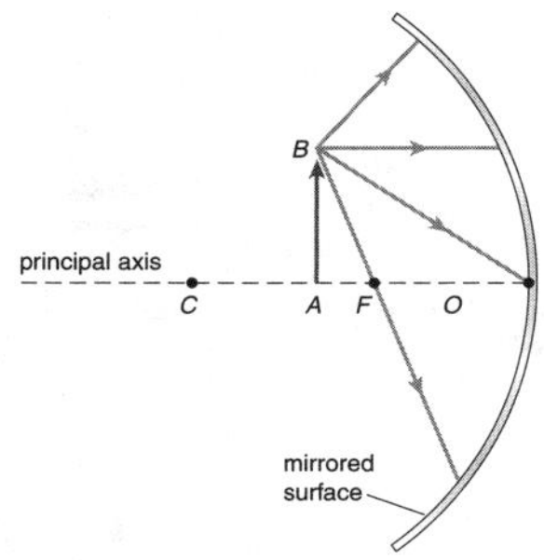

© It's About Time

Activity 5 A

Low-Tech Reflection: Curved Mirror

FOR YOU TO DO

If you do not have lasers or light boxes available, give the students penlights to use as the light source. Build long-focal-length mirrors by gently curving cardboard and taping mylar to the cardboard surface. Investigate the reflection of parallel flashlight beams. See if a concave mirror makes the beams cross.

© It's About Time

For use with *Let Us Entertain You*, Chapter 5, Activity 5: Curved Mirrors

Bulb and Image Distances

Distance of bulb from mirror	Distance of image from mirror

Bulb and Image Distances

Distance of bulb from mirror	Distance of image from mirror

© It's About Time

For use with *Let Us Entertain You*, Chapter 5, Activity 5: Curved Mirrors

© It's About Time

For use with *Let Us Entertain You*, Chapter 5, Activity 5: Curved Mirrors

ACTIVITY 6
Refraction of Light

Background Information

Refraction is the basis of the operation of lenses. Light rays bend as they enter and leave the convex lens, as shown in the drawing.

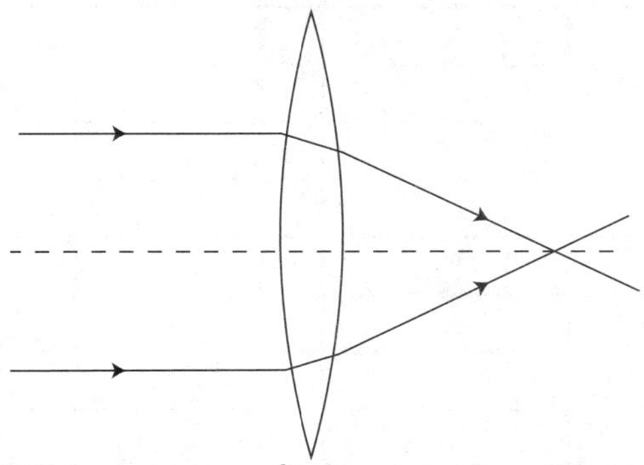

The angles of incidence (*i*) and refraction (*R*) specify the path of the light as it moves from one substance into another, as shown in the drawing.

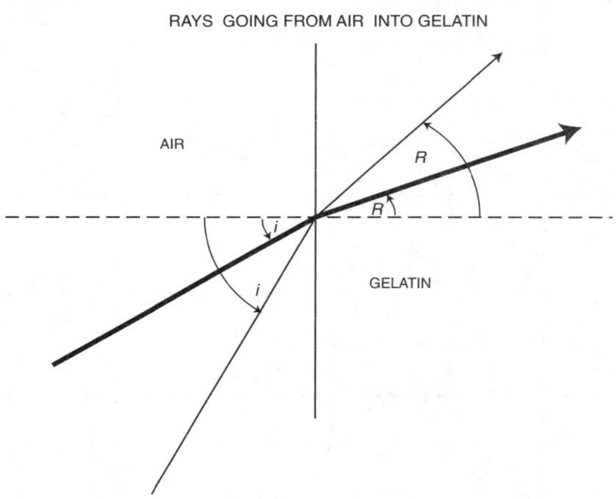

When light moves from air into gelatin, the equation relating the angles *i* and *R* can be written as

$\sin(i) = n(\text{gelatin})\ 3\ \sin(R)$

The quantity *n*(gelatin) is the index of refraction of gelatin. (The value of the index of refraction of air is 1.000, so it is not shown in the equation above.) Going from air into gelatin, light bends towards the normal. Consequently, the ratio $\sin(i)/\sin(R)$ is greater than one (except for the special case of $i = 0°$).

Going from gelatin into air, light bends away from the normal. The angle of refraction is greater than the angle of incidence

The critical angle occurs as light leaves the medium with the higher index of refraction (in this activity, moving from gelatin into air). Look at the diagram.

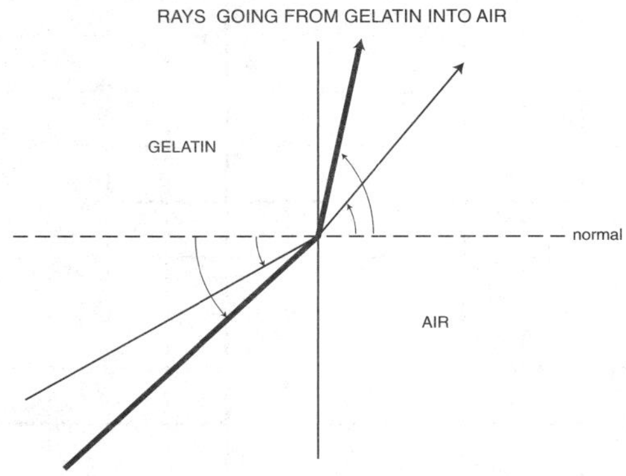

It shows how the angle of refraction changes as the angle of incidence, in the gelatin, increases from zero.

Eventually the angle of refraction reaches 90°. The corresponding angle of incidence is called the critical angle. What happens when the angle of incidence exceeds the critical angle? The light cannot be refracted, because the angle of refraction would be greater than 90°. It turns out that all the light is reflected back into the medium with the higher index of refraction. This effect is called total internal reflection. Although there appears to be an abrupt change at the critical angle, internal reflection begins considerably before reaching the critical angle and becoming total. Likewise, the intensity of the refracted beam diminishes progressively as the critical angle is approached, and this effect can make the critical angle challenging to measure.

Total internal reflection is the basis of the operation of the light pipe. A light beam enters a small diameter, flexible glass rod. When the beam hits the wall of the pipe, total internal reflection occurs and the beam reflects entirely within the pipe. In this way, light signals can carry information through a system of bundled glass rods called fiber optic cables.

Active-ating the Physics InfoMall

Many of the searches done previously will provide great information for refraction, since most optics topics are closely related. For example, "An investigation of student understanding of the real

© It's About Time

image formed by a converging lens or concave mirror," from the *American Journal of Physics*, vol. 55, issue 2, 1987, will work here, as well as in **Activity** 7. Additional searches you should try would include "critical angle" and "internal reflection." Although quite simple, these are effective.

In **For You To Read**, the index of refraction (another good search term?) is described as a ratio. In his book, *A Guide to Introductory Physics Teaching*, (in the Book Basement) Arons says "One of the most severe and widely prevalent gaps in cognitive development of students at secondary and early college levels is the failure to have mastered reasoning involving ratios." The discussion continues, and you are urged to read this section (it appears early in the "Underpinnings" chapter). Arons also addresses this issue in "Student patterns of thinking and reasoning, part one of three parts," from *The Physics Teacher*, vol. 21, issue 9 (1983).

Again, Snell's Law is a great search item, and the equation can again be found in the *Equations Dictionary* (although this one requires a few more clicks, as it is buried relatively deep). Don't forget to search for more information on fiber optics and prisms while you are looking on the InfoMall.

Planning for the Activity

Time Requirements:

one and a half class periods

Materials Needed

For each group:

- acrylic block
- cardboard sheet, 14" 3 14" square
- laser pointer
- cosmetic mirror
- glass rod
- protractor
- right angle prism
- ruler, metric
- unlined index cards (100)
- wooden blocks to raise light source (3)

For each student:

- 1 copy of the Blackline Master, Angles of Incidence and Refraction

Advance Preparation and Setup

Make one copy of the Blackline Master for each student. There are many kinds of gelatin you can use. Let the gelatin harden in a rectangular tray and cut out the blocks in advance. Take care to cut the blocks as precisely as possible. You can run hot water on the bottom of the tray to soften the bottom of the gelatin so the blocks will come out easily. Keep them wrapped and refrigerated until the lab so no microorganisms will grow on their surface. Be sure you have supports for the gelatin and the light source so they are at the proper height to permit light beams to go through the gelatin block. If your students are using lasers, find large pieces of cardboard for laser backstops. If you have neither a laser nor a raybox, each group can use a candle and a narrow (2 mm to 3 mm) slit in a piece of cardboard. With this arrangement the students can at least get semiquantitative data and observe total internal reflection.

Teaching Notes

Suggest that the students use a file card to follow the beam after it hits the glass rod. If the students say that they have difficulty seeing the refracted beam as they approach the critical angle, point out that less and less of the beam is refracted and more and more is reflected (as they approach the critical angle). You can introduce **Steps** 7 and 8 by pointing out that **Steps** 1 through 6 produced relatively small angles of incidence inside the gelatin block (as the light approached the gelatin-air surface). This occurs because the light is bent towards the normal upon entering the block, and the angle of refraction on the front surface is equal to the angle of incidence on the rear surface. However, in **Step** 8, the angle of incidence inside the block is the complement of the angle of refraction in the block, so a small angle of refraction entering the block makes a large angle of incidence leaving.

If the students use lasers, be sure they understand the importance of keeping their laser beam within their group's work area (by building some sort of backstop).

© It's About Time

If your class makes observations with laser beams, have each group build backstops of cardboard to absorb the beam. Also, explain to the students that they must not eat the gelatin. Keep the gelatin blocks covered and refrigerated before use. If your light source is a candle, have the students remove any loose papers from the area around the candle.

The concept that reflected light fans out from each point on an object is extremely difficult for students. Consequently, when asked to explain examples of refraction, they typically make a broad statement like "The water bent the light," rather than considering the path of light reflecting from the object in the water. For this reason it is particularly effective to have them trace the path of the laser beam. Asking a qualitative question, like "How does the laser beam bend when it goes from air into the gelatin," can help students organize their observations and contrast what happens in going from air to block with what happens going the other way.

Air to Acrylic Block

angle of incidence	angle of refraction
15 degrees	10 degrees
35 degrees	22 degrees
42 degrees	28 degrees

For the largest angle of incidence

$$1 \times \sin 42 \text{ degrees} = n_{gelatin} \times \sin 28 \text{ degrees}$$

$$n_{gelatin} = 1.4$$

© It's About Time

Activity 6 Refraction of Light

GOALS

In this activity you will:

- Observe refraction.
- Measure angles of incidence and refraction.
- Measure the critical angle.
- Observe total internal reflection.

What Do You Think?

The Hope Diamond is valued at about 100 million dollars. A piece of cut glass of about the same size is worth only a few dollars.

- **How can a jeweler tell the difference between a diamond and cut glass?**

Record your ideas about this question in your *Active Physics* log. Be prepared to discuss your responses with your small group and with your class.

For You To Do

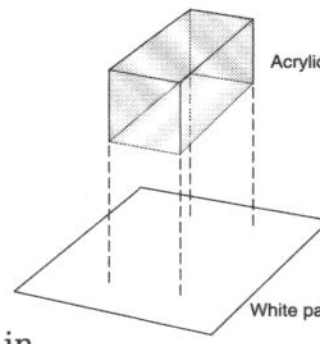

1. Place an acrylic block on a piece of white paper on your desk.
2. Carefully aim a laser pointer, or the light from a ray box, so the light beam moves horizontally, as you did in

Activity Overview

To prepare for the study of lenses, students investigate refraction. By shining a beam of light into a rectangular block of clear gelatin, they observe refraction as the light enters and leaves the gelatin. The students measure the angles of incidence and refraction and tabulate the results. They also observe total internal reflection and measure the critical angle.

Student Objectives

Students will:

- Observe refraction.
- Measure angles of incidence and refraction.
- Measure the critical angle.
- Observe total internal reflection.

ANSWERS FOR THE TEACHER ONLY

What Do You Think?

A jeweler with a trained eye can identify a diamond with a careful look. A diamond reflects more light than cut glass, because of the diamond's greater index of refraction. The diamond also has a range of color not found in glass and also particular small imperfections.

Despite these visual differences, many jewelers rely on an electrical device that applies a small amount of heat to an unknown stone and identifies a diamond from the resulting temperature change.

© It's About Time

ANSWERS

For You To Do

1. Be sure the front and back of the acrylic block are paralled.

2.-3. The students do not need to mark the path of the beam inside the block They find the angles inside by connecting the points where the light enters and leaves.

4. The students can measure every ten or fifteen degrees.

5. Remind the students to look straight down when marking the path of the beam to avoid parallax.

6. a) Remember that angles are measured from the normal.

b) Each ratio is >1 since ∡i > ∡R

c) This ratio is (approximately) a constant.

7. Be sure the side of the block is perpendicular to the front.

Never look directly at a laser beam or shine a laser beam into someone's eyes. Always work above the plane of the beam and beware of reflections from shiny surfaces.

previous activities. Place a glass rod in the light beam so that the beam spreads up and down.

3. Shine the laser pointer or light from the ray box through the acrylic block. Be sure the beam leaves the acrylic block on the side opposite the side the beam enters. Mark the path of each beam. You may wish to use a series of dots as you did before. Label each path on both sides of the acrylic block so you will know that they go together.

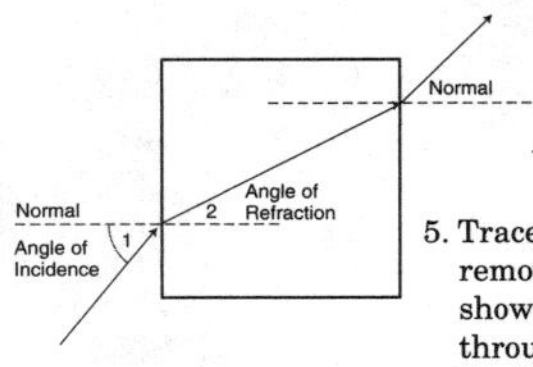

4. The angle of incidence is the angle between the incident laser beam and the normal, as shown in the diagram. Choose two other angles of incidence and again mark the path of the light, as you did in **Step 3**. As before, label each pair of paths.

5. Trace the outline of the acrylic block on the paper and remove the acrylic block. Connect the paths you traced to show the light beam entering the acrylic block, traveling through the acrylic block, and emerging from the acrylic block. Draw a perpendicular line at the point where a ray enters or leaves the acrylic block. Label this line the normal.

6. Measure the angles of incidence (the angle in the air) and refraction (the angle in the acrylic block).

a) Record your measurements in tables like the one shown.

Angle of incidence	Angle of refraction	Sine of angle of incidence	Sine of angle of refraction	Sin ∠i / Sin ∠R

b) Use a calculator to complete the chart by finding the sines of the angles (sin button on calculator).

c) Is the value of $\frac{\sin \angle i}{\sin \angle R}$ a constant? This value is called the index of refraction for the acrylic block.

7. Set up the acrylic block on a clean sheet of white paper. This time, as shown in the drawing (next page), aim the beam so it leaves the acrylic block on the side, rather than at the back.

Physics Words

critical angle: the angle of incidence for which a light ray passing from one medium to another has an angle of refraction of 90°.

index of refraction: a property of a medium that is related to the speed of light through it; it is calculated by dividing the speed of light in vacuum by the speed of light in the medium.

Snell's Law: describes the relationship between the index of refraction and the ratio of the sine of the angle of incidence and the sine of the angle of refraction.

© It's About Time

Activity 6 Refraction of Light

8. Make the first angle of incidence (angle 1) as small as possible, so the second angle of incidence (angle 2) will be as large as possible. Adjust angle 1 so that the beam leaves the acrylic block parallel to the side of the acrylic block, as shown. Measure the value of angle 2.

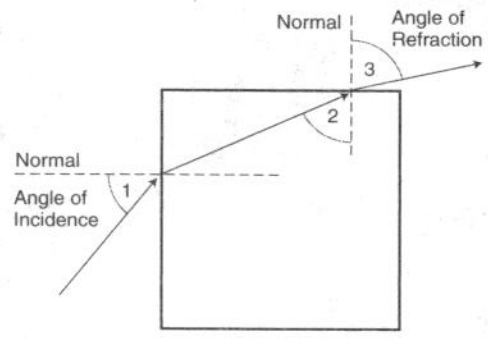

a) Record the value of angle 2. It is called the **critical angle**.

b) What happens to the beam if you make angle 2 greater than the critical angle?

c) What you observed in **(b)** is called "total internal reflection." What is reflected totally, and where?

9. It is possible to bend a long, rectangular acrylic block so the light enters the narrow end of the acrylic block, reflects off one side of the acrylic block, then reflects off the other and back again to finally emerge from the other narrow end. Try to bend an acrylic block rectangle so that the light is reflected as described.

FOR YOU TO READ

Snell's Law

Light refracts (bends) when it goes from air into another substance. This is true whether the other substance is gelatin, glass, water, or diamond. The amount of bending is dependent on the material that the light enters. Each material has a specific **index of refraction**, *n*. This index of refraction is a property of the material and is one way in which a diamond (very high index of refraction—lots of bending) can be distinguished from glass (lower index of refraction—less bending). The index of refraction is a ratio of the sine of the angle of incidence and the sine of the angle of refraction.

Index of refraction: $n = \dfrac{\sin \angle i}{\sin \angle R}$

This equation is referred to as **Snell's Law.**

As light enters a substance from air, the light bends toward the normal. When light leaves a substance and enters the air, it bends away from the normal. If the light is entering the air from a substance, the angle in that substance may be such that the angle of refraction is 90°. In this special case, the angle in the substance is called the critical angle. If the angle in the substance is greater than this critical angle, then the light does not enter the air but reflects back into the substance as if the surface were a perfect mirror. This is the basis for light fibers where laser light reflects off the inner walls of glass and travels down the fiber, regardless of the bend in the fiber.

ANSWERS

For You To Do *(continued)*

8. a) Note that as the critical angle is approached, the refracted intensity diminishes and the reflected intensiy grows.

b) If angle 2 is greater than the critical angle, the refracted ray disappears.

c) The incident ray (inside the acrylic block) is totally reflected, with an angle of reflection equal to angle 2.

9. This is a model of a light pipe. When the light hits the walls, there is total internal reflection.

© It's About Time

ANSWERS

Physics To Go

1. The light is bent towards the normal, so the angle of incidence is larger (except when the angle of incidence is zero).

2. a) See drawing on page 373.

   b) Yes

3. See drawing on page 373.

4. See drawing in **Background Information**.

5. a) 90°

   b) The angle of incidence equals 90°. The angle of refraction equals 42°, the critical angle.

Reflecting on the Activity and the Challenge

The bending of light as it goes from air into a substance or from a substance into air is called refraction. It is mathematically expressed by Snell's Law. When light enters the substance at an angle, it bends towards the normal. When light leaves the substance at an angle, it bends away from the normal. As you create your light show for the chapter challenge, you may find creative uses of refraction. You may decide to have light bending in such a way that it spells out a letter or word or creates a picture. You may wish to have the light travel from air into glass to change its direction. You may have it bend by different amounts by replacing one material with another. Regardless of how you use refraction effects, you can now explain the physics principles behind them.

Physics To Go

1. A light ray goes from the air into an acrylic block. In general, which is larger, the angle of incidence or the angle of refraction?

2. a) Make a sketch of a ray of light as it enters a piece of acrylic block and is refracted.

   b) Now turn the ray around so it goes backward. What was the angle of refraction is now the angle of incidence. Does the turned-around ray follow the path of the original ray?

3. A light ray enters an acrylic block from the air. Make a diagram to show the angle of incidence, the angle of refraction, and the normal at the edge of the acrylic block.

4. Light rays enter an acrylic block from the air. Make drawings to show rays with angles of incidence of 30° and 60°. For each incident ray, sketch the refracted ray that passes through the acrylic block.

5. a) Light is passing from the air into an acrylic block. What is the maximum possible angle of incidence that will permit light to pass into the acrylic block?

   b) Make a sketch to show your answer for **Part (a)**. Include the refracted ray (inside the acrylic block) in your sketch.

© It's About Time

6. a) A ray of light is already inside an acrylic block and is heading out. What is the name of the maximum possible angle of incidence that will permit the light to pass out of the acrylic block?
   b) If you make the angle of incidence in **Part (a)** greater than this special angle, what happens to the light?
   c) Make a sketch to show your answer for **Part (b)**. Be sure to show what happened to the light.

7. a) Make a drawing of a light ray that enters the front side of a rectangular piece of acrylic block and leaves through the back side.
   b) What is the relationship between the direction of the ray that enters the acrylic block and the direction of the ray that leaves the acrylic block?
   c) Use geometry and your answer to **Question 2 (b)**, to prove your answer to **Question 7 (b)**.

8. You have seen the colored bands that a prism or cut glass or water produce from sunlight. Light that you see as different colors has different wavelengths. Since refraction makes these bands, what can you say about the way light of different wavelengths refracts?

ANSWERS

Physics To Go
(continued)

6. a) The critical angle.
   b) All the light is reflected inside the block.
   c) See drawing on page 373. (Look at the incident ray inside the acrylic block.) With a slightly larger value for angle 2, the refracted ray in air will have an angle of refraction of 90°.

7. a) See drawing on page 373.
   b) These rays are parallel.
   c) By alternate interior angles, inside the gelatin the angle of refraction is equal to the angle of incidence at the rear of the block. Since the light paths can be reversed, the angle of refraction in the air equals the angle of incidence in the air. (It may help to rotate the diagram at the back side of the block by 180° and compare it with the diagram at the front.)

8. The amount of refraction depends on the wavelength. Equivantly, the angle of incidences is a function of the wavelength.

© It's About Time

Stretching Exercises

1. Cover the acrylic block with a red filter. Shine a red laser beam into the acrylic block, as you did in **For You To Do, Steps 1 through 3**. What happens? How can you explain what happens?

2. Find some $\frac{1}{2}$" diameter clear tubing, about 2 m long. Plug one end. Pour clear gelatin in the other end, through a funnel, before the gelatin has had time to set. Arrange the tubing into an interesting shape and let the gelatin set. You may wish to mount your tube on a support or a sturdy piece of cardboard, which can be covered with interesting reflective material. Fasten one end of the tube so laser light can easily shine straight into it. When the gelatin has set, turn on the laser. What do you see? This phenomena is called total internal reflection.

3. Place a penny in the bottom of a dish or glass. Position your eye so you can just see the penny over the rim of the glass. Predict what will happen when you fill the glass with water. Then try it and see what happens. How can you explain the results?

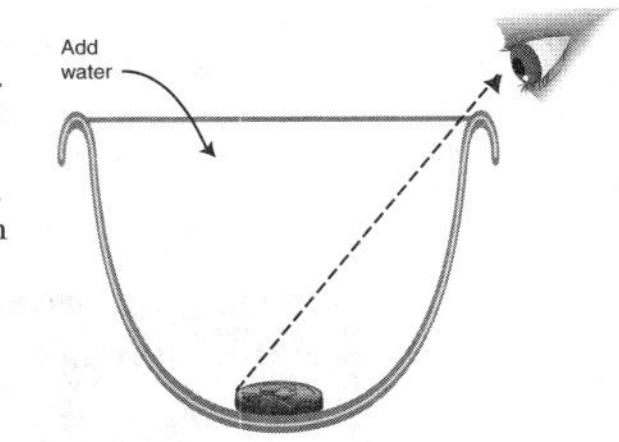

4. Place an empty, clear drinking glass over a piece of a newspaper. When you look through the side of the glass near the bottom, you can see the printing on the newspaper. What do you think will happen if you fill the glass with water? Try it and see. How can you explain the result? Does it help to hold your fingers over the back of the glass?

ANSWERS

Stretching Exercises

1. The red laser light passes through the acrylic block, which absorbs light of other colors.

2. The laser beam will pass through the tubing both times it reaches the outside at the gelatin, total internal reflection occurs, so this is a model of a light pipe.

3. When the light leaves the water, the light bends away from the normal (the vertical). If you find the ray from the water that enters your eye, extend this ray backward to find where you will see the coin.

4. The print disappears. Suggest that students draw diagrams to explain what happens.

© It's About Time

Angle of incidence	Angle of refraction	Sine of angle of incidence	Sine of angle of refraction	Sin< *i* Sin< *R*

For use with *Let Us Entertain You*, Chapter 5, Activity 6: Refraction of Light

Angle of incidence	Angle of refraction	Sine of angle of incidence	Sine of angle of refraction	Sin< *i* Sin< *R*

For use with *Let Us Entertain You*, Chapter 5, Activity 6: Refraction of Light

© It's About Time

ACTIVITY 7
Effect of Lenses on Light

Background Information

A light bulb sends out light in all directions. If a few of these rays pass through a pinhole, and the rest are blocked, a dim image of the bulb can be seen on a screen. Opening up the pinhole makes the image brighter but also less sharp. Only if the light is refracted can the image be both bright and sharp. A convex lens bends light to make such an image. Look at the lower ray diagram. It shows two of the countless rays that fan out from each point on the object.

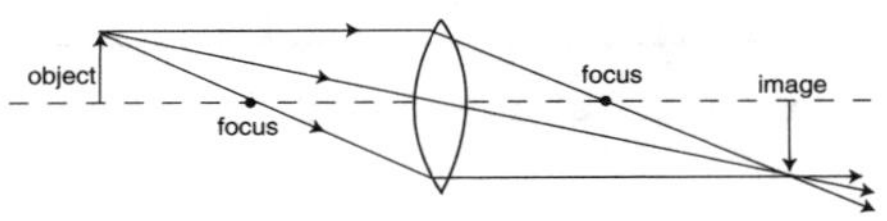

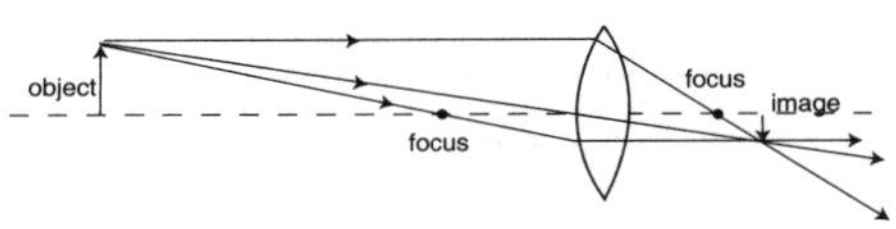

One ray enters the lens parallel to its axis, so this ray is refracted through the focus, the position of the image when the object is extremely distant. The other ray goes through the center of the lens. If the lens is thin, and the object is small, this ray maintains its original direction, since the front and back of the lens are parallel right at the center. Where these two rays meet is the image of the arrowhead. The image is upside down. Also, the image is smaller than the object and is located between $2f$ and f (f is the focal length) from the lens. The object is beyond $2f$. Notice that the object and image could be interchanged, without changing the ray diagram. When the object is close to the lens, the image is far away and larger than the object. When the image is close to the lens, the object is far away and larger than the image. The object and image can be equidistant from the lens, when each is located at $2f$. (See the upper ray diagram.) In this symmetrical arrangement, the object is the same size as the image.

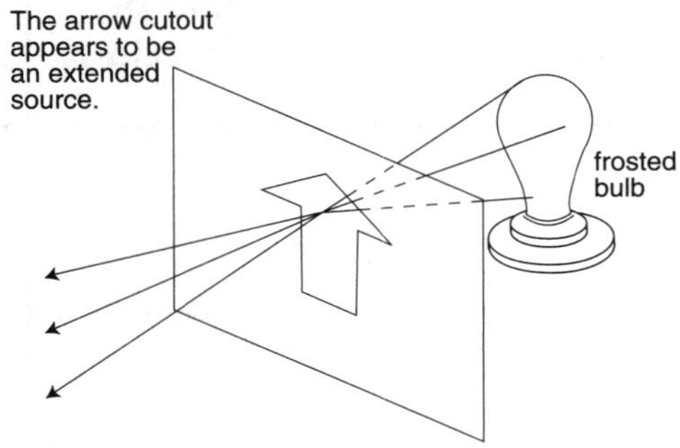

Projecting a slide is challenging to analyze. Think about **Steps** 7 and 8, where a student holds a card with an arrow-shaped cutout between the bulb and the lens. The lens projects the arrow on the screen.

Since the light bulb is an extended source, light rays from the bulb cross at each point in the cutout. In effect, the cutout arrow acts like an extended light source too. This extended source has a sharp edge right at the card, so a sharp arrow can be projected on the screen. See below for notes on changing the size of the projected image.

Active-ating the Physics InfoMall

Search using keyword "Convex lens*" and you will get a short list. Don't forget; using an asterisk in your search words tells the InfoMall to look for words that are similar ("lens*" could be "lens", "lenses", or "lensless"). Some of these include rays diagrams, labs, or homework problems. You may wish to check these out if you do not know what to expect from a convex lens. You may also want to read "Lenses, pinholes, screens, and the eye," from *The Physics Teacher*, vol. 29, issue 4 (1991), which also discusses student difficulties; and don't forget "An investigation of student understanding of the real image formed by a converging lens or concave mirror," from the *American Journal of Physics*, vol. 55, issue 2, 1987.

Another search to perform here might be "converging lens*". Indeed. that is the search that produced the two articles just mentioned. From the same search, we find that "The Camera" (found in the Pamphlet Parlor) is high on the list of search hits. We also find that the textbook *Modern College Physics* has an entire chapter on lenses.

Repeat these searches for concave lenses, also called diverging lenses.

© It's About Time

Planning for the Activity

Time Requirements:

one and a half class periods

Materials Needed

For the class:

- overhead projector (several, if possible)

For each group:

- 40-W bulb and socket, ceramic w/switch
- acetate sheet, about 8 cm square
- bulb, clear 100-W
- convex lens with focal length about 10 cm
- felt-tip marker
- ruler, metric
- steel scissors
- stick, meter 100 cm, hardwood

For each student:

- 1 copy of the Blackline Master: Table of Object and Image Distances

Advance Preparation and Setup

Make a copy of the Table of Object and Image Distance following this activity in the Teacher's Edition. Darken the room as much as possible so students can see the images easily. If you use 110-V light bulbs, be sure there are no exposed wires. Use the kind of sockets that have the wiring connections inside a plastic or ceramic cover. If the socket is ceramic, then the connections are probably exposed underneath. Screw each ceramic socket down to a piece of wood, so the wiring is safely out of reach.

Teaching Notes

Most students need help when they first make a real image with a lens. They often have difficulty getting started because they are unable to line up the lens, the file card, and the object (with the lens facing the object). To help them, you can refer them to the drawing in the book on page 378. Projecting the slide (**Step 9**) can be challenging. Many students make an image of the light bulb as seen through the arrow, but it is in fact the bright arrow that is the image. To enlarge the size of the image, they must bring the lens and arrow closer together. You may need to help them move both lens and arrow to find the position that gives a sharp image.

Keep the groups as well separated as possible. Even with the room dark, stray light from one group's bulb can easily interfere with other group's images.

If the students use a 110-V light bulb for the light source, be sure the students cannot touch any exposed wiring. If you use ceramic sockets with exposed screw terminals underneath, screw down the socket to a board so the terminals are inaccessible. See **Advance Preparation and Setup**. If the students use a candle, be sure no flammable materials are nearby.

Light is a challenging topic for students. Students do not easily understand the fanning out of reflected light nor, for that matter, even the idea that light travels from one place to another yet is not material. You can help your students by asking them to explain how they hold the lens when they make an image (facing the object) and how that orientation produces an image (the lens focuses the light that hits its surface, and the light, which fans out from the object, hits all of the surface).

Focusing Parallel Beams: If you have lasers available, set up the lens and laser so you can move the laser back and forth in front of the lens, with the beam parallel to the lens axis. Observe what happens to the beam after it passes through the lens. Measure the focal length by making an image of a distant object and compare the value with what you found from the laser beam.

© It's About Time

NOTES

© It's About Time

Activity 7 Effect of Lenses on Light

GOALS

In this activity you will:

- Observe real images.
- Project a slide.
- Relate image size and position.

What Do You Think?

Engineers have created special lenses that can photograph movie scenes lit only by candlelight.

- **How is a lens able to project movies, take photographs, or help people with vision problems?**

Record your ideas about this question in your *Active Physics* log. Be prepared to discuss your responses with your small group and with your class.

For You To Do

1. Look at the lens your teacher has given you.

a) Make a side-view drawing of this lens in your log. This is a convex lens.

ANSWERS

For You To Do

1. a) See lens drawing in **Physics To Go, Question 3**, in the Student Edition.

Activity Overview

Having investigated refraction in the last activity, students now move ahead to convex lenses. They observe the real image made by a convex lens and how its size and position change as the object distance changes. To prepare for the light show, the students project the image of a slide onto a wall and vary the size of the projected image.

Student Objectives

Students will:

- Observe real images.
- Project a slide.
- Relate image size and position.

ANSWERS FOR THE TEACHER ONLY

What Do You Think?

The lens does all these things by bending light, a process called refraction. Without refraction, cameras could take only pinhole images, as though the lens let in light through only a tiny opening. The images would be very dim. But by bending light, the lens can direct all the light that strikes the lens onto the proper place in the image, so the image is far brighter than a pinhole image.

To make an image, a lens must be convex. If you make a ray diagram of a light beam passing through a convex lens, you will see that it bends towards the lens axis as it enters the glass and then again towards the axis as it leaves the glass. The result is that the ray can be part of an image, which is located near the axis.

© It's About Time

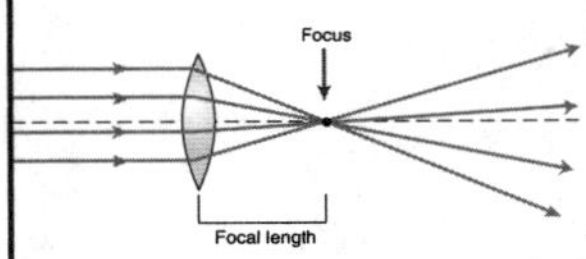

2. Point the lens at a window or at something distant outside. Use a file card as a screen. Look for the image on the file card. Move it back and forth until you see a sharp image of the distant object.
   a) Describe what you see. Is the image large or small? Is it right side up or upside down? Is it reversed left to right? This image is called "real" because you can project it on the screen.

3. Measure the distance between the image and the lens. If the object is very far away, this distance is the focal length of the lens. The position of this image is the focus of the lens. It is the same location at which parallel rays of light would converge.
   a) Approximate the object distance.
   b) Measure the image distance.
   c) Record your object and image distance. Note that the image distance is also the focal length of the lens.

Do not use lenses with chipped edges. Mount lenses securely in a holder. Use only light sources with enclosed or covered electrical contacts. Keep flammables/combustibles away from the candleholder.

4. Set up a 40-W light bulb or a candle to be a light source. Mount the lens at the same height as the light source. If you are using a light bulb, point it right at the lens, as shown.

5. Place the light bulb about a meter away from the lens. Try to find the image of the light bulb on a screen. The screen can be a file card or a sheet of paper.
   a) Record your results in a table, including the distance and appearance of the image.

6. Adjust the position of the object to create a larger image.
   a) Describe how the position of the object, the image, and the size of the image have changed. Record the results in a table.

ANSWERS

For You To Do *(continued)*

2. a) The image is upside down. If the object is outside, the image is quite small. It is inverted and reversed left-to-right.

3.-5. If the object is far away, the image is at the focus (close to the lens). As the object moves toward the lens, the image moves away.

6. a) If the image is larger, the object is closer to the lens. The image is further from the lens.

© It's About Time

Activity 7 Effect of Lenses on Light

7. Create an object by carefully cutting a hole in the shape of an arrow in an index card. Have someone in your group hold the card close to the light bulb.
   a) Can you see the object on the screen? Describe what you see.
   b) Have the person holding the object move it around between the light bulb and the convex lens. What happens?

8. Project the object onto the wall. Can you make what you project larger or smaller?
   a) In your log indicate what you did to change the size of the image.

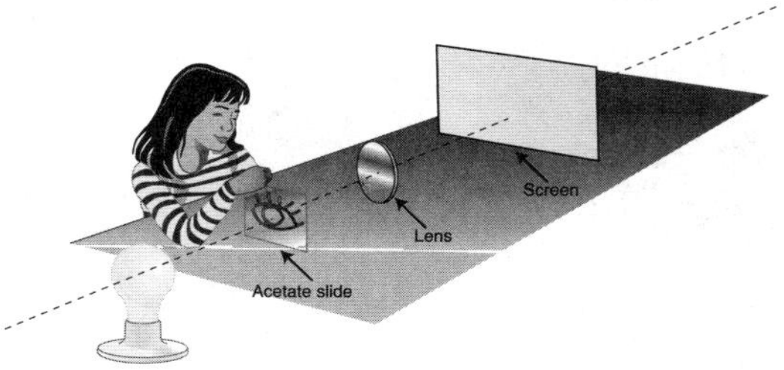

9. Create a slide by drawing with a marking pen on clear acetate. Try placing a 100-W light bulb and the slide in different positions.
   a) Describe how you can project a real, enlarged image of your slide onto a screen or wall.
   b) How can you use the lens to change the size of the image?
   c) Explore the effect of different lenses. In your log, record how you think this effect might be part of your light and sound show.

Caution: Lamps get very hot. Be careful not to touch the bulb or housing surrounding the bulb.

ANSWERS

For You To Do *(continued)*

7. a) Yes, you see a bright arrow.
   b) By moving the screen back and forth to the right place, the edge of the object can be seen sharply.

8. a) Yes, you can change the size of what you see on the wall by changing the distance between the lens and the object.

9. a) Line up the bulb, slide, lens, and wall. Move the lens toward and away from the slide until the image is sharp.
   b) If you move the lens closer to the slide, and then refocus by moving lens and slide together until the image is sharp, the size of the image will be larger.
   c) In your show you can project slides on the wall and change the size of the projected image.

© It's About Time

FOR YOU TO READ

Lens Ray Diagrams

You are probably more familiar with images produced by lenses than you are with images from curved mirrors. The lens is responsible for images of slides, overhead projectors, cameras, microscopes, and binoculars.

Light bends as it enters glass and bends again when it leaves the glass. The **convex converging lens** is constructed so that all parallel rays of light will bend in such a way that they meet at a location past the lens. This place is the focal point.

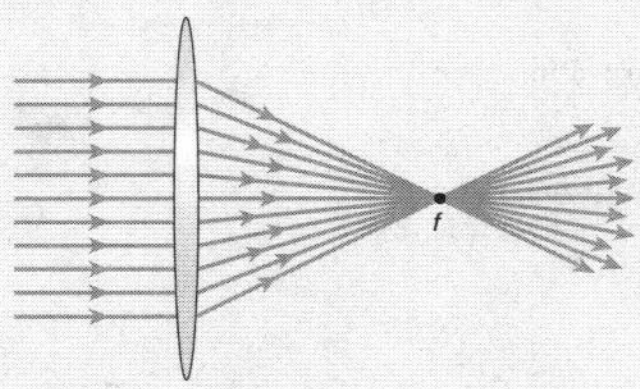

If an object is illuminated, it reflects light in all directions. If these rays of light pass through a lens, an image is formed.

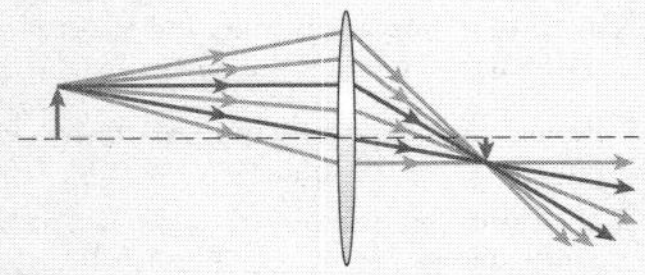

Although all of the light rays from the object help to form the image, you can locate an image by looking at two easy rays to draw—the ray that is parallel to the principal axis and travels through the focal point and the ray that travels through the center of the lens undeflected. (These rays are in red in the diagram.)

You can use this technique to see how images that are larger (movie projector), smaller (camera), and the same size (copy machine) as the object can be created with the same lens.

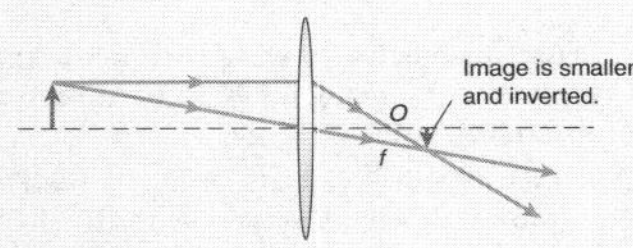

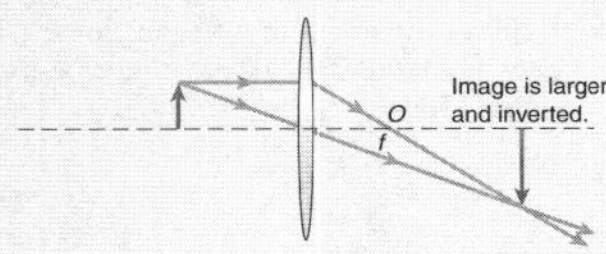

© It's About Time

If the object is close to the lens (an object distance smaller than the focal distance), then an image is not formed. However, if you were to view the rays emerging, they would appear to have come from a place on the same side of the lens as the object. To view this image, you put your eye on the side of the lens opposite the object and peer through it—it's a magnifying glass!

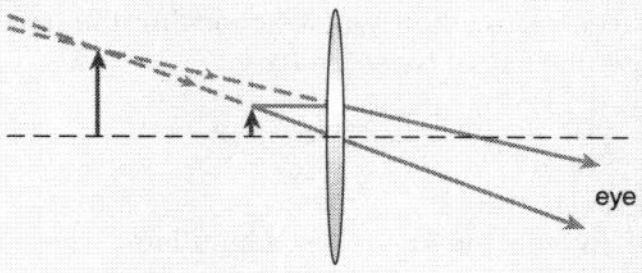

Sample Problem

The diagram shows a lens and an object.

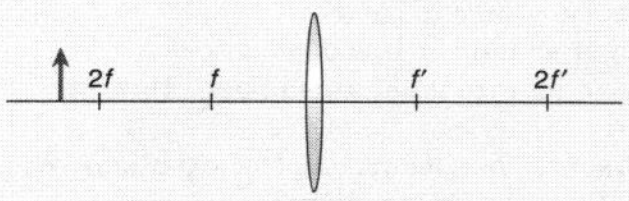

a) Using a ray diagram, locate the image of the object shown.

b) Describe the image completely.

Strategy: Choose a location on the object to be the origin of the rays. A simple choice would be the tip of the arrow. At least two rays must be drawn to locate the image.

Givens:

See the diagram.

Solution:

a)

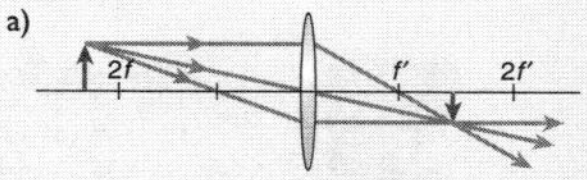

b) The image is real, reduced, and inverted.

As the object moves closer to the lens, its size will increase. At $d_o = f$ there will be no image and at $d_o < f$ the image will be virtual and upright.

Physics Words

converging lens: parallel beams of light passing through the lens are brought to a real point or focus (if the outside index of refraction is less than that of the lens material); also called a convex lens.

© It's About Time

ANSWERS

Physics To Go

1.a) The focus is the position of the image of a very distant object.

b) The object is very far away (in fact, it's at infinity).

c) The focal length is the distance between the focus and the lens.

d) See **For You To Do, Steps 2** and 3.

2. a) Yes.

b) Upside-down.

c) Yes. Light spreads out as it moves from the object to the lens (to see this, imagine the object is a tiny light). If the light were not bent, it would continue spreading after it went through the lens, so there could not be an image.

d) Move the screen back from the lens.

3.a) Move the object away from the lens (more than twice the focal length away).

b) Bring the object close to the lens (closer than twice the focal length), but not less than the focal length for a real image.

4.a) The image of the closer light is blurred.

b) The image of the more distant light is blurred.

Reflecting on the Activity and the Challenge

You have explored how convex lenses make real images. You have found these images on a screen by moving a card back and forth until the image was sharp and clear, so you know that they occur at a particular place. Bringing the object near the lens moves the image away from the lens and enlarges the image, but if the object is too close to the lens, there is no real image. These images are also reversed left to right and are upside down. You may be able to use this kind of image in your sound and light show. You have also projected images of slides on a wall. You may be able to add interest by moving the lens and screen to change the size of these images.

Physics To Go

1. a) What is the focus of a lens?
   b) If the image of an object is at the focus on a lens, where is the object located?
   c) What is the focal length of a lens?
   d) How can you measure the focal length of a lens?
2. a) You set up a lens and screen to make an image of a distant light. Is the image in color?
   b) Is the image right side up or upside down?
   c) Did the lens bend light to make this image? How can you tell?
   d) A distant light source begins moving toward a lens. What must you do to keep the image sharp?
3. a) You make an image of a light bulb. What can you do to make the image smaller than the light bulb?
   b) What can you do to make the image larger than the light bulb?
4. a) You have two lights, a lens, and a screen, as shown on opposite page. One light is at a great distance from the lens. The other light is much closer. If you see a sharp image of the distant light, describe the image of the closer light.
   b) If you see a sharp image of the closer light, describe the image of the more distant light.

© It's About Time

c) Could you see a sharp image of both lights at the same time? Explain how you found your answer.

5. Research how a camera works. Find out where the image is located. Also find out how the lens changes so that you can photograph a distant landscape and also photograph people close up.

6. Using a ray diagram, locate the image formed by the lens below.

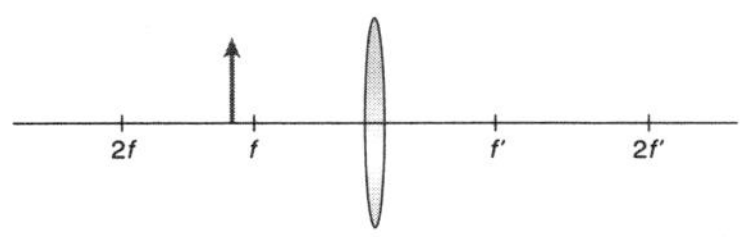

7. An object 1.5 cm tall is placed 5.0 cm in front of a converging lens of focal length 8.0 cm.
   a) Determine the location of the image.
   b) Completely describe the image.

8. A relative wants to show you slides from her wedding in 1972. She brings out her slide projector and screen.
   a) If she puts the screen 2.8 m from the projector and the lens has a focal length of 10.0 cm, how far from the lens will the slide be so that her pictures are in focus?
   b) If each slide is 3.0 cm tall, how big will the image be on the screen?

Answers

Physics To Go *(continued)*

4. c) Not if one object is very distant and the other is very close. But if you find a sharp image and then move the object a small distance, the image may remain sharp. That means there is a range of object distances that will produce a sharp image. (In photography, this range of object distances is called depth of field.)

5. The image is located on the film. Part or all of the lens moves in and out to change the focus. Notes: As the aperture—the opening—of the lens decreases (this corresponds to increasing the *f*-number), the range of sharp focus increases. For some cameras, the aperture is kept small to maintain a large range of object distances that produce a reasonably sharp image. For these cameras, the focus of the lens is fixed.

6.

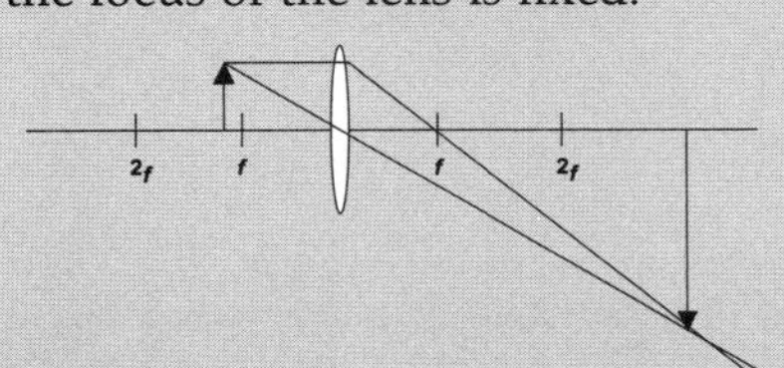

Answers

Physics To Go
(continued)

7. a) $\frac{1}{f} = \frac{1}{d_o} + \frac{1}{d_i}$

$$\frac{1}{d_i} = \frac{1}{f} + \frac{1}{d_o}$$

$$d_i = \left(\frac{1}{f} - \frac{1}{d_o}\right)^{-1} =$$

$$\left(\frac{1}{8cm} - \frac{1}{5cm}\right)^{-1} =$$

$$(-0.075cm^{-1})^{-1} = -13.3cm$$

b) The image is virtual, upright, and enlarged.

8. Apply the lens equation for **Part (a)**. For **Part (b)**, as with mirrors, there is a proportion between the heights (sizes) of the image and object and their respective distances from the lens.

a) $1/f = 1/d_o + 1/d_i$

Rearrange:

$1/d_o = 1/f - 1/d_i$

$= 1/0.1\text{ m} - 1/2.8\text{ m}$

$= 10\text{ m}^{-1} - 0.357\text{ m}^{-1}$

$= 9.643\text{ m}^{-1}$

$d_o = 0.103\text{ m}$

b) $h_i/h_o = d_i/d_o$

Rearrange:

$h_i = h_o d_i/d_o$

$= (3.0\text{ cm})$

$(2.8\text{ m})/(0.103\text{ m})$

$= 81.5\text{ cm}$

You can change the 3.0 cm to meters before using the equation, but the distance units will divide out and the unit of your answer will be whatever you had as the unit for your height. It could be centimeters, inches, etc. As long as the students carry the units through the problem, they should have no trouble identifying the units of the answer.

© It's About Time

ANSWERS

Physics To Go

(continued)

9. The image is real, reduced, and inverted.

10. The image will be located at $2f$.

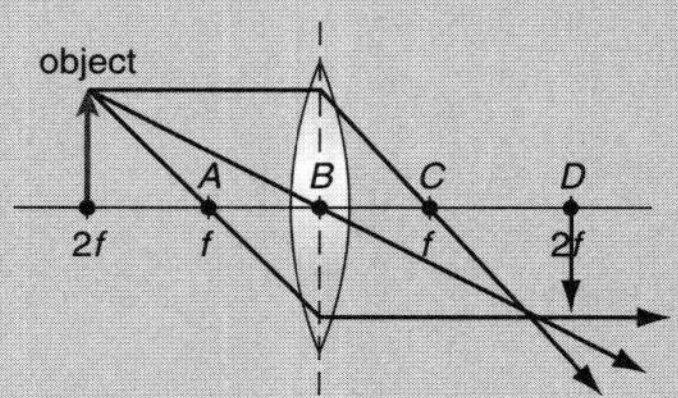

11. As the object moves closer to the lens, its size will increase.

9. The diagram shows an object 0.030 m high placed at point X, 0.60 m from the center of the lens. An image is formed at point Y, 0.30 m from the center of the lens. Completely describe the image.

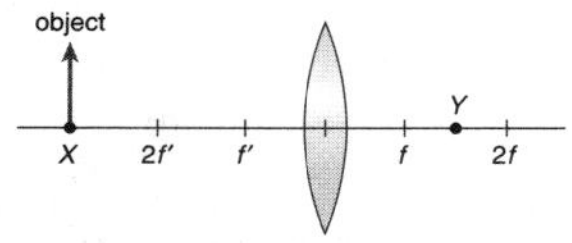

10. The diagram represents an object placed two focal length from a converging lens. At which point will the image be located?

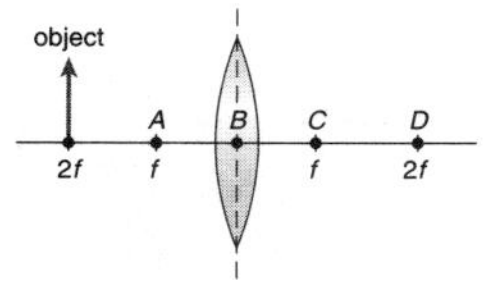

11. The diagram shows a lens with an object located at position A. Describe what will happen to the image formed as the object is moved from position A to position B.

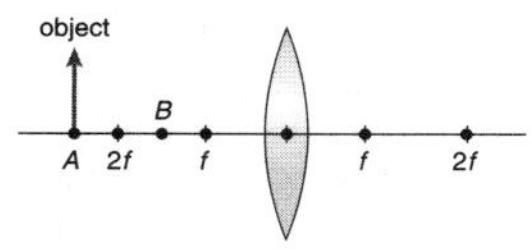

© It's About Time

Activity 7 Effect of Lenses on Light

Stretching Exercises

1. To investigate how the image position depends on the object position, find a convex lens, a white card, and a light source. Find the image of the light source, and measure the image and object distance from the lens. Make these measurements for as wide a range of object distances as you can. In addition, make an image of an object outside, such as a tree. Estimate the distance to the tree. The image of a distant object, like the tree, is located very near the focus of the lens. Draw a graph of the results. Compare the graph with the equation

$$\frac{1}{f} = \frac{1}{D_o} + \frac{1}{D_i}$$

2. Find a camera with a shutter that you can keep open (with a bulb or time setting). Place a piece of waxed paper or a piece of a plastic bag behind the lens, where the film would be if you took a picture. Find the image and compare it to the images you made in this activity. Focus the lens for objects at different distances. Investigate how well the object and image location fit the lens equation $\frac{1}{f} = \frac{1}{D_o} + \frac{1}{D_i}$.

   Remember that the focal length of the lens is typically printed on the lens.

3. Research how the concept of "depth of field" is important in photography. Report to the class on what you learn.

Answers

Stretching Exercises

1. To compare the graph with the equation, students can calculate $(1/D_o + 1/D_i)$ for each pair of images and object distances. If the values of $(1/D_o + 1/D_i)$ are approximately equal, then the equation shown describes their data (and they can calculate f).
2. The real image will lie on the wax paper.
3. Depth of fluid is the range of object distances that will produce a sharp image. It depends on the lens' focal length aperture (f - number) and the object distance.

© It's About Time

Object Distance	Image Distance	Appearance of Image

For use with *Let Us Entertain You*, Chapter 5, Activity 7: Effect of Lenses on Light

Object Distance	Image Distance	Appearance of Image

For use with *Let Us Entertain You*, Chapter 5, Activity 7: Effect of Lenses on Light

© It's About Time

NOTES

© It's About Time

ACTIVITY 8
Color

Background Information

Shadows are produced when one object blocks some of the light that is hitting another object. Outside on a sunny day, you can see a sharp shadow of your foot on the ground. You can also see shadows of leaves on the ground, but notice how sharp—or rather how fuzzy—is the border of these shadows. You can see the same effect in a dark room by turning on a single frosted bulb. Because the bulb and the sun are extended light sources, they make shadows with a penumbra region, a fuzzy border around the shadow illuminated by some, but not all, of the light source.

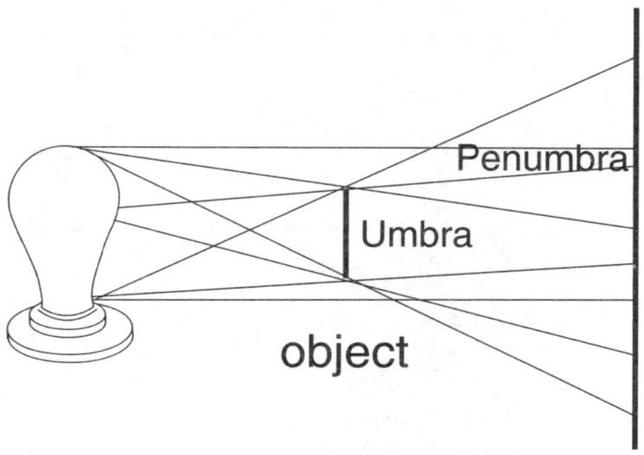

The size of this fuzzy border is directly proportional to the distance from the object to the screen. The top of a flagpole, or even better a tall building, casts a fuzzy shadow indeed.

Color is produced by the eye and the brain. The corresponding physical property of light is the wavelength. Visible wavelengths range from about $0.4 \times 10[-6]$ m to $0.7 \times 10[-6]$ m. Light with a wavelength in this range stimulates the eye and brain to produce a spectral color, as shown by the sprectral colors of sunlight formed by a prism (the index of refraction of the prism glass depends on the wavelength of light). These spectral colors are also produced by combinations of lights with different wavelengths. Red light plus green light produces light we see as yellow. Sometimes these color combinations contain white, and the result is an unsaturated color. Such a combination would be a red, green, and blue light, with a second blue light added, which would be seen as a pale blue. In a saturated color, there is no white. Saturated colors are often called "strong" or "intense." As different colored lights are progressively turned on to illuminate a screen, more and more light falls on the screen, which looks brighter and brighter. In the drawing of the three overlapping disks on page 389, the central area, illuminated by light of all three colors, would be the brightest area and would be white.

Overlapping colored filters gives quite different results. As more and more filters are overlapped, the light transmitted is less and less, so the stack of filters looks dark. Typically a filter passes a range of wavelengths, which correspond to a range of colors, so two overlapped filters pass only those wavelengths that both pass in common.

Active-ating the Physics InfoMall

Many of the searches to try here are probably obvious now. A warning may be helpful; if you search for "color" you will need to limit the search to only a few stores at a time, as the Articles & Abstracts Attic alone provides over 2000 hits. You should search for "pigments" to find the differences between light and pigment mixing. There are many useful searches you could perform here. Don't limit yourself; use your imagination. In other words, be colorful.

Planning for the Activity

Time Requirements:

two class periods

Materials Needed

For the class:

- overhead projector

For each group:

- 40-W bulb and socket, ceramic w/switch
- bulbs: white, red, blue, and green (4 each)
- file folder
- lamp, clip-on w/shade
- stand for clip-on lamp
- steel scissors
- white screen, about 2' × 3'

© It's About Time

Advance Preparation and Setup

Make a copy of the two Blackline Masters for each student. If possible, buy inexpensive "cans," which can be aimed and which have a slot to hold a filter (get these at a theatrical supply company). Set up the three colored lights so the beams all overlap on the screen.

Be sure all sockets have the wiring safely covered. If you use ceramic sockets with screw terminals underneath, screw the sockets down to a board so these terminals will be inaccessible.

Teaching Notes

The geometry of the shadows can be challenging. It may help to point a meter stick from a light to the puppet and to ask where that light will reach the screen. Also, with all three lights on, you can ask the students what will happen if you block the lights one at a time (for example, by covering one with a book or clipboard). Encourage the students to compare their observations for the colors they observe with their expectations. You might ask how the unexpected colors might become part of the student's light show.

If you have only one set of colored lights, at any give time you will have to keep the students busy and productive on other parts of the activity. If you have prisms available, you might have each group do a simple prism activity at this time.

SAFETY PRECAUTION: Be extremely careful with 110-V bulbs. Explain to the students that they must not touch the bulbs. Be sure all sockets have the wiring safely covered. If you use ceramic sockets with screw terminals underneath, screw the sockets down to a board so these terminals will be inaccessible.

Most students have little experience with combining colored lights. They probably will be surprised that red light plus green light looks yellow. It may help to explain that what is compicated here is not the physics but rather the functioning of the eye and brain. In fact, the way we commonly refer to light as "red" or "green" may create the preconception that the color is somehow embedded in the light, rather than being a sensory response to the physical characteristic of wavelength.

© It's About Time

Activity Overview

To investigate color addition, students make shadows of colored bulbs. First they observe shadows of an object illuminated by one bulb. They make a drawing to show the size of the shadow and the relationship of size to the positions of bulb, object, and screen. Then they progress to combinations of colored lights, and for each combination they make a drawing to record the pattern of the shadow colors.

Student Objectives

Students will:

- Analyze shadow patterns.
- Explain the size of shadows.
- Predict pattern of colored shadows.
- Observe combinations of colored lights.

ANSWERS FOR THE TEACHER ONLY

What Do You Think?

The answer to these questions lies in the field of both physiology and physics. The human eye and brain respond to the wavelength of light by producing the sensation of color, and each wavelength produces a distinct color, as seen in the spectrum of sunlight formed by a prism. (Recall that the index of refraction of the prism is a function of the wavelength of light.)

The nature of the light-sensitive cells in the eye, together with the structure of the brain, determine in a complex way the color we see. Mixing paints produces color subtraction, because only light that can be transmitted by both paints is seen (the other colors are subtracted out). Overlapping lights in a theater produce color addition, since more light is reflected. The eye and brain do the rest to produce the colors we see.

Let Us Entertain You

Activity 8 Color

GOALS

In this activity you will:

- Analyze shadow patterns.
- Explain the size of shadows.
- Predict pattern of colored shadows.
- Observe combinations of colored lights.

What Do You Think?

When a painter mixes red and green paint, the result is a dull brown. But when a lighting designer in a theater shines a red and a green light on an actress, the actress's skin looks bright yellow.

- **How could these two results be so different?**
- **How are the colors you see produced?**

Record your ideas about these questions in your *Active Physics* log. Be prepared to discuss your responses with your small group and with your class.

For You To Do

1. Carefully cut out a cardboard puppet that you will use to make shadows.

386

Active Physics

ANSWERS

For You To Do

1. The puppet can have any shape.

© It's About Time

2. Turn on a white light bulb only. Move the puppet around and observe the shadow.
   a) Describe the shadow you see.
   b) What happens to the shadow if you move the puppet sideways or up and down?
   c) What happens to the shadow if you move the puppet close to the screen?
   d) What happens to the shadow if you move the puppet close to the bulb?

Caution: Lamps get very hot. Be careful not to touch the bulb or housing surrounding the bulb.

3. Look at the drawing. It shows a top view of a puppet halfway between the light and the screen.
   a) Make a copy of this drawing in your log. Draw light rays going from the light to the screen. (Light rays will go in straight lines in all directions from all parts of the light.)

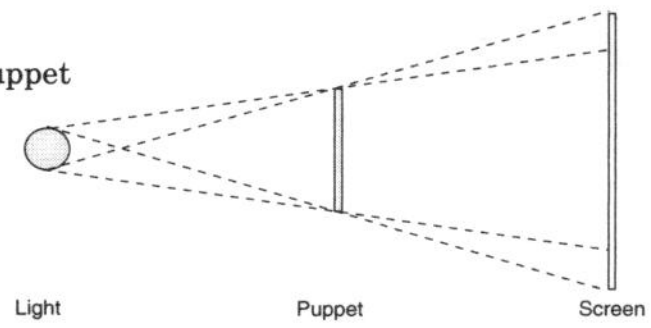

   b) Use the top-view drawing you drew to answer these questions: Which part of the screen receives light? Which part receives no light? Which part receives some light?
   c) Is the shadow larger or smaller than the puppet? Explain how you found your answer.
   d) Now copy the other two top-view drawings and show the path of the light rays.

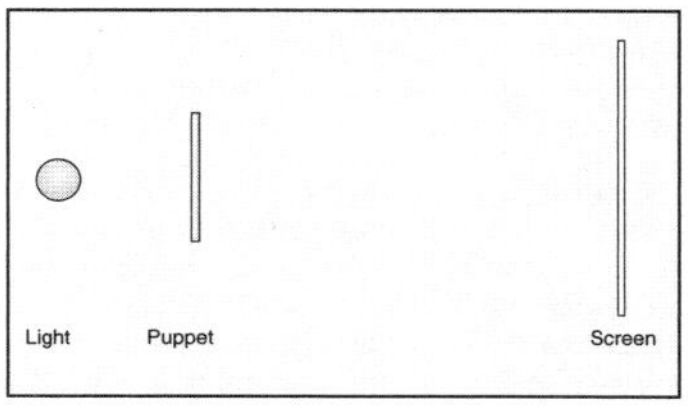

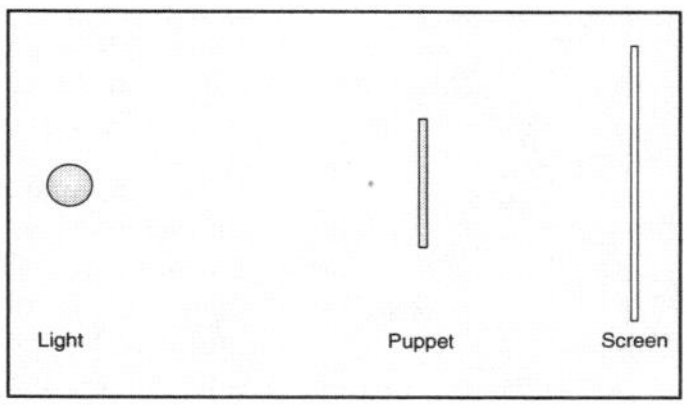

   e) On your drawings, show which part of the screen does receive light and which part does not receive light and which part receives some light.
   f) For each of these two drawings, tell whether the shadow is larger or smaller than the puppet. For each one, explain how you found your answer.

ANSWERS

For You To Do
(continued)

2. a) The sharpness of the edge of the shadow depends on the distance from the puppet to the screen.
   b) The shadow moves in the same direction.
   c) The shadow becomes sharp and is the same size as the puppet.
   d) The shadow becomes quite large, and the edge of the shadow will become fuzzy if the light bulb is frosted.
3. a) Notice that you have to draw lines from only the top and bottom of the bulb.
   b) The part of the screen between the inner dashed lines receives no light. The part of the screen between the inner and outer dashed lines receives some light (since it is possible to draw a line from some part of the bulb to this part of the screen). No part of the screen receives light from all of the bulb.
   c) The shadow is larger than the puppet, since light from the bulb continues to spread out after it passes the puppet.
   d) In the drawing on the left, the shadow is larger than the puppet. Imagine lines from the light source to the ends of the puppet. Extend these lines to the screen. Since these lines spread way out as they go towards the screen, the shadow will be much larger than the puppet. For the drawing on the right, the shadow will be only slightly larger than the puppet.
   e) Draw dashed lines from the top and bottom of the bulb, as in the drawing in **Step 3**.
   f) See answer to **Step d)** above.

© It's About Time

ANSWERS

For You To Do

(continued)

4. a) Where both lights strike the screen, you will see yellow (but it may appear white).

   b) Where the red light is blocked but the green is not, you will see a green shadow. Where the green light is blocked but the red is not, you will see a red shadow.

   c)-d) See diagram below. Also, note that where light from both bulbs hits the screen, you see yellow. Moving toward the center of the screen, the yellow gradually changes to red (above) and to green (below).

5.- 6. See answer to 4. above.

7. Where all three lights strike the screen, you will see white. Where one light is blocked, you will see a shadow with the color of the combination of the other lights. See **Steps 4** through **6** for these colors. Where two lights are blocked, you will see a shadow with the color of the unblocked light.

4. Turn off the white bulb. Turn on red and green bulbs. They should be aimed directly at the center of the screen.
   a) What color do you see on the screen?
   b) Predict what color the shadows will be if you bring your puppet between the bulbs and the screen. Record your prediction, and give a reason for it.
   c) Make a top-view drawing to show the path of the light rays from the red and green bulbs.
   d) On your drawing, label the color you will see on each part of the screen.

5. Turn off the green bulb and turn on a blue one. Repeat what you did in **Step 4**, but with the blue and red bulbs lit.

6. Turn off the red bulb and turn on the green one. Repeat what you did in **Step 4**, but with the blue and green bulbs lit.

7. Turn on the red bulb so all three—red, blue, and green—are lit. Repeat what you did in **Steps 5** and **6**.

Reflecting on the Activity and the Challenge

Different colored lights can combine to make white light. When an object blocks all light, it creates a dark shadow. Since some light comes from all parts of the bulb, there are places where the shadow is black (no light) and places where the shadow is gray (some light reaches this area). An object illuminated by different colored lights can create shadows that prevent certain colors from reaching the wall and allowing other colors to pass by.

In your light show creation, you may choose to use the ideas of colored shadows to show how lights can be added to produce interesting combinations of colors. By moving the object or the lights during the show, you may be able to produce some interesting effects. Lighting design is used in all theater productions. It requires a knowledge and understanding of how lights work, as well as an aesthetic sense of what creates an enjoyable display.

4. c - d)

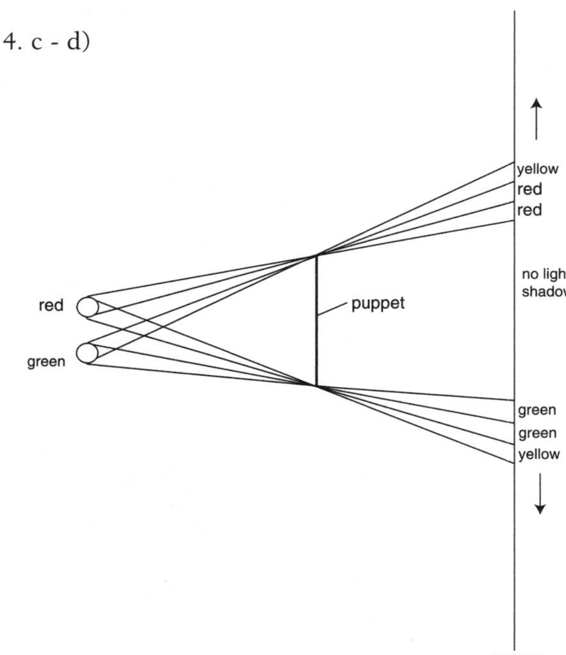

© It's About Time

Physics To Go

1. Show how a shadow is created.
2. How can moving the light, the object, and the screen all produce the effect of enlarging the shadow?
3. Explain why a gray halo surrounds a dark shadow made by a light bulb and an object.
4. a) Why is your shadow different at different times of the day?
   b) What is the position of the Sun when your shadow is the longest? The shortest?
5. Why is the gray halo about your shadow so thin when you are illuminated by the Sun?
6. a) Suppose you shine a red light on a screen in a dark room. The result is a disk of red light. Now you turn on a green light and a blue light. The three disks of light overlap as shown. Copy the diagram into your journal. Label the color you will see in each part of the diagram.

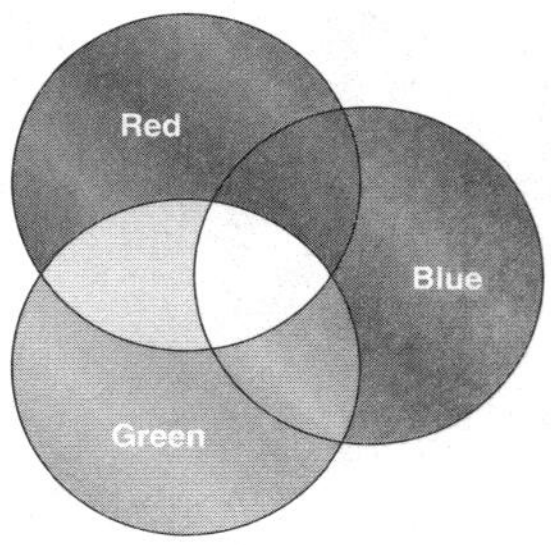

   b) Add the labels "bright," "brighter," and "brightest" to describe what you would see in each part of your diagram.

Answers

Physics To Go

1. When an object blocks light from hitting part of a screen, the result is a shadow.
2. The shadow becomes larger when light can spread out more after it passes the object. Light can spread out more when the screen is moved away from the light, the object is moved towards the light, or the light is moved towards the object.
3. The halo is a region of intermediate brightness at the edge of the shadow. The effect is to make the shadow edge appear fuzzy. The gray halo occurs when the light bulb is frosted (an "extended source"). Since the light comes from different places on the bulb, the area at the edge of the shadow is illuminated by light from only part of the bulb.

Answers

Physics To Go *(continued)*

4. a) The length of the shadow is determined by the altitude of the Sun (the angle between the horizon and the Sun). Since the light rays of sunlight are approximately parallel, the orientation of your body and also the orientation of the ground determine the length of the shadow.

   b) Your shadow is longest at sunrise and sunset. It is shortest at noon.

5. Because the Sun, although quite large, is so very far away.

6. a) Where all three lights overlap, in the center, you see white. Where red and green overlap, you see yellow. Where red and blue overlap, you see purple. Where blue and green overlap, you see blue-green.

   b) The central region is the brightest. Three regions with two overlapping lights (yellow, blue-green, and purple) are the next brightest. The red, blue, and green regions are the least bright.

© It's About Time

ANSWERS

Physics To Go
(continued)

7. a)-b) See drawing on p. 387. The rest of the screen is red.
   c) See drawings on previous pages.
8. Answers will vary.

7. a) Make a drawing of an object in red light. The object casts a shadow on the screen. Label the color of the shadow and the rest of the screen.
   b) Repeat **Part (a)** for an object in green light.
   c) Now make a copy of your drawing for **Part (b)**. Add a red light, as in **Part (a)**. Label the color of all the shadows.
8. List some imaginative ways that you can add colors to your light show.

Stretching Exercises

1. With the room completely dark, shine a red light on various colored objects. Compare the way they look in red light with the way they look in ordinary room light.
2. View 3-D pictures with red and blue glasses. Explain how each eye sees a different picture.
3. Shine a white light and a red light on a small object in front of a screen. What colors are the shadows? How is this surprising?
4. Prepare a large drawing of the American flag but with blue-green in the place of red, and yellow in the place of blue. Stare at the drawing for 30 seconds and then look at a white surface.

ANSWERS

Stretching Exercises

1. A green object illuminated in red light will look black (since a green object reflects only green light). A red object will still appear red.
2. The inks in the 3-D pictures are matched to the filters in the glasses. The eye looking through the red glasses sees the blue lines, which appear very dark. The red lines blend into the background, since both appear red. The other eye sees only the red lines, which also appear very dark. The red and blue images are slightly displaced to produce the 3-D effect.
3. It would seem that the shadow of the red light should be white. But often this shadow is seen as blue-green. The screen is a pale red, but the eye and brain often take this for white. If that happens, then subtracting red from this color (in the shadow of the red light) produces what the brain takes to be white minus red, or blue-green, the complement of red (remember that red plus blue-green gives white). It is a remarkable illusion.
4. Staring at the strangely-colored flag tires the color receptor cells in the retina, so they cease responding in that color. They continue to respond in the complementary color, however, so the result is that the brain perceives the complementary colors when viewing a white surface.

© It's About Time

PHYSICS AT WORK

Alicja and Dennis Phipps

Alicja Phipps has always been interested in electronics. As a child she wanted to be a television repair person. She now works with her husband, Dennis Phipps, in their company, Light & Sound Entertainment, which designs original content programming in a variety of areas—from rock concerts to the Olympics.

Light & Sound Entertainment got its name because Dennis believes strongly that the two are and should be linked. "It's terrible when the music of a production does not match what you are seeing," he says. "More and more theaters have lighting specialists come in and set up the theater with everything preset on a computerized lighting board. There will be a setting for 'outdoor lighting' and 'nighttime' or 'sunset'. The problem with that is that every production will look the same, when in reality nighttime in Canada looks very different from nighttime in Florida. There are also different lighting considerations depending on your audience. A production in front of a live audience needs different lighting than one being recorded on video. For instance, the human eye picks up shadows a lot better than a camera will. If a production is intended for both a live audience and video, lighting needs to be arranged accordingly."

Dennis continues, "The sound of a production is only as good as its setup, and nothing can replace the actual setting. The Red Rocks Theater in Colorado is terrific, for example, because stone has a very high reverberation rate, which is great for guitars. A huge wooden room like Carnegie Hall also provides a unique sound. However, these spaces and materials are not readily available." The hardest projects, Dennis says, are those in which you cannot control the elements. "Sound elements include the size and shape of the space, reverberation, feedback, and temperature."

"New media has its own set of challenges," explains Alicja who oversees the conversion of live events into various other formats, such as CD-ROMs, virtual reality, and Web sites. "We have to think about how much information (sound and image) we will be able to fit on a disc or on to a Web page and how long it will take to load. If it takes too long, no one will ever see or hear what we've done."

"We enjoy the creative process of every production," claim Alicja and Dennis. "Each one is a unique challenge."

© It's About Time

Chapter 5 Assessment

With what you learned about sound and light in this chapter, you are now ready to dazzle the world. However, you have neither the funds nor the technology available to professionals. All sounds you use to capture the interest of the class must come from musical instruments that you build yourself, or from human voices. Some of these sounds may be prerecorded and then played in your show. If your teacher has a laser and is willing to allow you to use it, you may do so. All other light must come from conventional household lamps. Gather with your committee to design a two- to four-minute sound and light show to entertain other students your age.

Review the criteria by which you decided that your show will be evaluated. The following suggestions were provided at the beginning of the chapter:

1. The variety and number of physics concepts used to produce the light and sound effects.

2. Your understanding of the physics concepts:
   a) Name the physics concepts that you used.
   b) Explain each concept.
   c) Give an example of something that each concept explains or an example of how each concept is used.
   d) Explain why each concept is important.

3. Entertainment value

At this time you may wish to propose changes in the criteria. Also decide as a class if you wish to modify or keep the point value you established at the beginning of the chapter.

Enjoy the sound and light productions!

Physics You Learned

Compressional and transverse waves

Wave speed = wavelength × frequency

Standing waves

Pitch and frequency

Sound production in pipes and vibrating strings

Controlling frequency of sounds produced electronically

Angle of incidence and angle of reflection

Location of image in plane and curved mirrors

$\frac{1}{f} = \frac{1}{D_o} + \frac{1}{D_i}$ in curved mirrors

Real images

Angle of incidence and angle of refraction

Lenses and image formation

$\frac{1}{f} = \frac{1}{D_o} + \frac{1}{D_i}$ in lenses

Color addition

$n = \frac{\sin\angle i}{\sin\angle R}$

© It's About Time

Chapter 5

NOTES

© It's About Time

Alternative Chapter Assessment

1. You shake a Slinky back and forth. No wave moves along the spring. At a few places, the spring does not move at all. You have made a:

   a) pulse

   b) frequency

   c) standing wave

   d) reflection

2. In a wave, the number of cycles per second is called the:

   a) wavelength

   b) amplitude

   c) speed

   d) frequency

3. As you decrease the tension on a string, the pitch it makes:

   a) is unchanged

   b) increases

   c) decreases

4. As you decrease the length of a tube, the pitch it makes:

   a) is unchanged

   b) increases

   c) decreases

5. You built an electronic circuit that made sound. When you changed the value of a capacitor in this circuit, the sound:

   a) became louder

   b) became softer

   c) stopped

   d) changed its pitch

6. If you walk towards a mirror, your reflection:

   a) doesn't move

   b) moves away from the mirror

   c) moves towards the mirror

© It's About Time

7. The image in a convex mirror:

   a) is real

   b) is virtual

   c) can be real or virtual

8. At the critical angle, there is:

   a) total internal reflection

   b) no reflection

   c) partial reflection

9. You make an image on a card with a convex lens. As the object moves away from the lens, the image:

   a) moves away from the lens

   b) stays in the same place

   c) moves closer to the lens

10. One light shines on a screen. If you shine a second light on the screen, the screen looks:

    a) yellow

    b) dimmer

    c) brighter

11. You make a spring wave with a wavelength of 1.5 m. You shake the spring with a frequency of 2 Hz. The speed of the wave is ________________.

12. The distance from one wave crest to the next is called the ______________.

13. As a wave goes by on a spring, the spring moves from its rest position. The distance from the rest position is called the ______________.

14. A concave mirror makes an image of a distant light at one point. The location of the image is called the ________________ of the mirror.

15. The bending of light when it goes from air into gelatin is called ________________.

16. Explain how to measure the focal length of a lens.

© It's About Time

17. a) Explain how to measure the *frequency* of a wave.

    b) If you have the frequency of a wave, how could you find the *period*?

18. Explain the difference between a *transverse* wave and a *compressional* wave on a Slinky. You can use drawings to give your answer.

19. a) A beam of light hits a flat mirror. Make a drawing to show the angle of incidence, the angle of reflection, and the normal.

    b) What is the relationship between the angle of incidence and the angle of reflection?

20. a) You are building an organ with pipes that are 3 m, 1.5 m, and .5 m long. Order these tubes according to how high a pitch they will make.

    b) Label the tube with the highest pitch and the tube with the lowest pitch.

21. a) A beam of light goes from air into a piece of gelatin. Make a drawing to show the angle of incidence, the angle of refraction, and the normal.

    b) On your drawing, show which way the beam of light bends.

22. a) You shine a red light and a blue light on a screen (see drawing). The puppet casts shadows on the screen. Make a drawing of the lights, the puppets, and the screen to show the different shadows.

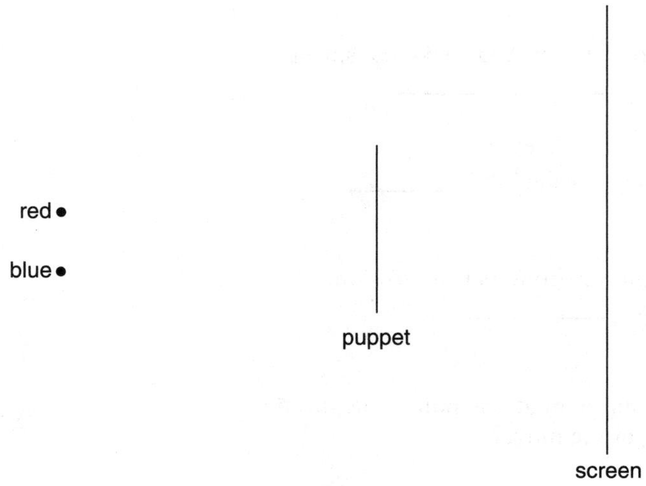

    b) Label the drawing to show the color of the screen and the color of each shadow.

    c) Are the shadows larger or smaller than the puppet?

    d) Explain your answer to part (c).

© It's About Time

Alternative Chapter Assessment Answers

1. c
2. d
3. c
4. b
5. d
6. c
7. b
8. a
9. c
10. c
11. 3 m/s
12. wavelength
13. amplitude or displacement
14. focus
15. refraction
16. Make an image of a distant object. Measure the distance from the image to the lens.
17. a) Count how many cycles of the wave there are in a certain time. Divide the time into the number of cycles.

    b) Invert the frequency (divide it into 1)
18. The difference is in the direction of the back-and-forth motion.

In a transverse wave, the Slinky goes back and forth perpendicular to the direction the wave moves. In a compressional wave, the Slinky goes back and forth in the same direction the wave moves. See drawings on p. 327 and 329.

19. a) See the drawing on page 356.

    b) the angles are equal
20. a) from longest to shortest

    b) The shortest makes the highest pitch; the longest makes the lowest pitch.
21. a) See angles 1 and 2 in the drawing on page 372.

    b) The light bends towards the normal (going into the gelatin).

© It's About Time

22. a) and b)

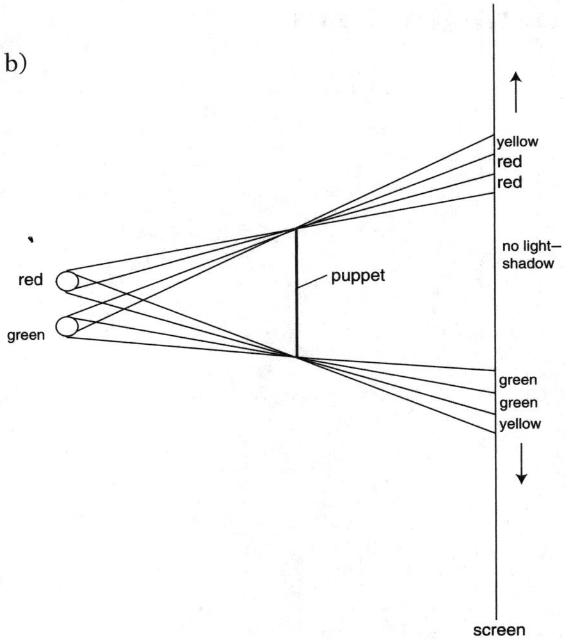

c) and d) The shadows are larger, because the light is small, and the light rays spread out to reach the puppet. After passing the puppet, they keep on spreading.

© It's About Time

Chapter 6

DESIGNING THE UNIVERSAL DWELLING

Chapter 6- Designing the Universal Dwelling

National Science Education Standards

Chapter Challenge

Homes for Everyone establishes the scenario with a quest for a universal dwelling that meets the need for homes in diverse environments. Students are challenged to design a home that can be constructed quickly and simply at a remote building site and should use the least amount of materials to create the most living space. And, it should be energy-efficient!

Chapter Summary

To gain understanding of the science principles necessary to meet this challenge, students work collaboratively on activities in which they learn about energy transfer, solar energy, the iterative process of engineering design, and how to use collected data to refine designs. These experiences engage students in the following content identified in the National Science Education Standards.

Content Standards

Unifying Concepts

- Evidence, models and explanations
- Constancy, change, and measurement
- Form and function

Science as Inquiry

- Identify questions and concepts that guide scientific investigations
- Use technology and mathematics to improve investigations
- Formulate and revise scientific explanations and models using logic and evidence

Science and Technology

- Abilities of technological design
- Understanding about science and technology

Science in Personal and Social Perspectives

- Natural resources
- Environmental quality
- Science and technology in local, national, and global challenges

Physical Science

- Conservation of energy and increase in disorder
- Interactions of energy and matter

© It's About Time

Key Physics Concepts and Skills

Activity Summaries	Physics Principles
Activity 1: Factors in Designing the "Universal Dwelling" Examining characteristics of houses designed for a variety of geographic locations introduces students to the concept of form and function in a home. They then list features necessary for a universal dwelling.	• Form and function • Analyzing and interpreting data
Activity 2: What is the "Right Size" for a Universal Dwelling? After researching and analyzing data on living space in sample homes, students plan the design of their universal dwelling and calculate the necessary dimensions of each room based on its function. They conclude by drawing a floor plan.	• Analyzing and interpreting data • Models, measurements, and scale
Activity 3: The Shape of the "Universal Dwelling" Students examine their floor plans, applying knowledge of area, perimeter, and volume to decide if the planned living space is maximized while building materials are minimized. They then refine their plans to reflect what was learned from investigating size as compared to form and function of rooms in other dwellings.	• Applying measurement and data to predictions
Activity 4: Solar Heat Flow in the "Universal Dwelling" This activity confronts students with implications of shape and size on occupants of the home. Students construct a model of the home, then use a heat lamp and temperature probe to investigate interior light and heat during a simulated day and night. This also introduces the use of passive solar designs for light and heat in the home.	• Radiant energy • Energy transfer • Passive solar heating
Activity 5: The Role of Insulation... Students expand their understanding of solar heat by exploring the affect of different types of insulation in an experiment with hot and cold water. They apply data collected to plan how to maintain a stable temperature inside the home. Students read to learn about transfer of radiant energy through conduction and convection.	• Energy transfer • Heat conduction • Heating and cooling curves
Activity 6: Investigating Insulation Placement... Students collect and compare temperature data during simulated 24-hour cycles for different types, thickness, and placements of insulation in the ceilings and walls of their model homes. Interpretation of this data combined with new understanding of heat loss and gain is used to further refine design of the home.	• Energy transfer • Heat loss through conduction
Activity 7: The Role of Windows... Light and ventilation are the focus of this activity as students plan and test placement, size, and materials of windows. As they continue experimenting with passive solar heating, they are challenged to again consider heat loss when an insulated wall is replaced with a window.	• Energy transfer • Heat loss through conduction • Passive solar heating • Passive solar lighting
Activity 8: Investigating Overhangs and Awnings Students investigate the affect of the angle of the Sun's rays in different geographic regions in a simulation with heat lamps and temperature probes. Overhangs and awnings are added to the house to compensate for increased interior temperature. Students then have the opportunity to further refine the design of their dwelling to maximize use of passive solar for light and heat.	• Energy transfer • Passive solar heating • Passive solar lighting
Activity 9: Too Hot, Too Cold, Just Right Students will be able to experimentally determine the final temperature when two liquids of different temperatures are mixed, determine the final temperature when a hot metal is added to cold water, and calculate the heat lost and the heat gained of two objects after they are placed in thermal contact. They will also determine if energy is conserved when two objects are placed in thermal contact and reach an equilibrium temperature, and explain the concept of entropy as it relates to objects placed in thermal contact.	• Heating and cooling curves • Irreversibility of heat • Specific heat

© It's About Time

Equipment List For Chapter 6

PART	ITEM	QTY	ACTIVITY	TO SERVE
CB-0912-H1	Corrugated Cardboard, 9 X 12	2	3	group
CB-7608-H1	Corrugated Cardboard, 1 X 2	6	6	group
CS-0024-H1	Cylinder, Graduated, 100 ml	1	9	group
CM-0207-H1	Set of 3 Metal Cans Varying Diameters	1	5	group
CM-1108-H1	Calculator, Basic	1	All	group
CS-3237-H1	Clock, Electric	1	4, 6, 7, 8	group
CS-6047-H1	Cup, Styrofoam	4	9	group
CS-7706-H1	Card, 3 X 5" W/1 Slit	1	1	group
CS-7707-H1	Card, 3 X 5" W/2 Slits	1	1	group
DS-6066-H1	Dish Pan, 10 qt.	1	5	group
FF-7704-H1	File Folders	24	4, 8	group
GS-7706-H1	Graph Paper Pad of 50 Sheets	1	3, 9	group
HL-0250-H1	250-Watt Reflective Heat Lamp	1	4, 6, 7, 8	group
KS-2814-H1	Knife, Retractable	1	4, 6, 7, 8	group
LS-4119-H1	Stand For Clip-On Lamp	1	4, 6, 7, 8	group
VC-0120-H1	120-Volt Clip On Fixture For Heat Lamp With Shade	1	4, 6, 7, 8	group
PM-0521-H1	Protractor	1	8	group
PP-6128-H1	Pushpins	100	3	group
RS-2826-H1	Ruler, Inch Scale	1	4, 6, 7, 8	group
SM-1676-H1	Stick, Meter, 100 cm, Hardwood	1	4, 6, 7, 8	group
SS-7722-H1	Ball of String, 225 ft.	1	3	group
TO-1600-H1	Tongs	1	9	group
TS-2662-H1	Tape, Masking, 3/4 X 60 yds.	1	4, 6, 7, 8	group
TS-3602-H1	Thermometer	3	4, 5, 6, 7, 8	group
VS-2143-H1	Insulating Material, 1qt.	3	5	group
WS-7546-H1	Wrap, Clear Plastic	1	7	group
ZZ-0066-H1	66 qt. Clear Plastic Storage Container	1		group
THINGS NEEDED NOT SUPPLIED				
	Hot water supply		5	class
	Ice water supply		5	class
	Pieces of Metal-Approximately 100 Grams each		9	group
	Paper Towels		9	class

© It's About Time

Organizer for Materials Available in Teacher's Edition

Activity in Student Text	Additional Material	Alternative / Optional Activities
ACTIVITY 1: Factors in Designing the "Universal Dwelling" p. 396	Floor Plan for a Home: Floor Area of a Home pp. 144-145	
ACTIVITY 2: What is the "Right Size" for a Universal Dwelling? p. 400	Floor Space for a Home pp. 153-154	
ACTIVITY 3: The Shape of the "Universal Dwelling" p. 405	Square Grid Paper p. 165	ACTIVITY 3 A: Exploring Surface Area-to-Volume Ratio p. 166
ACTIVITY 4: Solar Heat Flow in the "Universal Dwelling" p. 411	Grid for Heating/Cooling Data p. 176	ACTIVITY 4 A: How Does Location Affect Energy Demands? pp. 177-178 ACTIVITY 4 B: What Shape is Most Effective for Heat Storage? pp. 179-181
ACTIVITY 5: The Role of Insulation... p. 416		ACTIVITY 5 A: Transmission of Heat pp. 189-190
ACTIVITY 6: Investigating Insulation Placement p. 421	R Values for Insulation p. 199	
ACTIVITY 7: The Role of Windows... p. 425	R Values for Windows p. 207	
ACTIVITY 8: Investigating Overhangs and Awnings p. 429	Sun Location at Seasonal Times pp. 215-217	
ACTIVITY 9: Too Hot, Too Cold, Just Right p. 433		

© It's About Time

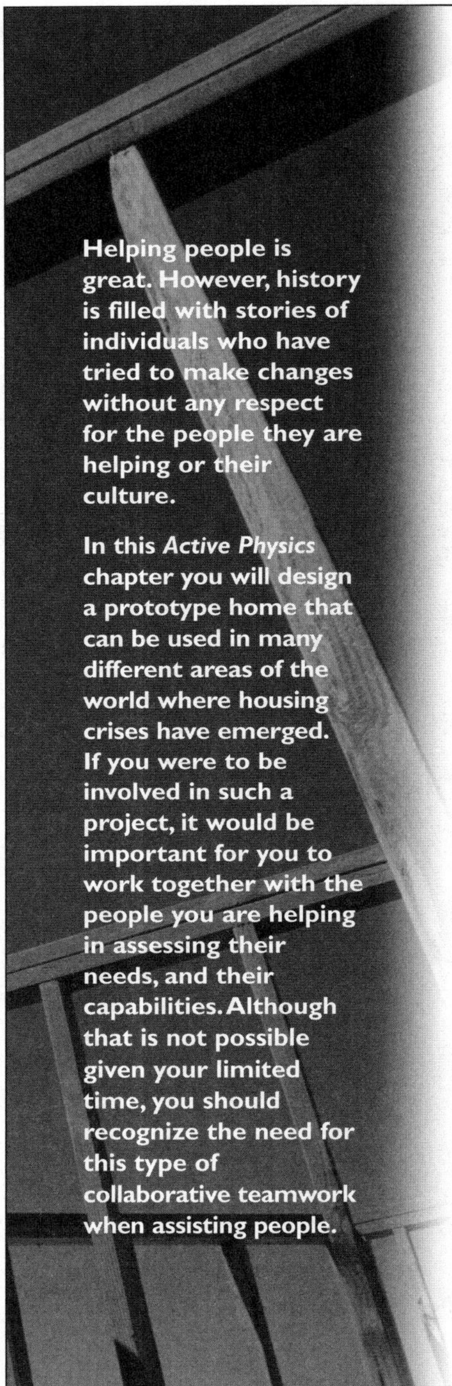

Helping people is great. However, history is filled with stories of individuals who have tried to make changes without any respect for the people they are helping or their culture.

In this *Active Physics* chapter you will design a prototype home that can be used in many different areas of the world where housing crises have emerged. If you were to be involved in such a project, it would be important for you to work together with the people you are helping in assessing their needs, and their capabilities. Although that is not possible given your limited time, you should recognize the need for this type of collaborative teamwork when assisting people.

Scenario

Imagine you and your team members are part of an international group called Homes For Everyone (HFE). The purpose of your organization is to address the growing housing shortage in many areas throughout the world. You have recently been sent to work with a self-help community group in a faraway area. Here is a letter you might write home.

Dear Mom and Dad,

Greetings! I've finally settled into my new home. Sometimes I think that I'm not in another country but on another planet. Everything here is so different. Sorry about the water splotches on this letter. It's pouring rain outside, and the roof is leaking. I have to remind myself that at least I have a roof over my head. Many people here have lost their homes from the last hurricane and are crowded into their relatives' already crowded spaces.

We're trying to figure out what we can do to help with the housing situation here. We're still not sure what the solution is. Right now I'm trying to learn more about these people who have been so generous and gracious to me. When we arrived the whole village got together to greet our group. I felt so honored! I expect the same kind of welcome from my family when I return home!

Tomorrow we are going to start looking around. I'll keep you posted on what we discover and on our progress. Signing off for now. Miss you all!

Love,

Kim

HFE plans to design a "universal dwelling" to meet the need for homes in diverse environments. The group needs a design that can be constructed quickly and simply at the building site. The design should also use the least amount of materials to create the most living space. To make mass production possible, the dwelling design should be uniform, but you should be able to make simple changes to meet local conditions. It should be energy efficient in any climate.

394

© It's About Time

Chapter and Challenge Overview

In this chapter, the students will be designing a home that will be energy-efficient, easily and inexpensively constructed, adaptable for many different climates, as well as being aesthetically pleasing.

In the beginning of the chapter the students will brainstorm as to the features, functions and other factors affecting the shape, size, and other attributes of a home. From there they will work their way through a designing process, while considering a number of different factors. Even though this is not meant to be an all-inclusive design, the students will get an appreciation for the need for looking at these issues when designing a home.

© It's About Time

Challenge

After completing the nine activities in this chapter you will be challenged to do the following to present to the HFE Architectural Committee:

1. Develop scale drawings of the floor plan and all side views of the universal dwelling. Following are specifications for the drawings:
   a) The drawings should be done on a scale of 1 inch = 4 feet (1:48).
   b) The scale drawings should include all sides of the house, showing roof lines, roof overhangs, and placement and dimensions of all windows for a selected climatic region and culture.
   c) The scale drawings should show the floor plans of the living spaces and include specifications for the thickness of the wall, types of insulation and their thickness, and kinds and types of windows or ventilation openings.
   d) The plans should show the geographic orientation of the house (in which direction the house should face).

2. Write a two-page explanation which gives your reasons for the following:
   a) your choices for the shape and dimensions of the dwelling
   b) the energy consideration that went into the design of the dwelling
   c) the changes that could be made to the basic design to take into account different climates and cultures
   d) the things you included to make your dwelling attractive.

Criteria

The HFE Architectural Committee will use criteria similiar to the ones below in evaluating your drawings and your written presentation. Discuss and decide as a class the exact criteria the committee should use.

- **(20%) The drawings and written presentation should meet all specifications listed in the Challenge.**
- **(10%) The drawings and written presentation should explain how you considered surface-to-volume ratio in your design.**
- **(20%) The written presentation should explain how the house will accommodate differing seasonal and climatic conditions in terms of heating and cooling. It should also show you have a basis for scaling the dwelling appropriate to the family and culture for which it is designed.**
- **(20%) The design of the roof and the placement and dimensions of windows should account for climatic, solar, and latitude considerations, and should show you have accounted for the advantages and disadvantages of windows.**
- **(20%) The placement and kind of insulation materials and thickness you choose should show you know the principles which make insulation effective.**
- **(10%) The appearance of your house and the interior layout should show consideration for fundamental human needs.**

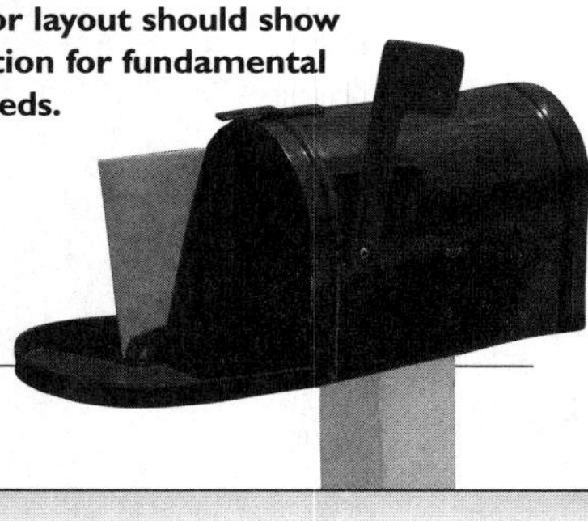

395

© It's About Time

Assessment Rubric for Challenge

The following rubric may be used to assess group work during Chapter 6. Each member of the group can evaluate the manner in which the group worked to solve problems with the challenge. This can constitute part of their mark, or can be incorporated in the criteria for evaluation.

This would be a peer evaluation. Emphasize that a mark with no comment is useless. Students must be able to justify their evaluation to the teacher.

Total = 12 marks

1. Low level – indicates minimum effort or effectiveness.
2. Average – acceptable standard has been achieved, but the group could have worked more effectively if better organized.
3. Good – this rating indicates a superior effort. Although improvements might have been made, the group was on task all of the time and completed all parts of the activity.

The following is a possible guideline for evaluation of the Challenge.

Place a check mark in the appropriate box. Each line may represent 5%. If you would rather, you could mark it holistically, where all the statements taken together earn a mark out of 30%, for example.

You may change the criteria, by adding or deleting any of the points determined by a class decision. For example, if you have a class that has had CAD training, those students might decide to put less emphasis on the actual scale drawings, and more emphasis on the written presentation.

Descriptor	Values		
1. The group worked cooperatively to design the HFE home.	1	2	3
2. Data was collected and recorded in an organized fashion in data tables in their logs.	1	2	3
3. The group was organized. Materials were collected and the problems were addressed by the entire group.	1	2	3
4. The two-page explanation was a good summary of the groups efforts, with input from all members.	1	2	3

© It's About Time

For use with *Designing the Universal Dwelling*, Chapter 6

Descriptor	**Yes**	**No**
Scale drawings (30%)		
scale of 1:48		
show all sides of house, roof lines, roof overhangs		
show placement of windows, including dimensions		
show floor plans of living space (thickness of walls, kinds of windows, etc.)		
geographic placement of home is clear		
drawing encompasses understanding of surface-to-volume ratio		
Written presentation (20%)		
clear statement for choice of shape and dimensions		
consideration of energy requirements for different climatic regions		
clear statement of changes to be considered, depending on different climates, cultures and family		
any additional solar energy trapping ideas included		
Window and roof design (20%)		
clear statements of climatic, solar and latitude regions chosen		
clear statements as to pros and cons of chosen roof design		
clear statements for and against the choice of windows		
diagram reflects accurate understanding of placement and sizes of windows, and the area where they are located relative to the climatic region		
Insulation (20%)		
clear statements of type of insulation used		
clear statements of choice of insulation based on climatic region		
clear understanding of principles of insulative materials		
diagrams show accurate thickness of insulation		
Appearance (10%)		
appearance of home is pleasant		
interior layout reflects consideration of fundamental human needs		

© It's About Time

For use with *Designing the Universal Dwelling*, Chapter 6

What is in The Sun's Joules CD-ROM for Chapter 6?

The Sun's Joules CD-ROM is a comprehensive collection of information resources developed by the Center for Renewable Energy and Sustainable Technology (CREST), funded by the United States Department of Energy and directed by the National Renewable Energy Laboratory (NREL). It provides easy access to a great deal of information about new technological advances in utilizing energy resources. As a supplemental resource, it will provide enjoyable enhancement materials to the basic concepts and activities of the chapter. Its features include: a glossary with definitions of energy-related terms; an index with information on its main topic areas, and a list of references for each of its topic areas.

One of the Main Menu items of the CD-ROM is entitled "Passive Solar" and contains a wealth of information that will directly help students in their designing of the Universal Home. Also, much of the information on the disc relates directly to the **Stretching Exercises** and **Inquiry Investigations** found in the chapter. For example, in **Activity** 7 the **Stretching Exercise** specifically asks students to find out about passive solar heating for homes and asks them a number of related questions.

What is in The Green Home CD-ROM for Chapter 6?

In The Green Home CD-ROM students are given the opportunity to walk through a digital home exploring important environmental and technology features that will help them in their Universal Home designs. They will therefore find this to be a valuable resource as they think about the many different types of design issues that they face in the chapter. For example one of the "topic" areas found in the walkthrough is entitled "Exterior Wall Construction" which is full of information and resources that will to help students in their inquires for **Activity 5** "The Role of Insulation."

There are also a number of interactive activities in which students manipulate various elements relating to one or more aspects of home design. Each activity has several input variables, for instance in the "Whole House Activity" students can explore the effects of house size, window configuration, window types, insulation, and location of the house (to name a few of the variables) on energy consumption. The Green Home CD-ROM is a wonderful enhancement resource to this chapter's challenge and activities.

© It's About Time

What is in the Physics InfoMall CD-ROM for Chapter 6?

Chapters 6, 7 and 8 in *Active Physical Science* concern the physics of the home. Many physics topics are discussed within this theme, and apply to these chapters as a whole more than to a single activity or chapter. These include energy conservation, alternate energy sources, and power. The Physics InfoMall CD-ROM contains many resources that address exactly these issues.

For example, look in the Textbook Trove at Madalyn Avery's *Household Physics: A Textbook for College Students in Home Economics*. Many topics from physics are discussed within the context of the home — ideal for this *Active Physical Science* book. Such topics include work, energy, and power; water supply and sewage disposal; heat and thermometers; heat transfer; refrigeration; air conditioning; and electricity. This book can provide information to augment *Active Physical Science*.

Another great resource on the CD-ROM is *Energy: Insights from Physics*, which contains sections specifically on home energy conservation and home heating. Other sections of this book should be equally useful. You can find out what a "degree-day" is, the effect on energy use, and how to calculate degree-days.

Chapter 11 of *The Fascination of Physics* describes why a fireplace can increase your home heating costs, not counting the cost of the firewood itself!

Articles describing use of energy can be found in the Articles and Abstracts Attic. For example, "Efficient Use of Energy," *Physics Today*, vol. 28, issue 8, discusses use of energy and what efficiency means. The topic is addressed again in *Physics Today*, vol. 33, issue 2 with "Efficient use of energy revisited." Keep in mind that these are not the only articles and information available on the InfoMall — this is just a place for you to start your search for information.

As you can well imagine, many of the resources on the InfoMall discuss the physics of the home, but often only as physics topics, leaving the home out of consideration. The references mentioned above are some of the resources that make the direct reference to the home, but the physics topics themselves are contained in most of the other resources on the CD-ROM; for example, energy is discussed in every physics textbook. However, some of the advanced books discuss the energy of subatomic particles without much development of the energy concept. With a little forethought, you can find any of these common topics discussed abundantly on the InfoMall.

Perhaps more than any other chapter *Designing the Universal Dwelling*, Chapter 6 concerns itself with solar energy. Perform a search of the entire CD-ROM with the words "solar energy" and you will find hundreds of hits (over 400 in the Articles and Abstract Attics alone!). These include discussions of pro-solar, anti-nuclear energy; conversion of solar energy to electrical energy; and much more. Do this search and see what you find.

© It's About Time

NOTES

Chapter 6

© It's About Time

ACTIVITY 1
Factors in Designing the "Universal Dwelling"

Background Information

The main concept involved in **Activity 1** is the calculation of the surface area of a two-dimensional object from measurements of the object's dimensions. Applied to determining the total surface area of the floor(s) of a home in **Activity 1**, the shape of the object (floor) involved usually encountered will be a rectangle:

Area of a rectangle $(ft)^2$ = Length (ft) × Width (ft)

The foot (ft) is used as the unit of length in the above equation because the standard unit of measurement for expressing the floor area of a home in the United States is square feet $(ft)^2$.

Complex floor areas of homes may be "broken up" into simple rectangular sub-areas and the dimensions (length and width) of each sub-area can be measured. The surface area of each sub-area then can be calculated and then the total area of the home can be calculated by addition of the surface areas of all of the sub-areas.

The outcome of the calculations will depend on "rounding off" rules used in the process of measuring the lengths and widths of rectangular areas in a home. Also, principles of precision, accuracy and significant figures could be applied to calculation of the floor area of a home. However, the overarching principle suggested to be applied with students at this stage is "keep it simple." The main idea is for students to get a gross estimate of the floor area of their home as part of the sample represented by the homes of class members.

Equations for calculating the surface areas of other two-dimensional shapes which may be needed now and probably will be needed later are:

Area of triangle = 1/2 × Base × Height

Area of circle = π × $(\text{Radius})^2$

The surface areas of very complex two-dimensional shapes for which equations may be difficult or impossible to find can be determined in several other ways:

- The shape can be superimposed on a grid paper and the number of grid squares bounded within the perimeter can be counted while estimating, if necessary, fractional parts of grid squares "cut" by the perimeter line at the edges of the shape. Knowing the scale of the grid (how many square feet, for example, each grid square is "worth") the total surface area can be determined.
- The shape can be superimposed on a grid paper, "cut out" along the perimeter with scissors and then weighed on a sensitive balance (perhaps available from a chemistry teacher). Similarly, a grid square can be cut out from the paper and weighed (or, for more accuracy, 10 grid squares, dividing the weight by 10). Dividing the weight of the complex shape by the weight of one grid square will give the area of the shape in units of "grid squares."
- An instrument called a polar planimeter (not often found in schools) exists which gives a readout of the surface area of any shape when the instrument is used to trace the perimeter (some computer software may provide similar capability).

Active-ating the Physics InfoMall

In **Physics To Go**, the concept of "area" is important. Students sometimes have difficulty with this basic concept. A Compound Search on the entire InfoMall (which took less than 30 seconds!) using the words "area" and "student difficulties" uncovers area numbers of good references.

SAFETY PRECAUTION: The term "area" is often used to describe topics, as in "this area if physics" and will often result in search hits that do not apply to what you may be looking for.

For example, in the introduction to Arnold B. Arons' *A Guide to Introductory Physics Teaching* (in the Book Basement) it is said that students may "have never been asked to define area." All they have ever done is calculate areas of regular figures such as squares, rectangles, parallelograms, or triangles, using memorized formulas that they no longer connect with the operation of counting the unit squares, even though this connection may have been originally asserted. As a teacher, you cannot assume that students fully understand a concept such as area, but with the knowledge that some have difficulty, you may be prepared to address such difficulties.

As you may guess, most physics textbooks assume that students can calculate and understand areas, and for this reason most texts use the concept without explaining it. Nor do many of the books provide problems that explicitly allow students to practice calculations.

Planning for the Activity

Time Requirements

Allow approximately 30 to 40 minutes to carry the brainstorming activity to completion.

© It's About Time

Materials Needed

For each group:

- calculator, basic
- 3 × 5 card with 1 slit
- 3 × 5 card with 2 slits

Advance Preparation and Setup

None is required.

Teaching Notes

Before proceeding with the brainstorming activity, plan to reserve about 10 minutes at the end of the class period to give instructions for **Physics To Go**; the data which students will gather during **Physics To Go** is critical for tomorrow's class.

The success of the brainstorming activity strongly depends on each student understanding the rules and purpose. Therefore, it is suggested that the teacher read the procedure to the class and provide any necessary clarification. Further, the teacher should suggest that, within each group, one student should serve as recorder and another student should serve as leader. The leader's responsibility is to keep the group on task, observing the rules. The recorder's responsibility is to write down each characteristic given by group members. During the brainstorming session, you should "circulate," monitoring interactions within groups and, if necessary, clarifying rules. But you should not interject your ideas about characteristics of dwellings.

Set a time limit (perhaps 15 minutes) for brainstorming the initial list of 100 items, and have the students begin. When the time limit is reached, stop the brainstorming activity and direct groups to proceed to narrowing and separating the list into two lists, and to answer the questions; at the same time, announce that you will "cut off" work on **For You To Do** 10 minutes before the end of class to allow time to give directions for **Physics To Go.**

SAFETY PRECAUTION: Some students may be sensitive about having their homes compared to others (**Question 8** in **Physics To Go**). Emphasize that the information will be kept anonymous.

Distribute a 3 × 5 card to each student, and point out that the card is to be used to record data about each student's home as shown on Page 399 of the Student Edition. Stress that the data is needed for tomorrow's class. Use the overhead transparency of a house floor plan to show an example of how to measure and calculate the floor areas of living spaces. Be sure that all students have a ruler to use for measurements at home — if needed, a string knotted at each foot of length could be provided.

© It's About Time

Activity Overview

This activity begins the chapter with the students examining different styles of houses from different areas of the world in order to get ideas of how they will design their HFE homes. When building a home, there are many different criteria to consider. This activity will get them thinking about these criteria.

Student Objectives

Students will:

- Identify essential characteristics which all human dwellings have in common.
- Identify characteristics of human dwellings which vary with environment.
- Measure the dimensions of living spaces in the student's home.
- Calculate the floor areas of living spaces in the student's home.

ANSWERS FOR THE TEACHER ONLY

What Do You Think?

Students will have an opportunity to be creative in their answers. Some of the things they should notice that are common are they provide shelter from water, Sun and wind, all have roof openings, windows, supports for the roof, walls, prominent entrance, two stories, etc.

They are alike in that they have supports for the roof, walls, etc. They are different in that they are built of different materials. One is built on stilts, one is built into the side of a mountain, thatched roof in one, the size of each seems to be different, etc.

Designing the Universal Dwelling

Activity 1 Factors in Designing the "Universal Dwelling"

GOALS

In this activity you will:

- Identify essential characteristics that all human dwellings have in common.
- Identify characteristics of human dwellings that vary with environment.
- Measure the dimensions of living spaces in your home.
- Calculate the floor areas of living spaces in your home.

What Do You Think?

All humans have a basic need for shelter. Examine the pictures of dwellings shown on this page.

- **What are some common characteristics of all these dwellings?**
- **How are they alike? How are they different?**

Record your ideas about these questions in your *Active Physics* log. Be prepared to discuss your responses with your small group and the class.

Active Physics 396

© It's About Time

For You To Do

1. "Brainstorming" is a process in which you simply generate a large number of ideas. The rule of brainstorming is that all ideas should be accepted and no idea should be evaluated or thrown out. With your group, brainstorm a list of the characteristics of a "universal dwelling" for people anywhere in the world. In this case, set your team's goal at 100 different characteristics. Use the following questions to guide your brainstorming:
   a) What are the functions of a home?
   b) What features are essential to any home?
   c) What are some of the factors that determine the size and shape of a home?
2. After brainstorming, narrow the lists down and develop two lists of ten or fewer items each. The first list should identify essential characteristics that should be present in all homes. The second list should identify special characteristics that are absolutely necessary for some environments, but that are non-essential in other environments.
   a) Record the lists in your log using these two headings:
      - Essential characteristics of the universal dwelling.
      - Characteristics that must be modified for various climatic and cultural conditions.
3. Return to your group and answer the following questions in your log:
   a) What things do you need to know about people, family sizes, and lifestyles in order to design a universal dwelling?
   b) How large should a universal dwelling be? How could you make an educated guess?
   c) How many of your "essential characteristics of the universal dwelling" require an energy input of some kind?

Answers

For You To Do

1. a) Students' answers will vary. Functions may include: shelter from cold/warm environment; safety; protection from elements and undesirables; comfort; status; etc.
   b) Some of the features essential to any home might include: kitchen; bedrooms; able to cook, clean, heat, use warm water, etc.
   c) Many factors can be included. Some are: cost, location, local bylaws, types of material, climate, environment, type of ground, etc.
2. a) Student data
3. a) Students' answers will vary greatly, but expect some of the following: shift work, pets, number of people in family, whether there is an extended family, do they go on vacation a lot, are they involved in community activities, etc.
   b) Student will probably assign based on his/her own experience. Expect much variation.
   c) Some of the essential characteristics which may require energy input of some kind are: pumping of water, cooling/heating the house, hot water, electricity, gas or other forms of heating energy, etc.

© It's About Time

ANSWERS

Physics To Go

1. a) Student input.

   b) If the students live in cold climates, some of the features would include, double-pane windows, storm windows, insulated walls and ceilings, no windows on north side of house, heating of some kind, storm doors.... Some of the features seen on houses in hot climates would be insulation, but not as much; single pane windows, or only screens; weeping tiles in wet areas, house on stilts near swamps or large bodies of water, etc.

2. Student answers will vary, continue with the theme from previous question.

3. Student activity.

4. Student activity.

5. a) Student activity: e.g.: 16 feet by 26 feet

   b) Using above example
   Area = $l \times w$ therefore,
   Area = $16 \times 26 = 416$ sq. ft.

Reflecting on the Activity and the Challenge

You may think that this activity didn't move you very far toward meeting the **Challenge**, but it really has gotten you started. Through interaction with others, you have identified and shared ideas about the two basic aspects of dwellings for humans: function and form. Function involves the many things that a dwelling must do for people, and you have identified basic functions. Form involves the physical characteristics that a dwelling must have to support necessary functions, and you have started thinking about the size of a dwelling as perhaps the most basic part of form. Congratulations! You're on the way to meeting the **Challenge**.

As you progress through this chapter, you and your group may find that you will need to make modifications to your original plans. Don't be concerned about this, as this is typical of the planning and research process in which you are involved.

Physics To Go

1. a) Describe the weather in the area of the country in which you live.
   b) Describe some features of homes in your area that provide protection against the weather.
2. Choose another area of the country, or in the world, which has weather different from yours. Describe the weather in that area. Describe some features of homes in that area that provide protection against the weather.
3. A square foot (ft^2) is a unit of measurement of surface area. Using a ruler, draw an accurate diagram of a square foot. (You may have to tape two pieces of paper together.) Label the dimensions of width and length.
4. Use your diagram of a square foot. Estimate the number of square feet in the room in which you are presently located.
5. a) Measure the length and width (in feet) of the room in which you are presently located.
   b) To calculate the area of a rectangle you multiply the length by the width.
   $A = l \times w$
   Calculate the area of the room you are in.

© It's About Time

c) How close was your estimate in **Question 4** to the area you calculated?

6. A family has decided to install a new floor in the kitchen of their home. The kitchen floor is a rectangle 8 feet wide and 12 feet long. The new floor will be covered with tiles which are square pieces measuring 1 foot on each side. How many tiles will be needed?

7. The family described in **Question 6** has changed its mind and has decided to use a type of flooring material that is available in square pieces measuring 2 feet on each side. How many pieces will be needed?

8. To prepare for **Activity 2**, determine the total floor area of your own or someone else's living spaces. Measure the amount of floor space in the bedrooms, the living areas, the kitchen, bathroom(s), and any other areas. Include any closets in your measurement. Place the data you collect on a 3 × 5 card, using the format shown here. All the information will be kept anonymous.

The Total Floor Area of a Home

Combined area of bedrooms in square feet:	______
Combined area of kitchens/eating areas in square feet:	______
Combined area of living room, den, etc., in square feet:	______
Total area of all the above spaces in square feet:	______
Combined area of bathroom(s) in your home in square feet:	______
Total areas of all other living spaces in square feet:	______
GRAND TOTAL of all living spaces in square feet:	______
The number of people occupying the space:	______

Stretching Exercises

Do a survey in your neighborhood. Make a list of, or draw features of homes that are well-suited to the local climate.

ANSWERS

Physics To Go
(continued)

c) Student observation.

6. Each tile is 1 square foot, and, the total area is 96 square feet, therefore, the number of tiles is 96.

7. If each tile is now 2 ft on each side, each tile is now 4 sq. ft. Therefore, 96 sq. ft / 4 sq. ft, is 24 tiles needed to tile the floor.

8. Student activity.

© It's About Time

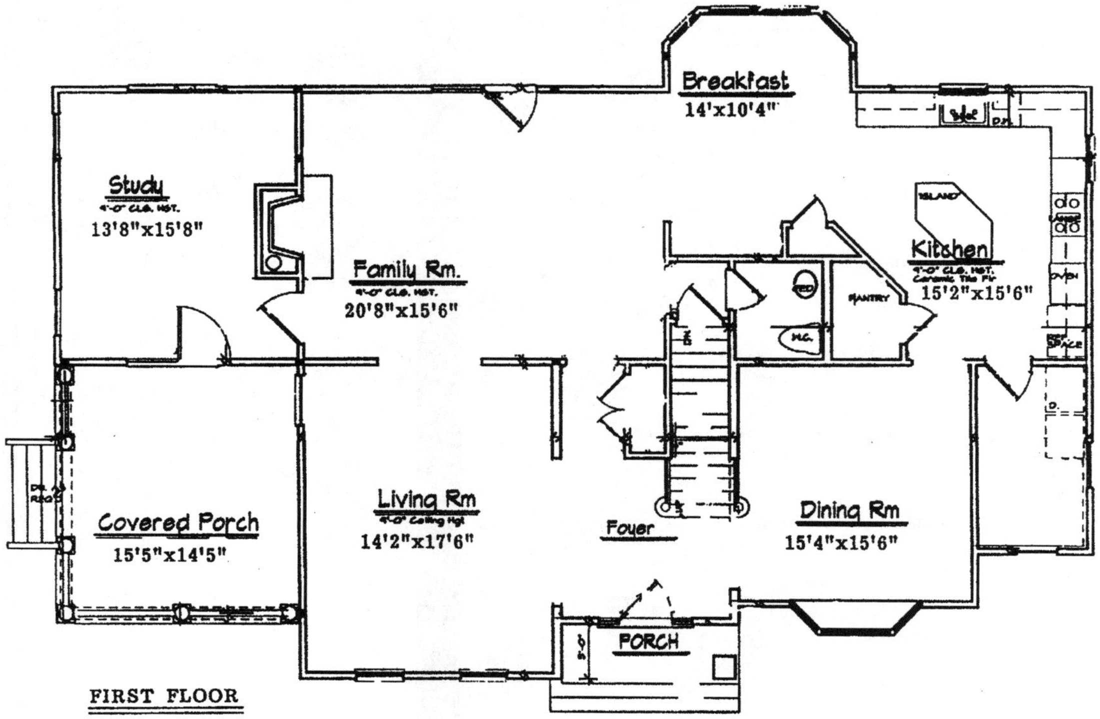

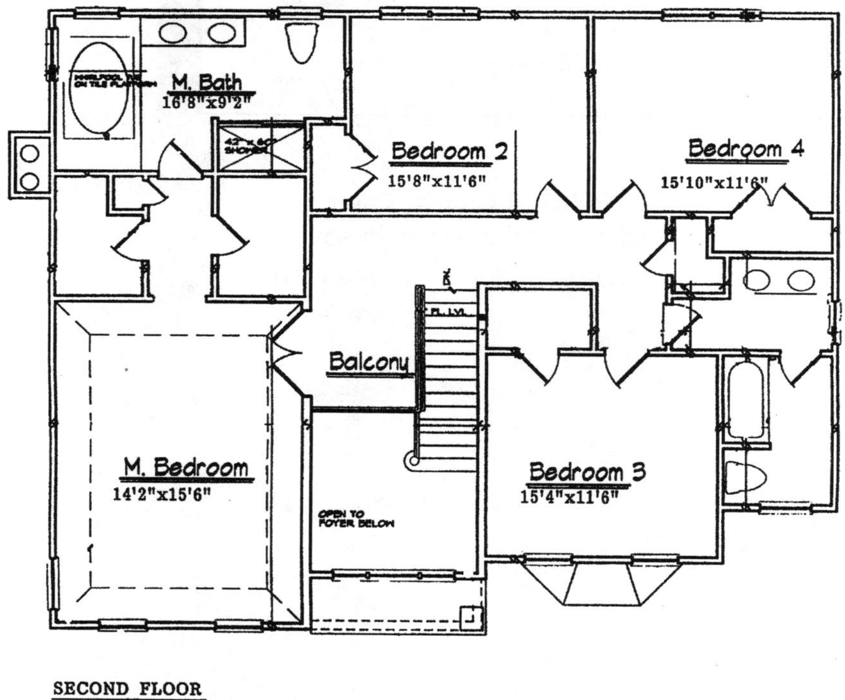

© It's About Time

For use with *Designing the Universal Dwelling*, Chapter 6, Activity 1: Factors in Designing the Universal Dwelling

Floor Area of a Home

The Total Floor Area of a Home

Combined area of bedrooms in square feet: ______
Combined area of kitchens/eating areas in square feet: ______
Combined area of living room, den, etc., in square feet: ______
Total area of all the above spaces in square feet: ______

Combined area of bathroom(s) in your home in square feet: ______
Total areas of all other living spaces in square feet: ______

GRAND TOTAL of all the living spaces in square feet: ______
The number of people occupying the space: ______

For use with *Designing the Universal Dwelling*, Chapter 6, Activity 1: Factors in Designing the Universal Dwelling

The Total Floor Area of a Home

Combined area of bedrooms in square feet: ______
Combined area of kitchens/eating areas in square feet: ______
Combined area of living room, den, etc., in square feet: ______
Total area of all the above spaces in square feet: ______

Combined area of bathroom(s) in your home in square feet: ______
Total areas of all other living spaces in square feet: ______

GRAND TOTAL of all the living spaces in square feet: ______
The number of people occupying the space: ______

For use with *Designing the Universal Dwelling*, Chapter 6, Activity 1: Factors in Designing the Universal Dwelling

© It's About Time

ACTIVITY 2
What is the "Right Size" for a Universal Dwelling?

Background Information

The main concept involved in **Activity 2** is a ratio, the "area per person" calculated by dividing the total surface area of the floor(s) of a home by the number of persons living in the home.

Area per person (ft^2/person) = Floor area (ft^2) ÷ Number (person)

Keep in mind that this ratio involves the interplay of two quantities, the floor area and the number of persons living in the home, and may be tricky. For example, it is possible that a big home having a very large floor area and many family members may have a smaller "area per person" ratio than a small home with few family members.

The class average of this ratio also is involved in this activity, and it is good to keep in mind that the average, or arithmetic mean (add all the values and divide by the number of values) is not always the best indicator of central tendency. It is possible, for example, that the class's average value of area per person may not be at all representative of the typical home for class members. This could happen if, for example, only a few homes had extremely large values compared to the rest.

If the class's average value of area per person does not seem meaningful, you could explore the possibility of using other indicators of central tendency, such as the median (halfway between the smallest and largest values) or the mode (the most frequent value).

Once a decision is made about the desired value for the area per person ratio for a home, the total floor area needed can of course be calculated by multiplying the value of the ratio by the number of persons who will live in the home. Similarly, the range of floor areas of homes needed for various family sizes can be calculated by multiplying the area per person by the smallest and largest number of family members.

Active-ating the Physics InfoMall

One of the concepts in this activity is the concept of scale. Again, this is an area where students have known difficulty. Arons has done much of the work on student understanding, and this concept is not excepted. Section IV of Arons' article "Student patterns of thinking and reasoning, part one of three parts," (*The Physics Teacher*, vol. 21, issue 9) is on scaling and functional reasoning, such as how area and volume depend on linear dimensions. Arons returns to this topic in *A Guide to Introductory Physics Teaching* (in the Book Basement), in "Underpinnings" section 1.12.

For discussions on interpolation and extrapolation, you can perform a search using either of these as keywords. One of the hits is Chapter 11 of the textbook *Physics for the Inquiring Mind.* This is a short passage, but discusses some of the pros and cons of using limited data to form predictions.

Planning for the Activity

Time Requirements

Allow one class period for this activity. If you do the **Stretching Exercise**, allow at least one more class, or possibly two, depending on the resources available to you.

Materials Needed

For each group:

- calculator, basic

Advance Preparation and Setup

Have some sample cards made up, with the calculation of the average number of square feet per person. Ave = Total sq. ft. / number of people.

© It's About Time

Teaching Notes

Ask students to be ready to hand in the 3 × 5 card on which they have recorded data about their individual homes (be sure to announce that anonymity will be preserved). Collect the cards, and mix them up. Read the grand total for the living areas in square feet, and the number of people living in those homes. Guide the students through **For You To Do**.

© It's About Time

Activity Overview

In this activity, the students will be looking at floor plans, and calculating the space required for individuals. They will be applying mathematical understanding of area, and exploring the relationship between total living area or space, and number of individuals in the home.

Student Objectives

Students will:

- Calculate the average number of square feet per person for a sample of homes represented in the student's class.
- Recognize ranges in family size and the number of square feet per person in the class's sample of homes.
- Use data on the class's sample of homes and family sizes to decide upon the size needed for a "basic" universal dwelling, and decide how the size of the dwelling will vary for large and small families.

ANSWERS FOR THE TEACHER ONLY

What Do You Think?

Answers will vary according to the students' environment. For example, students who have lived in apartments all their life will have a different concept of space than the students raised on the wide open prairie on a ranch or a farm. This is a good opportunity to initiate a discussion regarding their preconceptions of living space.

Designing the Universal Dwelling

Activity 2 What is the "Right Size" for a Universal Dwelling?

GOALS

In this activity you will:

- Calculate the average number of square feet per person for a sample of homes represented in your class.
- Recognize ranges in family size and the number of square feet per person in the class's sample of homes.
- Use data on the class's sample of homes and family sizes to decide upon the size needed for a "basic" universal dwelling, and decide how the size of the dwelling will vary for large and small families.

What Do You Think?

In some prisons in the United States, two inmates occupy an 8-foot by 10-foot cell. On the other hand, some people live in mansions of 10,000 to 20,000 square feet!

- **How much space do people need in their shelters to have a decent, humane life?**

Record your ideas about this question in your *Active Physics* log. Be prepared to discuss your responses with your small group and the class.

For You To Do

1. Give the "Total Floor Area of a Home" index cards that you completed for today's class to your teacher. He or she will then read aloud the Grand Totals of the living spaces in square feet and the number of people who live in some of the homes.

400

Active Physics

© It's About Time

a) Enter these numbers in your log.

b) Calculate the average number of square feet per person for at least five of the homes in your community.

2. Discuss the answers to the questions below with your group and then record the answers in your log.

a) How much total living space for all functions should your universal dwelling have per person?

b) Could you design a universal dwelling that would combine some living space functions? If so, how might you do this?

c) Will bathroom space be necessary for your universal dwelling? If not, explain how you might modify your dwelling if no indoor plumbing will be available. (Indoor plumbing is not available to many people in many countries.)

d) What size family (number of people) and what total square footage will you use for designing your "basic" universal dwelling? Explain your group's reasoning for choosing these family and dwelling sizes.

e) Decide on two additional sizes for your universal dwelling, the first to accommodate a larger family and the second to accommodate a smaller family. What will the square footage and family sizes of these versions be? What is your group's reasoning for deciding upon these specific larger and smaller sizes?

Reflecting on the Activity and the Challenge

Based upon the data that your class has collected, and your group discussions about what is needed in terms of living space for a universal dwelling, your group and other groups in the class have probably decided on universal dwellings of various sizes. This is another step toward completing the **Chapter Challenge**. You didn't just draw a size out of a hat. Your decision on size can be defended in terms of the data that you have collected about the sizes of homes of class members. You have also included necessary functions and various family sizes as factors in your decision.

There are probably a range of appropriate answers, but there is also opportunity to make changes and improvements as you proceed.

ANSWERS

For You To Do

1. a) Student data.

   b) Ave = Total sq. ft. / number of people (for example: total square feet for one house is 1780 sq. ft and five people live there. Therefore, the average square feet per person is 1780 / 5. The average is 356 sq. ft. per person.

2. a) Student data.

   b) Student data; Some students might have the bedroom and living room as one space (hide-a-bed).

   c) Student data; Some may realize that the toilet must be outside, as the smell inside the house might be rather noxious.

   d) Student data.

   e) Student data.

© It's About Time

Physics To Go

1. Make a floor plan drawing of a room in your home, or someone else's home, or in your school. A floor plan drawing shows what the room looks like when viewed from above. It should show the shape of the room and the positions and sizes of the doors and windows. You will need a scale for your drawing, for example, one square on a sheet of graph paper could equal one square foot.

2. Make a vertical cross-section drawing of the same room you used in **Question 1**. A vertical cross-section drawing should show what the room looks like when viewed from the side, including how high it is, and the shape and size of the windows and doors.

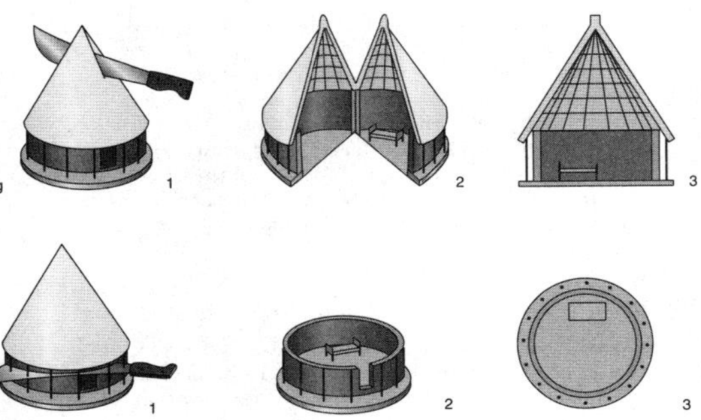

3. A house lived in by an elderly man and woman has 1500 square feet of floor area. Calculate the individual living space in square feet per person.

4. The house described in **Question 3** is sold to a young man and woman who have three children. Calculate the individual living space for the new family.

ANSWERS

Physics To Go

1. Students' sketches.

2. Student activity.

3. 1500 sq. ft. / 2 people = 750 sq. ft. per person.

4. 1500 sq. ft. / 5 people = 300 sq. ft. per person.

© It's About Time

Activity 2 What is the "Right Size" for a Universal Dwelling?

5. Here is a floor plan for houses being built for some people in southern India.
   a) Calculate the number of square feet of living space the house has.
   b) If the average family size is 6 to 8 persons, what is the average living space per person?

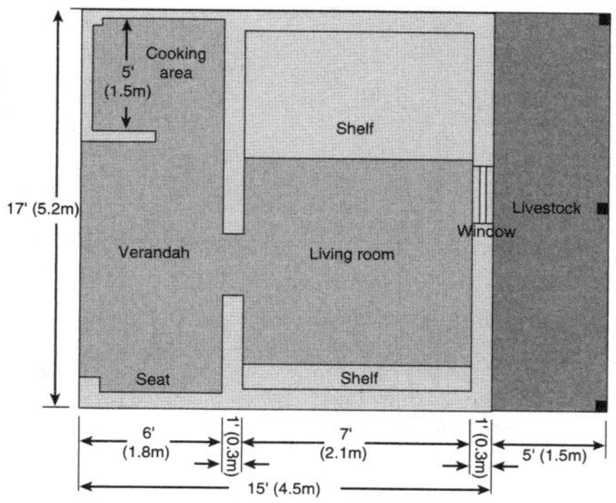

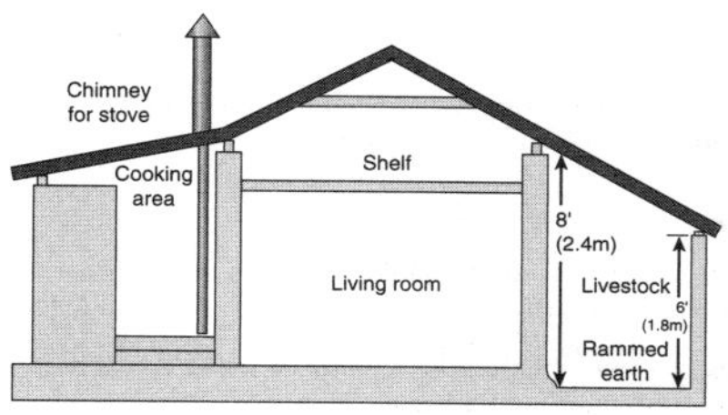

Typical family house—vertical cross-section drawing.

ANSWERS

Physics To Go *(continued)*

5. a) Area = $l \times w$; Therefore, Area = 17' × 15' = 255 sq. ft.

   b) 221 sq. ft. / 6 people = 37 sq. ft. per person;
   255 sq. ft. / 6 people = 43 sq. ft. per person;
   255 sq. ft. / 8 people = 32 sq. ft. per person.

© It's About Time

6. Look at the floor plan from Malawi of "a house that grows." It allows the homeowner to add rooms to the house as the household income or the family grows.

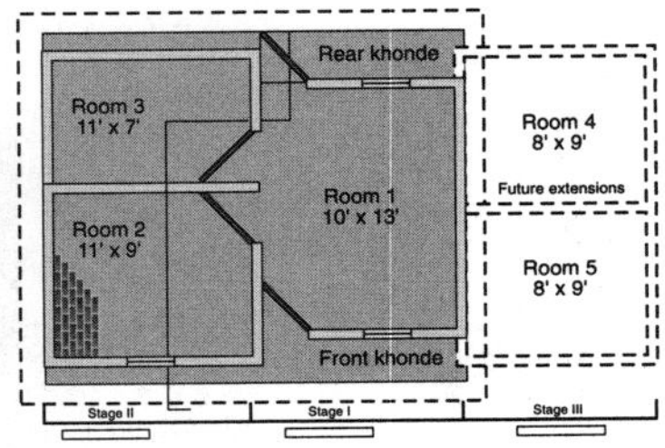

a) Calculate the number of square feet per person for a newly married couple who builds Stage I (Room 1).
b) The couple has two children and decides to build Stage II (Rooms 2 and 3). Calculate how many square feet there are per person for this family of four.
c) The couple has two more children and then builds Stage III (Rooms 4 and 5). Calculate the number of square feet per person for this family of six.

7. Make a floor plan drawing and a vertical cross-section drawing for your universal dwelling. Show the total square footage of each room and the total square footage of the dwelling. Include a scale for your drawing.

Stretching Exercises

1. Do research in your school library or on the Internet to find information on the size of the "average American home."
2. Locate information on the amount of living space used by an average person in another country.

ANSWERS

Physics To Go *(continued)*

6. a) Area of Stage I = $l \times w$ = 13' × 10' = 130 sq. ft.

Ave = 130 sq. ft. / 2 = 65 sq. ft. per person.

b) Add to the total area, Area of Stage II = $(l \times w)$ + $(l \times w)$ = (11' × 7') + (11' × 9') = 306 sq. ft.

Ave = 306 sq. ft. / 4 = 79 sq. ft. per person.)

c) Add to the total area, Area of Stage III = $(l \times w)$ + $(l \times w)$ = (8' × 9') + (8' × 9') = 450 sq. ft.

Ave = 450 sq. ft. / 6 = 75 sq. ft. per person.

7. Student activity

© It's About Time

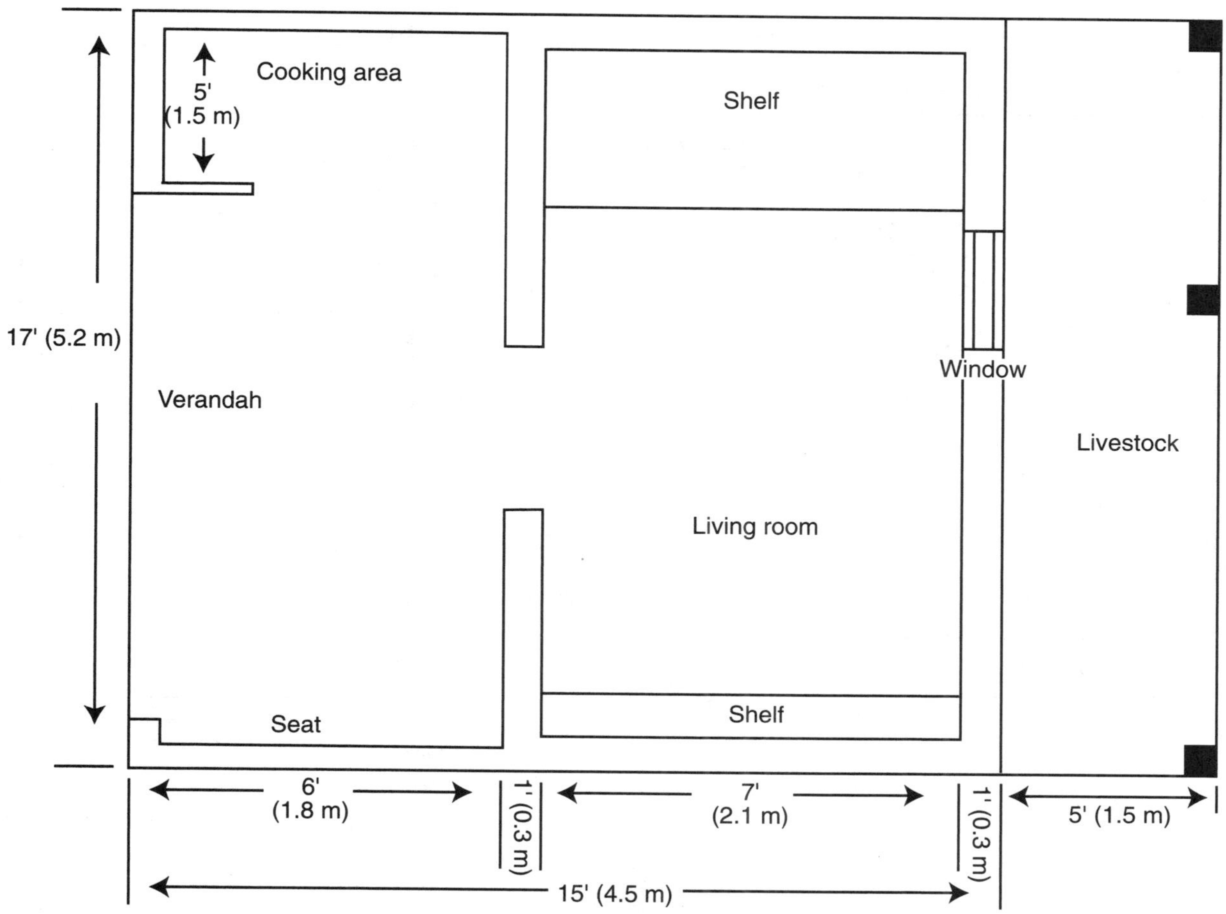

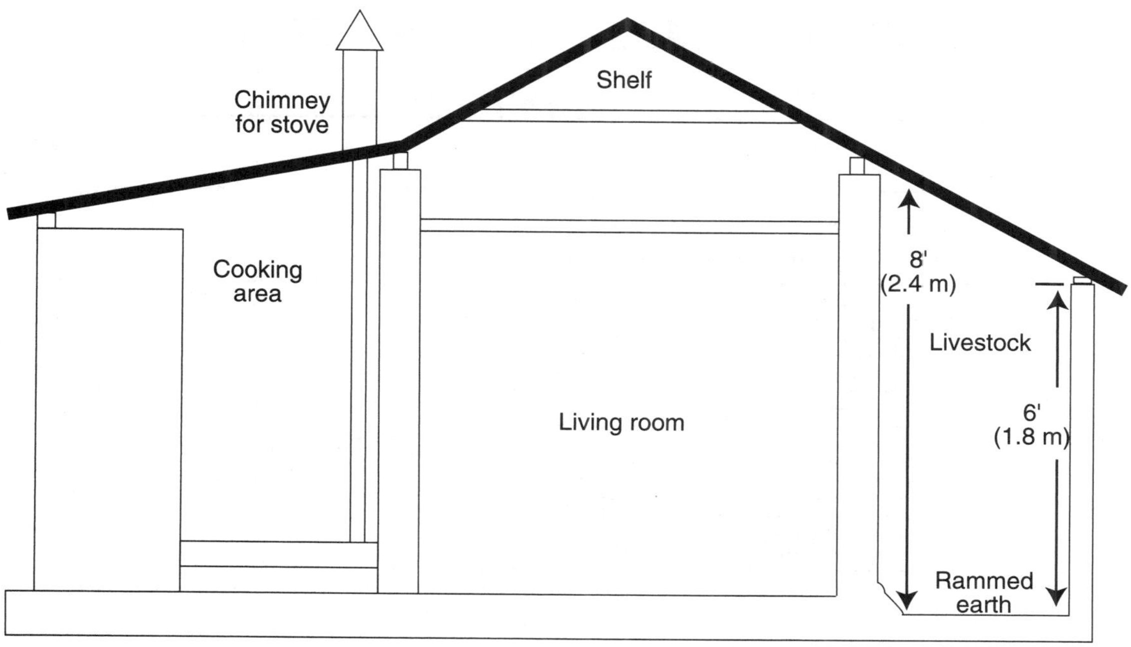

© It's About Time

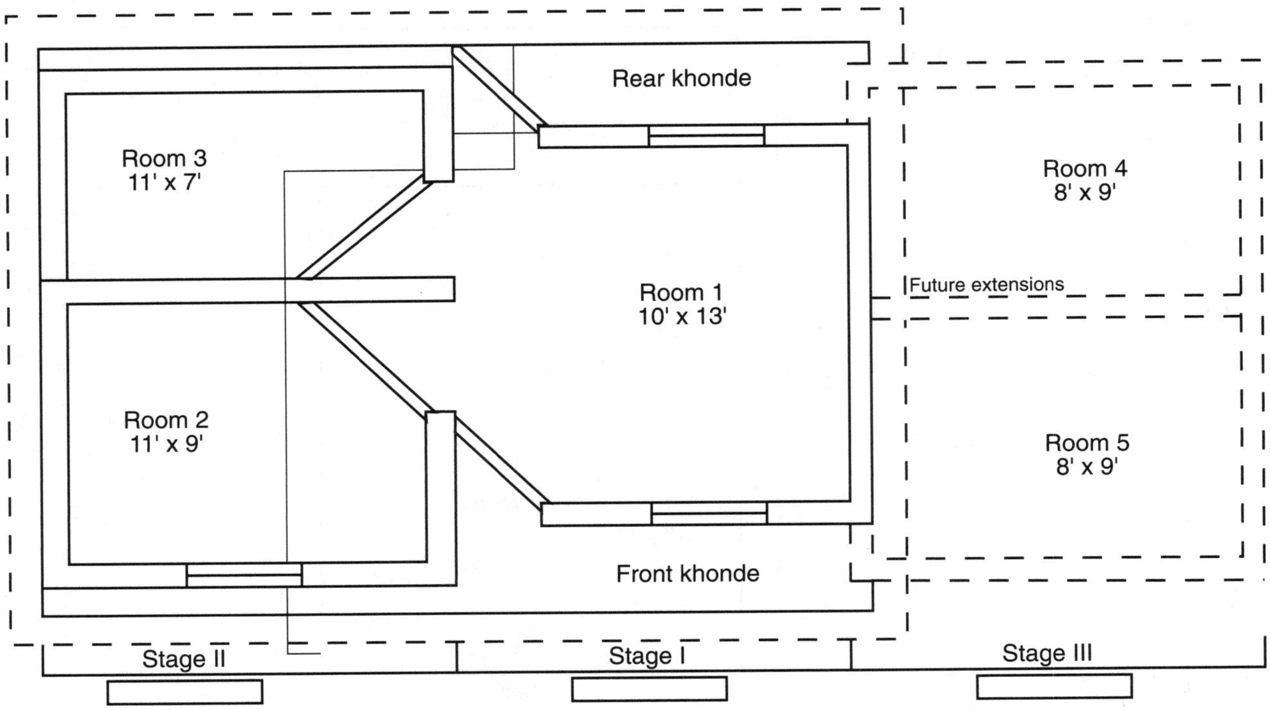

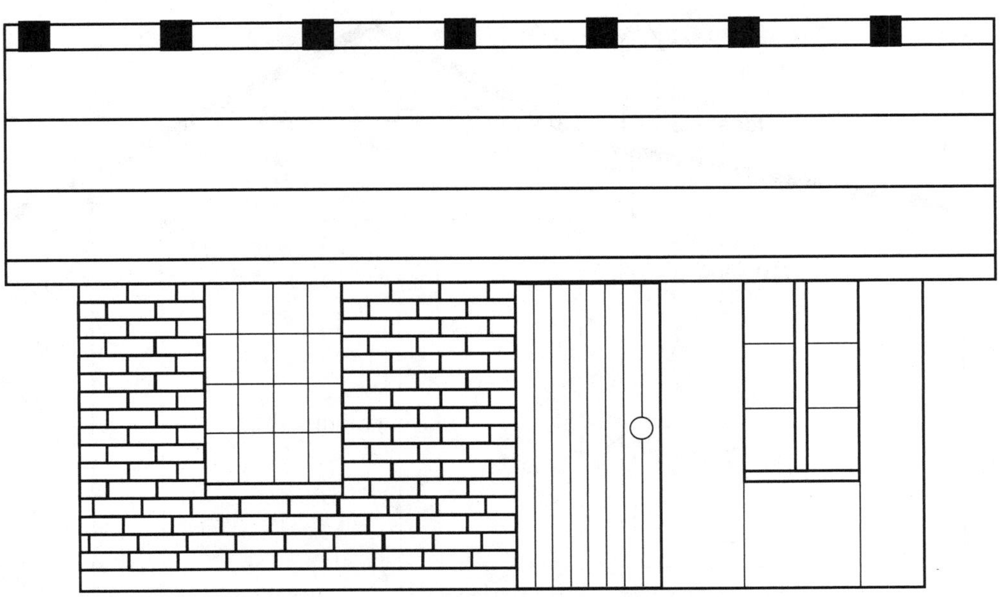

© It's About Time

NOTES

© It's About Time

ACTIVITY 3
The Shape of the "Universal Dwelling"

Background Information

Three mathematical relationships which affect the cost and performance of homes are introduced in **Activity 3**:

If the total floor area of a home is held constant, the total perimeter of the home (the total length of exterior walls) depends on the two-dimensional shape of the floor area.

The ratio of the exterior surface area of a home to the three-dimensional volume enclosed by the home, called the surface-to-volume ratio, depends on the three-dimensional shape of the home.

If the three-dimensional shape of a home is held constant, the surface-to-volume ratio of the home depends on the overall size of the home.

The dependence of shape on perimeter (area constant). For home designs limited to rectangular floor plans (four sides, all corners right angles) the smallest perimeter results for a square floor plan. For unlimited shapes the smallest perimeter for a given area is provided by a circular floor. Ideally, a circular house would require the least building material for exterior walls and, for vertical walls of standard height and thickness, would provide minimum surface area for heat escape through the walls.

The dependence of shape on surface-to-volume ratio. For home designs limited to three-dimensional shapes resembling rectangular solids (rectangular surfaces on all sides) the smallest surface area for a particular enclosed volume (and, therefore, the smallest surface-to-volume ratio) results for a cubical shape. For unlimited shapes (assuming that a flat floor is desired) a hemisphere (dome) shape would have the least surface-to-volume ratio. Consideration of the three-dimensional shape of a home rather than only the two-dimensional shape of the floor plan allows optimization of building materials and heat loss; a dome, although complex to construct, would provide the least building materials and the least surface area for heat loss.

You may recall that the example in nature which demonstrates the sphere as having the least surface-to-volume ratio is the raindrop; the strong cohesive forces of water molecules cause the surface of a raindrop to shrink to the smallest possible size, resulting in a sphere.

The below equations apply to calculating the areas and volumes of three-dimensional shapes which are likely to be involved in this activity. The prism is included for cases where students may choose home designs which involve pitched or sloped roofs.

Volume of rectangular solid = Length × Width × Height

Surface area of sphere = 4 × π × (Radius)^2

[divide by 2 for hemisphere]

Volume of sphere = 4/3 × π × (Radius)^3

Surface area of cylinder including circular ends

= 2 × π (Radius) + [(Radius) + (Length)]

Volume of cylinder = π × (Radius)^2 × Length

Volume of prism = Area of triangular side × Length of prism

= (1/2 × Base × Height) × Length

If dimensions are measured in feet of length, calculated areas and volumes will be in units of, respectively, square feet (ft^2) and (ft^3). In this application surface-to-volume ratio would have dimensions of "square feet per cubic foot" (ft^2/ft^3); mathematically, this dimension could be reduced to reciprocal feet (1/ft), but it has more meaning to retain the entire statement "square feet per cubic foot."

The dependence of surface-to-volume ratio on size. As shown immediately above in the discussion of units for expressing surface-to-volume ratio, the value of the ratio varies inversely with the fundamental dimensions, or scale of size, of a three-dimensional object of a given shape (see the **Stretching Exercise** for this activity in which this dependence is illustrated as a cube is scaled upward in size). Applied to a home design, this means, for example, that doubling all of the dimensions of the home (scaling upward by a factor of two) would decrease the home's surface-to-volume ratio by a factor of two. The "doubled" home would require more building materials, but not twice as much. Further, the larger home would have more heat loss through its surfaces than would a home half that size, but not twice as much. Therefore, some efficiency is gained with increased size. Similarly, there is a loss in efficiency upon scaling down; reducing the size of a home by a factor of two

© It's About Time

causes the surface-to-volume ratio to increase by the same factor.

The interplay between surface-to-volume ratio and scale of size is demonstrated in nature by the fact that small birds and mammals must ingest more food per unit of body volume each day—some consume many times their body weight each day—to offset the heat loss caused by the large surface-to-volume ratios of their small bodies. We humans, by comparison, don't need to eat as much on a "pound for pound" basis to maintain similar body temperature because, being of larger scale, our bodies have a comparatively smaller surface-to-volume ratio.

Active-ating the Physics InfoMall

In addition to the references mentioned earlier for area calculations, there is information on the InfoMall regarding student problems with area-to-volume ratios. If you perform a compound search for "area," "volume" and "ratio," you get too many hits. A good idea, as a first try, is to limit the search by indicating that the terms must appear in the same paragraph (alternately, you could limit which stores on the InfoMall are searched and reduce the number of hits as well). Among the hits is another from Arons, "Thinking, reasoning and understanding in introductory physics courses," *The Physics Teacher*, vol. 19, issue 3. Arons indicates that "The majority of U.S. college students in introductory physics courses have difficulty with this problem" of ratios.

Another hit from this list (*Physics: Foundations and Frontiers* in the Textbook Trove, look in the Biophysics section) tells how this area-to-volume ratio affects animals; smaller animals must have a high metabolic rate to keep their small volumes heated due to large heat loss through relatively large surface area. This compared well with **Physics To Go, Step 9.**

Planning for the Activity

Time Requirements

Allow at least 40 minutes for this activity, depending on how many different shapes are to be looked at and calculated.

Materials Needed

For each group:

- ball of string (225 ft.)
- calculator, basic
- 9 × 12 corrugated cardboard (2)
- graph paper pad of 50 sheets
- pushpins (100)

Advance Preparation and Setup

Have the materials available for the students, and depending on time restraints, have the 20 inches loops of string already closed and ready to use.

Teaching Notes

Introduce the term "optimize" and explain that the purpose of the **For You To Do** activity is to optimize the shape of the dwelling. Give the idea to the students that they should be "making the most with the least."

Some students may not understand the concept of surface area-to-volume ratio, and how it affects the design of the house. To help them understand, have them think of tents and how much material they would need as they increased the height of the walls, without increasing the floor space inside. They will come to see, as the tent gets higher and higher, there is greater surface area, and greater volume, but no difference in the floor space.

If 1-inch grid paper is not available, 1/4-inch grid or graph paper is okay. If using different grid paper, make the students aware of this change, and show how they can make 1-inch grid paper from different graph paper available.

© It's About Time

NOTES

© It's About Time

Activity 3 The Shape of the "Universal Dwelling"

GOALS

In this activity you will:

- Explain how the surface area of a two-dimensional figure is related to its shape when the perimeter is held constant.
- Calculate the surface-to-volume ratio of simple three-dimensional shapes.
- Relate the shape of a dwelling to optimization of living space for a limited amount of building materials.

What Do You Think?

Throughout the world, homes take on many different shapes.

- **Why are most homes in the United States built in rectangular shapes?**

Record your ideas about this question in your *Active Physics* log. Be prepared to discuss your responses with your small group and the class.

For You To Do

1. Make a closed 20-inch loop of string. You will also need pins and 1-inch square grid graph paper attached to a sheet of cardboard as shown on the next page.

2. Carefully place four pins at the intersections of grid lines. Run the closed loop of string around the pins. Adjust the

ANSWERS

For You To Do

1. Student activity.

2. a) Student activity and observations.

Activity Overview

In this activity, the students will look at the area of a house, and then compare it to the area needed for each individual in the house. They will determine the optimal size of floor space, while keeping the cost of the materials as low as possible.

Once they have looked at the floor space, they will look at the volume, and surface area. This is to aid them in determining more accurate costs for the building.

Chapter 6

Student Objectives

Students will:

- Explain how the surface area of a two-dimensional figure is related to its shape when the perimeter is held constant.
- Calculate the surface-to-volume ratio of simple three-dimensional shapes.
- Relate the shape of a dwelling to optimization of living space for a limited amount of building materials.

ANSWERS FOR THE TEACHER ONLY

What Do You Think?

Answers will vary. Generally, when building with straight lumber, the easiest structure to build is one with 90° angles. Hence the "box" we are familiar with. An interesting discussion can develop from this if there are students from other countries, or who have traveled to other countries. Ask the students what is the most common structure, and what is the most common building material.

© It's About Time

ANSWERS

For You To Do

(continued)

3. a) Student activity and observations.

4. a) All the rectangles should have equal perimeters.

   b) As the sides get longer and the perimeter stays the same, the area gets smaller.

   c) The size of rectangle which has the largest area of given perimeter is the square (5 × 5 × 5 × 5).

5. a) Student activity and observations.

   b) As the number of the sides increases, the enclosed area gets larger.

   c) Students' answers will vary. Some students will name large polygons (> 10 sides) as being the shape with the greatest area. However, it would become evident that as the shape approaches a circle, it becomes the shape with the largest area.

Perimeter	Length of Long Sides	Length of Short Sides	Area Inside Perimeter

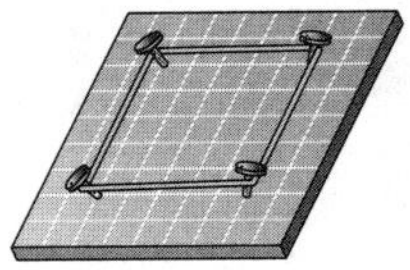

pins until you have constructed a rectangle on your grid paper. The loop should form a tight fit around the perimeter.

a) Count and record the number of 1-inch squares that lie within this perimeter. Record your data in a table similar to the one shown on the following page.

3. By carefully moving the pins, construct at least five more rectangles.

   a) Record the results in the table in your log.

4. Use the data you collected to help you answer the following questions:

   a) Which rectangle has the greatest perimeter, or do they all have equal perimeters?

   b) What happens to the area of the rectangle as one pair of sides gets longer while the perimeter stays the same?

   c) Which rectangle seems to have the largest internal area?

5. With your partner, use your loop of string and graph paper to determine how triangular (3-sided), pentagonal (5-sided), hexagonal (6-sided), octagonal (8-sided), and other shapes compare to that of a square of the same perimeter.

   a) Record your results in a table. (You can determine the area of the shapes by counting the number of boxes in the graph paper enclosed by the string. You will probably need to count half-boxes as well.)

   b) What appears to happen to the enclosed area as the number of sides increase?

   c) What shape do you think will generate the largest area for a given perimeter?

6. Extend your two-dimensional knowledge to three dimensions. Since homes are surrounded by walls and not string, you must consider how the information about enclosing the largest possible area with a loop of string can be used in designing your home.

© It's About Time

a) If you only had a limited quantity of material for walls to enclose a home, which shape would give the most living area? Why?

7. Meet with the other members of your group and, using the information you have gathered from these activities, choose a design shape for your universal dwelling. Your group will also have to determine the ceiling height, number of stories, and surface area-to-volume ratio for your basic dwelling design.

a) Record your group's decisions regarding these specifications in your log.

The shape of the universal dwelling will be:

The ceiling height of the rooms will be:

The living space of the dwelling in square feet will be:

The number of stories in the dwelling will be:

The volume of the dwelling will be:

Reflecting on the Activity and the Challenge

Your group has reached some tentative decisions about how many square feet of living space per person your universal dwelling should have. You have to be concerned about making sure that you provide enough living space. You also need to make sure that the building is not too expensive to produce. This means that a good design will not require a large building to contain the floor space.

In this activity, you explored how you may construct a dwelling with the required floor space, while keeping the dimensions of the building small and the amount of materials needed to construct the building as small as possible.

ANSWERS

For You To Do

(continued)

6. a) Students' answers will vary. The shape chosen should be a circle, as it takes the least amount of materials per maximum unit of floor space.

7. a) Student activity and observations.

© It's About Time

Now you have addressed the two most basic questions about the form of the universal dwelling: size and shape. You can defend your choice of shape in terms of keeping the cost of materials low. You are on your way to being able to present your plan to the Architectural Committee with confidence.

Physics To Go

1. The total area of all the faces of an object is called the surface area. To calculate the total surface area, find the area of each of the faces, and then add these areas.

   To find the surface area of the following cube, you can imagine unfolding the cube to form a flat pattern.

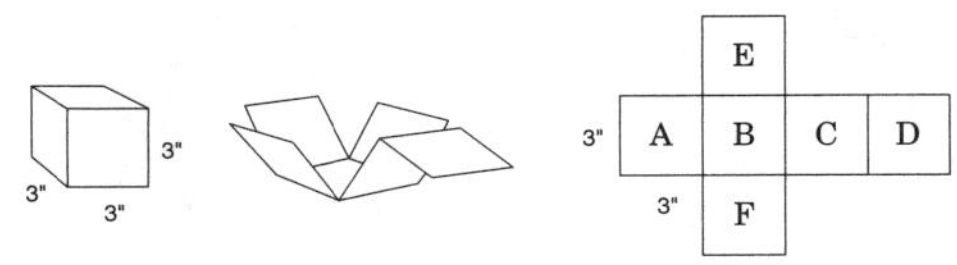

   a) What is the area of face A?
   b) What is the total surface area of the cube?

2. To find the surface area of the shape shown in the diagram below (a rectangular prism) you can once again imagine unfolding it to form a flat pattern.

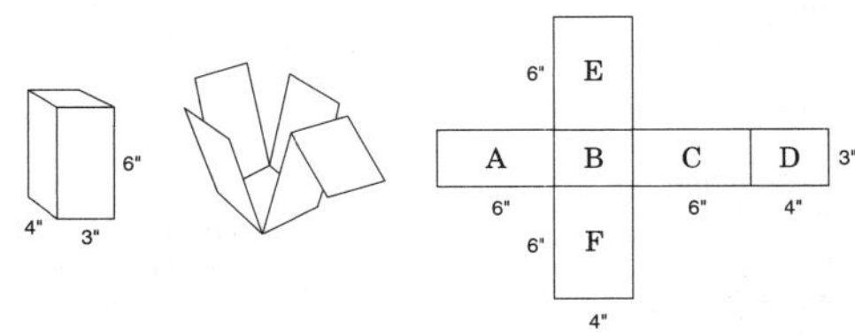

ANSWERS

Physics To Go

1. a) Face A has an area of 9 square inches.

   b) Total surface area is 54 square inches.

© It's About Time

a) Calculate the areas of A, B, C, D, E, and F.
b) What is the surface area of the shape?

3. Find the surface areas of the following:

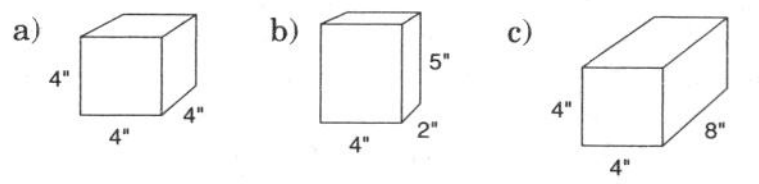

4. To calculate the volume of a rectangular shape you can use the following relationship:

Volume = length × width × height

Calculate the volume of each figure in **Question 3**.

5. Find the surface area-to-volume ratio of each figure in **Question 3**. (Divide the surface area by its volume.) Which figure has the greatest surface area-to-volume ratio? The lowest?

6. A house was designed with outside walls that form a square base 27 feet on each side.
   a) Calculate the perimeter (distance around) of the house in feet.
   b) Calculate the total floor area of the house in square feet.

7. A house was designed with an outside wall that forms a circle that has a radius of 17.2 feet.
   a) Calculate the circumference (distance around) of the house in feet. The equation for the circumference of a circle is:
   Circumference = 2 × π × radius.
   You can use a value of 3.14 for π.
   b) Calculate the total floor area of the house in square feet. The equation for the area of a circle is
   Area = π × radius × radius.

8. Use numbers to compare the houses in **Questions 6** and **7**.
   a) If the height of the outside walls of the two houses were equal, how would the amount of paint needed to cover the walls compare? How significant is the difference?

ANSWERS

Physics To Go
(continued)

2. a) Area A = 18 in^2;
   Area B = 12 in^2;
   Area C = 18 in^2;
   Area D = 12 in^2;
   Area E = 24 in^2;
   Area F = 24 in^2.

   b) Total Area = 108 in^2.

3. a) 6 sides × 4" × 4" = 96 in^2.

   b) 2 sides × 4" × 2" = 16 in^2;
   2 sides × 5" × 2" = 20 in^2;
   2 sides × 4" × 5" = 40 in^2;
   Total = 76 in^2

   c) 2 sides × 4" × 4" = 32 in^2;
   4 sides × 4" × 8" = 128 in^2;
   Total = 160 in^2

4. a) 4" × 4" × 4" = 64 in^3.

   b) 4" × 2" × 5" = 40 in^3.

   c) 4" × 4" × 8" = 128 in^3.

5. a) The surface area-to-volume ratio is surface area/volume = 96/64 = 1.5 to 1

   b) The surface area-to-volume ratio is surface area/volume = 76/40 = 1.9 to 1

   c) The surface area-to-volume ratio is surface area/volume = 160/128 = 1.3 to 1. **Figure b)** has the greatest (1.9 to 1) and **Figure c)** has the least (1.3 to 1).

6. a) Perimeter is sum of all sides, therefore the perimeter is 27' × 4 sides = 108'.

   b) The total floor area is l × w = 27' × 27' = 729 square feet.

7. a) Circumference = 2 × π × r =
   2 × 3.14 × 17.2 feet = 109.9 or 110 feet.

   b) Total floor area = π × r^2 =
   3.14 × $(17.2 \text{ feet})^2$ = 928.9 or 929 square feet.

8. a) If we use the height of the walls to be 8 feet, then the area of the walls will be the circumference × height = 108 × 8 = 864 sq. ft., and the circular house will be circumference × height, 110 × 8 = 880 sq.ft. There is not a significant difference between the two.

   b) There are 200 more square feet in the circular house than in the square house. Yes, this is significant.

Chapter 6

© It's About Time

c) As there is about 180 sq. ft. per person, there would be about four people in the square house and five people in the round house. It is very significant to the fifth person!

9. a) hut: 8 ft. × 8 ft. = 64 sq. ft., cabin: 16 ft. × 16 ft. × two floors = 512 sq.ft.

b) hut: 8 ft. × 8 ft. × 6 surfaces = 384 sq. ft.
cabin: 16 ft. × 16 ft. × 10 surfaces = 2560 sq. ft.

Therefore, the ratio is 2560/384 or approximately 6.6 times (cabin to hut).

c) Answer: hut: 8 ft. × 8 ft. × 8 ft. = 512 cu. ft.
cabin: 16 ft. × 16 ft. × 32 ft. = 8192 cu. ft.

Therefore, the ratio is 8192/512 or approximately 16 times (cabin to hut).

d) hut: 384/512 = 0.75
cabin: 2560/8192 = 0.3125
Therefore, the hut would have the greatest possibility of losing heat through its surfaces.

10. Student activity.

c) If living space of at least 180 square feet per person is needed, how many persons could live in each house? How significant is this difference?

9. A research outpost for a scientist in a remote part of Alaska has a hut shelter which is a cube eight feet on each side. The scientist sometimes visits a nearby family who live in a two-story cabin also shaped as a cube. The cabin's dimensions are double the dimensions of the hut, measuring 16 feet on each side. Use the numbers above to make the following comparisons:

a) Compare the total floor areas of the two dwellings as a ratio. (The cabin has an upstairs level.)
b) Compare the outside surface areas of the two dwellings including the bottom floor of each. Express the comparison as a ratio.
c) Compare the volumes of the two dwellings as a ratio.
d) Calculate the surface area-to-volume ratio for each dwelling. (Divide the outside surface area of each dwelling by its volume. The answers will be in units of square feet per cubic foot.) Which dwelling has the greatest possibility for heat loss through its surfaces per cubic foot of inside volume?

10. Using the information you gathered in this activity, modify your floor plan drawing and section drawing for your universal dwelling. Show the total square footage of each room and the total square footage of the dwelling. Include a scale for your drawing.

© It's About Time

Square Grid Paper

For use with *Designing the Universal Dwelling*, Chapter 6, Activity 3: The Shape of the "Universal Dwelling"

© It's About Time

Activity 3 A

Exploring Surface Area-to-Volume Ratio

FOR YOU TO DO

1. The surface area-to-volume ratio changes with the scale of an object. Begin with a unit cube (1 × 1 × 1). This cube is one unit long on each side.

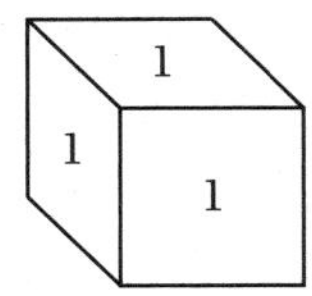

a) Calculate its volume.

b) Calculate its surface area.

c) What is its surface area-to-volume ratio?

2. This cube is two units long on each side.

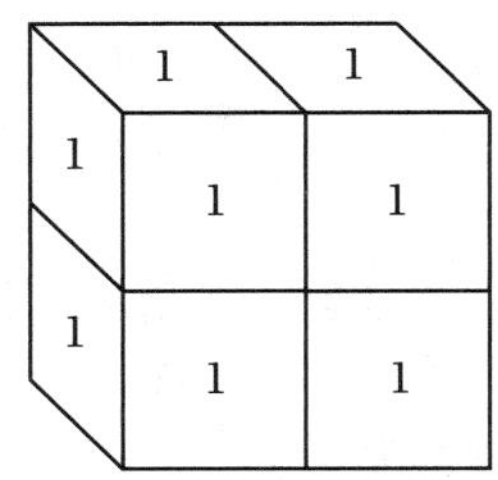

a) Calculate its volume.

b) Calculate its surface area.

c) What is its surface area-to-volume ratio?

3. Calculate the surface areas, volumes, and surface area-to-volume ratios of three-, four-, and five-unit cubes.

a) 3-unit cube: surface area: _______ volume: _______
surface area-to-volume ratio: _________

b) 4-unit cube: surface area: _______ volume: _______
surface area-to-volume ratio: _________

c) 5-unit cube: surface area: _______ volume: _______
surface area-to-volume ratio: _________

4. On a separate graph, plot the change in surface area-to-volume ratios against unit cube size.

5. If heat energy radiates equally out of each side of a cube, how does heat loss change as the cube increases in unit size? Why do small birds and mammals spend so much of their time eating?

© It's About Time

For use with *Designing the Universal Dwelling*, Chapter 6, Activity 3: The Shape of the "Universal Dwelling"

NOTES

© It's About Time

ACTIVITY 4
Solar Heat Flow in the "Universal Dwelling"

Background Information

The concepts and skills introduced in **Activity 4** fall into two areas:

- producing a model of a three-dimensional object to a specified reduced scale of size.
- producing and interpreting a temperature versus time graph from temperature measurements made inside a model home which is first heated by an external source over time and then allowed to cool over time.

Modeling to a specified scale. The 1:48 scale specified for students to use for constructing their model homes in **For You To Do** was chosen to provide models of reasonable size for this and subsequent activities. It is the "shrink factor" which would allow, for example, that a home 48 feet long would have its length reduced to 1 foot in a model made this specified scale. As explained to students in the text, this also means that one inch on the model would represent 4 feet on the real home. The same scale can be expressed as 1/4 inch representing one foot:

1 foot: 48 feet = 12 inches: 48 feet = 1 inch : 4 feet = 1/4 inch : 1 foot

This is a standard choice of scale often used by architects and home designers who refer to it as a "one-fourth inch scale," carrying the silent meaning that one-fourth inch represents one foot in the real world. This particular scale offers the advantage that scale drawings of many homes, or rooms within large homes, will fit on standard sized sheets of paper; universally available 8-1/2" × 11" paper printed with a 1/4" grid is very handy for sketching plans to this particular scale. Other scales commonly used for home design include 1/8" and 1/16". Inexpensive computer software for do-it-yourself home designers are available for producing drawings to these common scales and may be useful for your students to render and refine plans for their universal dwellings. The technical education teachers at your school may be able to provide help in the area of computer-assisted drawing; this would be a good opportunity for interdisciplinary cooperation.

Heating/cooling curves. The background information presented here will apply to all of the activities remaining in this chapter. It is critical for you as the teacher to understand the principles presented here.

Before proceeding, is suggested that you read the **For You To Read** for **Activity 5** in the student text to refresh your knowledge of conduction, convection and radiation as heat transfer mechanisms; please do that reading now.

Having reviewed the three means by which heat is transferred in the reading suggested above, you, as the teacher, should have awareness that all three mechanisms will be involved when students heat and cool their model homes. During the heating phase, the heat from the lamp will be transferred primarily through infrared radiation to the outside surfaces of the model home which face the lamp . The heat next will conduct through the tagboard walls and roof to warm the inside surfaces which, in turn, will warm the air in contact with the surfaces. Finally, heat will be delivered to the thermometer at the center of the model home by convection through the air inside the model. The cooling phase will involve a similarly complex reverse chain of heat transfers involving all three mechanisms. Don't be fooled into believing that the heating and cooling processes in this exercise are simple, because they aren't. More than one mechanism is sure to be active at any given time; for example, final cooling of the outside surfaces of the home will be accomplished as the surfaces transfer heat to the air through simultaneous conduction by contact and emission of radiation.

For your background only, and not suggested to be treated with students, the rate of conduction of heat through a solid which would apply only to that part of the chain of heat transfers in this activity which involves the conduction of heat through the tagboard walls/roof is governed by the equation:

Rate of heat transfer through a material = $kA(\Delta T)/d$

where k represents the heat conductivity of the material, and A is the common area across which a temperature difference ΔT exists between surfaces of the material separated by thickness d.

The main factor which will affect the rate of heat transfer to and from the model home will be the difference in the temperature of the illuminated outside surfaces of the model and the temperatures of the inside surfaces and the air inside the model. The greater the temperature differences are at a particular time, the greater will be the rate of heat

© It's About Time

transfer at that time; the less the temperature difference, the less the rate of transfer.

Because the temperature difference between the outside and the inside of the model home diminishes during heating, the rate of heat transfer also can be expected to diminish. Therefore, the temperature versus time graph can be expected to climb rapidly at first, continue to climb with decreasing slope and then level off to plateau at an equilibrium temperature.

The equilibrium temperature, of course, is explained by the fact that, once the interior of the model home has been heated, it simultaneously cools, or loses heat, through the surfaces not exposed to the lamp because, for those surfaces, the inside temperature is higher than the outside temperature The equilibrium temperature is reached when the model is losing heat to the surrounding cool air at the same rate that it is receiving heated air from the lamp.

The cooling phase depends on the difference in the temperature of the air inside the model and the temperature of the air surrounding the model (the ambient temperature of the room). This temperature difference will diminish during cooling, so the cooling curve can be expected to fall rapidly at first, continue to fall with decreasing negative slope, and finally plateau at room temperature.

The key to interpreting the behavior of rates of heating and cooling over time is to focus on the slope of each temperature versus time graph. How the slope changes with time tells the story about how the rate of heating or cooling changes with time.

Be sure to have students carefully measure and record the position and orientation (distance and angle) of the heat lamp relative to the model home so that the setup can be replicated as called for in activities later in this chapter. It will be desirable that the heat input provided by the lamp will remain constant (not enter as a variable) in those future activities.

Active-ating the Physics InfoMall

The physics topics (solar energy and heat flow) in this activity were already discussed above. However, students are asked to construct graphs now. Search the InfoMall with the keywords "student difficult*" and "graph*". Note that the asterisks are wild-characters - "graph*" can be "graph", "graphs" and "graphing", for example. You will find several articles that discuss the difficulties students have interpreting graphs. However, most of the research done in this area (in physics, that is; math research may also report on general graphing difficulties, but math research does not appear on the InfoMall) is on kinematics. Nonetheless, the results may be enlightening. Students do not always understand what lines on a graph mean, and what slopes indicate, for example.

Planning for the Activity

Time Requirements

This activity can be completed within one period (40 - 50 minutes). Allow an extra 15 - 20 minutes if the students need to construct their houses. The graphing of the data may take up to 20 minutes to complete, or may be assigned for homework. If it is assigned for homework, ensure the students know how to label and plot the points on a graph. Allow 20 minutes to 30 minutes to guide the students in producing a graph from common data for the whole class if needed.

Materials Needed

For each group:

- 120-Volt clip-on fixture for heat lamp with shade
- 250-Watt reflective heat lamp
- calculator, basic
- electric clock
- file folders (24)
- inch scale ruler
- hardwood meter stick (100 cm)
- masking tape (3/4" × 60 yds.)
- retractable knife
- stand for clip-on lamp
- thermometer

Advance Preparation and Setup

Almost any heat lamp will do. A commonly available 250-W clear infrared heat lamp with a built-in reflector is recommended; a 125-W bulb of the same design will work, but will provide a lower rate of heating. If lamp fixtures with parabolic metal reflectors are available at the school, "Crystal Clear" (unfrosted) light bulbs available in 150-, 200-, and 300-W sizes also will work (these bulbs, which have

© It's About Time

a straight filament aligned coincident with the bulb's axis of symmetry also are useful as intense "point" sources of light (filament viewed on end) for other laboratory activities.

Teaching Notes

The activities for this class session are very intensive. Student teams construct their scale model homes according to their prior specifications and then conduct the heating and cooling experiment on their model. You should have all materials they will need available and well-organized to ensure that the teams will be able to complete the activity. Begin the class by immediately having the student groups discuss **What Do You Think?** Next, focus their attention on the model-building and heating/cooling phases of **For You To Do.** Check that the 1:48 scale of size is understood and encourage student to monitor time as they work so that they will complete the tasks.

SAFETY PRECAUTION: Care should be taken when cutting the holes in the cardboard house. Also, remind the students to handle the thermometer carefully, and if it should break, to inform the teacher immediately.

© It's About Time

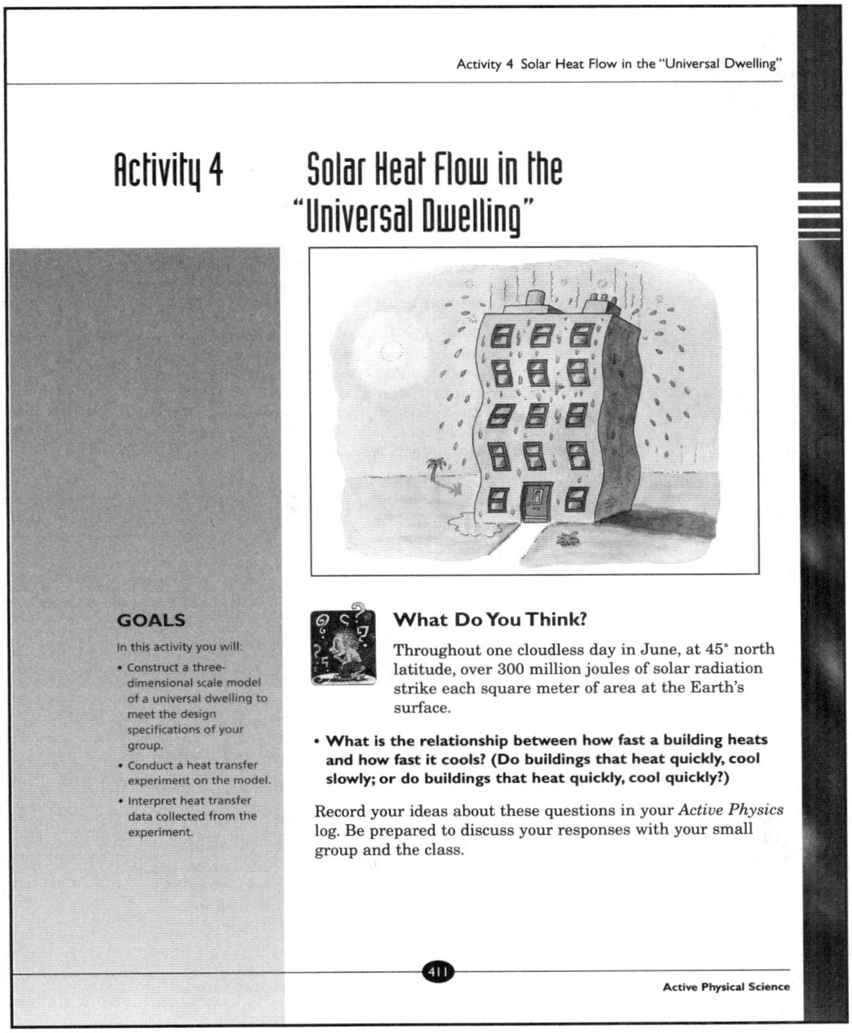

Activity 4 Solar Heat Flow in the "Universal Dwelling"

Activity 4 Solar Heat Flow in the "Universal Dwelling"

GOALS

In this activity you will:

- Construct a three-dimensional scale model of a universal dwelling to meet the design specifications of your group.
- Conduct a heat transfer experiment on the model.
- Interpret heat transfer data collected from the experiment.

What Do You Think?

Throughout one cloudless day in June, at 45° north latitude, over 300 million joules of solar radiation strike each square meter of area at the Earth's surface.

- **What is the relationship between how fast a building heats and how fast it cools? (Do buildings that heat quickly, cool slowly; or do buildings that heat quickly, cool quickly?)**

Record your ideas about these questions in your *Active Physics* log. Be prepared to discuss your responses with your small group and the class.

411 Active Physical Science

Activity Overview

In this activity the students will be performing experiments to simulate the heating of a building during the noonday Sun, and cooling of the building at night (when the heat source is removed).

Student Objectives

Students will:

- Construct a three-dimensional scale model of a universal dwelling to meet the design specifications of student groups.
- Conduct a heat transfer experiment on the model.
- Interpret heat transfer data collected from the experiment.

Chapter 6

ANSWERS FOR THE TEACHER ONLY

What Do You Think?

Some students may think that some buildings will heat up quickly and cool down slowly, and vice versa. However, the rate of heating/cooling has to do with the change in temperature between the building materials, and the environment. The rate of absorption is also dependent upon the types of materials with which the building is constructed, the texture of the building, how much glass is in the south side of the building, etc.

© It's About Time

ANSWERS

For You To Do

1. Student activity.

2. Student activity.

3. Student activity.

For You To Do

1. Using the resources that have been provided, construct a scale model of your universal dwelling using the design specifications for the basic unit. The model home should be built on a scale of 1:48. That means that one inch on the model will equal four feet, or 48", on the real home. In order to build the model quickly, build only the walls and roof of your home. Don't worry about putting a bottom (floor) on the model or placing windows and doors at this time.

2. Using the figure below, assemble a heat lamp, ring stand support and clamp, timer, measuring tools and thermometers. Conduct a timed heat-transfer experiment of 20 min. Place the nearest edge of the model 20 cm from the bulb of the heat lamp, with the lamp at approximately 45° from the center of the model home. (A 45° angle is halfway between the vertical and the horizontal.)

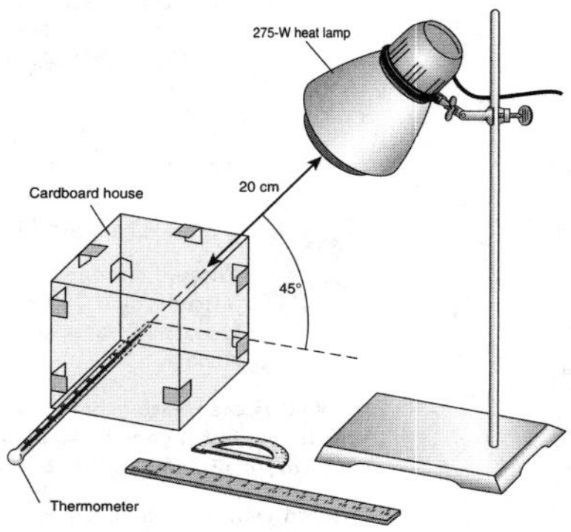

If the thermometer should break, *immediately* notify your instructor.

3. Carefully make a small hole in one wall. After making the hole, place the thermometer (or temperature probe) into one side of your model so that the bulb or probe is located in the dead center of the model. Be sure you can read the thermometer once it is in place!

© It's About Time

Activity 4 Solar Heat Flow in the "Universal Dwelling"

4. When you have your model, the thermometer, and the heat lamp placed properly, have your teacher check it before you begin heating.
   a) Record the temperature of the thermometer before heating and record this as the temperature at 0 time.
   b) Turn on the heat lamp and, after 30 s, record the temperature.
   c) Continue this procedure for 10 min, recording temperature at each 30-s interval.

Caution: Heat lamps get very hot. Be careful not to touch the bulb or housing surrounding the bulb.

5. Turn off the lamp at the end of the 10-min heating phase.
   a) Continue to record the temperature every minute after the lamp is turned off for another 10 min.
   b) Calculate and record the change in temperature during each timed interval of heating and cooling.
   c) Graph your recorded data (temperatures against time) on the graph paper provided by your teacher. Your time axis should extend from 0 min (when you began) to 20 min (after the experiment was completed), and be placed on the *x*-axis.

6. Use the graph and the data you recorded to answer the following questions:
   a) What was the total temperature increase in your model home during the heating phase?
   b) What was the total temperature decrease in your model home during the cooling phase?
   c) Was the rate of temperature increase constant? How do you know?
   d) Was the rate of temperature decrease constant? How do you know?
   e) What time of day did the heating phase represent?
   f) What time of day did the cooling phase represent?
   g) What does your experiment suggest in regard to the potential of your model home to be heated by solar energy?

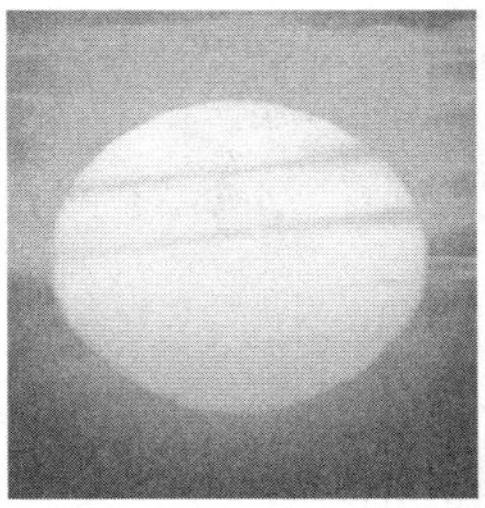

413 Active Physical Science

Answers

For You To D (continued)

4\. a)-c) Student activity and observations.

5\. a)-b) Student activity and observations.

c) Student activity and observations. The graph should have fairly even heating and cooling. Depending on the types of materials, the heating curve and cooling curve should be approximately the same.

6\. a)-b) Student data.

c) Student data. If the line is straight, it is a constant change. If it is a curve, it is not a constant change.

d) Student data. If the line is straight, it is a constant change. If it is a curve, it is not a constant change.

e) Heating phase represents the daytime heating.

f) Cooling phase represents the night.

g) Construction of the home can aid in increasing the amount of solar heating that can be utilized. Depending on the construction, the rate of heating will be increased or decreased, as needed depending on the climatic region in which the home is located.

© It's About Time

ANSWERS

Physics To Go

1. Using a four feet to one inch scale, then one foot is represented by 1/4 inch on the scale.

2. Using the 1:48 inch scale, the size of the house would be 8.5 × 4 ft. by 11 × 4 ft. or 34 ft. × 44 ft.

3. One advantage is that the size of the scale drawings would be one-half the size. This would allow a larger house to be put on ordinary paper.

Reflecting on the Activity and the Challenge

Over the past three activities you have made good progress in deciding upon the design specifications for your universal dwelling. In the last activity, you explored finding the "best fit" between living space and minimum surface area for your universal dwelling.

There are several reasons why you would want to keep surface area of the dwelling small, while keeping the interior as large as possible. One reason is that the materials for construction of the surface area of a building are expensive and a smaller surface area requires fewer materials. Another reason has to do with controlling the inside temperature of the building. In this activity, you constructed a model of your universal dwelling and investigated its heating and cooling properties.

The Architectural Committee probably will have some hard questions for you about your plans for heating and cooling the dwelling you are designing. That's because the greatest cost in operating most homes is for heating and cooling the home. This activity has given you some baseline information about how your dwelling design responds to the best kind of energy: free energy from the Sun. It will be important for you to learn how to use solar energy to best advantage in the next activities. You do not want to design a building that is expensive to heat or cool.

Physics To Go

1. If four feet of length on a real home is represented by one inch on a scale model of the home, what length on the scale model would represent one foot on the real home?

2. A 1:48 scale drawing of the floor plan of a home must fit one $8\frac{1}{2}$ × 11-inch sheet of paper. What maximum length and width in feet can the home have?

3. Home designers sometimes use a 1:96 scale ($\frac{1}{8}$ inch represents 1 foot) when making drawings of home plans. What advantages would a 1:96 scale have over a 1:48 scale ($\frac{1}{4}$ inch represents 1 foot)?

© It's About Time

Activity 4 Solar Heat Flow in the "Universal Dwelling"

4. If the heat lamp represented the Sun at its noon position during the heating phase of the activity, which side of the model dwelling—north, south, east, or west—must have been facing the Sun?

5. Describe how you would need to change the position of the heat lamp relative to the model dwelling to have the lamp represent the Sun during the morning or afternoon.

6. Do you think that all of the radiation from the heat lamp that hit the model dwelling was absorbed by the dwelling? Give evidence to support your answer.

7. Explain how you think heat travelled from the inside surface of the model dwelling to the thermometer at the center of the model during the heating phase. Obviously, the heat had to travel through the air inside the model. How did it do that?

STRETCHING EXERCISES

1. Write down three questions that you would ask someone who uses solar heating or solar cooling for their home.

2. If you know of any people who use solar energy to heat their homes, or have designed ways to keep solar energy out to keep their homes cool, you may wish to talk with them about how they have done this. Ask them about the special things that they have done to control the effects of solar energy on their homes.

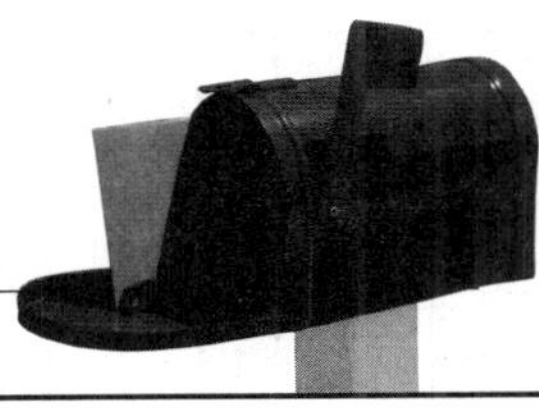

415

ANSWERS

Physics To Go

(continued)

4. The side of the model dwelling facing the Sun would be south.

5. If you were facing the house, moving the lamp to the right would represent the morning Sun, and moving the lamp to the left would represent the evening Sun.

6. Not all of the radiation hit the house or was absorbed by the house, as the lamp shade heated up, as well as the area around the house (table top, air, any objects close by, etc.).

7. As particles are heated up, they increase their kinetic energy. This causes them to move faster and faster, until they bump into other particles. As the bumping increases, the energy is transferred from particle to particle. There is also convection currents which are set up inside the house, in which the heat is moved in "streams" or currents around the house.

Chapter 6

ANSWERS

Stretching Exercises

1. Students' answers will vary. Some might include:
   - How much does it cost?
   - Describe how it works.
   - How does it work on cloudy days?
   - How does it work in the winter/summer?
   - How do you heat at night?

2. Student activity. Encourage students to seek out homes which are passively or actively solar heated. It is an opportunity to see how well it works in the area in which they live. Encourage the students to explore with the owners of the homes suggestions on how they might improve their home or ways in which it would work more efficiently in different climates.

© It's About Time

Grid For Heating/Cooling Data

HEATING AND COOLING DATA

Time (min)	Empty house °C	Insulation Ceiling only °C	Insulation Full house °C	Windows °C	Summer °C	with Awnings °C	Winter °C
0.5							
1							
1.5							
2							
2.5							
3							
3.5							
4							
4.5							
5							
5.5							
6							
6.5							
7							
7.5							
8							
8.5							
9							
9.5							
10							
11							
12							
13							
14							
15							
16							
17							
18							
19							
20							

© It's About Time

For use with *Designing the Universal Dwelling*, Chapter 6, Activity 4: Solar Heat Flow in the "Universal Dwelling"

Activity 4 A

How Does Location Affect Energy Demands?

WHAT DO YOU THINK?

Some places have nearly the same temperature year-round. This may be hot, cold, or temperate. Other places have seasonal temperature changes. Where it is always cold and where there are noticeable seasonal changes, heating and/or cooling a dwelling is necessary. The amount of heat that must be used varies from day to night and from day to day. The range of temperatures and other conditions, such as amount of rainfall and sunshine, help determine how much heat is necessary.

- Using your assigned or chosen location, make a prediction about summer and winter climates. Imagine the people in the lowest economic category in this community.
- Do you think that they heat or air-condition their homes? What would they use as a source of fuel?

Record your ideas about these questions in your *Active Physical Science* log. Be prepared to discuss your responses with your small group and the class.

FOR YOU TO DO

1. Determine answers to the following questions for your team's location. Record your answers in your log.

a) What is the average high temperature in the middle of summer?

b) What is the average low temperature in the middle of summer?

c) What is the average high temperature in the middle of winter?

d) What is the average low temperature in the middle of winter?

e) What source of fuel for heat is most likely used by the poorest of the people there?

f) How many heating 'degree-days' are there in an average year? (There is one degree-day for every degree Fahrenheit the average daily temperature is below 600° F.) If you do any calculations, show your work in your log.

g) 'Cooling degree-days' are also kept track of, especially in more developed countries. How would you define one cooling degree-day? How many cooling degree-days are there in an average year?

h) Does the location you chose require a heating or cooling system? Write your answer in your log.

© It's About Time

PHYSICS TO GO

1. Find out how many heating degree-days there were in the last heating season in your local home town. (This information is available from the gas company that serves your area or from the people that supply the energy your family uses for heat.)

2. Determine how much energy your family used for heat last year. (If your family hasn't saved last year's bills, the company that supplies your energy can tell you, or they can give you a value for a 'typical' family.)

3. Calculate how much energy per degree-day your family used for heating last year.

4. How much energy might be used annually for heat, by a family like yours, in the location your team is researching?

© It's About Time

For use with *Designing the Universal Dwelling*, Chapter 6, Activity 4: Solar Heat Flow in the "Universal Dwelling"

Activity 4 B

What Shape is Most Effective for Heat Storage?

WHAT DO YOU THINK?

- Which shape of hot metal will cool more quickly: a sphere or a pancake?

Record your ideas about this question in your *Active Physical Science* log. Be prepared to discuss your responses with your small group and the class.

FOR YOU TO DO

1. Measure the mass and dimensions of a spherically and flat cylindrically shaped piece of the same type of metal. Record their values.

a) Mass of spherically shaped metal: _________g

b) Radius of spherically shaped metal: _________mm

c) Mass of flat, cylindrically shaped metal: _________g

d) Radius of flat, cylindrically shaped metal: _________mm

e) Thickness of flat, cylindrically shaped metal: _________mm

2. Put on a pair of goggles. Place both samples in a 250-mL beaker with enough water to cover both samples.

3. Obtain two polystyrene cups. Place 50 mL of room-temperature water in each cup. (Note: The mass of 1 mL of water is 1 gm.) Measure the temperature of room-temperature water in each of your two polystyrene cups.

a) Record the temperature of the water.

Do not let your thermometer rest in the polystyrene cup without external support. Unsupported thermometers will tip polystyrene cups over and possibly break.

4. Place the metal samples and water on a hot plate and bring to a boil.

5. After the water has come to a boil, using tongs, carefully remove the spherically shaped metal sample from the 250-mL beaker and place it in the first polystyrene cup. Start the stopwatch at this time.

© It's About Time

6. Immerse your thermometer bulb in the water without letting it touch the metal. Stir the mixture in the polystyrene cup gently until the water increases by three degrees Celsius. Stop the stopwatch and record the time.

   a) Calculate the amount of heat lost by the spherically shaped metal to the water and the amount of heat lost per minute per gram of material. (Heat lost by metal = heat gained by water = mass of water × specific heat of water × change in water temperature). The heat lost per minute per gram is found by dividing the heat lost by the mass in grams and the time in minutes.

7. Repeat **Steps 5** and **6** for the flat, cylindrically shaped metal.

8. From your previous measurements, calculate the volumes, surface area, and surface area-to-volume ratio for each piece of metal. Show all your calculations in your log.

   a) Volume of spherically shaped metal:

   b) Surface area of spherically shaped metal:

   c) Surface area-to-volume ratio of spherically shaped metal:

   d) Volume of flat, cylindrically shaped metal:

   e) Surface area of flat, cylindrically shaped metal:

   f) Surface area-to-volume ratio of flat, cylindrically shaped metal:

   g) Describe how the surface area-to-volume ratio affects the rate at which heat is transferred. Write in your log how this applies to keeping heat from leaking away too fast.

PHYSICS TO GO

Do as many of the following as you can:

1. By treating your arms, legs, and torso as cylinders and your head as a sphere, estimate the dimensions, then calculate the surface area-to-volume ratio for your body. Repeat this for a baby (make approximations if you need to). Discuss the implications of your findings for retaining body heat. (Should a baby be kept more warmly dressed than an adult?)

2. Look at several water heaters at an appliance store or in advertising brochures. Take the necessary data to calculate surface area-to-volume ratios. Are some more effective at retaining heat than others? Is this related to brand? To capacity? (If you can get one or two other people to cooperate with you, you can get more data in less time!)

© It's About Time

For use with *Designing the Universal Dwelling*, Chapter 6, Activity 4: Solar Heat Flow in the "Universal Dwelling"

3. Look at several air-cooled, gasoline-powered engines, such as lawn mowers and chain saws. What is the biggest one (horsepower or cubic inches or liters) you can find? Do you know of any cars that have air-cooled engines? Do any airplanes have air-cooled engines? In your journal write about your discoveries and explain them in terms of this activity.

STRETCHING EXERCISES

1. Obtain a tall, cylindrical container such as a graduated cylinder or drinking glass. Fill it with different amounts of hot water and measure how fast the water cools. (Be sure to have a control to account for the fact that some of the heat will go to warming the container itself, and the more water you put in it the more energy goes into warming the container.)

To avoid breaking glass due to extreme temperature changes, do not put hot water in glass containers unless they are made of heat-proof material such as Pyrex.

2. Make a graph showing how heat loss depends on the surface area-to-volume ratio of the water. If you know how, determine the equation of this relationship.

© It's About Time

ACTIVITY 5
The Role of Insulation: Investigating Insulation Types

Background Information

If you did not consult the teacher **Background Information** for **Activity 4** it is suggested that you do so now. The **Background Information** presented for **Activity 4** is needed for this activity and the remaining activities in this chapter.

The focus of **Activity 5** is thermal insulation, which in the thermodynamics context of this chapter, applies to the ability of materials to resist heat transfer by conduction. It is important that you and your students not confuse use of the term here with the electrical context in which the same term, insulation, applies to the ability of materials to conduct electricity.

The term insulation is used only informally among scientists and engineers who work in the area of thermodynamics. Instead, scientists use a physical property called "heat conductivity" to describe the ability of a material to conduct heat. The heat conductivity of a material expresses the amount of heat energy which flows per unit of time through a unit of cross-sectional area when a one-degree temperature difference exists between surfaces of one unit of thickness of the material (see the equation governing the rate of heat transfer through materials in the **Background Information** for **Activity 4**; heat conductivity is symbolized by "*k*" in that equation).

Values of heat conductivity have been measured for virtually all materials, and, in general, metals are found to be good conductors of heat while nonmetals are poor conductors. A scientist desiring to find a good thermal insulator would seek a poor conductor—a material having a low value of heat conductivity.

"R values" are included in one of the items for **Physics To Go** in this activity, and it is likely that you will need some knowledge about what R values mean. The industries which supply thermal insulating materials for the building trades have developed a relative scale of resistance to heat conduction called "R values," wherein "R" represents resistance in absolute (dimensionless) numbers. The higher the R value, the better the insulating quality. Insulating materials such as fiberglass wool batts and Styrofoam® are manufactured in thicknesses tailored to the needs of builders (for example, batts of thickness to match the hollow space between 2 × 4 and 2 × 6 boards used to construct walls).

Homeowners and builders today can consult tables which recommend the R values needed in the walls, ceilings and floors of homes to provide efficiency for heating and cooling homes according to climatic region.

R values offer the easy-to-use advantage for consumers of being additive. For example, a homeowner in the northern part of the United States. wishing to upgrade ceiling insulation from R-19 to the recommended value of R-38, can place a six-inch thickness of fiberglass wool batting having insulating value R-19 over the top of the existing insulation in the attic of the home.

Comparison of the heating and cooling curves generated by students in this activity should lead to two findings:

- Different materials have different insulating effects.
- Increasing the thickness of a particular insulating material has a generally inverse effect on the rate of flow of heat energy and, therefore, an increased insulating effect.

The inverse relationship between the rate of heat flow and the thickness of the material through which the heat is flowing suggests that, theoretically, no thickness of insulation will absolutely stop heat flow. Applying this to insulating a home, it is clear that cost and available space for insulation cause need for a trade-off: an acceptable level of heat loss must be decided upon in relation to the cost and ability to prevent heat loss through insulating.

Active-ating the Physics InfoMall

For alternate activities or demos, you can browse through the Demo & Lab Shop. Simply choose a book and find the section on heat. There are plenty from which to choose. Many of these have explanatory figures. Take a look.

Again, graphs may be a problem. Check out the references for **Activity 4**.

A search of the InfoMall discovers several good references on insulation. For example, *American Journal of Physics*, vol. 52, issue 7, contains "Cost per million BTU of solar heat, insulation, and

© It's About Time

conventional fuels." *The Physics Teacher*, vol. 16, issue 6, has "Insulation and the rate of heat transfer." Not to be outdone, *Physics Today* has two articles on the efficient use of energy, as discussed above (vol. 28, issue 8, and vol. 33, issue 2).

Simply search the entire CD-ROM with keyword "insulation" to find these and many other hits. One more example is the textbook *Physics Including Human Applications*, with Chapter 11 on Thermal Transport. This addresses conduction and convection. Do the search and browse some of the hits for yourself. You may find more than you can easily use!

Planning for the Activity

Time Requirements

Allow about 10 to 20 minutes for the discussion regarding the results from the previous activity.

Allow about 10 minutes to set up the apparatus, and the hot water bath. Twenty minutes is needed for the activity, and about another 10 to 20 minutes to collect and interpret data. If desired, the class may share the results of the various kinds of insulating materials from each of the other groups.

Materials Needed

For the class:

- ice water supply
- hot water supply

For each group:

- 3 metal cans, varying diameters
- calculator, basic
- dish pan (10 qt)
- insulating material (1 qt), 3
- graph paper pad of 50 sheets
- thermometer (3)

Advance Preparation and Setup

In order to get reliable results, collect the number of cans needed for your particular class size. Three different size identical cans per group are needed. Before the students arrive, have a model set up and ready to go to show the students how to do this activity.

Teaching Notes

In the previous class, heating and cooling experiments were conducted on the students' model homes. Begin this class by having each group report its results in terms of heat transfer (heating and cooling) for their model home. Discuss with the entire class the similarities and differences of results among groups.

Ask: "To what extent do you think the shape, styles and volumes of the houses were factors affecting the heating and cooling rates? What other factors might influence heating and cooling rates in your model homes?"

Suggest that some groups may wish to consider design changes in light of the comparative results of groups.

Scientific inquiry can also be aided by working with the students to develop a proper procedure for this activity in order to achieve true scientific results. This can be done by having a discussion before the class, and establish the independent and dependent variables and which variables are being controlled in the experiment.

Discuss the heat transfer mechanisms described in the **For You To Read** with the class, and then have students complete **What Do You Think?** within their groups. Next, have students conduct **For You To Do.**

Remind the students of the safety concerns in the textbook.

The transfer of heat is not dependent on whether there is heating or cooling as much as on the change in the temperature. Heat always moves from high temperatures to low temperatures. Therefore, the rate of heating or cooling is dependent on the temperature difference, and that it is not always exactly the same in heating in the hot water bath, and cooling in the ice bath.

At the end of class you may want to tabulate the class data so that the students can examine the comparative effects of different insulating materials. Discuss the results with the class and attempt to arrive at a generalization about the effects of two variables: insulation thickness and insulation quality. If the cans were not identical, ask the question, "If the sets of cans were not identical for all groups, would it be "fair" to compare the effects of the different kinds of insulation used by different groups?"

© It's About Time

Activity Overview

The students will be evaluating properties of insulation, and their effect on temperature cooling with given variables (thickness, type of material, cooling/heating, etc.). They will be looking at the three ways in which heat energy is transferred — radiation, convection and conduction.

Student Objectives

Students will:

- Explain how insulation retards heat flow in terms of conduction.
- Explain how insulation retards heat flow in terms of convection.
- Conduct an experiment to test how the thickness and kind of insulating material influence the rate of heat transfer.

ANSWERS FOR THE TEACHER ONLY

What Do You Think?

Students may give different answers based on preconceptions. For example, a student living in Arizona, will say that it is more important for summer cooling, however, a student from Alaska may say that it is more important for winter heating. Generally, the importance of insulation for heating is about as important as for cooling.

Activity 5

The Role of Insulation: Investigating Insulation Types

GOALS

In this activity you will:

- Explain how insulation retards heat flow in terms of conduction.
- Explain how insulation retards heat flow in terms of convection.
- Conduct an experiment to test how the thickness and kind of insulating material influence the rate of heat transfer.

What Do You Think?

Insulation used in homes is usually a lightweight material that is designed to reduce the flow of thermal energy through the walls or ceiling.

- **Is insulation in a home more important in the heat of summer or the cool of winter?**

Record your ideas about this question in your *Active Physics* log. Be prepared to discuss your responses with your small group and the class.

© It's About Time

For You To Do

1. Place the insulating material supplied to your team in each of the three cans. Insert a thermometer in the absolute center of each material.
   - a) Measure the thickness of the insulation surrounding the thermometer in each can.
   - b) Record the temperature shown by each thermometer.
2. Place the three cans in a hot water bath.
   - a) Record the temperature of each can every minute for 10 min.

Do not change the temperature setting on the hot water bath.

3. Carefully check that the cans are not too hot to touch. Carefully remove the cans from the bath.
   - a) Record the temperature each minute for another 10 min.

Use a cloth or paper towel to grasp the cans when removing them from the baths. Moisture may form on the outside—be careful that the cans do not slip from your grasp.

4. Now place the three cans in an ice water bath and repeat the steps above.

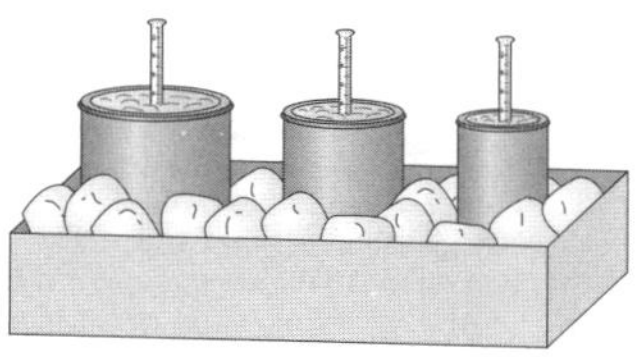

   - a) Record the temperature in each can every minute for 10 min of cooling in the bath and for another 10 min after removing them from the bath.

If the thermometer should break, *immediately* notify your instructor.

5. Display the data on a graph.
   - a) Graph your heating and cooling data for each of the three cans on a single graph, labeling each appropriately.
6. Use your graph to answer the following questions:
   - a) Compare the three graphs that you made. What differences do you note among the graphs?
   - b) What do you conclude about the effectiveness of insulation in heating as opposed to cooling situations?
   - c) What do you conclude about the effectiveness of using varying thickness of insulating material?

ANSWERS

For You To Do

1. a)-b) Student activity.

2. a) Student activity and observation.

3. a) Student activity and observation.

4. a) Student activity and observation.

5. a) Student activity and graphing.

6. a) Student data from their observations. They should note that there is a steeper curve with the least insulation.

   b) Students' answers will vary depending on the type of insulation. In general the students should notice that the effect of the insulation is about the same for heating and cooling.

   c) Generally, as you get thicker insulation, the effectiveness goes up. More is better. One variable to consider, is how much insulation is stuffed in, and students may note that if it is packed in too tightly, there is a reduction in the effectiveness. Discuss this with the students.

© It's About Time

FOR YOU TO READ

Transmission of Thermal Energy

You have most likely studied in prior science classes the three means by which heat is transmitted: **conduction**, **convection**, and **radiation**. As you read the descriptions of these means of transmission, try to recall experiments or investigations that you pursued if you studied this earlier.

In the case of conduction, the thermal energy possessed by a material is transmitted to another material by direct contact of the materials with one another. The rapidly moving (and hence high energy) atoms in a red hot iron bar come in direct contact with, let us say, cold water. The thermal energy of the iron atoms is conducted by physical contact directly to the water molecules, which, in turn, heat other adjacent molecules and so on, causing the heat to dissipate throughout the water, warming it in the process. In convection, the molecules or atoms of a fluid (a liquid or a gas) that have more thermal energy will move faster than the surrounding molecules of that material. As a result, they will move further apart and, therefore, become pushed by the colder, more closely packed molecules or atoms. As the hotter molecules spread out through the colder surrounding material, they carry their thermal energy (motion) with them, losing it along their path to other colder molecules by conduction. Convection then, is the transmission of heat from one place to another due to physical movement of a warmer fluid through a colder fluid (typically in response to the force of gravity).

Radiation is the third form of energy transmission. In this case, the electromagnetic radiation given off by high energy (hot) materials can be transmitted through objects, or even a vacuum, at the speed of light. When this energy strikes a material that can absorb it, the radiant energy causes the atoms or molecules of the material it strikes to move faster and the molecules now have more heat energy.

Reflecting on the Activity and the Challenge

Homes For Everyone (HFE) realizes that it is important to control the interior temperatures of their universal dwellings regardless of climate and location. It is important because of energy, comfort, and health concerns.

One of the criteria that the Architectural Committee will use to evaluate your plan is "the kind of insulation materials and thickness you choose should show you know the principles which make the insulation effective." That's exactly what this activity was about, so it will be very helpful in meeting the **Challenge**.

© It's About Time

Activity 5 The Role of Insulation: Investigating Insulation Types

Physics To Go

1. Explain how conduction, convection, and radiation were involved in both the heating and cooling phases of the model dwelling in **Activity 4**.
2. Which of the three ways of transferring thermal energy was most directly involved in the investigation in **Activity 4**? Explain your answer.
3. In this activity you tested the effects of two variables on the transfer of thermal energy. Identify the variables and explain what you learned about the effect of each on thermal energy transfer.
4. If you wanted to reduce the amount of thermal energy conducted through a particular kind of material to one-half the amount, what change in the thickness of the material would be needed? Also provide answers for reducing the transfer of thermal energy to one-third, and one-fourth the amount.
5. Would insulating a home be more, less, or about equally important for keeping a home warm in a cold climate or keeping a home cool in a hot climate? Use evidence from the data for your answer.
6. Years ago, the answer to keeping warm during cold weather was to put on more clothes. Today, the answer can be to put on different clothes. Comment on these two ways of keeping warm in terms of the variables tested in this activity, and the evolution of clothing available for winter.
7. Find out what kinds and thickness of insulating materials are used in the walls and ceiling of your home. Your parents or building supervisor may have this information.
8. Go to a building supply store and examine various insulating materials. The insulating properties of these materials are given in "R values." Find out what an R value is.

Physics Words

conduction: (of heat) the energy transfer from one material or particle to another when the materials or particles are in direct contact.

convection: the heat transfer resulting from the movement of the heated substance, like air or water currents.

radiation: (heat transfer) electromagnetic radiation strikes a material that can absorb it, causing the particles in the material to have more energy often resulting in a higher temperature.

ANSWERS

Physics To Go

1. Conduction is the transfer of heat from one particle to another. In **Activity 4**, the heat is transferred by conduction. This is when the outside molecules of the box transfer the heat to other molecules of the box, and eventually to the particles of air inside the box. Convection is the movement of heated particles, as some of the molecules are moving faster (more heat energy) than surrounding molecules. The heat is transferred from one area in the box to other areas in the box similar to the movement of water in a stream. Radiation is the transfer of heat in the form of electromagnetic radiation. Heat is transferred from one object to the next when electromagnetic radiation is absorbed by some materials. In this case the bulb is light and heat (electromagnetic radiation) and the energy is absorbed by the box.
2. Conduction is most directly involved in the transfer of heat as there is direct molecule to molecule transfer (hot water in direct contact with the can and other molecules in the insulation, etc.).
3. One variable was the thickness of insulation. Generally, the thicker the insulation, the greater the insulative properties. The other variable is the temperature (either heating up or cooling down).
4. Student data. Probably reduce the amount of insulation to half. Students will be able to surmise from their data, based on the time it took to reach certain temperatures while heating or cooling.
5. Insulating the home would be about equally important for keeping the home warm in winter and cool in summer. One factor would be the cost of insulating the home. If the heat was bearable, the cost of insulating would too expensive.
6. The students, after comparing the results, might notice that some materials insulate better than others. Before some of the modern insulative materials were invented, all we could do was put on more layers. With modern insulation, the amount of mass is reduced by the type of material. Therefore, you can have better insulation with thinner insulation or smaller amounts of the materials.
7. In colder climates, walls generally have between 4 and 6 inches, and ceilings have between 6 and 20 inches.
8. The "R value" of a material is the relative insulative properties of different materials compared to a standard.

Chapter 6

© It's About Time

ANSWERS

Physics To Go
(continued)

9. The aluminum foil will increase the R value as it reflects the heat from the shiny surface. Therefore, the aluminum foil side should be facing the heat source. In colder climates the foil is facing inward, and in hotter climates the foil is facing outward.

10. Depends on the area. Check with local contractor or building inspector.

11. The amount of insulation is largest in the ceilings as hot air rises, and there will be a large amount of heat lost through the ceiling. The walls have the next largest amount, and the floors the least.

9. Some rigid insulating material has an aluminum foil surface bonded to the sheet. What purpose does this aluminum foil surface serve? How does it influence the "R value" of the material?

10. How much insulation is recommended for a home in your area of the country?

11. Is the amount of insulation recommended the same for floors, walls, and ceilings? Why?

Stretching Exercises

1. Design a beverage container to keep a cold drink cold, and a hot drink hot.

2. Design an experiment to test the insulating effectiveness of natural material such as grass, dry mud, adobe brick, or stone. After your teacher approves your design, conduct the experiment.

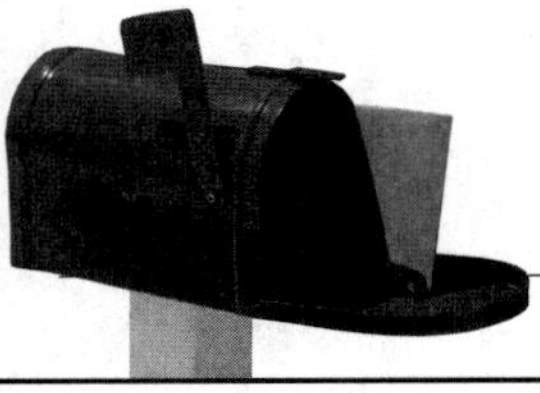

ANSWERS

Stretching Exercise

Student activity. Look for innovative designs. Some of the considerations are: size, mass, portability, durability, cost, etc.

© It's About Time

Activity 5 A

Transmission of Heat

FOR YOU TO DO

1. Drip candle wax onto one end of a long nail. Put on a heat-proof glove and hold the end of the nail with tongs.

2. Carefully hold the other end of the nail in the candle flame. Observe what happens to the candle wax at the opposite end of the nail, as the nail heats up.

a) Record your observations.

Keep flammable materials away from open flame.

b) Speculate on how the heat was transmitted from one end of the nail to the other.

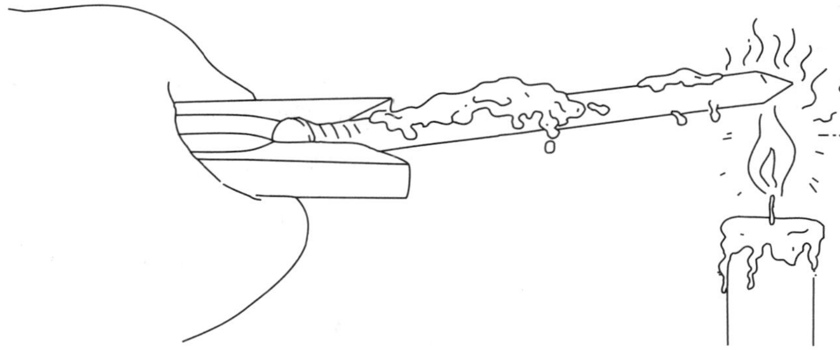

3. Obtain two identical-sized beakers or jars. Pour an equal amount of water into each beaker.

4. Cover one beaker with black paper, and the other with aluminum foil, with the shiny silver side outward.

5. Measure the initial temperature of the water in each beaker.

a) Record the initial temperature in your log

© It's About Time

6. Set up both beakers equidistant from a heat lamp, or place the beakers in a sunny window. Allow the beakers to stand for one hour. Measure the temperature of the water in each beaker after the hour.

a) Record the final temperature of the water.

b) Speculate on how the heat from the lamp or the Sun was transmitted to the water in the beakers.

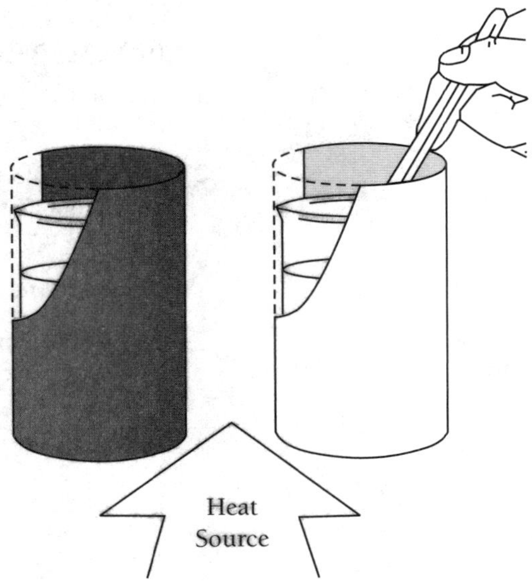

7. Fill a small plastic bottle with hot water. Add a few drops of food color to the bottle.

Seal the mouth of the bottle with aluminum foil attached to a string.

8. Fill a large tank with cold water. Carefully lower the plastic bottle into one corner of the tank. Let the water settle and then pull the foil lid off the bottle using the string. Observe what happens.

a) Record your observations in your log. You may wish to use a diagram.

b) Speculate on how the hot water was transmitted through the tank.

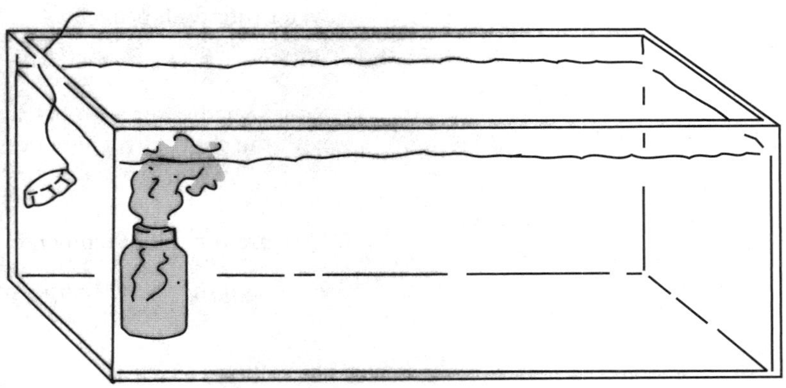

© It's About Time

NOTES

© It's About Time

ACTIVITY 6
Investigating Insulation Placement in Your Universal Dwelling

Background Information

No new physics principles are introduced in **Activity 6** as students conduct an experiment to explore the effect of adding insulation to their model homes and repeat the heating/cooling procedure conducted on uninsulated models in **Activity 4**. The results of this activity should be able to be directly related to outcomes in **Activity 5** in which thickness was explored as a variable governing the effectiveness of insulating materials. However, some subtleties are involved which need your attention as the teacher.

Any departure from setting up the heat lamp in the same way as it was set up for **Activity 4** (same lamp intensity at the same distance and angle relative to the model home) will limit ability to compare results of this activity to the results of **Activity 4**. If a variety of heat lamps were used by different groups in **Activity 4**, each group should use the same kind (wattage and style) in this activity so that the heat output of the lamp will not enter as a variable. Since the angle of illumination also affects the heat delivered to a surface, it is desirable to use the same angle as in **Activity 4**. Finally, since the intensity of illumination provided by the lamp varies inversely with the square of distance, a small change in the distance betweenthe lamp and the model home will cause a dramatic change in the heat delivered to the surface of the home. Therefore, it will be important to maintain the distance used in **Activity 4**. The need to replicate the setup of the heat lamp, and the cautions about changes in the setup just noted, will also apply to later activities in this chapter.

The sage saying "heat rises" has strong implications for insulating homes. While most homes have insulation in both exterior walls and ceilings (and sometimes in floors), ceiling insulation is generally regarded as the most important and effective procedure for preventing heat loss in homes. It is common practice today in residential home design to provide two or more times the insulating effect in ceilings than in walls. For example, for homes in the northern United States, recommended minimum insulation values are R-19 for walls and R-38 or more for ceilings. It will be interesting to see if similar trends are reflected in students' data as they progressively insulate the ceiling and walls of their model homes.

The thickness of cardboard used as insulation in this activity probably would not be realistic in terms of the 1:48 scale of the model home. For example, a 1/4 inch thickness of cardboard scaled up to real world size would correspond to a 12-inch thickness. This would be a reasonable thickness for ceiling insulation in a home, but would be unrealistically large for wall thickness in conventional construction. Only in recent years have home designers begun using 2" × 6" lumber as stud members in exterior walls to allow placing a greater thickness of insulation in the hollow spaces between studs than is allowed by 2" × 4" studs. Of course, a lesser thickness of insulation for the model which would scale up to a reasonable real-world wall thickness could be provided by choosing a more effective insulating material than cardboard.

Active-ating the Physics InfoMall

For alternate activities or demos, you can browse through the Demo & Lab Shop. Simply choose a book and find the section on heat. There are plenty from which to choose. Many of these have explanatory figures. Take a look.

Again, graphs may be a problem. Check out the references for **Activity 4**.

A search of the InfoMall discovers several good references on insulation. For example, *American Journal of Physics*, vol. 52, issue 7, contains "Cost per million BTU of solar heat, insulation, and conventional fuels." *The Physics Teacher*, vol. 16, issue 6, has "Insulation and the rate of heat transfer." Not to be outdone, *Physics Today* has two articles on the efficient use of energy, as discussed above (vol. 28, issue 8, and vol. 33, issue 2).

Simply search the entire CD-ROM with keyword "insulation" to find these and many other hits. One more example is the textbook *Physics Including Human Applications*, with Chapter 11 on Thermal Transport. This addresses conduction and convection. You may find more than you can easily use!

© It's About Time

Planning for the Activity

Time Requirements

Allow about 40 minutes for this activity.

Materials Needed

For each group:

- 120-Volt clip-on fixture for heat lamp with shade
- 250-Watt reflective heat lamp
- 1 × 2 corrugated cardboard (6)
- calculator, basic
- electric clock
- inch scale ruler
- hardwood meter stick (100 cm)
- masking tape (3/4" × 60 yds.)
- retractable knife
- stand for clip-on lamp
- thermometer

Advance Preparation and Setup

Have the same setup ready from **Activity 4**. Have several pieces of cardboard ready so that the students are not waiting for one piece to be cut.

Teaching Notes

Start this class session by reminding students that they should seriously examine the **Criteria for Evaluation** at the beginning of the chapter. Point out that they can refine their plans for their universal dwelling throughout the next few lessons, and that the last day of the unit will be devoted to their presentations.

Have all student teams compare their findings from **Activity 5**. Each of the different materials used will show a different insulating effect. Two important ideas to develop in discussion of the findings are that:
1) those materials which trap air in small spaces within the insulating material retard conduction and convection best, and 2) there are diminishing returns past an optimum thickness of insulation. Elicit feedback on **Physics To Go** from **Activity 5**. Those students who completed the assignment should be able to relay information on R values and recommended thickness' of insulation for homes in your location.

Before having students begin **For You To Do**, discuss the importance of replicating the conditions used in **Activity 4** (the same heat lamp, at the same distance and angle) so that the only new variable introduced is the insulation added to the model dwelling. Also point out that data on window and wall areas of all students' homes as described in **Physics To Go** will be needed for tomorrow's class.

© It's About Time

NOTES

© It's About Time

Activity 6 Investigating Insulation Placement in Your Universal Dwelling

GOALS

In this activity you will:

- Identify and explain which surface in a house is vulnerable to the greatest heat loss in terms of conduction and convection.
- Infer an optimum thickness for home insulation.
- Identify limitations of using scale models in conducting experiments.

What Do You Think?

The density (mass per unit volume) of air decreases with temperature, leading to the common knowledge expression, "heat rises."

- **Why are attics of homes so hot on a summer day?**
- **What are these same attics like on a cold winter day?**

Record your ideas about these questions in your *Active Physics* log. Be prepared to discuss your responses with your small group and the class.

Activity Overview

In this activity, the students will be repeating the heating and cooling of their model homes, but using insulation this time. The students should notice a difference immediately, but if not, have them repeat the experiment using more cardboard.

Student Objectives

Students will:

- Identify and explain which surface in a house is vulnerable to the greatest heat loss in terms of conduction and convection.
- Infer an optimum thickness for home insulation.
- Identify limitations of using scale models in conducting experiments.

ANSWERS FOR THE TEACHER ONLY

What Do You Think?

Attics are hot due to the solar heating during the day. If there is little insulation, then the Sun will heat up the attic quickly. Also, there is generally little circulation in the attic, and the air continues to heat up.

The same attics are cold because of the same reasons. That is, the attic is insulated against the heat from inside the home, but not the cold outside. Because the air outside is cold the heat that is trapped in the attic very quickly leaves the attic to the outside.

© It's About Time

ANSWERS

For You To Do

1. Student activity.

2. a) Student activity.

   b) Student activity.

   c) Students should notice that there is a reduction in the transfer of heat to the inside during heating, and a reduction in the transfer of the heat to the outside from the inside during the cooling.

3. a) Again, there should be a reduction in the heat transfer.

4. a) Students' observations.

   b) Some of the differences may be due to the angles of the roofs, reflective properties of the roofs, quality of construction, etc.

   c) If 1 inch = 48 inches then 1:48 = 1/4:12. Therefore, the thickness on your house should be 12 inches.

   d) If you double the thickness, you should probably double the insulative value. Tripling the thickness should triple the insulative value. The students' answers will vary depending on their answers from **Activity 5**.

For You To Do

1. Cut the cardboard as insulation so that it will fit snugly within the ceiling of your model home. Use masking tape to fix it in place.

2. Set up your experiment in the same manner as you did for **Activity 4**. Be sure to keep the lamp at the same distance (20 cm) and at the same 45° angle as in the first experiment.

   a) Construct a temperature vs. time graph of your data as you did in **Activity 4**.

   b) Compare this heat transfer experiment data with data from **Activity 4** and interpret the results.

   c) How did insulating the ceiling affect the heating and cooling of your model?

3. Insulate the walls of your model in the same way as you did the ceiling. Repeat the heat transfer experiment.

   a) Compare and interpret these results with the first two experiments.

4. Use the results of **Activities 4** and **5** to answer the following questions:

   a) Compare the results of your team's findings with those of the other teams.

   b) How do you explain the similarities and differences?

   c) Given the scale of your model home, to what thickness of insulation in a real home does your $\frac{1}{4}$ inch of cardboard insulation correspond?

   d) What effect do you predict would occur if you used twice the thickness of cardboard in your model? Three times the thickness? Defend your responses with data from **Activity 5**.

© It's About Time

Reflecting on the Activity and the Challenge

In this activity, you insulated your model home and then tested it to see what effects insulating will have on controlling the internal temperature of the home. The Architectural Committee will expect you to be able to show that you know where insulation is most effective in trapping energy. You now have gathered scientific data which you can use to show that you know where to place insulation most effectively in a home.

Physics To Go

1. In insulating a real home you probably wouldn't use cardboard as the insulating material as you did for the model dwelling. What materials might you use, and why?
2. What considerations, other than the effect of the insulation, might guide your choices of insulating materials for a home? Think about fire prevention.
3. Homes are usually designed to have about twice as much insulating effect in ceilings than in exterior walls. Why?
4. Could a vacuum be a good insulator? Explain why or why not, and under what conditions.
5. Compare and contrast the purpose of insulation in hot weather versus cold weather. Should placement of insulation be different for these different purposes?
6. Insulating materials for homes are rated on a relative scale of R values. Sheets or rolls of different kinds of materials in various thickness are available, and each has an assigned R value. The R values of two or more thickness can be added together to find the total insulating value of combined layers. What combinations of the following materials could be used to provide R-19 insulation for a wall and R-38 insulation for a ceiling?
   - R-8 expanded polystyrene board 2 inches thick
   - R-11 fiberglass wool $3\frac{1}{2}$ inches thick
   - R-30 fiberglass wool $9\frac{1}{2}$ inches thick

Answers

Physics To Go

1. Cardboard is flammable, and would constitute a fire hazard. Also, the amount of cardboard needed would be rather costly, and it loses its insulative properties when wet. Materials used would be fiberglass insulation, Styrofoam®, plastics, etc. These are used because they are fire retarding, relatively waterproof, resistant to mold, mildew and other odor-producing growths, relatively inexpensive, resistant to packing. Even when wet, they retain their insulative properties.

2. See previous answer, as well as resistant to insects and small rodents. Ease of installation is also a good criteria.

3. The heat from the Sun is more direct on the roof than on the walls. From the inside of the house, the hot air rises, therefore, having insulation in the ceiling will prevent heat loss.

4. A vacuum is a good insulator when the heat is transferred only by conduction (molecule to molecule) or convection (moving currents of heat energy). If there is also the possibility of radiation (as in light or other forms of EMR), then a vacuum will do no good. Space is a vacuum, and there is radiation in space.

5. Students' answer will vary. The primary purpose of insulation in hot weather is to keep the heat out, and the cool in. In cold weather, it is to keep the heat in and the cold out. Placement of the insulation should be the same for both situations.

6. For the wall: in order to get at least R-19, the walls must be at least 6" thick. You could use 3 layers of R-8 polystyrene (R-value of 24); or R-11 fiberglass wool and R-8 polystyrene. For R-38, use R-30 fiberglass wool, and R-8 polystyrene, or 2 layers of R-8 plus 2 layers of R-11 fiberglass wool.

© It's About Time

ANSWERS

Physics To Go

(continued)

7. Student activity and data for each a), b) and c).

7. In preparation for tomorrow's class work, collect the following information tonight at your home:
   a) Measure all the windows in your home and calculate the total number of square feet of window area of your home. Record your calculation.
   b) Measure all the exterior wall surface areas of your home, and calculate the total number of square feet (including the window areas) of exterior walls of your home. Record your calculation.
   c) Divide the total window area by the total wall surface area. Record your calculation.

Stretching Exercises

1. In constructing homes today, high, open (cathedral) ceilings are often part of the design plans for the living space. What advantages and disadvantages do high ceilings have with respect to heating and cooling considerations for the home?
2. Contemporary homes in America today are built with full basements, half-buried basements, or are constructed on top of concrete slabs cast on the surface. What heating advantages and disadvantages do you believe would exist for each of these house foundation types? Explain your responses for each type of foundation.

ANSWERS

Stretching Exercises

1. One advantage is that the heat is pulled to the ceiling away from the living area during summer months. One disadvantage is that the heating of the space is more expensive as the heat rises to the ceiling during the winter months.

   Often, many windows are associated with the high ceilings, and this causes problems during summer as more heat is entering the home, and in winter more heat is leaving the home.

2. Students' answers will vary. Homes built with in-ground basements, will have a greater living space, with less home above ground to insulate. Basements do not need as much insulation, because the ground acts as an insulator. Basements are also necessary for the plumbing to remain in a heated environment.

© It's About Time

R Values for Insulation: Manufacturers and Government Information

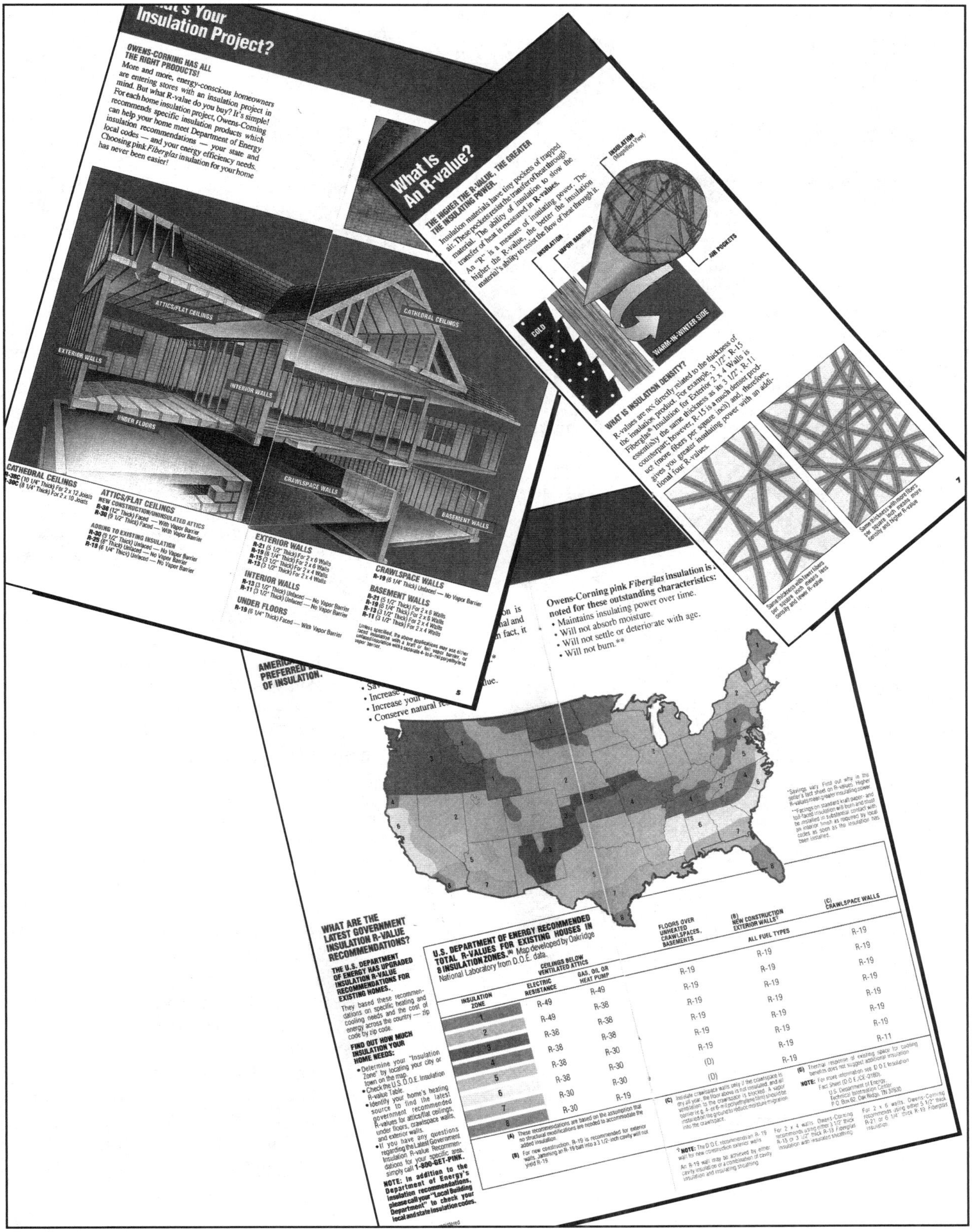

© It's About Time

For use with *Designing the Universal Dwelling*, Chapter 6,
Activity 6: Investigating Insulation Placement in Your Universal Dwelling

ACTIVITY 7
The Role of Windows... Placing Windows in Your Universal Dwelling

Background Information

The orientation of the model home with respect to geographic direction and the orientation of the Sun in the sky is introduced in **Activity 7**. The main thrust of this activity is that students add windows to their model homes and repeat the heating/cooling procedure used in **Activity 4** and **Activity 6**. The added specification is made that the model home will be oriented so that its south side faces the sun-simulating heat lamp. As in **Activity 6** which tested the effect of adding insulation to the home, the temperature versus time graphs produced from measurements taken during the heating and cooling phases are used to infer the effect of including windows in the model home.

As explained in the **Background Information** provided for **Activity 6**, it remains important for comparison of results across activities to maintain a constant setup for the heat lamp.

The 45° angle used for the orientation of the heat lamp in this series of activities takes on greater significance for this activity. That particular orientation matches the orientation of the Sun in the sky at noon on the spring and autumn equinoxes (March 21 and September 23) for locations at 45° north latitude; on those dates at such locations, the Sun appears 45° above the south point of horizon at local noon. This is the median noon altitude of the Sun throughout the year (the variation of the Sun's altitude throughout the year will be discussed in the **Background Information** for **Activity 8**). In the southern hemisphere, the Sun appears in the northern sky at noon and the seasons are reversed.

The main property of windows which can be expected to affect heating and cooling results of this activity is the surface area of the windows. The radiant heat transmitted into the home through windows facing the Sun should, in principal, be directly proportional to the window area. Similarly, heat loss through windows via conduction and radiation should be directly proportional to the total surface of windows regardless of the side of the house on which they are located. It is important to keep in mind that, during the heating phase of this activity, simultaneous gains and losses of heat will be occurring through windows.

This activity can be expected to produce the onset of greater diversity of results across groups than was the case for **Activity 6**. Even though a variety of sizes and shapes of model homes corresponding to groups' designs were used in **Activity 6**, the kind and amount of insulation added to the variety of models was the same across groups. Therefore, some uniformity in effects on heating and cooling rates probably was realized. **Activity 7**, however, allows freedom for each group to choose the number, size and locations of windows to be added to their model and, therefore, the effects on heating and cooling rates should be expected to be as diverse as the choices made by students for windows. Overall, it can be expected students will find that windows on a southern exposure should offset some of the effect of the insulation added to the home in the previous activity to cause an increase the rate of heating; similarly, the rate of cooling can be expected to increase due to loss of some insulating effect.

Obviously, windows exposed to the Sun during daylight hours can provide a profound heating effect through the process of radiation. This is true for windows not only on the south side of a home where the Sun shines for most of each day, but also to a lesser, yet significant extent for windows placed on the east and west sides.

Less obvious to most people is that windows also cause very significant heat loss from homes through conduction and radiation. Despite tremendous efforts within the window industry to prevent heat loss (e.g., multiple panes of glass separated by layers of gas, thin coatings which reflect infrared radiation that would otherwise escape back into the home), it remains true that the window has not been made which matches the insulating quality of a well-insulated wall. A typical insulated wall is rated at R-19; windows range in insulating effect from about R-1 for a single pane of glass to less than R-5 for the best triple-glazed windows money can buy. Therefore, windows are a two-edged sword; they provide views, light, and solar heating, but their low insulating effects also cause problems with heating and air conditioning efforts.

© It's About Time

Active-ating the Physics InfoMall

The textbooks mentioned at the beginning of this book (*Household Physics, and Energy: Insights From Physics*) contain chapters on heat, home energy conservation, home heating, solar radiation, solar heating, and home illumination. You are encouraged to look at these sections to see just how much information is available on this single CD-ROM.

Planning for the Activity

Time Requirements

Allow about 40 – 50 minutes for this activity.

Materials Needed

For each group:

- 120-Volt clip-on fixture for heat lamp with shade
- 250-Watt reflective heat lamp
- calculator, basic
- clear plastic wrap
- electric clock
- inch scale ruler
- hardwood meter stick (100 cm)
- masking tape (3/4" x 60 yds.)
- retractable knife
- stand for clip-on lamp
- thermometer (3)

Advance Preparation and Setup

Have the students work within their teams to calculate the ratio of window-to-wall areas represented by the homes of students within each team. Circulate among the teams as the students work, moving the process along. Then turn the students' attention to need to make a decision about windows for each group's model home.

Teaching Notes

Before moving on to **For You To Do**, ask each group to give a brief report (anonymous, no names) on the ratios found for students' homes and the group's decision on windows for the universal dwelling. Point out in the instructions for **For You To Do** the conditions (cold winters, hot summers in the northern hemisphere) and allow that some groups may wish to adjust their initial decision about the number of windows and their placement. Also, remind students about the need to replicate the same heating conditions used in **Activities** 4 and 5, this time introducing windows as a variable. Ask the entire class the question, "When you have completed this activity to gather data on the heating and cooling effects for a model home with windows, what data will you use for comparison? Will you use the data from **Activity** 4, a windowless home without insulation, or the data from **Activity** 5, a windowless home with insulation?" Discuss and decide as a class.

SAFETY PRECAUTION: Remind students to be careful while cutting their windows.

Chapter 6

© It's About Time

NOTES

© It's About Time

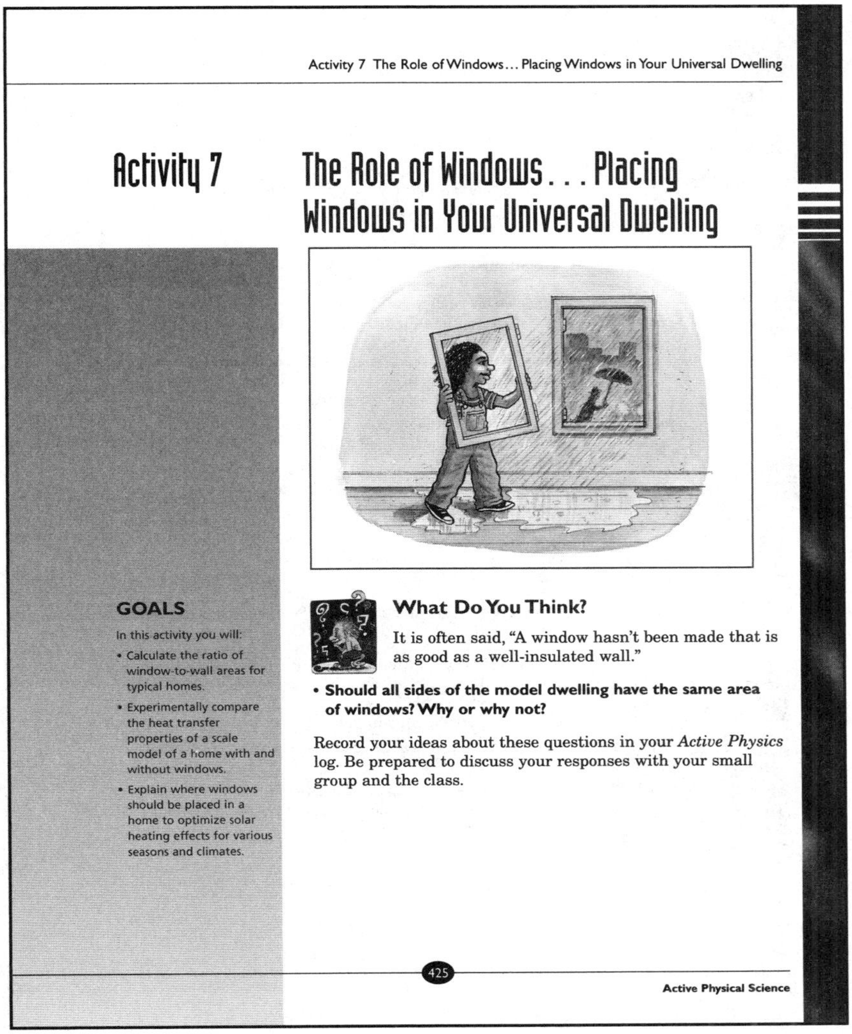

Activity 7 The Role of Windows... Placing Windows in Your Universal Dwelling

GOALS

In this activity you will:

- Calculate the ratio of window-to-wall areas for typical homes.
- Experimentally compare the heat transfer properties of a scale model of a home with and without windows.
- Explain where windows should be placed in a home to optimize solar heating effects for various seasons and climates.

What Do You Think?

It is often said, "A window hasn't been made that is as good as a well-insulated wall."

- **Should all sides of the model dwelling have the same area of windows? Why or why not?**

Record your ideas about these questions in your *Active Physics* log. Be prepared to discuss your responses with your small group and the class.

Activity Overview

The students will continue their exploration into the heating and cooling of their home. In this activity, the students will place windows with a predetermined window-to-wall area ratio from their calculations in Activity 6.

Student Objectives

Students will:

- Calculate the ratio of window-to-wall areas for typical homes.
- Experimentally compare the heat transfer properties of a scale model of a home with and without windows.
- Explain where windows should be placed in a home to optimize solar heating effects for various seasons and climates.

ANSWERS FOR THE TEACHER ONLY

What Do You Think?

In colder climates, there would be fewer windows on the north side of building than in a warmer country. However, in the warmer climate the opposite is true. There would be fewer windows on the south side than on the north side.

© It's About Time

Designing the Universal Dwelling

For You To Do

1. Compile a chart in your group to summarize your calculations of window area to wall area ratios in your homes.
   a) What is the average window to wall area ratio for your group?

2. Use all the information and resources available to you. Decide upon the number, sizes, and placement of the windows for your model home. Remember, you want to provide for optimal lighting and ventilation and you also want to allow solar energy to enter the home and heat it when needed. However, you do not want to transfer large amounts of energy from the home to the outside when direct sunlight does not enter the window.
   a) Approximate the window to wall area ratio you plan to use in your model home.
   b) How does the window to wall area ratio for your model home compare to the average window to wall area ratio you calculated for your group? Explain why you chose the ratio you did.

Caution: The cutting tools are very sharp. Be careful and get help from your teacher as needed.

3. Draw outlines of the windows to scale on the walls of your model and carefully cut out the windows with a cutting tool.

4. Use masking tape to tape transparent food wrap over the window cutouts.

5. Repeat the heat transfer experiment with the heat lamp on your model home. Place the heat lamp and model home so that the lamp is shining on the home at the same angle and at the same distance as in previous activities.
   a) Record and graph your findings as you did in previous activities.
   b) Compare your heating and cooling graphs for this activity with those in **Activities 4** and **6**. How do the graphs differ? Explain your results.

ANSWERS

For You To Do

1. a) Student data.

2. a) Student data.

   b) Student data. Students' answers will vary, depending on their location, climate, apartment dwellers, people who live on farms, whether they like lots of natural lighting, etc.

3. Student activity.

4. Student activity.

5. a) Student activity.

   b) Students should notice that there is an increase in the heating and cooling curves, showing that there is a lot of energy lost through the windows.

© It's About Time

Reflecting on the Activity and the Challenge

Virtually all dwellings have windows of one type or another to provide interior light and to provide for ventilation and temperature control. The windows in a home built today are extremely well-designed and engineered in comparison to those of 30 years ago. The problem your group faces is a complicated one. You wish to design and place windows in your Homes For Everyone (HFE) universal dwelling to provide for optimal lighting and ventilation, but you also know that windows have energy advantages and disadvantages. Windows, which can allow solar energy to enter the home and heat it when needed, lose heat through conduction, convection, and radiation to a much greater degree than the insulated walls.

You will need to convince the Architectural Committee that you understand that windows offer both advantages and disadvantages. You have gathered evidence by comparing the heating and cooling curves from this activity to the corresponding curves from **Activity 6**.

Physics To Go

1. Heat can enter and leave a home through windows. How are conduction, convection, and radiation involved in:
   a) The transfer of heat into a home through its windows?
   b) The transfer of heat out of a home through its windows?
2. The amount of heat gained or lost through a window is directly proportional to the surface area of the window.
   a) Compare the amount of heat expected to be transferred through a window 2 feet wide by 2 feet high to the amount expected for a window 4 feet wide by 4 feet high.
   b) Glass doors behave about the same as windows regarding heat transfer. Compare the amount of heat expected to be transferred through a glass patio door 8 feet wide by 6 feet high to the amount expected for a 2-feet-wide by 2-feet-high window.

ANSWERS

Physics To Go

1. a) Conduction: Sun heats up the window glass, and window pane, then is transferred from the window to the inside of the home.

   Convection: Wind moves air by window and transfers heat to the window and air around the window.

   Radiation: Sunlight radiates through the window, and heats up the inside of the home.

   b) Conduction: The heat inside the home is transferred to the window and frame. It is then transferred by molecule to molecule contact to the outside.

   Convection: The warm air inside the home is moved to the cold air close to the windows, and then the heat is transferred outside.

   Radiation: Heat is radiated from the inside to the outside through the glass.

2. a) Find the area of both windows, then one would expect that the larger window would lose or gain proportionately. $2 \times 2 =$ 4 sq. ft. and the larger $4 \times 4 =$ 16 sq. ft. Therefore, the larger window loses (or gains) 4 times as much heat.

   b) $6 \times 8 = 48$ sq. ft. compared to the 4 sq. ft. window, therefore, the patio doors will lose (or gain) 48/4 or 12 times as much heat.

© It's About Time

ANSWERS

Physics To Go

(continued)

3. Visit your local building supply store for this information. Generally, as you increase the number of panes of glass in a window you increase the R value. Thermopane windows have different functions and construction depending on the climate in your region.

4. There is more dead air space in a wall as compared to the space in a window. The air in the window is free to move around inside the panes, and therefore can heat up more quickly than the dead air spaces in the walls. The walls also will not radiate heat as quickly as windows, as the heat energy must pass through the wall itself before getting to the trapped air. In a window the air is directly heated by radiation.

5. Curtains and shades provide a barrier to the different temperatures of the inside to the outside, or vice versa. They provide a barrier to the radiation of the heat energy, as well as minimize the movement of air with different temperatures.

6. The Sun does not shine on the north side of the home, so there is no advantage using those windows for heating. Also, the prevailing winds are often from the north, and therefore will speed the process of heat loss.

7. The problems are associated with keeping the warm air out, and cool air in. This is just the opposite problem of the cold climate window problem. Windows are poor heat insulators, and therefore whether cooling or heating, the greater the surface area of windows, the greater the heat gain or loss.

3. Visit a building supply store in your community. Examine the windows available. Find out about the following:
   a) The R value of the cheapest and most expensive windows.
   b) The difference between thermopane and single pane windows.
   c) Why thermopane windows are built the way they are.
   d) What "low E" windows are, and how they function.

4. Even the highest quality windows do not have as much insulating effect as a well-insulated wall. Explain why.

5. Window drapes and shades can provide privacy and beauty in a home. Explain how they may also be used to control heat transfer through windows.

6. Designs for energy-efficient homes in parts of the United States that have cold winters have very few if any windows on the north side of the home. Why?

7. People who live in warm climates often use air conditioners to cool their homes. What, if any, problems do windows present related to air conditioners?

Stretching Exercises

Find out about passive solar heating for homes. Answer the following:

a) How can adding more windows to a house serve to warm the house during the day?
b) How can you prevent heat from escaping from these extra windows during the night?
c) What can be used to store the heat that is collected during the day?
d) Where should windows for passive solar heating be placed to be most effective?

ANSWERS

Stretching Exercises

a) The solar heating of the carpets, floors and furniture.

b) Using shutters or blinds or curtains will prevent the heat loss to the outside at night.

c) Water is the best way to store heat. It has a very high specific heat capacity, and is easy to handle.

d) Windows used for passive solar heating should be placed on the south side of the home.

© It's About Time

R Values for Windows: Manufacturers and Government Information

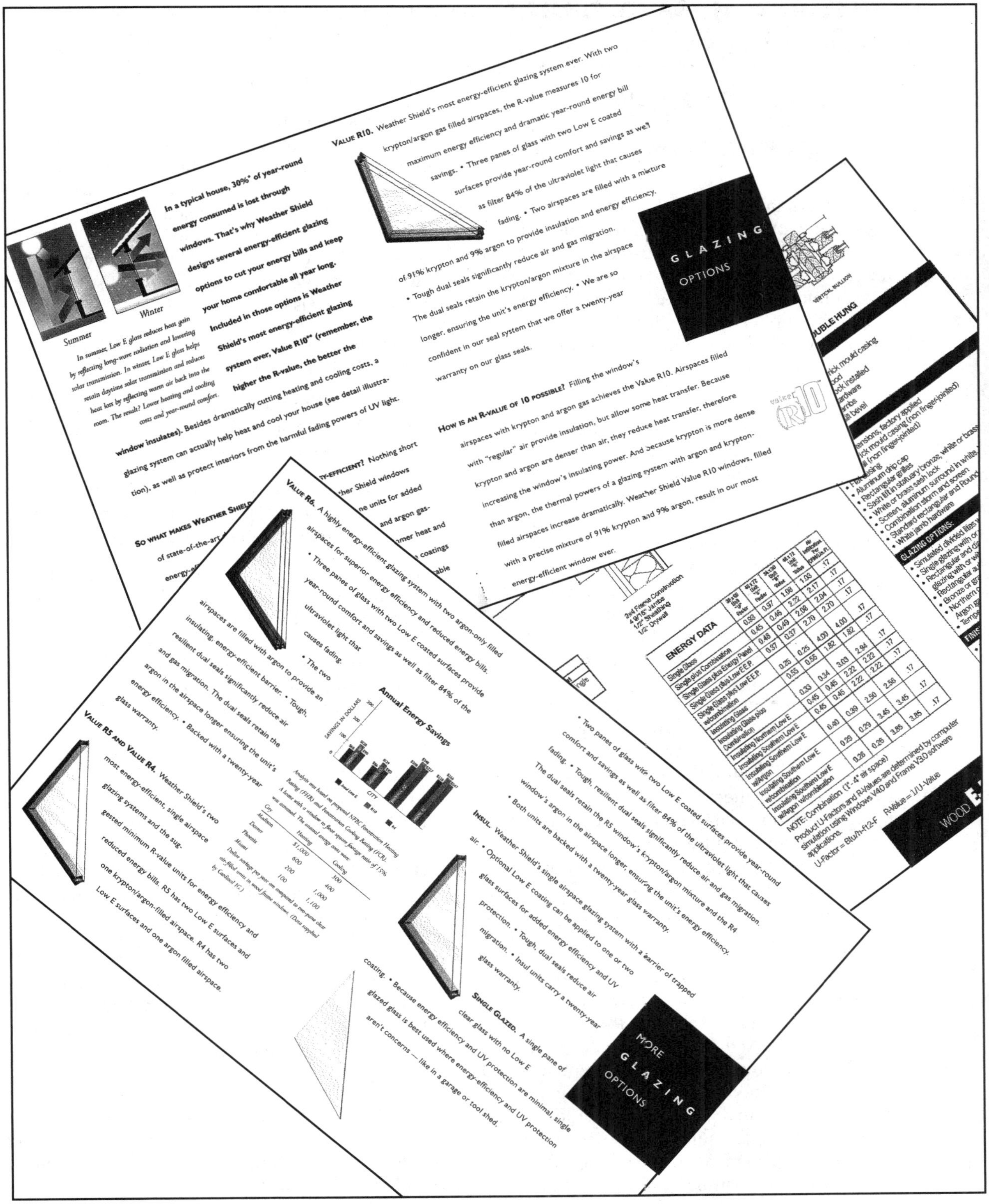

For use with *Designing the Universal Dwelling*, Chapter 6,
Activity 7: The Role of Windows... Placing Windows in Your Universal Dwelling

Chapter 6

© It's About Time

ACTIVITY 8
Investigating Overhangs and Awnings

Background Information

Activity 8 has students explore how seasonal changes in the altitude of the Sun may affect the heating and cooling of homes. Using the heat lamp to simulate the Sun at its extreme noon altitudes during summer and winter, students again heat and cool their model homes with and without the addition of window awnings and/or roof overhangs to control the amount of light allowed to enter windows.

The Sun's annual cycle of changing noon altitudes is caused by the fact that the Earth's spin axis is not perpendicular to the plane of Earth's orbit around the Sun. Instead, the spin axis is tilted 23.5° away from the perpendicular to the plane of the orbit. The Earth maintains its tilted orientation, rotating daily as it completes its orbit in an amount of time which serves as the definition of one year on Earth. The tilt of Earth's orbit causes seasons. When the tilt causes the northern hemisphere to "lean" toward the Sun it is summer; the Sun appears higher in the sky at noon than it does at other times of year and radiation from the Sun strike the Earth's surface most directly, causing maximum heating of the surface for the year. When the northern hemisphere leans away from the Sun, the Sun appears lower in the sky, the heating effect is less and it is winter. Spring and autumn occur when the axis is tilted neither toward nor away from the Sun, but sideways. The seasons are exactly reversed in the southern hemisphere, and observers at mid-southern latitudes see the Sun residing above the north point on the horizon at noon.

The four special dates which mark the beginning of each season are helpful for keeping track of the Sun's altitude throughout the year: the spring (or vernal) equinox on March 21, the summer solstice on June 21, the autumnal equinox on September 23 and the winter equinox on December 21. For observers in the United States, noon altitude of the Sun is maximum on the summer solstice and is minimum on the winter solstice, varying 47° (2 × 23.5°) in altitude (measured from the south point on the horizon) during the six-month time interval from one solstice to the next. The altitude of the Sun is a its median value on both equinoxes. The noon altitude of the Sun varies sinusoidally throughout the year and, if necessary, can be predicted for any day of the year for any observer.

For purposes of home design involved in this activity, it is necessary to be able to calculate the extreme noon altitudes of the Sun throughout the year (the altitude on each solstice) as viewed from the particular latitude at which the home is located. A diagram and narrative which explain how to accomplish the calculation is included in **For You To Do** for **Activity 8**; a condensed version for you as the teacher is:

- Calculate the complement of the latitude of the home (90° minus latitude).
- dding 23.5° to the complement of the latitude gives the Sun's altitude above the horizon at noon on the summer solstice.
- Subtracting 23.5° from the complement of the latitude gives the Sun's altitude above the horizon at noon on the winter solstice.

By this discussion, it may be apparent that the 45° angle of the heat lamp used in prior activities can be thought of as having modeled the Sun at noon on the equinoxes shining on a home located at 45° north latitude. This was an arbitrary choice by the authors and happens to correspond to locations in the northern contiguous United States.

To allow comparison of results of this activity to the results of **Activity** 7, it will be necessary for students to be careful not to introduce an undesired variable by changing the distance between the heat lamp and the model home when the lamp is being repositioned to illuminate the home at the solstice angles. The distance should be held constant at the same value used in earlier activities.

If your latitude is greatly different from 45° north, be aware that the 45° angle of the heat lamp used in prior activities did not represent the Sun's noon altitude on the equinoxes for your location, but on dates other than the equinoxes.

Students should find it possible, as do home designers, to manipulate roof overhangs, awnings or other methods to shield windows from the Sun during summer while allowing sunlight to enter windows during winter.

© It's About Time

Active-ating the Physics InfoMall

Search the InfoMall with keywords "overhang*" or "awning*". You will find several hits with good information. One of these, again, is *Household Physics*. The section on Solar Heating (found, oddly, in the Air Conditioning the Home chapter) has an excellent graphic showing how awnings allow the winter Sun to pass into the house, but prevent the summer Sun from causing unwanted heating.

Planning for the Activity

Time Requirements

Again, this will be about 40 – 50 minutes. Allow extra time for students to try different kinds of awnings or overhangs if desired.

Materials Needed

For each group:

- 120-Volt clip-on fixture for heat lamp with shade
- 250-Watt reflective heat lamp
- calculator, basic
- clear plastic wrap
- electric clock
- file folders (24)
- inch scale ruler
- hardwood meter stick (100 cm)
- masking tape (3/4" x 60 yds.)
- protractor
- retractable knife
- stand for clip-on lamp
- thermometer (3)

Advance Preparation and Setup

The data from **Activity 6** should point to the potential for using passive solar heating during winter months. However, a problem during summer months develops when south-facing windows may allow more solar energy to enter the home than would be desired. Elicit those conditions from students in a discussion of the results of **Activity 6.**

Teaching Notes

Ask for brief reports on student findings from **Physics To Go** for **Activity 7** regarding single pane, thermopane and low e glass windows.

Announce that, in this activity, students will design roof overhangs or awnings for their universal dwelling which will shade the south-facing windows in the summer and yet allow maximum solar energy to enter the home during winter months. Use a world globe to point out the local latitude and determine the angle of the Sun at summer and winter solstices for your location. Discuss the importance of maintaining the same distance between the heat lamp and the model home when adjusting the angle of the heat lamp during the **For You To Do** procedure, and begin the activity.

SAFETY PRECAUTION: Be careful when using sharp instruments while cutting the overhangs.

© It's About Time

NOTES

© It's About Time

Activity 8 Investigating Overhangs and Awnings

GOALS

In this activity you will:

- Explain how the altitude and heating effect of the Sun varies with seasons.
- Design overhangs and awnings to control solar heating through the windows of a home during summer and winter.
- Conduct a heat transfer experiment on a scale model of a home to test the effectiveness of overhangs and awnings for controlling solar heating during summer and winter.

What Do You Think?

Heating and air conditioning are expensive items for operating the average American home.

- **What can be done so that windows will let Sun enter the dwelling when it is cold outside, and block sunlight from entering when it is hot outside?**

Record your ideas about these questions in your *Active Physics* log. Be prepared to discuss your responses with your small group and the class.

For You To Do

1. Find the latitude of the location for which you are designing your model home.
2. In the temperate regions, the Sun is low in the sky at noon in winter and high in the sky at noon

Activity Overview

This activity is the culmination of this chapter, where students will put the final variable on their home. Again, they are to be reminded of the importance of keeping all the other variables the same and only changing the one variable for this experiment.

Student Objectives

Students will:

- Explain how the altitude and heating effect of the Sun varies with seasons.
- Design overhangs and awnings to control solar heating through the windows of a home during summer and winter.
- Conduct a heat transfer experiment on a scale model of a home to test the effectiveness of overhangs and awnings for controlling solar heating with during summer and winter.

ANSWERS FOR THE TEACHER ONLY

What Do You Think?

Letting the Sun in during the winter months is valuable in utilizing the Sun's heat. However, it can be uncomfortable during the summer months. Therefore, people put up awnings, or have large overhangs on their roofs.

ANSWERS

For You To Do

1. If you want an accurate reading, you can go to a good quality Atlas, or phone the local weather station. You could also ask any of the students, if they know how to get a reading of the latitude from the North Star at night.

© It's About Time

ANSWERS

For You To Do
(continued)

2. a) Student activity.
   b) Student activity.
3. Student activity.
4. a) Student activity and data.

in summer. In fact, it is this changing elevation of the Sun which causes the winter and summer temperature differences.

a) Use the example below and the figure to the left to help you calculate the angle of the Sun at noon at the winter and summer solstices in your area.

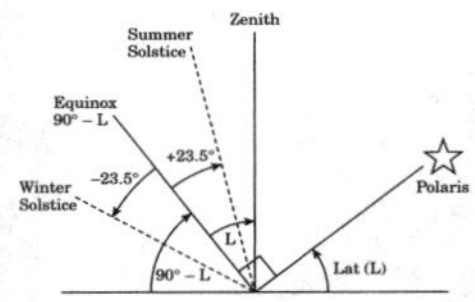

Finding Sun angles for a given latitude

Example:
The latitude of St. Louis, Missouri, is 38° north. Calculate the noon altitude of the Sun as viewed from St. Louis on the equinoxes and solstices.

Sun angle on equinoxes = (90°) minus (latitude angle)
= 90° –38°
= 52°

Sun angle on summer solstice = (Sun angle on equinoxes) plus (23.5°)
= 52° + 23.5°
= 75.5°

Sun angle on winter solstice = (Sun angle on equinoxes) minus (23.5°)
= 52° –23.5°
= 28.5°

b) Use the same figure and example to help you calculate the angle of the Sun at noon at the winter and summer solstices in the area for which you are designing a model home.

3. Set up your model home and the heat lamp as in previous activities.

4. Use a protractor and the Sun angle data to adjust the heat lamp so that it matches the angle position of the Sun at noon at the summer solstice.

a) Measure, record, and graph the heating and cooling curves for your model home.

© It's About Time

5. With the lamp turned on, place a small piece of tag board (or other lightweight cardboard) over one of the windows on the side of the model facing the heat lamp. Try different widths of tag board and the angles of attachment until you find the width and/or angle that will block the lamp's light entirely but still provide someone inside with a view. Use masking tape to attach awnings and/or roof overhangs to your model.
6. Repeat the heat transfer experiment on your model home that is now fitted with awnings or overhangs for summer solstice conditions.
   a) Record and graph your data.
7. Do not change the awnings or overhangs on your model, but adjust the lamp so its position matches the angle of the Sun at the winter solstice. Repeat the heat transfer experiment.
   a) Record and graph this data on the same graph.
   b) What differences do you note in the summer and winter solstice heating graphs? How do you explain the differences?

Reflecting on the Activity and the Challenge

One of the main requirements is that the universal dwelling must be able to be used in a wide variety of locations on Earth. Since the dwelling also depends on energy from the Sun, you must include in your design the knowledge about how solar energy varies with location on the Earth. This activity has prepared you to answer questions that the Architectural Committee will ask about how your design allows for solar energy at different locations.

Physics To Go

1. Imagine yourself standing outdoors at noon at your home latitude. You are facing south straight toward the horizon. You are doing this on four special days during the year, whose titles are listed on the next page. For each special day give its date, explain its meaning in terms of seasons, and then predict the angle of the Sun above the horizon.

ANSWERS

For You To Do
(continued)

5. Student activity.

6. a) Student data.

7. a) Student activity and data.

   b) There should be an increase in the heating of the home in the winter scenario. This is due to more direct Sun entering the window, where the inside is heated.

ANSWERS

Physics To Go *(see next page for Questions a, b, c, d and e)*

1. a) The vernal equinox is the position of Earth when the Sun is directly over the equator. The plane of Earth's orbit is perpendicular to its axis. This is the spring, and in the northern hemisphere the days are getting longer and they are heating up. The date is close to March 21.

   b) The summer solstice is when the axis is tilted at a 23.5° angle where the northern hemisphere is tilted towards the Sun. This gives us the longest days of the year, when the northern hemisphere is receiving more direct sunshine. The Sun is directly over the Tropic of Cancer. The date is close to June 21.

   c) The autumnal equinox is the position of Earth when the Sun is directly over the equator. The plane of Earth's orbit is perpendicular to its axis. This is the fall, and in the northern hemisphere the days are getting shorter and they are cooling down. The date is close to September 21.

   d) The winter solstice is when the axis is tilted at a 23.5° angle where the northern hemisphere is tilted away from the Sun. This gives us the shortest days of the year, when the southern hemisphere is receiving more direct sunshine. The Sun is directly over the Tropic of Capricorn. The date is close to December 21.

© It's About Time

ANSWERS

Physics To Go
(continued)

2. a) 90° - 25° = 65°
   b) 90° - 30° = 60°
   c) 90° - 35° = 55°
   d) 90° - 40° = 50°
   e) 90° - 45° = 45°

3. 26° + 23.5° = 49.5°

4. On the summer solstice, the Sun is directly over the Tropic of Cancer. The latitude of the Tropic of Cancer is 23.5°. Therefore, at noon on the summer solstice, the Sun will be shining on the back of the house (see diagram lower right).

5. Students provide rough sketch. (See diagram lower right.)

6. Awnings that can be retracted are good for those sides of the home. Therefore, in summer they can be extended, and in winter they can be drawn up. Another way is having trees or bushes that partially shade the window in summer. In winter when the leaves drop off, more light and more heat will get in. From the inside, curtains or shutters can be used in the hottest parts of the year.

7. It will depend on the area that the home is going to be built. If it is built in hot areas, then it would be worth the added expense.

8. Because the Sun is close to the horizon, the awnings would not be very effective. Therefore, shades can be drawn, to let only as much as you want, or to shut out the heat altogether.

a) Vernal equinox
b) Summer solstice
c) Autumnal equinox
d) Winter solstice

2. Latitudes in the continental United States vary from about 25° north to 50° north. Calculate the angle of the Sun above the south horizon on the summer solstice for the latitudes listed below:
   a) 25° North
   b) 30° North
   c) 35° North
   d) 40° North
   e) 45° North

3. One town in the United States is 2000 miles west of another town. At noon on the winter solstice at each town, the sun is observed to have an altitude of 26°. What is the latitude of each town?

4. The latitude of Mexico City, Mexico is 19° north. Do a calculation to show yourself that the Sun will shine on the north side of a home in Mexico City at noon on the summer solstice. State a rule for locations in Earth's northern hemisphere where this can happen.

5. Sketch a design for a roof overhang or window awning that will always shade a south-facing window from the noontime sun during the six months between the vernal and autumnal equinoxes, but will allow the noon Sun to shine on the entire window during the other six months of the year.

6. The Sun shines on windows on the east and west sides of a home, too, but for only part of the day. What could be done to control heat gains and losses through those windows?

7. Do you think the added cost of building a roof overhang or using awnings on your model dwelling would be worth the additional expense? Why or why not?

8. Shades are more effective than awnings for blocking sunlight from entering windows on the east and west sides of buildings. Why is this so?

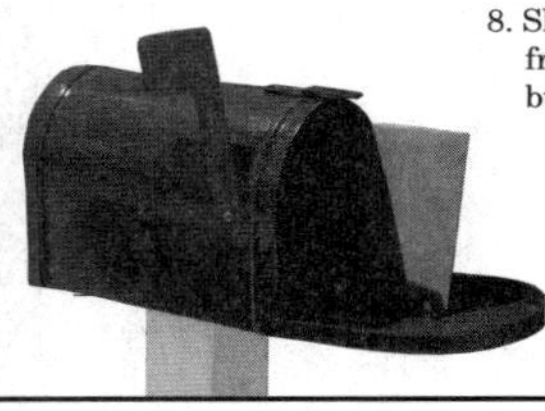

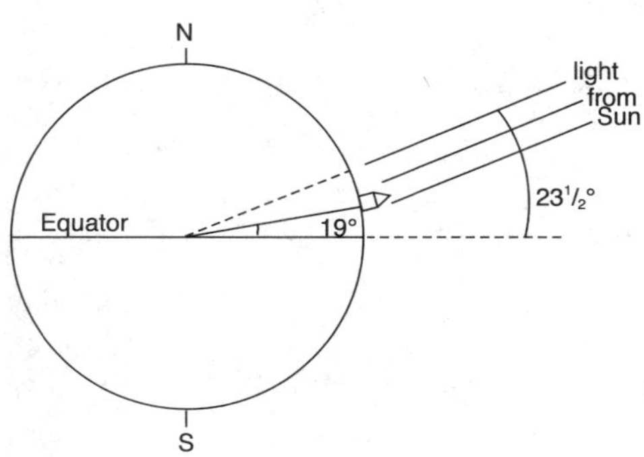

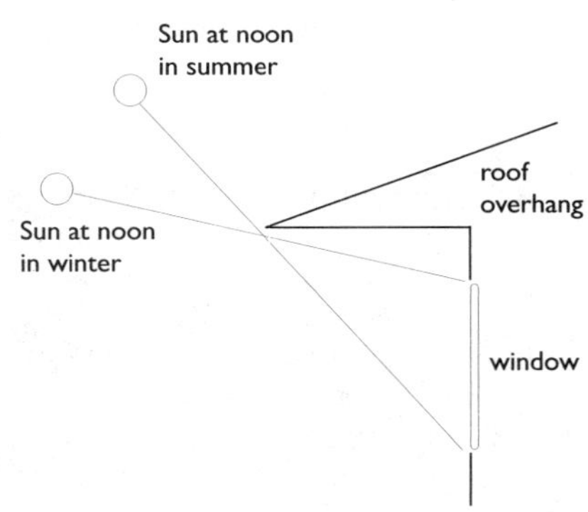

© It's About Time

Sun Location at Seasonal Times

Fall & Spring Equinoxes
40°N Latitude
Sun is directly over the Equator

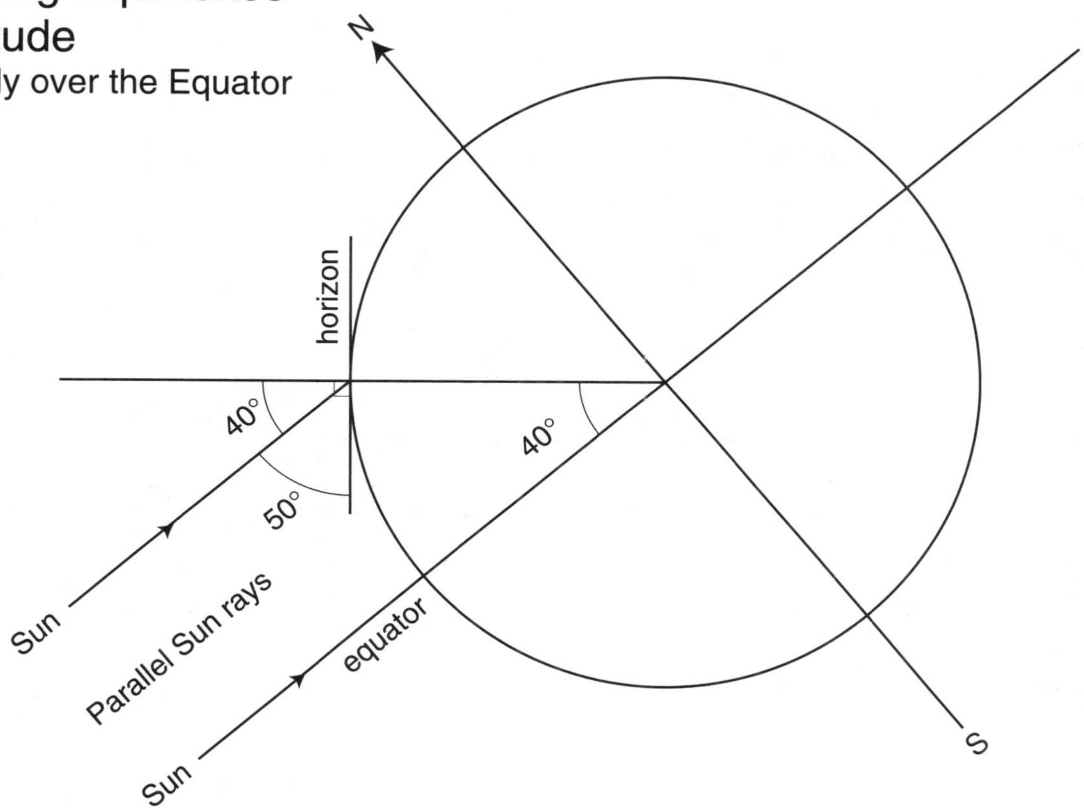

© It's About Time

For use with *Designing the Universal Dwelling*, Chapter 6, Activity 8: Investigating Overhangs and Awnings

Sun Location at Seasonal Times *(continued)*

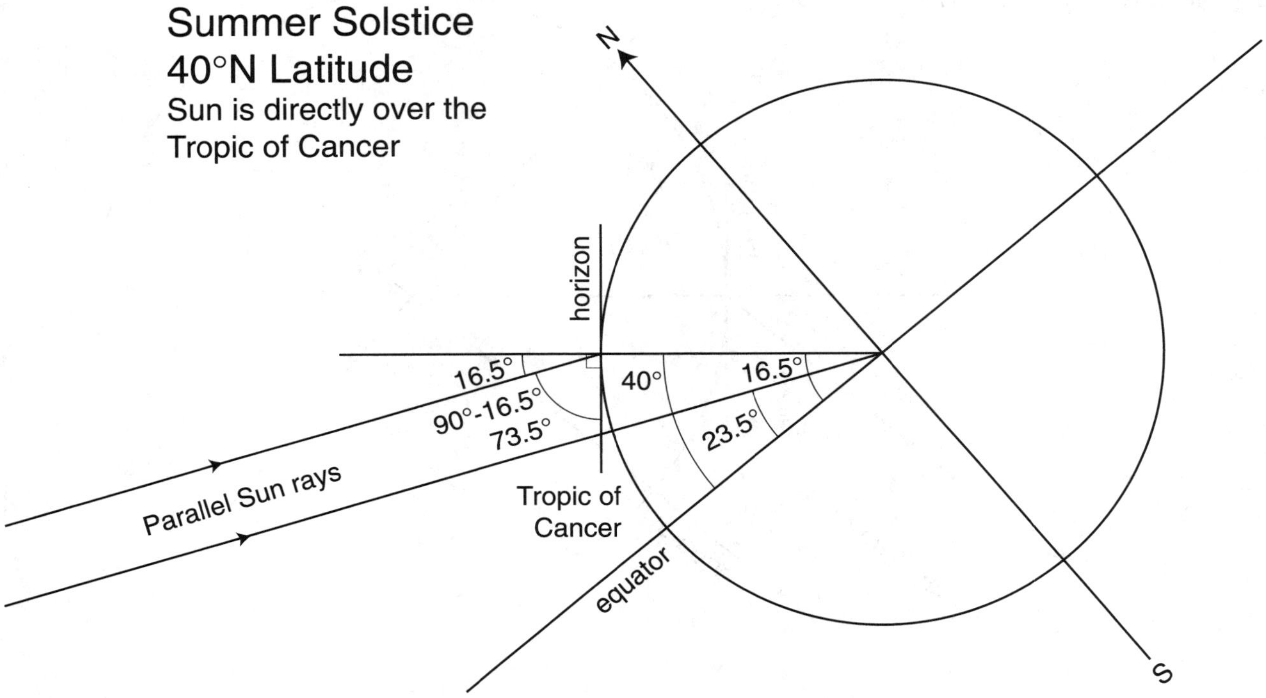

© It's About Time

For use with *Designing the Universal Dwelling*, Chapter 6, Activity 8: Investigating Overhangs and Awnings

Sun Location at Seasonal Times *(continued)*

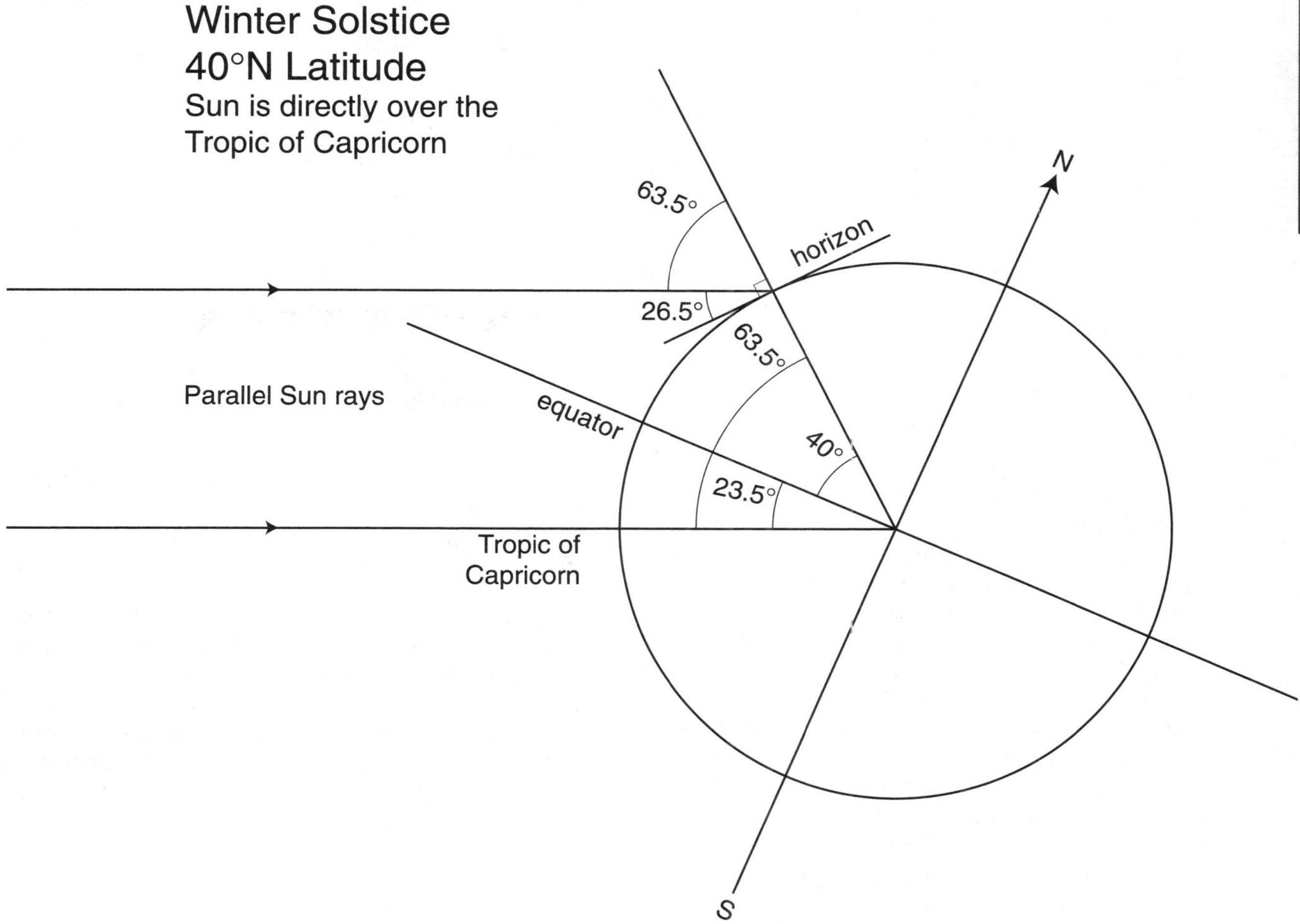

© It's About Time

For use with *Designing the Universal Dwelling*, Chapter 6, Activity 8: Investigating Overhangs and Awnings

ACTIVITY 9
Too Hot, Too Cold, Just Right

Background Information

It is suggested that the the **For You to Read** section of the student text for this activity be read before proceeding in this section.

There are well-documented studies that show the confusion students have in differentiating between temperature and heat. Temperature is a property of the material. On the macroscopic level, it is related to how hot or cold an object feels and is measured with a thermometer. On the microscopic level, it is related to the average kinetic energy of the molecules.

The heat energy is related to changes in temperature when placed in contact with an object of a different temperature. Large amounts of cold water mixed with hot water can lower the temperature considerably more than small amounts of cold water.

Hot objects placed in contact with cold objects never get hotter. Conservation of energy does not preclude this. The energy gained by the hot object could be equal to the energy lost by the cold object. Since it never occurs, we look to a restriction that nature places on these interactions. Entropy accounts for the direction of flow of the energy when objects are placed in thermal contact. Entropy also accounts for why all of the air in the room in which you are sitting does not all move to the other half of the room. The mathematics of entropy is too difficult for students in *Active Physical Science*. The concept of a directionality of time, or a means by which to distinguish past and future is something that will intrigue your students. They should be encouraged to research the topic of entropy and report to the class.

Planning for the Activity

Time Requirements

About one class period is required to complete this activity.

Materials Needed

For the class:

- paper towels
- pieces of metal (approximately 100 grams each)

For each group:

- calculator, basic
- graduated cylinder (100 mL)
- graph paper pad of 50 sheets
- Styrofoam® cup (4)
- tongs

Teaching Notes

Care should be taken when using the hot water. Impress upon the students how severe a burn can be.

Care should be taken when the hot metal is removed from the hot water. The metal should not have any water droplets on it before being submerged in the cold water. Drops of water will affect the final temperature of the water.

The activity gives students an opportunity to design their experiment of measuring the final temperature when hot metal is submerged into the cold water. The students should be able to extend their design to compare different metals. If different groups work with different metals, students will be able to draw conclusions concerning the specific heat of the metals.

© It's About Time

Activity 9 Too Hot, Too Cold, Just Right

GOALS

In this activity you will:

- Experimentally determine the final temperature when two liquids of different temperatures are mixed.
- Experimentally determine the final temperature when a hot metal is added to cold water.
- Calculate the heat lost and the heat gained of two objects after they are placed in thermal contact.
- Determine if energy is conserved when two objects are placed in thermal contact and reach an equilibrium temperature.
- Explain the concept of entropy as it relates to objects placed in thermal contact.

What Do You Think?

As you add cold milk to hot coffee, you expect that the milk will get a bit warmer and the coffee will get a bit colder.

- **What determines the final temperature of the coffee and milk?**

Record your ideas about this question in your *Active Physics* log. Be prepared to discuss your responses with your small group and the class.

Activity Overview

Students will predict the final temperature of mixtures of various amounts of hot and cold water. They will then conduct experiments to test their predictions. They will then design an experiment to test whether a hot metal of 100 g will heat up water as well as 100 g of water. From the results of this experiment, they will then be introduced to specific heat. Conservation of energy and entropy are described to provide a structure to understand the heat.

Student Objectives

Students will:

- Explain how the altitude and heating effect of the Sun varies with seasons.
- Experimentally determine the final temperature of a mixture of hot and cold water.
- Experimentally determine the final temperature of a mixture of a hot piece of metal and cold water.
- Calculate the heat lost and the heat gained of two objects after they are placed in thermal contact.
- Determine if energy is conserved when two objects are placed in thermal contact and reach an equilibrium temperature.
- Explain the concept of entropy as it relates to objects placed in thermal contact.

ANSWERS FOR THE TEACHER ONLY

What Do You Think?

The final temperature of the coffee-milk mixture will depend on:

- The initial temperature of the milk
- The initial temperature of the coffee
- The amount of coffee
- The amount of milk
- The type of milk (fat content – skim, whole, half & half)

Other factors that will also affect the final temperature is the container holding the coffee and how insulated it is, as well as air currents that can increase the evaporation rate of the liquid.

© It's About Time

ANSWERS

For You To Do

1. a) Answers will vary.

   b) Answers will vary.

   c) The prediction should be between the two values of the hot and cold water.

   d) The final temperature should be the average of the temperatures of the hot and cold water.

For You To Do

1. In this activity, you will determine the final temperature of a cold water and hot water mixture. Styrofoam® cups work well as containers in this activity. The insulation "protects" the experiment from the environment.

   Pour 100 mL of hot water into a Styrofoam cup. Measure its temperature.

   a) Record the temperature of the hot water. Measure the temperature of 100 mL of cold water.

   b) Record the temperature of the cold water.

   c) Predict the final temperature of the mixture of hot and cold water. Add the cold water to the hot water. Measure the final temperature.

   d) Record the final temperature.

Use a heat-proof holder while pouring, such as a glove or tongs.

2. Vary the experiment by changing the amount of cold water. Mix 100 mL of hot water with 50 mL of cold water; 75 mL of cold water; 125 mL of cold water; and 150 mL of cold water.

   a) Create a data table. Record your observations.

   b) Construct a graph of the results. Plot the final temperature on the y-axis and the amount of cold water added on the y-axis.

   c) Use your graph to predict the final temperature when 108 mL of cold water is added to 100 mL of hot water.

Clean up any spilled water immediately, especially off the floor so that no one slips on it.

3. The law of conservation of energy informs you that if the cold water gained heat energy (as indicated by its rise in temperature), then the hot water must have lost an equal amount of heat energy.

   The amount of energy change can be expressed using the following equation:

   $$\Delta Q = mc\Delta t$$

   where ΔQ is a measure of heat energy in joules;
   m is the mass of the substance in grams;
   c is the specific heat of the substance (the specific heat of water is 4.18 J/g°C)
   Δt is the change in temperature.

2. a) Answers will vary. The final temperature should decrease as the amount of cold water increases. Sample data:

Hot water volume (mL)	Cold water volume (mL)	Initial temperature of hot water (°C)	Initial temperature of cold water (°C)	Final temperature (°C)
100	50	80	10	57
100	75	80	10	50
100	100	80	10	45
100	125	80	10	41

   c) Using the sample graph the predicted final temperature would be 44° C.

© It's About Time

The equation requires you to use the mass of the water. For water, a volume of 50 mL has a mass of 50 g. Converting from volume to mass is easy for water, since the density of water is 1 g/mL.

$$D=\frac{m}{V} \text{ or } m=DV$$

Heat lost by hot water = Heat gained by cold water

$$(mc\Delta t)_{\text{hot water}} = (mc\Delta t)_{\text{cold water}}$$

Sample Problem

100.0 g of hot water at 80.0°C is mixed with 60.0 g of cold water at 20.0°C. The final temperature is 57.5°C. Show that energy was conserved.

Strategy: You can use the Law of Conservation of Energy to solve this problem.

Givens:

$m_h = 100.0$ g
$m_c = 80.0$ g
$c = 4.18$ J/g°C
$t_h = 80.0$°C
$t_c = 20.0$°C
$t_f = 57.5$°C

Solution:

Heat lost by hot water = Heat gained by cold water

$$(mc\Delta t)_{\text{hot water}} = (mc\Delta t)_{\text{cold water}}$$

$$(100.0\text{ g})(4.18\text{ J/g°C})(80.0\text{°C} - 57.5\text{°C}) = (60.0\text{g})(4.18\text{ J/g°C})(57.5\text{°C} - 20.0\text{°C})$$

$$9405\text{ J} = 9405\text{ J}$$

The energy lost by the hot water is gained by the cold water.

4. Cool water can be heated with the addition of hot water. How well would a piece of hot metal heat the cool water?

Design an experiment to compare the effect of adding 100 g of hot water and 100 g of heated metal to separate Styrofoam cups of cool water.

ANSWERS

For You To Do
(continued)

3. Sample problem is provided in the textbook.

4. Students are required to design an experiment.

   a) Student predictions will vary.

   b) Different metals do have an impact on the final temperature.

© It's About Time

ANSWERS

For You To Do
(continued)

5. Experimental design should take into account safety considerations.

6. The metal will not heat up the water as much as an equal amount of water. The value of *c* for the metal is less than the value of *c* for the water.

One way to heat the metal is to place it in a bath of hot water for three to five minutes. The length of time you need to keep the metal in the water bath depends upon the size of the metal. You can then use tongs to gently lift the metal from the hot water. As soon as the metal is out of the water, you will need to hold it over several pieces of paper towel folded to make a small mat and shake off drops of the hot water so that none of the hot water enters the beaker with the cold water. You want to try to place only the metal gently into the cold water.

Physics Words

temperature: a measure of the average kinetic energy of the molecules of a material.

a) Predict whether the individual cups of cool water will reach the same final temperature.

b) Do you think it matters what kind of metal you use? For example, will copper or zinc produce the same final temperature? Explain your answer.

5. Ask your teacher to comment on your experimental design. Make sure that your design is approved before you continue with the experiment.

6. Conduct the experiment.

a) Record your results.

b) Energy is conserved whether the cold water is mixed with hot water or hot metal. Look at this equation again:

Heat lost by metal = Heat gained by water

$$(mc\Delta t)_{\text{metal}} = (mc\Delta t)_{\text{water}}$$

The value of *c* for the metal may be different from the value of *c* for the water. Determine the value of the specific heat, *c*, for the metal that you used to heat the water.

c) Predict what would happen to the temperature of the cold water if the same amount of a different hot material with a lower specific heat than the metal you used were placed in the same volume of cold water.

d) Explain in your own words what the specific heat allows you to predict.

© It's About Time

FOR YOU TO READ

Conservation of Energy

Energy is conserved. Kinetic energy can become gravitational potential energy and vice versa. When an object like a book is dropped to the ground, the gravitational

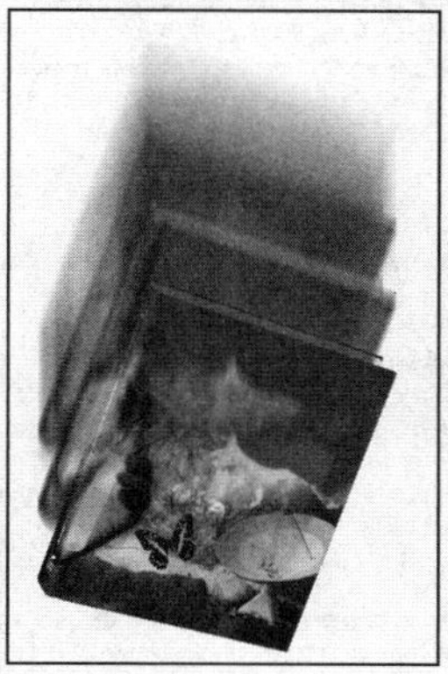

potential energy becomes kinetic energy as the book gains speed. As the book hits the ground and stops, some of the kinetic energy is converted to sound, as you hear a "thump." The rest of the kinetic energy of the book becomes heat energy and the temperatures of the book and of the ground both rise a bit.

You can calculate gravitational potential energy and kinetic energy. You also now know how to calculate heat energy. Heat energy is part of the total energy picture. Conservation of energy demands that the sum of all of the energies must remain constant.

Temperature and Heat

Temperature and heat energy are not the same. Temperature is a measure of the kinetic energy of the molecules of the material. You can measure this energy with a thermometer. The temperature and the kinetic energy of the molecules change when the object touches a material of a higher or lower temperature.

Heat energy has to do with the nature of the material, the mass of the material, and the temperature of the material. For example, 100 g of hot water has more energy than 100 g of cold water because of a difference in temperature. A swimming pool of 10,000 kg of cold water will have more energy than 1 kg of hot water, mainly because of a difference in mass. If the 1 kg of hot water is poured into the swimming pool of 10,000 kg of cold water, the temperature of the pool water will rise a tiny amount. The temperature of the hot water will drop considerably.

When you mix cold milk with hot coffee, you expect the cold milk to warm a bit and the coffee to cool a bit. They will soon arrive at the same temperature. This can be explained using the conservation of energy. The milk gained some energy and the hot coffee lost some energy. It might be clearer if you look at some numbers. If the cold milk is at 5°C and the hot coffee is at 90°C, the final temperature of the milk-coffee

→

Chapter 6

© It's About Time

mixture could be 88°C. In this case the temperature of the milk rose 83°C and the temperature of the coffee fell 2°C. Energy was conserved. If you knew the mass of the coffee and the milk, you could compute the gain and loss of the energy by each substance using the relation $Q = mc\Delta t$. The change in energy of both the milk and the coffee would be the same.

This situation is quite common. If you put a piece of cold metal into the coffee, there would be a similar effect. The coffee could cool from 90°C to 88°C and the metal could warm from 5°C to 88°C. If you knew the masses of the metal and the coffee and the specific heat of the metal, you could again compute the gain and loss of the energy with the relation $Q = mc\Delta t$. Again, the change in energy would be the same for the metal and for the coffee.

When a piece of cold metal is placed in hot coffee, it never happens that the metal gets even colder and the coffee gets even hotter. It never happens that the coffee heats up from 90°C to 92°C and the metal cools from 5°C to 3°C. The conservation of energy would be satisfied if the cold metal lost energy and the coffee gained energy.

If something never happens, you must assume that nature has placed a restriction on it. This restriction is called **entropy.** It informs you that the two materials in contact will reach a common equilibrium temperature. The transfer of heat energy can only take place in one direction. A cooler metal will heat up when placed in contact with the hot coffee, but the cooler metal will never become cooler when placed in contact with the hot coffee.

This irreversibility of heat and the related concept of entropy help to distinguish the past from the future. If you watch a movie of a pendulum moving back and forth, you may not be able to tell whether the film is being played forward or backward. If you watch someone break an egg and fry it, the film would look quite silly when played backward. It doesn't make sense that the egg could get un-fried and then return to its shell.

© It's About Time

Reflecting on the Activity and the Challenge

As you design your universal dwelling, you must be very aware of the temperature. You have investigated what happens when you change the shape of the dwelling or add windows. You also know that a cold drink will warm up if it sits on the table and that a hot drink will cool down if it sits on the same table. All objects in the dwelling will reach the same final temperature – the equilibrium temperature. Since water requires lots of energy to heat up or cool down, some people have used water or ice to provide comfort in their dwellings. You can now calculate the energy changes when cold objects and warm objects are put in contact. You may want to consider cold water or ice as a type of air conditioning unit. Similarly, you may want to consider the idea of heating water by solar energy and then moving that hot water through the house to heat up the air.

Physics Words

entropy: a measure of the degree of disorder in a system or a substance.

Physics To Go

1. A hot cup of coffee at 90°C is mixed with an equal amount of milk at 80°C. What would be the final temperature if you assume that coffee and milk have identical specific heats?
2. Explain why heating up a pot of water when you only need enough for one cup of tea is wasteful of time and wasteful of energy consumption.

ANSWERS

Physics To Go

1. Since there are equal amounts of coffee and milk, the temperature will be the average of the two initial temperatures. The final temperature will be 85°C.
2. If you only need one cup of water, heating up more than this will require more energy. The more water heated, the more energy is required. It will also take more time to heat the additional water. If it takes more time and more energy, why do people heat up a kettle of water to make a single cup of tea?

© It's About Time

3. A container of water can be heated with the addition of hot water or the addition of a piece of hot metal. If the mass of the water is equal to the mass of the metal, which material will have the greatest effect on the water temperature? Explain your answer.

4. Suppose 200 g of water at 50°C is mixed with 200 g of water at 30°C.
   a) What will be the final temperature?
   b) Calculate the energy gained by the cold water.
   c) Calculate the energy lost by the hot water.

5. Suppose 200 g of water at 50°C is placed in contact with 200 g of iron at 30°C. The final temperature is 48°C.
   a) Calculate the energy gained by the iron. The specific heat *c* of iron is 0.45 J/g°C.
   b) Calculate the energy lost by the hot water.
   c) If the final temperature could have been measured more accurately, would you expect that it would have been a bit more or less than 48°C? Why?

6. Suppose 100 g of water at 50°C is placed in contact with 200 g of iron at 30°C. The final temperature is 46.5°C.
   a) Calculate the energy gained by the iron. The specific heat *c* of iron is 0.45 J/g°C.
   b) Calculate the energy lost by the hot water.
   c) If the final temperature could have been measured more accurately, would you expect that it would have been a bit more or less than 46.5°C? Why?

7. Suppose 300 g of water at 50°C must be cooled to 40°C by adding cold water. The temperature of the cold water is 10°C. How much of the cold water must be added to the hot water to bring the temperature down to 40°C?

ANSWERS

Physics To Go

(continued)

3. The hot water will have a larger effect on the final temperature. Water has a larger specific heat. It takes more energy to heat the water in the first place than to heat the metal.

4. a) Since both the hot and cold waters have equal masses (200 g) and since they have the same specific heats (they are both water), the final temperature will be the average of the two temperatures. The final temperature will be 40° C.
   b) Heat gained by cold water
   $= mc\Delta T$
   $= (200\text{ g})(4.18\text{ J/g°C})(10°\text{C})$
   $= 8360\text{ J}$
   c) Heat lost by warm water
   = Heat gained by cold water
   Heat lost by warm water
   = 8360 J

5. a) Heat energy gained
   $= mc\Delta T$
   $= (200\text{ g})(0.45\text{ J/g°C})(18°\text{C})$
   $= 1620\text{ J}$
   b) Heat energy lost
   $= mc\Delta T$
   $= (200\text{ g})(4.18\text{ J/g°C})(2°\text{C})$
   $= 1672\text{ J}$
   c) The final temperature would have been a bit higher so that the energy lost would be equal to the energy gained.

6. a) Heat energy gained $= mc\Delta T = (200\text{ g})(0.45\text{ J/g°C})(16.5°\text{C}) = 1485\text{ J}$
   b) Heat energy lost $= mc\Delta T = (100\text{ g})(4.18\text{ J/g°C})(3.5°\text{C}) = 1463\text{ J}$
   c) The final temperature would have been a bit less so that the energy lost would be equal to the energy gained.

7. Heat energy lost $= mc\Delta T = (300\text{ g})(4.18\text{ J/g°C})(10°\text{C}) = 12\,540\text{ J}$

Heat energy lost = Heat energy gained

$$mc\Delta T = mc\Delta T$$

$$mc(Tf - Ti) = mc(Ti - Tf)$$

$$(300\text{ g})(4.18\text{ J/g°C})(10°\text{C}) = (m)(4.18\text{ J/g°C})(30°\text{C})$$

$$m = 100\text{ g}$$

© It's About Time

PHYSICS AT WORK

Ray Aguilera

HABITAT FOR HUMANITY

Ray is the construction manager at the Valley in the Sun, Arizona, Habitat for Humanity branch. Habitat for Humanity International is a non-profit organization that seeks to eliminate poverty, housing, and homelessness from the world. They do this by building and selling homes for no profit, to families who cannot get conventional financing. Homeowners also become partners in the process by contributing 500 hours of "sweat equity" toward the construction of their own home.

Our former president, Jimmy Carter, has been deeply committed to Habitat since 1984. Each year former President Carter and his wife, Rosalynn, join Habitat volunteers to build homes and raise awareness of the critical need for affordable housing.

Ray Aguilera was on his way to becoming a lawyer, when one summer he got a job building homes, and he has never stopped. Ray gets a great deal of satisfaction from his work with Habitat. "I enjoy going from an empty lot and watching something magical grow out of it," he says.

Arizona has a different climate than many other places in the country and takes special considerations when planning homes. In this southwestern desert, the temperatures are often in the 100s and rarely very cold. For example, the foundations and footings for houses do not need as much concrete as other places because the soil conditions are so different. And, in Arizona, you never have to worry about winter frosts. "The most essential characteristic of building a house in Arizona," states Ray, "is to keep it energy efficient. We also strive to design houses that will blend in with the existing environment. In Arizona, we make more of an effort to keep the hot out and the cold in. To do this we use double-paned windows, and as much insulation as we can fit between the walls and in the attic."

Chapter 6

© It's About Time

Chapter 6 Assessment

You and your group have been investigating the design of your model universal dwelling for this entire chapter. You have learned a great deal about this problem and have explored possible solutions. You now have reached the point where you must finalize your plans and prepare your presentation for the hearing before the Homes for Everyone (HFE) Architectural Committee. Good luck!

Using the resources that you have been provided, develop scale drawings of your model universal dwelling. Refer to the **Challenge** given at the beginning of this chapter to guide your group's activities. Your presentations will be strictly limited to five minutes, so make every drawing and sentence count!

Review the criteria that you decided the HFE Architectural Committee should use in evaluating your drawings and your written presentation.

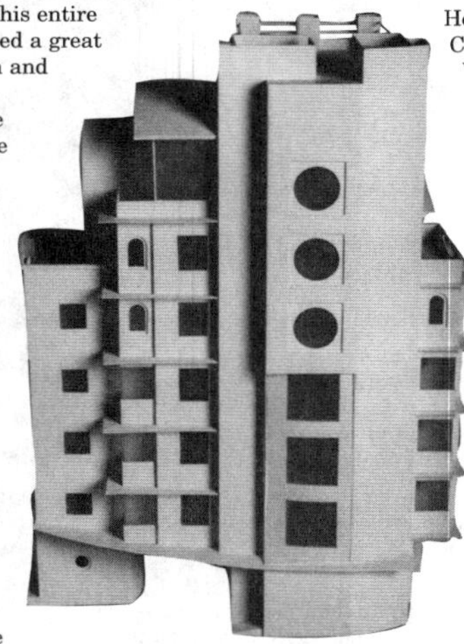

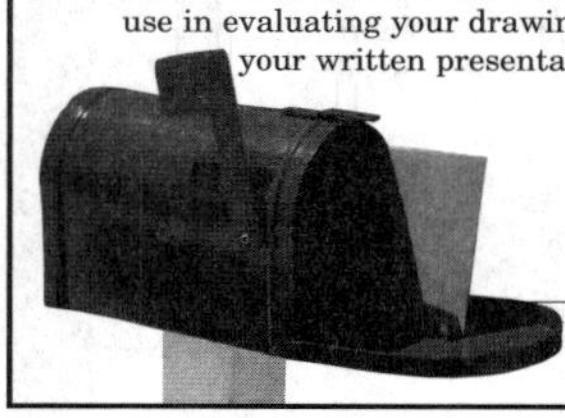

Physics You Learned

Surface area-to-volume ratio

Brief statements of conduction, convection, radiation

Heating curves and Cooling curves
- With opaque backgrounds
- With windows
- With insulation
- With awnings

Properties of insulation

Effect of thickness on insulation effectiveness

Brief review of latitude and solar position at equinox

© It's About Time

Alternative Chapter Assessment

Select the best response for each statement or question.

1. Use the dimensions shown in the drawings of the outside walls of Houses A, B, and C to compare the total floor areas of the houses in square feet. The drawings show views from above the houses.

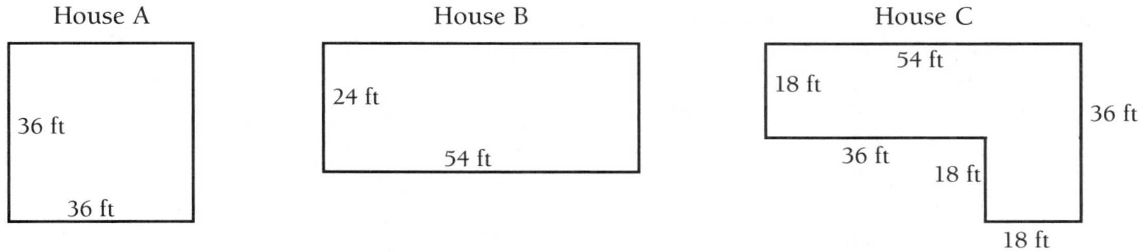

   a) House A has the greatest floor area.
   b) House C has the smallest floor area.
   c) House B's floor area is less than House A and greater than House C.
   d) The floor areas of the three houses are equal.

2. In six months the angle at which sunlight strikes a location on Earth's surface at noon can change by a maximum of:
   a) 23.5°
   b) 45°
   c) 47°
   d) 90°

3. The vernal equinox occurs on:
   a) March 21
   b) June 21
   c) September 23
   d) December 21

4. The summer solstice occurs on:
   a) June 1
   b) June 21
   c) July 4
   d) July 21

5. The autumnal equinox occurs on:
   a) September 1
   b) September 23
   c) October 1
   d) October 31

© It's About Time

6. The winter solstice occurs on:
   a) December 1
   b) December 21
   c) December 31
   d) January 1

7. During one minute in full sunlight the amount of solar energy striking a glass patio door 6 feet wide and 6 feet high compared to a window 2 feet wide and 2 feet high is:
   a) 3 times greater than the window.
   b) 4 times greater than the window.
   c) 9 times greater than the window.
   d) 16 times greater than the window.

8. To prevent heat loss from most homes the most insulation is needed in the:
   a) ceiling.
   b) outside walls.
   c) floor.
   d) doors and windows.

9. Three ways in which heat is transferred are:
   a) covention, convection and radiation.
   b) conduction, conversion and radiation.
   c) conduction, convection and irritation.
   d) conduction, convection and radiation.

10. Which unit of measurement would be correct expressing surface area?
   a) feet
   b) square feet
   c) cubic feet
   d) square feet per cubic foot

11. Which unit of measurement would be correct expressing volume?
   a) feet
   b) square feet
   c) cubic feet
   d) square feet per cubic foot

12. Which unit of measurement would be correct for expressing individual living space in a home having two or more family members?
   a) square feet per cubic foot
   b) persons per foot
   c) square feet per person
   d) cubic feet

© It's About Time

13. Which unit of measurement would be correct for expressing the ratio of surface area-to-volume?
    a) square feet
    b) cubic feet
    c) square feet per person
    d) square feet per cubic foot

14. If the total distance around the outside walls of a house is limited to 120 feet, the shape formed by the walls to give the largest floor area at ground level would be a:
    a) square.
    b) rectangle.
    c) hexagon.
    d) circle.

15. If the total surface area of the ground floor of a home has been decided to be 1000 square feet, the three-dimensional shape for the home which would have the smallest surface area-to-volume ratio would be a:
    a) cube.
    b) rectangular solid.
    c) cylinder.
    d) hemisphere.

16. One-fourth of an inch on a 1:48 scale model represents how much real length?
    a) 3 inches
    b) 1 foot
    c) 4 feet
    d) 12 feet

17. A length of 8 feet should be made on a 1:48 scale model as a length of:
    a) 2 inches.
    b) 6 inches.
    c) 1/4 inch.
    d) 1/6 inch.

18. Radiation is the only way heat can transfer through:
    a) a glass window.
    b) outer space.
    c) a steel door.
    d) an insulated wall.

19. If other factors which affect the flow of heat through insulation stay constant, the effect of doubling the thickness of insulation in a wall or ceiling of a house should be to reduce the rate of heat flow through the insulation:
    a) by a factor of 2.
    b) by a factor of 4.
    c) to zero.
    d) not at all (no effect).

© It's About Time

20. If other factors which affect the flow of heat through insulation stay constant, the effect of doubling the difference in temperature between the inside and the outside surfaces of an insulated wall should be to:
    a) have no effect on the rate of heat flow through the wall.
    b) increase the rate of heat flow through the wall by a factor of 4.
    c) reduce the rate of heat flow through the wall by a factor of 2.
    d) increase the rate of heat flow through the wall by a factor of 2.

© It's About Time

Alternative Chapter Assessment Answers

1. d
2. c
3. a
4. b
5. b
6. b
7. c
8. a
9. d
10. b
11. c
12. c
13. d
14. d
15. d
16. b
17. a
18. b
19. a
20. d

© It's About Time

NOTES

© It's About Time

Chapter 7

ELECTRICITY FOR EVERYONE

Chapter 7- Electricity for Everyone

National Science Education Standards

Chapter Challenge

Homes for Everyone (HFE) again establishes the scenario, this time with the need for electricity for all homes. Wind is given as the universally available source for generating electricity. Students are challenged to write a manual that describes how to obtain electricity from the generator. They then must determine which electrical appliances can and should be provided for use in all homes built by Homes for Everyone.

Chapter Summary

To gain understanding of the content necessary to meet this challenge, students collaborate on activities in which they learn about electromagnetic motors, how electricity is generated, the use of wind to generate electricity, and the relative efficiency of different appliances. These experiences engage students in the following content identified in the National Science Education Standards.

Content Standards

Unifying Concepts

- Evidence, models and explanations
- Constancy, change and measurement

Science as Inquiry

- Identify questions and concepts that guide scientific investigations
- Use technology and mathematics to improve investigations
- Formulate and revise scientific explanations and models using logic and evidence

Science and Technology

- Abilities of technological design
- Understanding about science and technology

Science in Personal and Social Perspectives

- Natural resources
- Environmental quality
- Science and technology in local, national, and global challenges

Physical Science

- Motion and force
- Conservation of energy and increase in disorder
- Interactions of energy and matter

© It's About Time

Key Physics Concepts and Skills

Activity Summaries	Physics Principles
Activity 1: Generate Students investigate electrical energy and electric circuits with a simple hand generator, wires, and bulbs. Using the hand generator introduces the concept that electricity is the result of converting one form of energy into another.	• Energy conversions • Electricity • Mechanical energy • Simple circuits
Activity 2: Lighten Up An experiment in which students compare the relative brightness of light bulbs in series and parallel circuits and increases their understanding of energy flow necessary in generating electricity. They also learn that a generator has a limit of what it can support. Students also read to learn more about electrical energy.	• Energy conversions • Electricity • Mechanical energy • Series and parallel circuits
Activity 3: Ohm's Law Students calculate the resistance of an unknown resistor given the potential drop and current, construct a series circuit, properly use a voltmeter and ammeter in a series circuit, and graph the relationship between voltage and current for a resistor that obeys Ohm's Law.	• Ohm's Law • Voltage, Current, Resistance • Series and Parallel circuits
Activity 4: Load Limit To experience load limit, students add bulbs to their simple circuit until the load is exceeded. They then calculate the load limit of a household circuit and watts required by appliances, comparing these to the limits given in the challenge. This also introduces use of terms and equations for calculating power.	• Watts, voltage, currents • $P = IV$
Activity 5: Who's In Control? Students again assemble a parallel circuit to explore how switches control the flow of electricity to specific bulbs. They then propose and test strategies for using parallel circuits and switches to increase the number of bulbs on a circuit. Finally, they are introduced to circuit diagrams.	• Parallel circuits • Switches
Activity 6: Cold Shower Electricity used by water heaters is the focus of this activity, which also reinforces concepts of energy transfer. Students investigate the amount of energy in joules needed to raise the temperature of water, then calculate the efficiency of different water heaters. They also consider alternate solutions to the expectation for hot water in a home.	• Energy conversion • Power • Watts, joules • Specific heat
Activity 7: Pay Up Students compare different electric bills as they continue to explore concepts related to power consumption, efficiency of electrical appliances, and relative costs and benefits of specific electrical appliances. This activity also serves to review concepts units used to communicate about energy transformation.	• Heat energy • Electrical energy • Joules, watts • Energy flow and power • Electrical efficiency
Activity 8: More for Your Money Students conduct an experiment in which they determine and compare the power consumed and the efficiency of three systems that could be used to heat water. They apply data collected to confirm their response to the challenge in which they recommend appliances for the universal home.	• Energy conversion • Energy flow and power • Electrical efficiency

© It's About Time

Equipment List For Chapter 7

PART	ITEM	QTY	ACTIVITY	TO SERVE
AC-6004-H2	Alligator Clip Leads	14	1, 2, 5	group
BK-0250-H2	Beaker, 250 ml	1	8	group
BK-1000-H2	Beaker, 1L	2	8	group
BS-5870-H2	Blinking Bulb	3	1	group
BS-7705-H2	Set Sample Electric Bills	1	7	group
CM-1108-H2	Calculator, Basic	1	3, 5, 6, 7	group
CS-0024-H2	Cylinder, Graduated, 100 ml	1	6, 7	group
CS-6047-H2	Cup, Styrofoam	2	6	group
GH-7704-H2	DC Generator, Hand-Operated W/Wire Leads	1	1, 2, 5	group
HS-5630-H2	Hot Plate, Electric	1	8	group
HS-3875-H2	Heater, Electric Immersion	1	6, 7	group
LS-6959-H2	Miniature Bulb, Light	5	1, 2, 5	group
LS-6960-H2	Bulb Base For Miniature Screwbase	4	1, 2, 5	group
RH-6132-H2	Rod, Stirring	1	6, 7	group
RS-0310-H2	Resistors, ohm 5, 10 Watt	4	3	group
RS-1010-H2	Resistors, ohm 10, 10 Watt	4	3	group
RS-1510-H2	Resistors, ohm 15, 10 Watt	4	3	group
SS-2393-H2	Pad of Coarse Grade Steel Wool	1	1	group
SS-7778-H2	Stopwatch	1	6	group
SS-9019-H2	Switch, Spst.	3	5	group
TS-3602-H2	Thermometer	1	6, 7	group
VI-0001-H2	V/I Meter	2	3	group
ZZ-0027-H2	27 qt. Clear Plastic Storage Container	1		group
TEACHER DEMONSTRATION ITEMS				
CH-7703-H2	Circuit for Demonstrating Overload, 120 Volt	1	4	class
FU-0003-H2	Fuses, 3 Amp	5	4	class
FU-0008-H2	Fuses, 8 Amp	5	4	class
THINGS NEEDED NOT SUPPLIED				
	Microwave Oven	1	8	class
	Source of cold water	1	5, 8	class
	Source of hot water	1	5, 8	class
	1100 Watt Hair Dryer to blow fuses	1	5	class
	Electric Calorimeter	1	6	class

Organizer for Materials Available in Teacher's Edition

Activity in Student Text	Additional Material	Alternative/Optional Activities
ACTIVITY 1: Generate p.446		Activity 1A: Generator Principles p.256
ACTIVITY 2: Lighten Up p.450	A Water Model for Electricity p.265	
ACTIVITY 3: Ohm's Law p.455		
ACTIVITY 4: Load Limit p.463		
ACTIVITY 5: Who's In Control? p.472	Circuit Diagrams p.309	
ACTIVITY 6: Cold Shower p.486	Home Electrical Appliances: Power and Energy Use for a Family pp.324-325	
ACTIVITY 7: Pay Up p.498	Sample Electric Bills for New York and Connecticut pp.333-334	
ACTIVITY 8: More for Your Money p.503		

© It's About Time

Throughout history people have tried to help other people. However, changes often have been made without any respect for the personal and cultural needs of those who are being assisted.

If you ever become involved in a self-help community group, it would be important for you to work together with the people you are helping in assessing their needs and their capabilities. Although that is not possible given your limited time in class, you should recognize the need for this type of collaborative teamwork when assisting people.

Scenario

The Homes For Everyone (HFE) Architectural Committee has just accepted your design for a "universal dwelling" to meet the growing housing shortage in many diverse areas of the world. The organization would now like you to develop an appliance package that would help meet the basic needs for healthy, enjoyable living for the families who will reside in the universal dwellings.

The source of electrical energy chosen for this particular project is a wind generator. The following is a description of the wind-generator system chosen for HFE. Try to get a sense of the meaning of unfamiliar words. When the chapter is completed, you will understand these terms.

The wind-generator system chosen for HFE is a highly reliable, mass-produced model that has an output of 2400 W (2.4 kW). Experience has shown that in areas having only moderate average wind speed (6 to 8 km/h) the generator system will deliver a monthly energy output of about 90 kWh (kiloWatt-hours) to the home, or about 3 kWh per day.

Direct current (DC) from the wind-driven generator is stored in batteries that allow storage of electrical energy to keep the home going for four windless days. The batteries deliver DC electricity, but most home appliances are designed to use alternating current (AC). An inverter changes the DC from the batteries into AC before it enters the home. A circuit breaker rated at 2400 W protects the batteries from overheating if too much energy is asked for at any single time. Finally, a kiloWatt-hour meter is provided to keep track of the amount of electrical energy that has been used. The result is that the dwelling will have the same kind of electricity delivered to it as do most homes in the U.S., but less electrical power and energy will be available than for the average homes in the U.S.

444

© It's About Time

Chapter and Challenge Overview

Students will be studying electricity, first of all, through their own eyes by exploring how a light bulb works, and then producing a list of terms about electricity. This offers an opportunity for the teacher to catch any misconceptions that students may have regarding electricity.

Understanding about electricity will be the primary objective of the first few activities. The students will need to have a basic understanding of circuits, series and parallel circuits, safety and circuits, and switches. At this level, it is not necessary to go into a lot of detail when discussing electricity, but the teacher will be able to gauge what is too much for their students and what is a sufficient challenge for them.

The students will next have an opportunity to apply their understanding of electricity, in order to analyze transfer of energy in the heating of water, and ultimately the cost of using electricity. The hope is that they will be able to apply that knowledge and understanding to build their HFE home, and finally to their own lives.

Students will be analyzing the difference between an appliance that uses 1200 W of power, and one that uses 750 W of power. This is to enable them to understand cost-effectiveness, and areas in which they will be able to save electrical energy.

Finally, students will be analyzing electricity bills, in order to appreciate that the cost of electric energy is not simple, but affected by a variety of factors. Again, in understanding the cost of electricity, students will better appreciate the real cost of energy, and bring to their own lives some appreciation for the conservation of energy sources.

You may wish to point out that wind generators often were used to provide lighting for remote homes, especially farms, before rural electrification. You may also want to discuss the impact of the REA on life in rural America. If you wish to take the time, videos on wind generators are available; Hawkhill Associates, for example, has one. Data on the availability of wind in various locations is not necessary for executing the chapter. The availability of moderate wind is assumed for all locations. If you or your students wish to check on the wind in various locations, data is available. Data on annual mean wind speed for various locations in the United States is available in the *Climatic Atlas of the United States*. Daily wind data is available from telecommunications-based sources such as Weather Machine, available to schools on a subscription basis. Also, global wind data may be available from the same references used in earlier chapters.

© It's About Time

Challenge

You will use your experience with electricity in your home and what you learn in this chapter to decide which electrical appliances, powered by a wind generator, can and should be provided for the HFE dwellings.

1. Your first task is to decide what electrical appliances can and should be used to meet the basic needs of the people whose HFE dwelling will be served by a wind generator.
   - **Use the list of appliances on pages 496-497, any additional information that you can gather about appliances, and the characteristics of the wind-generator system to decide what appliances to include in an "appliance package" for HFE.**
   - **As part of your decision-making process, determine if it seems best to provide a basic appliance package that would be the same for all dwellings, or if packages should be adapted with "options" to allow for factors such as different family sizes, climates, or other local conditions.**
   - **Describe how each appliance in your package will contribute to the well-being of the people who live in the dwelling.**
2. Your second task is educational. The people will need to be instructed how to stay within the power and energy limits of their electrical system as they use their appliances.
   - **You must outline a training manual for volunteers who will be going into the field to teach the inhabitants about the HFE wind-generator system and the appliances. The volunteers have no special knowledge of electricity. Therefore, the volunteers need a "crash course" that will prepare them to teach the people to use their electrical system with success.**
   - **Two factors will be especially important to teach: the power demand of the combination of appliances being used at any one time may not exceed 2400 W, and the average daily total consumption of electrical energy should not exceed 3 kWh.**

Criteria

The criteria for this challenge will be judged on the basis of 100 points. Discuss the criteria below, add details to the criteria if it would be helpful, and agree as a class on the point allocation.

- **The list of appliances to be included in the HFE appliance package must be as comprehensive as possible, and it must be clear how each appliance will enhance the health or well-being of the people who live in the dwelling.**
- **The outline of the training manual for HFE volunteers must explain the difference between 2400 W and 3 kWh. It must also give clear examples of how use of the appliances in the package can be scheduled to stay within the power and energy limits of the electrical system on both a daily and a long-term basis.**

© It's About Time

Assessment Rubric for Challenge

This is a possible rubric for evaluating the challenge in this chapter. It has some criteria, but it would be advantageous for the teacher to go over the criteria with the students, and have the students add any that may have been overlooked. Try to have the number of points be a multiple of 10, for ease of calculation of a mark out of 100. However, whatever the criteria, a mark of 100 can be given using percent.

Descriptor	5	4	3	2	1
Appliance Package					
comprehensive list of appliances					
appliances are energy-efficient					
appliances are useful (not frivolous e.g., electric can opener)					
choice of appliances is within the maximum power usage					
appliance package has options for different climates, family sizes, etc.					
appliance package has a description of benefits to people's well-being					
package is aesthetically pleasing					
Training Manual					
concise, simple language					
explains circuits					
explains AC and DC current					
explains series and parallel circuits					
explains power, current, and voltage					
explains the difference between 2400 W and 3kWh					
explains the need to stay within the power and energy limits, both daily and long term					
Appearance of Manual					
easy to read					
professional quality					
well laid out					

© It's About Time

For use with *Electricity For Everyone*, Chapter 7

What is in the Physics InfoMall for Chapter 7?

When using the Physics InfoMall CD-ROM, it is easy to get in the "search" mode, where every time you want something, you engage the search engine. Don't forget that this InfoMall is also great for simply browsing. For example, this *Active Physical Science* book is all about the physics of the home. Is there something on the InfoMall that we might expect to provide plenty of information? There is; Madalyn Avery's *Household Physics*. This chapter of *Active Physical Science* deals with electricity. A look at the table of contents from *Household Physics* uncovers several chapters that might be useful: Sources and uses of electricity, Magnetism, Electrostatics, Sources of electrical energy, Simple electromagnetic generators, Electrical measurements, Resistance in series and in parallel, Electric heating devices, Electric lights, Electric motors, Chemical effects of a current, Transformers, and House wiring. We should not be surprised when these appear as search results later. However, using the powerful search engine on the InfoMall is still a great method for becoming familiar with the contents of the CD-ROM.

If you are already familiar with using the search engine, you may wish to skip this paragraph and the next. The search engine for the InfoMall enables the user to locate almost any word or passage included in the more than 3000 articles, 19 textbooks, etc. If you use a Macintosh (remember—the CD-ROM works on both PC and Mac), you can search the entire database (that is, all stores) at once. This discussion was prepared using a Mac. If you use a PC, there will be minor differences but the database is identical. The same information can be found on either platform. There are two categories for searches: Simple and Compound. Simple searches look for a single word or phrase. Compound searches can locate multiple words or phrases that may not occur together. The Compound search also provides search options that are not available in the Simple search. Since a Compound search can always be done for a single word, it provides the same function as a Simple search. All searches in this discussion will be done using the Compound search.

The search option can be found under the "Functions" menu. When you select "Compound Search", you will get a window that has all the options you need for just about any search you are likely to want. Among these choices are database selection (which store or stores you want searched), category (perhaps only certain articles), and search words. You can enter any words or words you want the engine to search for. These words can be combined with the logical operators AND, OR, and NOT. That is, you can find passages that contain, for example, "wind" AND "generator" but NOT "politician". In addition, you can use an asterisk * as a wild character. That is, search word "fun*" will search for any word beginning with "fun", such as "fun", "funny", "funding", "fundamental", etc. Wild characters allow great freedom in searching, but as this example shows, you need to be careful. And finally, if your search is too broad, you will get "Too Many Hits" and you will have to restrict your search parameters to something that will not be found so many times on the CD-ROM. Such restriction techniques include searching in fewer stores, using fewer wild characters, or searching for less common words followed by a search of only the search hits. With a little practice, you should have very little trouble finding the information you desire.

The **Scenario** mentions a wind generator. It may be a good idea to have some understanding of generators in general before jumping into this chapter. You never know what students will ask, but "how does a generator work?" seems likely. To find an answer (or several answers) search the InfoMall for "generator*". The list of hits includes articles and chapters on such things as Van de Graaff generators and function generators, but the list also includes information relevant to our needs. For example, there is a chapter on "Generation and Transmission of Electricity" in

© It's About Time

Philip DiLavore's *Energy: Insights from Physics* found in the Textbook Trove. Not surprisingly, Household Physics is also on this list. There are many items on this list, and you should check them out. Some will be relevant, others will simply be interesting.

The Physics InfoMall was created by physicists who are interested in how physics is taught and learned. Therefore, many of the references in the database relate to physics education. These can be especially helpful. Electricity is a great example of how helpful this can be. Perform a search using "student" AND "difficult*" AND "electric*" on the entire CD-ROM. This search initially provides "Too Many Hits" so it must be limited to Terms Must Appear in the Same Paragraph. This finds several references on the difficulties students have with concepts related to electricity. Knowing these difficulties will allow the teacher to expect them, and have ideas on how to overcome them. A similar search can be done using "misconcept*" in place of "difficult*". The results are not identical, so both searches should be tried. It may also be interesting to note that most of the search hits are journal articles. The remainder are from the Book Basement. None are from textbooks.

What is in the Sun's Joules CD-ROM for Chapter 7?

The Sun's Joules CD-ROM is a comprehensive collection of information resources developed by the Center for Renewable Energy and Sustainable Technology (CREST), funded by the United States Department of Energy and directed by the National Renewable Energy Laboratory (NREL). It provides easy access to a great deal of information about new technological advances in utilizing energy resources. As a supplemental resource, it will provide enjoyable enhancement materials to the basic concepts and activities of the chapter. Its features include: a glossary with definitions of energy-related terms; an index with information on its main topic areas and a list of references for each of its topic areas. Much of the information on the disc relates directly to the **Stretching Exercises** and **Inquiry Investigations** found in the chapter.

One of the main menu topic sections from the Sun's Joules is entitled "Wind Energy" and therefore corresponds very specifically to the context of the **Scenario** for **Chapter 7** in which students are challenged to write a manual that describes how to obtain electricity from a wind generator. It contains a wealth of information that will directly help students understand this energy source. Also, another nice enhancement for the chapter comes from one of the interactive activities found in the Photovoltaics section of the disc. In this exercise, students drag appliances into or out of a house and observe the different energy consumption used for each appliance, making it a complementary exercise for their **Chapter Challenge**.

What is in the Green Home CD-ROM for Chapter 7?

In the Green Home CD-ROM, students are given the opportunity to walk through a digital home exploring important environmental and technology features that will help them in their understanding of generating electricity and energy consumption. They will therefore find this to be a valuable resource as they think about the many different types of appliances and the relative efficiency of different appliances as they complete their **Chapter Challenge**.

There are also a number of interactive activities in this CD-ROM in which students can manipulate various aspects of a house to see how different appliances, fixtures, designs and practices lead to different amounts of resource consumption. For example, they have the opportunity to compare the energy and economic performance of conventional incandescent lighting and compact fluorescent lighting.

Also, much of the information on the disc relates directly to the **Stretching Exercises** and **Inquiry Investigations** found in the chapter. For example the "Lighting Activity" will help students in the **Stretching Exercise** of **Activity 1** which asks them to investigate different types of light bulbs. The "Indoor Water Use Activity" relates directly to the **Stretching Exercise** from **Activity 6** in which students are asked to research ways to reduce the amount of electrical energy needed to provide hot water to a dwelling.

© It's About Time

ACTIVITY 1
Generate

Background Information

The physics principles involved in **Activity 1** are:

- resistive heating by an electric current (light bulb effect).
- production of electricity using a generator.
- energy transformations including conservation of energy.

Resistive Heating. As students use a hand-operated generator to drive an electric current through light bulb filaments and a short strand of steel wool, the current heats the metal resulting in the release of heat and light. Although it is not intended to quantify this process with students at this stage, it is important for you to understand that the mechanical power input provided by the student to crank the generator [Power = work/time = (force × distance)/time] is transferred—less, of course, the power expended overcoming friction in the generator—to the filament as electrical power [Power = Current × Voltage = Current squared × Resistance]. In turn, the electrical power is released by the resistive filament in the forms of heat and light. Similar transformations will be treated quantitatively later in this chapter.

Generator. Throughout this chapter the generator is meant to be treated as a "black box" by students; that is, the generator is used to produce electricity without understanding how or why it works. You as the teacher will be more comfortable and better prepared to respond to student questions about the generator if you understand its principles of operation.

Background Information about AC and DC generators provided for students from another part of the *Active Physical Science* course materials labeled Generator Principles is reproduced for you as the last section of this background discussion. The kind of generator used in this chapter is a DC generator, meaning that it produces direct (not alternating) current and voltage.

If you are not familiar with the operating principles of generators it is recommended that you study the Generator Principles section now.

The particular generator recommended for use in the Teacher's Guide for this and activities in a later chapter has a power output (current and voltage) which is generally proportional to the rate of cranking the handle (below a cranking rate which would damage the gears). The direction of the current reverses as the direction of rotation of the handle is reversed.

Energy Transformations. Perhaps the most profound thing to be learned about the generation of electrical energy in **Activity 1** is found in the "feel" of operating the generator with and without it delivering current to an external device such as a light bulb. When the generator is cranked without delivering a current only minimal effort is needed. This is the effort needed to overcome mechanical friction within the moving parts of the generator. However, when the generator is cranked to deliver a current to the circuit, the effort becomes noticeably greater. This is where a subtle, yet highly profound phenomenon comes into action. The greater effort upon delivery of current to an external device happens because as soon as current begins to flow within the internal wire coil of the generator, that current, in the presence of magnets within the generator, causes the generator to act as an electric motor which wants to run backward against the cranking action. This is known as the "back emf" (back electromotive force, or backward voltage) which always causes a generator to "fight back" any effort to make it produce an electric current. The harder and faster one may try to crank an operating generator, the more the generator fights back. Therefore, it is necessary to maintain as much input power to a generator as the desired output power; in fact, since friction always is present to be overcome, input power must always exceed output power. Electric generators do not produce free energy. For fullest understanding of this phenomenon you may also wish to consult a reference on the operating principles of electric motors.

For your information, light bulbs are filled with an inert gas to prevent oxidation of the metal (usually tungsten) filament. The strand of steel wool used in this activity oxides immediately because it is exposed to air.

The Sun is the primary source of energy in the chain of transformations which culminates in a student cranking a generator to light a bulb.

Active-ating the Physics InfoMall

One of the strengths of the InfoMall is the enormous number of demonstrations that can be found in the Demo & Lab Shop. As an example, consider **Step 3** of **For You To Do**. Steel wool is used in a circuit. Are there demonstrations that could also be used? (Note that a demonstration should not replace the student activity, but can be used in addition.) Try a search for "steel wool". The second hit on the list includes one

© It's About Time

way to use steel wool as a fuse in a simple circuit with light bulbs. This appears in Arnold Arons' *A Guide to Introductory Physics Teaching*, found in the Book Basement. This was not in the Demo & Lab Shop. Suppose we want to force the issue and limit our search to only the Demo & Lab Shop, and search for "electricity". Try this, and you will not be disappointed. However, high on this list is the following: *LAB MANUAL: Recipes for Science: Misconceptions*. Quite by accident, we have found an item that did not appear in our search for student misconceptions about electricity (done previously, in the Scenario section). Why? The word "Student" was included in our search. If we eliminate that word, our search provides additional references for us. Try it.

Again, this illustrates one of the properties of the InfoMall—it is like a real mall. Sometimes you find just the right item for someone while shopping for something or someone else.

Physics To Go, Step 1 asks for a list of "electricity words." You can find a long list in the Keyword Kiosk. Simply go to the Kiosk and find the electricity and magnetism keywords. These are also good words for searching the InfoMall in order to find definitions.

Physics To Go also asks about light bulbs. A search for "light bulb*" provides many informative references that you may wish to read. Don't forget—the text on the InfoMall (together with graphics) can be copied and pasted directly into your own word processor for your own handouts.

Planning for the Activity

Time Requirements

This activity can be completed in less than one period. Allow the students to have sufficient time so that each student is allowed to experience what the cranking "feels" like under load and no load.

Materials Needed

For each group:

- alligator clip leads (14)
- manual electric generator: a hand-operated DC generator capable of a maximum output of about 300 milliamps at 6 volts is recommended for this and several other activities in the remaining part of this and the next chapter. The Genecon manufactured by Nakamura Scientific Co., Ltd. is recommended. It is available from several United States science supply houses. Generators are considered essential for student use in this and the next chapter, and batteries or power supplies should not be substituted for them.
- light bulb, miniature screwbase (flashlight bulbs): Just about any low voltage bulb will do; a No. 13 (3.6 volt, .30 amp) is a good choice.
- blinking bulb, miniature screwbase: a No. 406 or No. 407, available at electronics stores
- bulb base for miniature screwbase bulbs: any kind (4)
- steel wool: coarse grade is easiest to "grab" with electrical clips; a little goes a long way

Advance Preparation and Setup

Have one set up to show the students how to set up the generator and the bulbs.

Teaching Notes

Do not give extensive directions for the **FYTD** activity because it is meant to be exploratory. Reading the objectives to the students will suffice to indicate to the students what to think about as they engage in the **WDYT** and **FYTD** activities. Ask students not to worry about how the generator works, just use it as a "black box" source of electricity.

SAFETY PRECAUTION: Remind the students that electricity can be dangerous. Although the currents and voltages are not likely to be harmful, the possibility for harm is there for students who may have heart or otherbio-electrical problems.

Some students are not familiar with what electricity is. They may try to explain that electricity is a "something" (a particle, molecule, or other substance) that is converted into light and that the thing goes away once it is converted into light (or heat). Don't try to explain everything during this first activity, but allow the students to discover what is electricity. The activities allow for this discovery of the nature of electricity.

AC Generator

The following figure shows the position of the rotating coil of an AC generator at instants separated by one-fourth of a rotation of the coil. Students should analyze your observations during the AC generator demonstration, and record their understanding of the drawing and the subsequent explanation in their journals.

© It's About Time

It is easier to understand the creation of a current if we invent a set of invisible threads to signify the magnetic field of the permanent magnets. The very thin threads fill the space and connect the north pole of one magnet with the south pole of the other magnet. If the wire of the generator is imagined to be a very thin, sharp knife, the question you must ask is whether the knife (the wire) can "cut" the threads (the magnetic field lines). If the wire can "cut" the field lines, then a current is generated. If the wire moves in such a way that it does not "cut" the field lines, then NO current is generated.

For the purpose of analyzing the generator in the figure of the rotating coil, you may want to build a small model of the rectangular coil so that you can move the model as a way to make sense of the drawings. The coil model can be constructed with three pens, or pencils held together with rubber bands. Rest the coil between two pieces of paper: the left paper is the north pole of a magnet and the right paper is the south pole of a magnet.

For the purpose of analyzing the rotating coil figure, the four sides of the rectangular coil of the AC generator will be referred to as sides AB, BC, CD, and DA, corresponding to the lines from corners A to B, B to C, etc. Side DA is "broken" to allow extension of the coil to the rings. The "brushes" labeled 1 and 2, make sliding contact with the rings to provide a path for the induced current to travel to an external circuit (not shown) connected to the brushes. The magnetic field has a left-to-right direction (from the north pole to the south pole) in the space between the magnets in the rotating coil figure. It is assumed that the coil has a constant speed of rotation.

1. When the generator coil is in the position shown in the rotating coil figure (1), no current is being induced because all sides of the coil are moving parallel to the magnetic field. The AB wire and the CD wire do not "cut" the field lines. On Graph 1, plot a point at the origin of the graph, indicating the amount of induced current is zero at the instant corresponding to the beginning of one rotation of the coil.

I

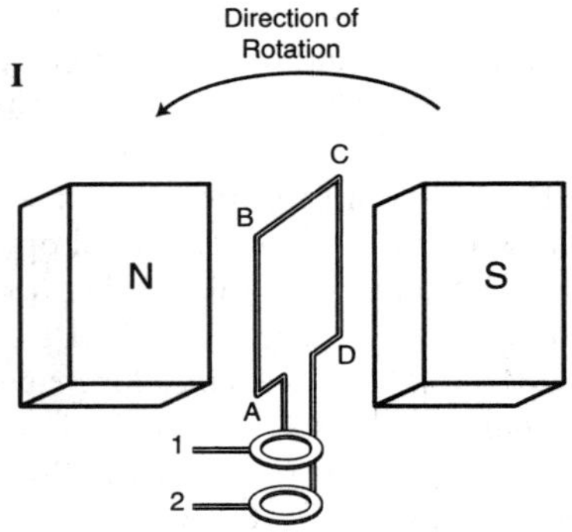

2. One-fourth turn later, at the instant shown in the rotating coil figure (II), the induced current is maximum because both sides AB and CD are "cutting" perpendicularly through the magnetic field. At this instant, the relative motion between the magnetic field and the coil is maximum. If the generator were connected to an external circuit, current would flow out of brush 1, travel around the circuit, and enter brush 2. The amount of current flowing in the circuit would be exactly the same as the induced current flowing around the coil as indicated by the arrows. On Graph 1, plot a point directly above the one-quarter turn mark at a height equal to the top of the vertical axis to represent maximum current flow in one direction.

II

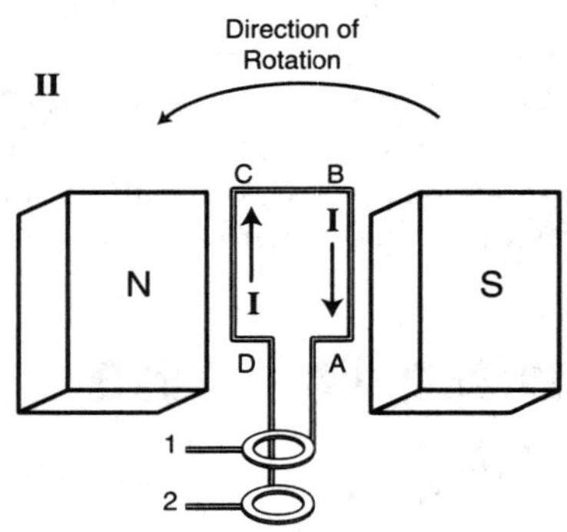

3. One-half turn into the rotation of the coil, at the instant shown in the rotating coil figure (III) on next page, the current again is zero because all sides of the coil are moving parallel to the magnetic field. Plot a point at the 1/2 mark on the horizontal axis to show that no current is being induced at that instant.

© It's About Time

III

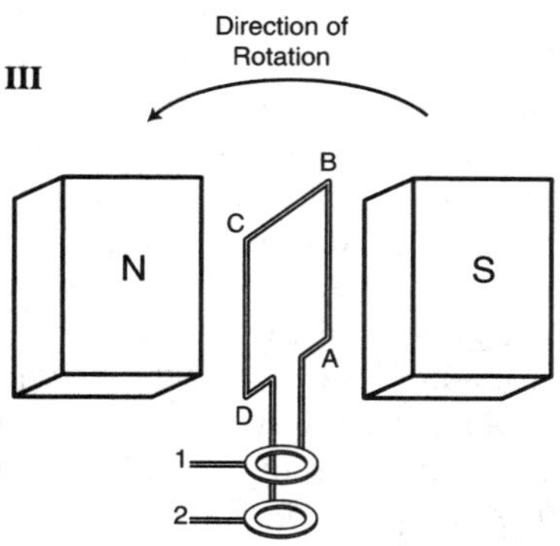

4. At the instant at which 3/4 of the rotation of the coil has been completed, shown in the rotating coil figure (IV), the induced current again is maximum because coil sides AB and CD again are moving across the magnetic field at maximum rate. However, this is not exactly the same situation as shown in the rotating coil figure (II); it is a different situation in one important way: the direction of the induced current has reversed. Follow the directions of the arrows which represent the direction of the current flow in the coil to notice that, at this instant, the current would flow to an external circuit out of brush 2 and would return through brush 1. On Graph 1, plot a point below the three-quarter turn mark at a distance as far below the horizontal axis as the bottom end of the vertical axis. This point will represent maximum current in the opposite, or "alternate," direction of the current shown earlier at one-quarter turn.

IV

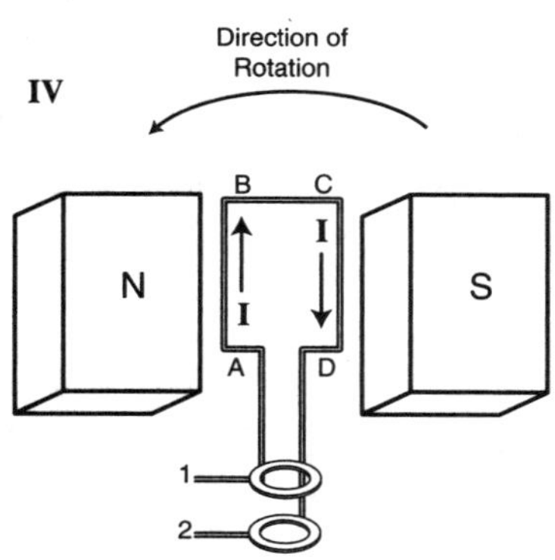

5. The rotating coil figure (1) is used again to show the instant at which one full rotation of the generator coil has been completed. Again, all sides of the coil are moving parallel to the magnetic field, and no current is being induced. Plot a point on the horizontal axis at the one-turn mark to show that the current at this instant is zero.

6. You have plotted only 5 points to represent the current induced during one complete cycle of an AC generator. What about the points to be plotted that would represent the amount of induced current at each instant during one complete rotation of the generator coil? What is the overall shape of the graph? Should the graph be smooth, or have sharp edges? Sketch it to connect the points plotted on Graph 1.

7. Additional rotations of the generator coil would, if the same speed and resistance in the external circuit were maintained, result in the same graph, over and over.

DC Generator

The DC generator figure below shows the important parts of a DC generator. Students should analyze your observations during the DC generator demonstration, and record their understanding of the drawing and the subsequent explanation in their journals.

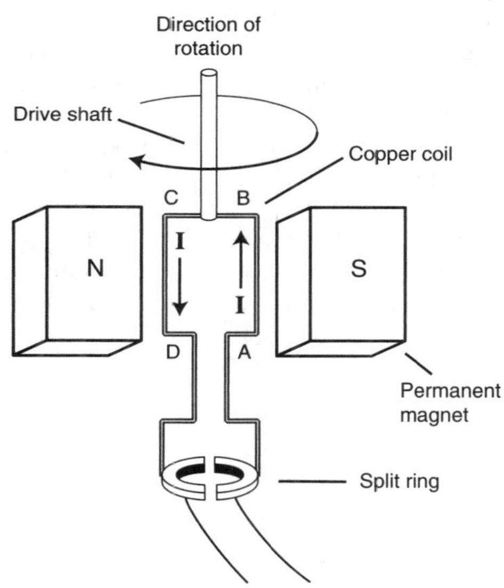

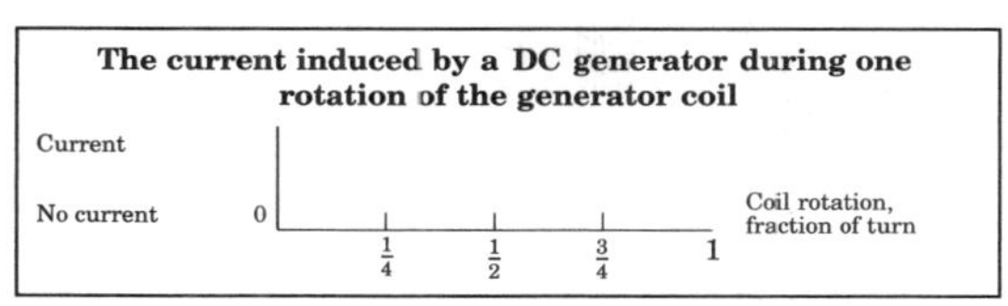

Graph 2

Graph 2 can be completed using the same pattern of analysis applied to the AC generator.

1. At the instant shown in the DC generator figure above, the induced current is maximum both in

© It's About Time

the generator coil and the external circuit to which the generator is connected (the external circuit is not shown). The instant corresponds to the rotating coil (II). Plot a point on Graph 2 directly above the one-quarter turn mark at a height equal to the top of the vertical axis to represent maximum current flow at that instant.

2. At the instant one-quarter turn earlier than the instant shown in the DC generator figure, corresponding to the zero mark of rotation, the current would have been zero because all sides of the coil would have been moving parallel to the direction of the magnetic field. Therefore, plot a point at the origin of Graph 2. Similarly, the induced current again would be zero at the instant one-quarter turn later than the instant shown in the DC generator figure; therefore, plot a point on the horizontal axis at the one-half turn mark.

3. Notice the arrangement used to transfer current from the generator to the external circuit for the DC generator. It is different from the arrangement used for the AC generator. The DC generator has a "split-ring commutator" for transferring the current to the external circuit. Notice that if the coil shown in the DC generator figure were rotated one-quarter turn in either direction, the "brush" ends that extend from the coil to make rubbing contact with each half of the split ring would reverse, or switch, the connection to the external circuit. Further, notice that the connection to the external circuit would be reversed at the same instant that the induced current in the coil reverses due to the change in direction in which the sides of the coil move through the magnetic field. The outcome is that while the current induced in the coil alternates, or changes direction each one-half rotation, the current delivered to the external circuit always flows in the same direction. Current that flows always in one direction is called direct current, or DC. Therefore, it is appropriate to plot a point on Graph 2 at a point directly above the three-quarter turn mark at the same height at the point plotted earlier for the one-quarter turn mark.

4. As done for the AC generator, find out how to connect the points plotted on Graph 2 to represent the amount of current delivered always in the same direction to the external circuit during the entire cycle.

Generators will supply electricity to the homes that we build through HFE. The voltage of an AC generator varies from zero to a maximum then back to zero and then to a maximum in the other direction. You will hear people give one value for AC voltage in the same way that they give one value for a battery. A 1.5-volt battery or a 9-volt battery provides a steady, direct voltage. The 120-volt AC in your home varies as we have shown; the 120 volts is a type of average that allows us to compare different AC voltages.

© It's About Time

NOTES

© It's About Time

Activity Overview

This activity introduces students to electricity in a way that may not be familiar to them. They will be generating electricity using a hand generator. This will allow them to see the effect of cranking quickly or slowly, and to experience the "feel" of the cranking.

Student Objectives

Students will:

- Trace energy transformations.
- Identify the Sun as Earth's ultimate energy source.
- Begin developing a personal model for electricity.

ANSWERS FOR THE TEACHER ONLY

What Do You Think?

Students may refer to solar photovoltaic cells, or wind generators as being "free." However, there is no such thing as "free" energy. The First Law of Thermodynamics states that energy can neither be created nor destroyed, only converted from one form into another. Therefore, energy is never "free."

Even with solar energy and wind energy (often referred to as renewable), the converter from solar (wind) to electrical energy for your home, must be built and costs money. The transmission from the point of generation to the point of use, costs money.

Activity 1 Generate

GOALS

In this activity you will:

- Trace energy transformations.
- Begin developing a personal model for electricity.

What Do You Think?

Electricity affects most parts of your life. You pay for it, over and over, in the form of electric bills and batteries. Also, most products that you purchase are manufactured by processes that use electricity, so you pay for electricity in indirect ways, too.

- **Is there any "free" electricity available and, if so, why pay for it?**

Record your ideas about this question in your *Active Physics* log. Be prepared to discuss your responses with your small group and the class.

© It's About Time

For You To Do

1. You will be provided with a bulb, bulb base, connecting wires and a generator. Assemble the bulb, bulb base, connecting wires, and hand generator, and turn the crank of the generator to make the bulb light. Never turn the crank too fast. You can strip the gears!
   - a) Draw a diagram of how you assembled the equipment for the bulb to light.
   - b) Under what conditions will the bulb not light? Use words and a diagram in your answer.
   - c) What are the effects of changing the speed or direction of cranking the generator?
   - d) What are the effects of reversing the connections of the wires to the bulb or to the generator?

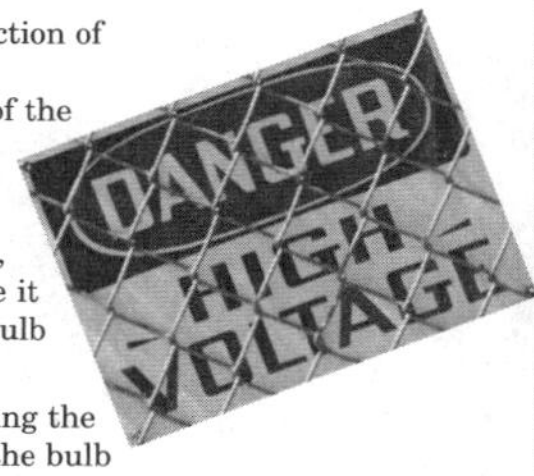

2. Replace the bulb that you have been using with a blinking bulb, the kind used in some toys, flashlights, and decorations. As before, use the generator to make it light, and keep cranking the generator to make the bulb go through several on and off cycles.
   - a) Describe any difference that you can feel in cranking the generator when the bulb is on compared to when the bulb is off.
   - b) How do you think that the blinking bulb works? What makes it go on and off?

3. Replace the blinking bulb with a strand of steel wool. Connect the ends of the steel wool to alligator clips on the generator. Crank the generator and observe what happens to the steel wool. Be careful not to touch the hot steel wool! You may push the steel wool with the point of a pencil to provide a better contact with the socket.
   - a) Describe the appearance of the steel wool.
   - b) What factors affect whether or not the steel wool glows, how much it glows, and for how long?
   - c) What were the similarities and differences between the steel wool and the light bulbs as used above?

⚠ **The steel wool will get very hot. Do not touch it while conducting the experiment. Allow it to cool before removing it.**

ANSWERS

For You To Do

1. a) Student activity and drawing.

   b) Students' answer will vary. However, the bulb will not light when the generator is not moving, or if they have the wires hooked up incorrectly. If there is a short the bulb will not light.

   c) The students should notice as they increase the speed of the generator, the bulb will get brighter. The direction should have no bearing on the brightness of the bulb. They will notice that when changing from one direction to the other, the bulb will go dim, out, and then bright again.

   d) There should be no effect.

2. a) The students will notice that the crank is harder to turn when the bulb is on than when the bulb is off.

   b) Students' answers will vary.

3. a) As the generator is cranked, the steel wool will start to glow.

   b) Students will notice that the harder they crank the generator, the more intense the glow of the steel wool. They may notice at some point that the steel wool starts to melt or break down. The steel wool will only glow for a short period of time.

   c) **similarities** – they both produced light and heat

   differences – the steel wool glowed for a short time, the bulb for a long time. The steel wool was exposed to air, the bulb filament is enclosed in glass.

© It's About Time

ANSWERS

For You To Do

(continued)

4\. a) Some students will not be able to answer this exactly. However, they should be able to understand that the cranking of the generator puts energy into the system. Understanding this process exactly is not necessary at this time. The energy is not free, but students may see it as free meaning rather that there is no monetary cost.

b) Students' answer will vary. There is stored chemical energy in the arm muscles. They produce kinetic energy of movement of the crank. This kinetic energy is converted to electrical energy in the generator (see **Background Information** for details). The electrical energy is then converted into light and heat energy of the light bulb. Some students may also refer to the "lost" energy of friction and heat lost to the environment. (In essence it is not lost energy but rather energy that is converted into useless energy.)

c) The source in this activity is the chemical energy stored in the arms. In a flashlight bulb, it is the chemical energy stored in the battery. The source of energy for the house lamp is dependent on the type of electric power plant (coal—chemical energy stored in the coal; nuclear power—nuclear energy in the fission or the breaking down of the atom into smaller particles, and giving off energy; gravitational potential energy that is stored in the water in a dam, mechanical energy of the wind; etc.).

4\. Was the electrical energy that you used to "light things up" in this activity "free"? Did you get something for nothing? Using your observations in this activity, write a short paragraph to answer each of the following questions:

a) What was the energy source for each part of the activity (bulb, blinking bulb, steel wool)? Was it free energy, at no cost?

b) What forms of energy were involved in the activity, and in what order did the forms appear?

c) How is the energy source used in this activity different from the source used to light a bulb in a flashlight, or in a house lamp?

Reflecting on the Activity and the Challenge

This activity has given you some experience with a process that is involved in the electrical system you will use for the HFE dwelling: using a generator to provide energy for electric light bulbs. The generator and the light bulbs used in this activity are scaled-down versions of the ones to be used for the dwelling, but they work in the same way. One additional feature will exist in the electrical system for the dwelling: the electrical energy from the generator will be able to be stored in batteries until it is needed to operate lights and other appliances.

Part of your challenge is to write a training manual to help instructors teach the inhabitants about their wind-generator system. You will probably want to include what you learned in this activity in your manual.

Word	Meaning
electricity	electrical energy
current	measure of amperage or flow of electrons
voltage	measure of potential difference (electromotive force)
ohms	measure of resistance
resistance	property of a substance which slows motion of electrons (sometimes referred to as non-conductivity)
conductor	allows the motion of electrons easily without resistance
insulator	impedes the movement of electrons
electrons	negatively charged particles which are free to move
switch	controls flow of electrons in a circuit
circuit	closed path where electrons flow from source (battery) to resistor (light bulb)

© It's About Time

Physics To Go

1. Make a chart with two columns, the first one labelled "Word" and the other labelled "Meaning."
   a) In the first column make a list of "electricity words"—words that you have heard used in connection with electrical units of measurement, parts of electrical systems, or how electricity behaves.
   b) In the second column write what *you* think each word means, or describes.
2. You know that electricity comes "out of the wall." You also know that it "starts" in a power plant. Draw a picture that shows how *you* think the electricity is "created" and how it gets to your home.
3. Explain what *you* think electricity is, how it behaves, and how it does what it does.
4. A variety of energy sources are used to operate light bulbs. Identify as many energy sources as you can that are used to power light sources.
5. The kind of light bulb you used in this activity is called "incandescent." Another kind of light bulb often used is called "fluorescent." Look up the meaning of the two words and explain how they are related to what glows to cause each kind of bulb to give off light.
6. "You don't get something for nothing." Explain how this expression applies to using a hand-operated generator to light a bulb.

Stretching Exercises

Some light bulbs are frosted on the inside, while others are clear. Some light bulbs have built-in reflectors to make them serve as a spotlight or a floodlight. Research different types of light bulbs. On a poster, describe how each is different, how each works, and where and why each is useful.

ANSWERS

Physics To Go

1. Answers will vary. There will be others. If you do not know the definition, check out any senior-level physics textbook.
2. Many possibilities for generation of electricity depending on the area. Hydroelectric dams, coal- and gas-fired power plants, nuclear plants are the most common. All of them will have similar paths from generator to home. The primary difference is how the electricity is generated. The students' diagrams should show how the electricity is generated, and then that it is transported to the consumer via power lines. Although most students have never been to a power generating station, a lot of them will have a basic understanding that the generator needs an input energy (water falling over a turbine), turning the generator, then passes through a series of invertors, and transformers increasing voltage to transmission levels (> 500 kilovolts), through a different series of transformers to step down the voltage to 240 V into our homes. With a coal-fired power plant, the coal is burned to heat up the water to steam, which is then used to drive the turbines. Nuclear power is using the nuclear reactor to generate heat to drive the turbines. Students at this level need not know the details, but if time or curiosity allows, a special project may be assigned as enrichment.
3. Students' answers will vary.
4. Students' answers will vary. The most common answers will be as listed previously, as well as batteries, car, RV batteries, gas-powered generators, solar panels, water wheels, geothermal power generators, etc.
5. Incandescent means that the light comes from the heating of filament which glows at certain temperatures. Fluorescence is the glowing of a substance when electromagnetic radiation of shorter wavelengths (such as UV light) hit the substance (tube lining) and emit light of a longer wavelength and greater visibility.
6. Although you are getting light without a monetary cost, you need to start out with some form of energy, (chemical stored energy in your muscles); use a converter (muscles), and convert the energy into a usable form (movement of the arm, and ultimately the crank of the generator). That kinetic energy is converted by the generator into electrical energy which is then passed through the light bulb which converts it to light and heat energy.

© It's About Time

Activity 1 A: Demonstration

Generator Principles

FOR YOU TO DO

Examine AC & DC Generators

1. Your teacher will explain and demonstrate a hand-operated, alternating current (AC) generator. During the demonstration, make the observations necessary to gain the information needed to answer these questions:

a) When the AC generator is used to light a bulb, describe the behavior in time of the brightness of the bulb when the generator is cranked.

(a) slowly, and (b) rapidly. Write your observations in your journal.

b) When the AC generator is connected to a galvanometer, describe the action of the galvanometer needle when the generator is cranked

(a) slowly, and (b) rapidly.

2. Your teacher will explain and demonstrate a hand-operated, direct current (DC) generator.

a) During the demonstration, make the observations necessary to gain the information needed to answer these questions:

b) When the DC generator is used to light a bulb, describe the behavior in time of the brightness of the bulb when the generator is cranked

(a) slowly, and (b) rapidly. Write your observations in your journal.

c) When the DC generator is connected to a galvanometer, describe the action of the galvanometer needle when the generator is cranked

(a) slowly, and (b) rapidly.

© It's About Time

For use with *Electricity For Everyone*, Chapter 7, Activity 1: Generate

ACTIVITY 2
Lighten Up

Background Information

The physics principles involved in **Activity 2** are:

- basic electrical theory
- basic electrical units of measurement
- series and parallel forms of electric circuits
- dynamic rules of series and parallel circuits

Basic electrical theory and basic electrical units. Explanations of basic electrical theory and units of measurement are presented in the **For You To Read**, of the student text for **Activity 2**. It is suggested that you complete that reading now.

Series and parallel forms of electric circuits. The diagrams of series and parallel circuits shown in **Activity 2** illustrate the most fundamental difference in the two kinds of circuits: only one possible path which a particular electron may take to traverse a series circuit, and multiple paths which an electron could take to complete a parallel circuit.

Students may come up with another possibility: a combination series/parallel circuit in which part of the circuit is series (one possible path) and another part of the circuit is parallel (two or more paths).

Both kinds of circuits have advantages which depend on the applications for which they are used. Parallel circuits are preferred for homes because, for example, when one light bulb on a circuit burns out the others are not affected. Also, parallel circuits provide the constant voltage needed for each parallel "branch" of a home circuit.

Dynamics and rules of series and parallel circuits. To enhance your background about what happens, and why, in series versus parallel circuits, it is suggested that you first read A Water Model for Electricity at the end of this **Background Information** and **Physics To Go** sections of the student text for **Activity 2**. When you have completed reading those sections, apply the rules below for series and parallel circuits to the quantitative examples presented in the sections.

Definition: 1 ohm is the amount of resistance which allows 1 amp of current to flow when 1 volt is applied.

Kirchoff's Law: All of the voltage (energy per unit of charge) applied to a circuit will be exactly dissipated in resistive parts of the circuit (that is, each unit of charge will have zero energy left when it has completed the circuit).

Ohm's Law: The total current (in amps) which will flow in a circuit may be calculated by dividing the voltage (in volts) applied to the circuit by the total resistance (in ohms) of the circuit.

Ohm's Law applies to both series and parallel (and other) circuits.

Rules applying to series circuits

The total resistance of a series circuit (R_T) is the sum of the individual resistances (R_1, R_2, R_3, . . .) in the circuit:

$$R_T = R_1 + R_2 + R_3$$

The total current in a series circuit (I_T) is equal to the current flowing in each resistor (I_1, I_2, I_3, . . .) in the circuit:

$$I_T = I_1 = I_2 = I_3$$

The total voltage (V_T, or energy per unit of charge) applied to a circuit is equal to the sum of the voltages dissipated in the individual resistors (V_1, V_2, V_3, . . .) in the circuit.

$$V_T = V_1 + V_2 + V_3$$

Further, Ohm's Law transposed as $V = IR$ may be applied to both entire circuit and to each resistor to show that the voltage dissipated by each resistor in a series circuit is proportional to its amount of resistance (that is, the larger resistors dissipate the most energy in this kind of circuit). Substituting IR for V in the right-hand terms of the above equation:

$$V_T = I_1R_1 + I_2R_2 + I_3R_3$$

Since $I_1 = I_2 = I_3$, the voltage dissipated at each resistor is proportional to the amount of resistance it represents in the circuit.

Rules applying to parallel circuits

The total current in a parallel circuit (I_T) is equal to the sum of the currents flowing in the individual resistors (I_1, I_2, I_3, . . .) in the circuit:

$$I_T = I_1 + I_2 + I_3$$

The total voltage (V_T, or energy per unit of charge) applied to a circuit is equal to the sum of the voltage dissipated by each resistor (V_1, V_2, V_3, . . .) in the circuit (see Kirchoff's Law above):

$$V_T = V_1 = V_2 = V_3$$

Chapter 7

© It's About Time

Substituting Ohm's Law in the form $I = V/R$ for each term in the equation $I_T = I_1 + I_2 + I_3$:

$$V_T/R_T = V_1/R_1 + V_2/R_2 + V_3/R_3$$

Since $V_T = V_1 = V_2 = V_3$ the above equation can be simplified through division by V to:

$$1/R_T = 1/R_1 + 1/R_2 + 1/R_3$$

Therefore, the reciprocal of the total resistance of a parallel circuit (R_T) is the sum of the reciprocals of the individual resistances (R_1, R_2, R_3, . . .) in the circuit.

Active-ating the Physics InfoMall

This activity immediately asks about light bulbs. We have already done the search for light bulbs in **Activity 1**. You may wish to repeat the search now to find items specifically for this activity.

For You To Do Step 2 mentions series and parallel circuits. A search for "series" AND "parallel" results in "Too Many Hits" again. If you limit this search, you can find many hits with wonderful information. For example, search for "misconcept*" AND "series" AND "parallel". The first hit is "Electricity Visualized: The Castle Project: Section 6: Circuits As Systems: What Makes Bulb Brightness Change?" from the Demo & Lab Shop. This contains activities that you may find useful. A little further down the list is "Potential difference and current in simple electric circuits: A study of students' concepts," from the *American Journal of Physics*, vol. 51, issue 5, 1983.

For You To Do Step 2 also asks the students to make a prediction. The importance of the prediction should not be overlooked; indeed, predictions force students to examine their understanding of a phenomena and actively engage thought. If you were to search the InfoMall to find more about the importance of predictions in learning, you would find that you need to limit your search. Sadly, not much information exists on the InfoMall regarding predictions about electric circuits. To find information on the importance of predictions, we must turn to a topic that has been researched extensively; dynamics. Let's borrow from a search we perform in *Active Physical Science: Sports* and *Active Physical Science: Transportation:* A search for "prediction*" AND "inertia" resulted in several hits; the first hit is from *A Guide to Introductory Physics Teaching: Elementary Dynamics*, Arnold B. Arons' Book Basement entry. Here is a quote from that book: "Because of the obvious conceptual importance of the subject matter, the preconceptions students bring with them when starting the study of dynamics, and the difficulties they encounter with the Law of Inertia and the concept of force, have attracted extensive investigation and generated a substantial literature. A sampling of useful papers, giving far more extensive detail than can be incorporated here, is cited in the bibliography [Champagne, Klopfer, and Anderson (1980); Clement (1982); di Sessa (1982); Gunstone, Champagne, and Klopfer (1981); Halloun and Hestenes (1985); McCloskey, Camarazza, and Green (1980); McCloskey (1983); McDermott (1984); Minstrell (1982); Viennot (1979); White (1983), (1984)]. Note that students' preconceptions can have a large effect on how they learn something. It is important that they are forced to consciously acknowledge their preconceptions by making predictions.

The InfoMall can also be used for more entertaining purposes. In **For You To Read**, we are told that Ben Franklin named charges as positive or negative. One of the underused parts of the InfoMall is the Calendar Cart. If you search the Calendar Cart for "Franklin" you will find that on "07/11/1747 In a letter to Peter Collinson, Benjamin Franklin described a discovery that would lead to the invention of the lightning rod and introduced the terms 'positive' and 'negative' and 'to electrise plus or minus.'" and "09/01/1747 In a letter to Peter Collinson, a Fellow of the Royal Society in London, Benjamin Franklin used his ideas about the 'two states of Electricity, the plus and minus' to explain how a Leiden jar stores electrical charge."

We are also told in **For You To Read** that a lightning bolt transfers approximately one coulomb of charge. If you look in the Utility Closet, in *Many Magnitudes: A Collection of Useful and Useless Numbers* (within Physics Pfact Potpourri) you will find that the "charge transfer in lightning stroke" is about 25 coulombs. Not very different, but it makes lightning that much more impressive.

Physics To Go asks if household wiring is series or parallel. For an answer, search the InfoMall for "house" AND "series" AND "parallel" but limit it to "same paragraph" to avoid too many hits.

The **Stretching Exercises** ask about Thomas Edison and his inventions. While this information does not seem to be on the InfoMall, a search for "Edison" AND "inventions" provides an article from the *American Journal of Physics* that contains the references for many books that you could use to

© It's About Time

find a list of his inventions. This article, and an update, are "Book-length biographies of physicists and astronomers," *American Journal of Physics*, vol. 12, issue 1, 1944, and "Book-length biographies of physicists and astronomers - addendum," *American Journal of Physics*, vol. 16, issue 3, 1948.

Planning for the Activity

Time Requirements

The students will be able to complete this activity in one 40-minute period or less.

Materials Needed

For each group:

- alligator clip leads (14)
- manual electric generator: a hand-operated DC generator capable of a maximum output of about 300 milliamps at 6 volts is recommended for this and several other activities in the remaining part of this and the next chapter. The Genecon manufactured by Nakamura Scientific Co., Ltd. is recommended. It is available from several United States science supply houses. Generators are considered essential for student use in this and the next chapter, and batteries or power supplies should not be substituted for them.
- light bulb, miniature screwbase (flashlight bulbs): Just about any low voltage bulb will do; a No. 13 (3.6 volt, .30 amp) is a good choice.
- bulb base for miniature screwbase bulbs: any kind (4)

Advance Preparation and Setup

Discuss the "electricity words" identified by students the previous day and ask some students to share some features of their personal models for electricity. You may wish to make notes about misconceptions which exist, for a later treatment.

Teaching Notes

The students should not need extensive instructions to proceed with the **FYTD** in this activity. The **FYTD** activity should not take much time, so plan to reserve time for with-class reading and discussion of **For You To Read**, and encourage students to begin **Physics To Go**.

Encourage students to try to turn the generator at the same speed throughout the activity, because it will help them to determine which variable is causing the brightness in the bulbs.

SAFETY PRECAUTION: Remind the students they are working with electricity, which is dangerous under certain circumstances. Watch to make sure the students are not trying to "zap" other students. Students with heart problems may be affected by the currents from the generator.

Because this lab is over quickly, it is advisable that you get to each of the student lab groups before they are finished. This is to ensure that they are on task, and also to watch for proper and safe lab technique.

Some students may refer to electricity as the transforming of electricity into heat and light energy. Emphasize that it is the transfer of energy in one form (electric kinetic energy) to another form (thermal energy—molecules are heated up as there are more and more collisions with the electrons). Students may ask for an explanation of what the difference is between series circuits, and parallel circuits. You may refer to the water provided, or another method may be the analogy of the stream. The potential difference in the water is the change in height. With a series circuit, it is analogous to the river with several rapids in a single line. As the water flows past each of the rapids it loses some of its potential energy at each rapid (the voltage in the electric circuit). The amount of water before and after each rapid is the same, regardless of the number of rapids (the current in the electric circuit). With a parallel circuit, the river has different channels, with the potential energy before each channel the same, and the potential after each is the same. Therefore, the potential difference in the stream is like the voltage in the circuit. However, the current through the channels may be different, and some may have a large flow (current in the electric circuit), and some may have a small flow, but the total flow will be the same before and after.

© It's About Time

Activity Overview

The students will be investigating what electricity is, and how it works, in a very simplified form. They do not need the depth of explanation in the **Background Information**.

Student Objectives

Students will:

- Qualitatively describe current, resistance, voltage.
- Define: coulomb, ampere, volt.
- Compare series and parallel circuits.
- Recognize generator output limit.
- Extend a personal model of electricity.

ANSWERS FOR THE TEACHER ONLY

What Do You Think?

Students' answers will vary. Electric energy is transformed into heat or light energy when electrons with a large amount of kinetic energy collide with the atoms in the wire. As more and more collisions occur, the wire heats up, and if it heats up enough, as in the filament of an electric light bulb, it will glow

Electricity for Everyone

Activity 2 Lighten Up

GOALS

In this activity you will:

- Qualitatively describe current, resistance, voltage.
- Define: coulomb, ampere, volt.
- Compare series and parallel circuits.
- Recognize generator output limit.
- Extend a personal model of electricity.

What Do You Think?

Lights were the first electric appliances for homes.

- **How do light bulbs, and the electricity that makes them glow, work?**

Record your ideas about this question in your *Active Physics* log. Be prepared to discuss your responses with your small group and the class.

For You To Do

1. Before you begin this activity, remind yourself about the "feel" of the generator and the brightness of the bulb when the generator was used to energize one bulb in **Activity 1**. Connect a single bulb to the generator and crank. Use your observations of a single bulb as a basis for comparison when you use two or more bulbs during this activity.

Active Physics 450

ANSWERS

For You To Do

1. Student activity.

© It's About Time

2. There are two distinct ways to connect more than one light bulb to the generator. Look at the two diagrams showing three bulbs connected in series and in parallel.

a) Describe in your log how the two circuits are different.

b) Make predictions about how each circuit operates.

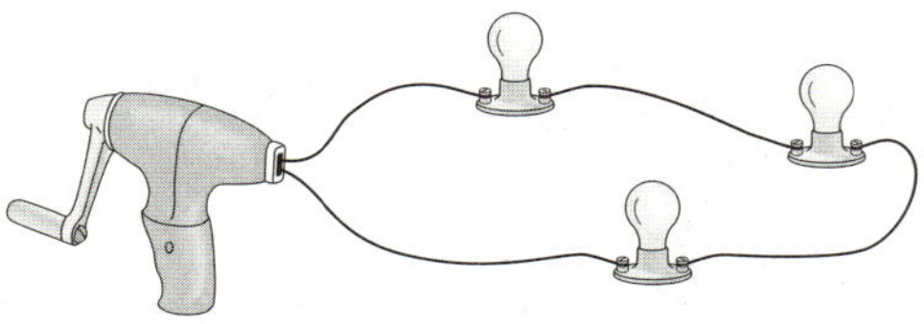

A series circuit.

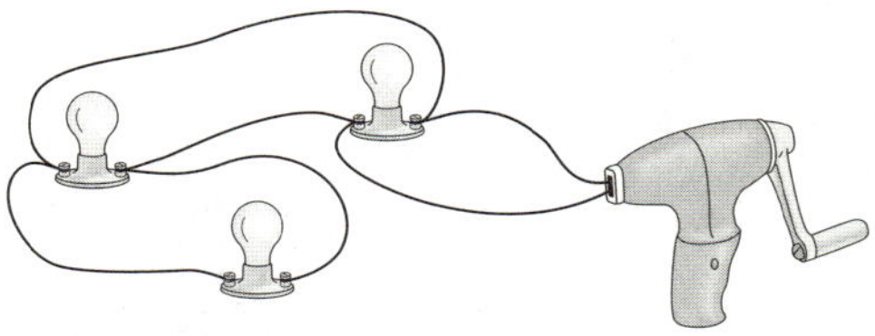

A parallel circuit.

3. Connect *two* bulbs in series with the generator. Use the diagram showing three bulbs connected in series to help you. Crank the generator, and notice the "feel" of the generator and the brightness of the bulbs. Repeat this for three bulbs, and four bulbs.

a) Describe what happens and try to explain why it happens.

4. What would happen if, in a series circuit of several bulbs, one bulb were to be disconnected, or burn out? Try it by unscrewing one bulb from its base while the circuit is operating.

a) Describe what happens, and try to explain why it happens.

Chapter 7

ANSWERS

For You To Do *(continued)*

2. a) Student response. Some students may observe that there is only one path in a series, and more paths in the parallel circuit.

b) Student response. Students' answers will vary.

3. a) In a series circuit, as you increase the number of the bulbs, the resistance increases. Therefore, as you increase the number of light bulbs, the cranking of the generator gets more difficult. You will also notice that you must crank the generator faster in order to light up more light bulbs. Again, the faster you crank the generator, the brighter the bulbs.

4. a) When you disconnect one of the bulbs, you actually break (open) the circuit, which stops the flow of electrons. No flow of electrons, no energy transformation.

© It's About Time

Physics Words

electric charge: a fundamental property of matter; charge is either positive or negative.

proton: a subatomic particle that is part of the structure of the atomic nucleus; a proton is positively charged.

electron: a negatively charged particle with a charge of 1.6×10^{-19} coulombs and a mass of 9.1×10^{-31} kg.

coulomb: the SI unit for electric charge; one coulomb (1 C) is equal to the charge of 6.25×10^{18} electrons.

5. Connect *two* bulbs in parallel with the generator and, again, observe the "feel" of the generator and the brightness of the bulbs. Repeat this for three bulbs, and four bulbs.
   a) Describe your observations and compare them to your predictions for a parallel circuit.

6. What would happen if one bulb were to fail in a parallel circuit? Try it by unscrewing one bulb.
   a) Describe what happens, and try to explain why it happens.

FOR YOU TO READ

The Language of Electricity

Now you are ready to become acquainted with some of the basic language of electricity. You are ready to learn the meanings of, and use of, some of the "electricity words" that you identified in **Activity 1**. Here are some theories and definitions about electricity to help you:

- There are two kinds of electric charges, positive and negative, named by Benjamin Franklin. Protons have positive charge, and electrons have negative charge. Like kinds of electric charges repel, and unlike kinds attract.

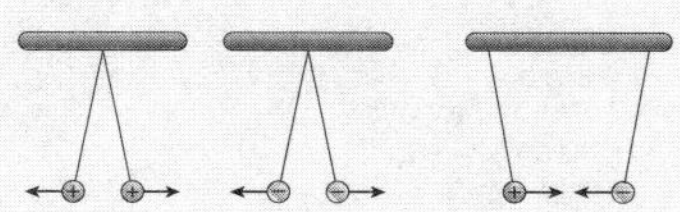

Like charges repel, and opposite charges attract.

- There is a smallest amount of the property called **electric charge**, the amount possessed by one proton and one electron. While **protons** and **electrons** differ in several ways (such as mass), an electron and a proton have an identical amount of charge.
- It is electrons that move in electric circuits of the kind you have been exploring. They flow through the circuit path, delivering energy, often in the form of heat and light, as they go. Protons, although present in the materials from which circuits are made, do not flow because they are locked within atoms. Positive particles sometimes do move in circuits where part of the path is liquid or gas.
- Scientists have agreed upon a standard "package" of electric charge, called the coulomb. Just as 12 eggs are called a dozen, 6.25 billion basic charges are called a **coulomb.** To provide a sense of how big this number is, one coulomb is approximately the charge transferred during a lightning bolt!
- Scientists have agreed upon a standard rate of flow of the **electric current** in circuits. When one coulomb of charge passes through a point in a circuit during each second of time, the current is said to be one **ampere**, symbol A, or often abbreviated to amp.

ANSWERS

For You To Do *(continued)*

5. a) The brightness should be about the same. Because the amperage will remain the same for each bulb as you add bulbs, then the brightness of each bulb will be the same. The students should not feel any difference in the cranking as they add bulbs.

6. a) If one bulb were to fail, there would be no change in the brightness, and there should be little, if any, difference in the cranking.

© It's About Time

Activity 2 Lighten Up

- Different materials offer different resistance, or opposition, to the flow of electric charge through them. Tungsten, from which light bulb filaments are made, has high electrical resistance. When electricity flows through it, the tungsten metal "robs" energy from the moving electrons and gets hot. Copper, by contrast, has low resistance; electrons transfer very little energy when flowing through copper. **Electrical resistance** is measured in **ohms** (Ω).
- Batteries or the hand generator provide energy to the electrons. These electrons are then able to light bulbs, heat wires, or make motors turn. The energy given to each coulomb of charge is measured in **volts** (V).

Reflecting on the Activity and the Challenge

In this activity you were introduced to the parallel and series circuits and to the electrical terms that you will need to know and be able to use for planning electric circuits to be used in the universal dwelling.

It is a fact that homes are "wired" using parallel circuits. Individual houses, apartments, mobile homes, or any other dwellings that receive electricity from a power company, have circuits of the parallel kind. Some older homes have as few as four circuits, and newer homes usually have many more. Each circuit in a home may have several light bulbs and other electrical appliances "plugged in," all in parallel. When electrical appliances are hooked up in parallel, if one is off or diconnected, the others can still be on. In a series circuit, if any appliance is disconnected, the other appliances cannot work. In your training manual, you will need to explain why the circuits in the universal dwelling are wired in parallel.

Physics Words

electric current: the flow of electric charges through a conductor; electric current is measured in amperes.

ampere: the SI unit for electric current; one ampere (1 A) is the flow of one coulomb of charge every second.

electrical resistance: opposition of a material to the flow of electrical charge through it: it is measured in ohms (Ω); the ratio of the potential difference to the current.

ohm: the SI unit of electrical resistance; the symbol for ohm is Ω.

volt: the SI unit of electric potential; one volt (1 V) is equal to one joule per coulomb (J/C).

Chapter 7

© It's About Time

ANSWERS

Physics To Go

1. Students answers will vary. Expect students to choose parallel circuits, because of the multiple paths the current is taking. If one bulb burns out on a series circuit, it will turn the whole circuit off. Also, on a series circuit, every time you turn on a new appliance, or light, there would be a "dip" in the current, and the lights would become dimmer each time a new resistor was added to the circuit.
2. Depending on the type of generator, you will notice it has an upper limit on the current it will produce. Before the students arrive, the teacher should establish what the upper limit is.
3. Students will see that there is a limit to the number of appliances and light bulbs a circuit can accommodate. As they go through the next few activities, they will discover the limit of the number of electric devices allowed with a limited electric output.

Physics To Go

1. What kind of circuit, series or parallel, would you choose for household wiring, and why? Write a short paragraph to explain your choice.
2. Did the generator used in this activity seem to have an "output limit"? In other words, did you arrive at conditions when the generator could not make the bulbs glow brightly even though you tried to crank the generator? Discuss this in a few sentences.
3. There is a great big generator at the power plant that sends electricity to your home. The wind generator chosen for HFE is much smaller than the generators used at power plants, but much larger than the one used for this activity. What implications might the output limit of the HFE electrical system have for the number of light bulbs and other electrical appliances that can be used in the HFE appliance package that you will recommend? Discuss this in a short paragraph.

Stretching Exercises

Thomas Edison is arguably one of the greatest inventors in world history. When we think of Edison, we think of the light bulb and the changes that this invention has made on the world. Edison dreamed of a world where we could read at night, where we could stroll down a lit street and where we could enjoy daytime all the time. Electricity and the light bulb have made that dream a reality. We live in Edison's dream!

Edison once said that genius is 1% inspiration and 99% perspiration. Explain the meaning of this phrase.

Construct a list of Edison's major inventions. (Edison had 1903 patents in his name!)

ANSWERS

Stretching Exercises

As an extension to this activity, the students may want to choose one of Edison's inventions and give a report to the class. This can be done, by simply a written report, or a visual, or even having the students reconstruct an invention. Students should submit a plan and discuss it with their teacher before building.

© It's About Time

A Water Model for Electricity

Imagine a pump that can lift water from the ground to a height of 4 m (12 feet). This water can then be dropped to the ground and, on its way, it can turn the paddles of a water wheel. A rotating water wheel is a motor that can be used for many purposes. The water can be lifted again and, once again, can turn the water wheel. In a more complex system, the water can be lifted 16 meters and dropped through a series of different-sized water wheels. Some of the wheels would spin fast and some would spin slowly—their speed would depend on their sizes. But all could be used as motors. The water would have to be lifted as high as 16 m again, but the energy we would get in our water wheel motors would be very useful.

In our electrical circuit, the hand generator gave the electrons energy. The electrons then "dropped" their energy in the three light bulbs. The electrons could, just as easily, have "dropped" their energy into electric motors that spin.

These models help you become more familiar with some of the electrical terms described earlier. If you turn the hand generator faster, you are providing more voltage to the electrons in the circuit. This is similar to raising the water to 16 m instead of 4 m. The number of amps is similar to the amount of water. Some light bulbs require more voltage to light (they have greater resistance) in the same way that some water wheels are harder to turn.

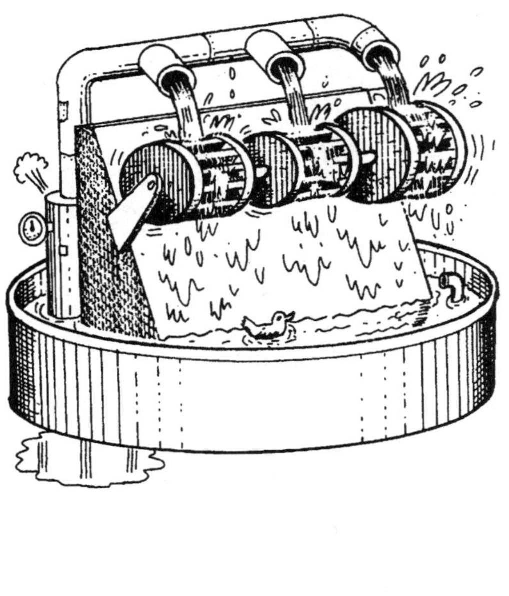

© It's About Time

ACTIVITY 3
Ohm's Law

Background Information

It is suggested that the **Physics Talk** and the **For You to Read** sections of the student text for this activity be read before proceeding in this section.

Ohm's Law is a fundamental relation between the voltage and current in a resistor. Certain resistors have a constant resistance. Ohm's Law $V = IR$ shows that the voltage is directly proportional to the current through the resistor.

Planning for the Activity

Time Requirements

About one and a half class periods are required to complete this activity.

Materials Needed

For each group:

- calculator
- resistors, ohm 5, 10 Watt (4)
- resistors, ohm 10, 10 Watt (4)
- resistors, ohm 15, 10 Watt (4)
- V/I meter (2)

Advance Preparation and Setup

Set up a sample circuit for the students before they come to class.

Teaching Notes

This activity is a combination of a black-box experiment and an inquiry activity. Students are required to design an experiment that will reveal the relationship between voltage and current for a resistor (Ohm's Law: $V = IR$)

They then use Ohm's Law to determine the value of an unknown resistor from the values of voltage and current given to them when the resistor is hidden from them.

Students should be encouraged to investigate different strategies but should not be given an answer to how to set this up. Inquiry requires students to struggle. Through that struggle, students learn much more than if they merely follow the teacher's directions.

© It's About Time

Activity 3 Ohm's Law

GOALS

In this activity you will:

- Calculate the resistance of an unknown resistor given the potential drop and current.
- Construct a series circuit.
- Properly use a voltmeter and ammeter in a series circuit.
- Graph the relationship between voltage and current for a resistor that obeys Ohm's Law.

What Do You Think?

Lighting makes some rooms conducive to work and other rooms more relaxing.

- **What determines the brightness of a light bulb?**
- **How can a dimmer switch make the same light bulb appear dim or bright?**

Record your ideas about these questions in your *Active Physics* log. Be prepared to discuss your responses with your small group and the class.

455 Active Physical Science

Activity Overview

Students design an experiment to determine Ohm's Law. They are then able to use their discovery of Ohm's Law to predict the resistance of an unknown resistor in a black box.

Student Objectives

Students will:

- Construct a series circuit.
- Properly use a voltmeter and ammeter in a series circuit.
- Graph the relationship between voltage and current for a resistor that obeys Ohm's Law.
- Calculate the resistance of an unknown resistor given the potential drop and current.
- Design an experiment for determining the resistance of an unknown resistor.

ANSWERS FOR THE TEACHER ONLY

What Do You Think?

The wattage rating of the bulb determines the brightness of the bulb if there is an ample supply of electricity. During a "brownout" the bulb's brightness can be diminished.

The dimmer switch can limit the electricity that gets to the bulb.

Chapter 7

© It's About Time

For You To Do

1. Imagine a closed box with something inside it. If you wanted to discover the contents, you could shake the box, weigh it, roll it, or twirl it. The sounds that you hear could help you decide what is inside.

   In this activity, you will be given a "black box." Inside the box is an electrical resistor. Outside the box are two terminals. The box will be set up as a part of an electrical circuit.

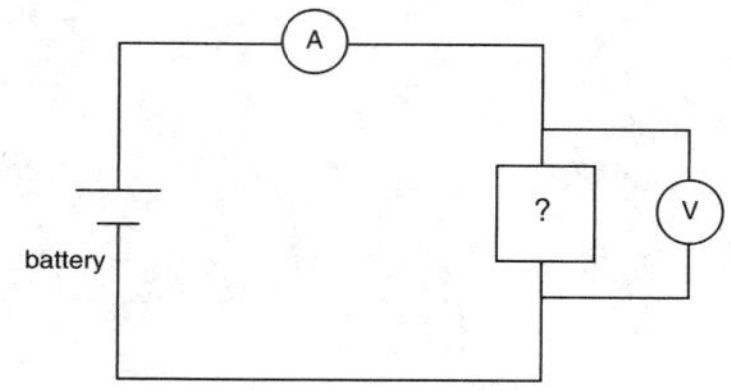

   You will be told the voltage across the black box and the current in the ammeter.

2. The box contains either a 3-Ω (ohm), a 10-Ω, or a 15-Ω resistor. By varying the voltage of the battery, you can learn how the current of the circuit and the voltage across a 3-Ω resistor changes. You can then conduct similar studies of the 10-Ω and 15-Ω resistor.
3. After you gather your data, your teacher will give you the values of the ammeter and voltmeter of the circuit with the black box. From the information, you will have to determine the contents of the black box.
4. Before beginning your study, please note the following circuit rules:
   - The ammeter is always placed in series in the circuit.
   - The positive terminal of the ammeter is always closest to the positive terminal of the battery.

ANSWERS

For You To Do

1. – 4. These steps provide instructions for the students

© It's About Time

- The voltmeter is always placed in parallel in the circuit. The simplest way to measure the voltage drop is to take the two leads from the voltmeter and touch the two ends of the resistor simultaneously.
- The positive terminal of the voltmeter is always closest to the positive terminal of the battery.
- Keep the connection in the circuit on for as long as it takes to read the ammeter and voltmeter and no longer.
- You can vary the voltage with a variable voltage supply or by adding additional batteries into the circuit.

5. Plan your experiment.
   a) Write down your plan in your *Active Physics* log.
   b) Create a table that will summarize the data you intend to collect.
   c) Describe how you will predict the contents of the black box from your data and the values given by the teacher.

6. Share your plan with your teacher.

7. After your teacher has approved your plan, conduct your experiment.
   a) Record your data in your *Active Physics* log.
   b) Record in your log the values given by your teacher for the black box.
   c) Which resistor is inside the black box?
   d) Explain how you know.
   e) How confident are you about the contents of the black box? (1 = I have no idea; 10 = I am completely sure; nothing else is possible.)

ANSWERS

For You To Do

(continued)

5. - 6. Students will design their own experiment. It should allow them to measure the voltage and current across the different resistors. By creating a data chart for each resistor, they will then be able to match the given data with one of their charts.

7. Sample data:

Resistor value (ohms)	Voltage (volts)	Current (amps)
3	1.5	0.5
3	3	1
3	4.5	1.5
10	1.5	0.15
10	3	0.3
10	4.5	0.45
15	1.5	0.1
15	3	0.2
15	4.5	0.3

d) The teacher will provide data that will have a value different from any of those measured. The students will have to realize that the ratio of the voltage to the current is the clue to determining the resistance of the resistor.

A graph of the voltage vs. the current for the three resistors will show three slopes.

© It's About Time

Physics Words

Ohm's Law: the relationship among resistance, voltage, and current.

PHYSICS TALK

Ohm's Law Relates Resistance to Voltage and Current

The measurements of voltage and current showed that increasing the voltage increased the current in the circuit. This was true for all three resistors. These resistors also obey **Ohm's Law**, which states that as the voltage increases in a fixed rate, the current increases at the same rate. The ratio of the voltage to the current is a constant for any one of these resistors. This is expressed mathematically:

$$R = \frac{V}{I}$$

where R is the resistance in ohms (Ω),

V is the voltage in volts (V), and

I is the current in amperes (A).

The equation is known as Ohm's law. This equation may be rearranged to calculate the value of any of the terms.

$$V = IR \qquad I = \frac{V}{R}$$

Some resistors obey Ohm's Law over a wide range of voltages. For these resistors, the value of R always remains the same.

Sample Problem

A 2-Ω resistor is placed in a circuit. Record the currents corresponding to voltage measurements of 10 V and 30 V.

Strategy: You are asked to calculate current for a known resistor. You can use Ohm's Law, but you will need to rearrange the equation to solve for current.

© It's About Time

Activity 3 Ohm's Law

Givens:

$R = 2\ \Omega$

$V = 10\ \text{V}$ and $30\ \text{V}$

Solution:

$$I = \frac{V}{R}$$

$$I = \frac{10\ \text{V}}{2\ \Omega} = 5\ \text{A} \qquad I = \frac{30\ \text{V}}{2\ \Omega} = 15\ \text{A}$$

FOR YOU TO READ

Graphing Ohm's Law

Ohm's Law expresses the relationship between the voltage and current ($R=V/I$). For certain resistors, the ratio of the voltage and current is a constant. Recording the varying voltage and the corresponding current would allow you to make a graph. If the resistor obeys Ohm's Law, the ratio of the voltage and current remains the same and the graph would be a straight line. If the current is on the *x*-axis and the voltage on the *y*-axis, the slope of the line is equal to the resistance. (If the current is on the *y*-axis and the voltage on the *x*-axis, the slope of the line is equal to $1/R$.)

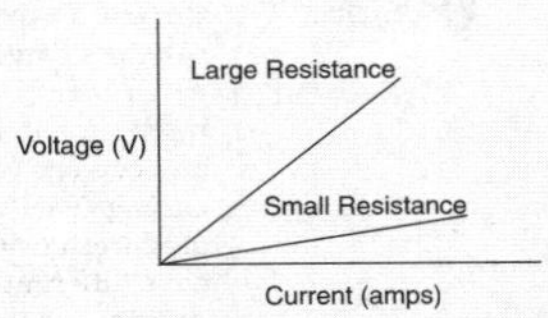

Many electrical devices do not obey Ohm's Law. A graph of the voltage versus current for a neon gas lamp is shown below.

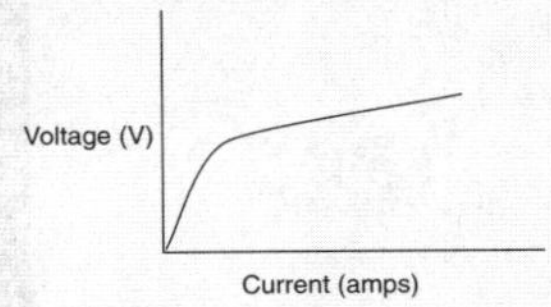

→

Chapter 7

© It's About Time

The resistance has a high value (large slope) for small currents and a low value (small slope) for large currents. The neon gas lamp does not obey Ohm's Law because the resistance is not constant.

The brightness of a bulb is dependent on the power (the rate of energy) supplied to the bulb. The power, as you will learn in the next activity, is equal to the product of the voltage and the current ($P = VI$). You have learned in this activity that increasing the voltage increases the current. This, in turn, increases the power to the light bulb. More power is more energy every second and that will make the bulb brighter.

The dimmer switch is a variable resistor. It controls the voltage and the current available to the light bulb and affects its brightness.

Reflecting on the Activity and the Challenge

Ohm's Law expresses the relationship between voltage and current for many resistors. Your appliances are all electrical resistors. These resistors can get hot (in toasters) or help a motor turn (in fans) or create light (in bulbs). Knowing the resistance allows you to use Ohm's Law to compute the voltage and current required in a circuit. As you develop your appliance package, you will need to consider the voltage and currents used in the circuit. You may wish to choose appliances that have different resistances to limit the power required in the circuit.

© It's About Time

Physics To Go

1. A resistor is placed in a circuit. Calculate the resistance in each of the following cases.
   a) The current in the circuit is 3 A and the voltage drop across the resistor is 12 V.
   b) The current in the circuit is 2 A and the voltage drop across the resistor is 6 V.
2. A resistor of 5 Ω is placed in a circuit. The voltage drop across the resistor is 12 V. What is the current through the resistor?
3. A resistor is placed in a circuit. The current in the circuit is 2 A and the voltage drop across the resistor is 8 V. The voltage is then increased to 12 V. What will be the new current?
4. Using the information in the data table, construct a graph following the directions below.
   - Mark an appropriate scale on the x-axis labeled "Current (A)."
   - Mark an appropriate scale on the y-axis labeled "Voltage (V)."
   - Plot the data points for voltage versus current.
   - Draw the best-fit line.

Current (A)	Voltage (V)
0.010	2.3
0.020	5.2
0.030	7.4
0.040	9.9
0.050	12.7

   a) Using your graph, find the slope of the best-fit line.
   b) What physical quantity does the slope of the graph represent?
5. Your hair dryer has a resistance of 9.6 Ω and you plug it into the bathroom outlet. Assume household voltage to be 120 V.
   a) What current will it draw?
   b) Suppose your brother has an identical hair dryer and plugs it into the same part of the circuit. What current will the two hair dryers draw?

ANSWERS

Physics To Go

1. a) $R = V/I = 12\text{ V}/3\text{ A} = 4\ \Omega$

   b) $R = V/I = 6\text{ V}/2\text{ A} = 3\ \Omega$

2. $I = V/R = 12\text{ V}/ 5\ \Omega = 2.4\text{ A}$

3. We can find the resistance from Ohm's Law and then use that resistance to calculate the new current.

   $R = V/I = 8\text{ V}/2\text{ A} = 4\ \Omega$

   $I = V/R = 12\text{ V}/ 4\ \Omega = 3\text{ A}$

4. See below.

5. a) $I = V/R$
   $= 120\text{ V}/ 9.6\ \Omega$
   $= 12.5\text{ A}$

   b) 12.5 A + 12.5 A = 25 A

   c) The fuse will blow or the circuit breaker will trip.

Chapter 7

4.

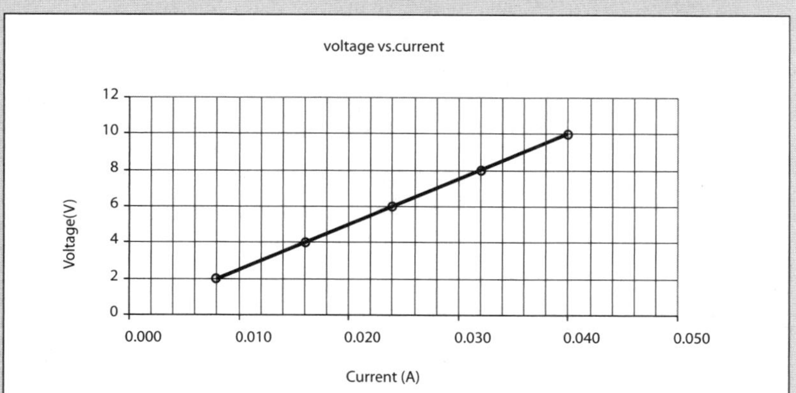

a) Slope = 250 V/A = 250 W

b) The slope represents the resistance.

© It's About Time

ANSWERS

Physics To Go
(continued)

6. Wire A has the greater resistance as you can see from the larger slope. On a voltage vs. current graph, the slope is equal to the resistance.

c) If the maximum current the circuit breaker in the system can handle is 20 A, what do you think will happen?

6. Two wires are tested in a lab setting. Current was measured as the voltage across the wire was varied. The results of the experiment are shown in the graph below. Both wire A and B obey Ohm's Law. Which wire has the greatest resistance?

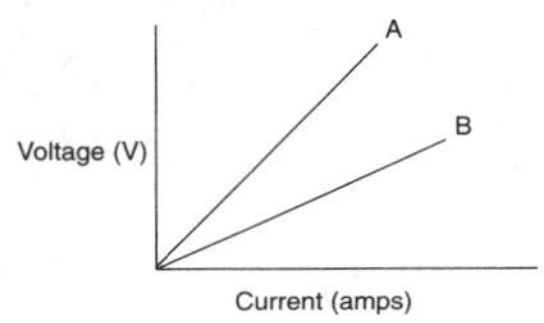

© It's About Time

NOTES

© It's About Time

ACTIVITY 4
Load Limit

Background Information

Principles involving power in a 120-volt AC household electric circuit are introduced in **Activity 4**. In particular, power calculations are applied to:

- the power consumed by individual appliances connected in a household circuit
- the power limit, or "load limit," of a household circuit which is protected by a fuse or circuit breaker

It is suggested that you first read the entire student text for **Activity 4** before proceeding in this background narrative. Please do that reading now.

The main focus of **Activity 4** is centered on the calculation of electric power presented in **For You To Do** as $P = IV$. Satisfy yourself that the units of measurement for the equation $P = IV$ are valid:

1 watt is defined as 1 joule/second

1 amp is defined as 1 coulomb/second

1 volt is defined as 1 joule/coulomb

Substituting the above definitions in the equation $P = IV$ gives:

$$\text{joule/second} = (\text{coulomb/second}) \times (\text{joule/coulomb}) = \text{joule/second}$$

An alternate equation for calculating power is good for you to know in cases where current and resistance are known but voltage is not known. Need for the alternate equation probably will not arise in this activity, but may be useful later:

$$P = I^2R$$

As discussed in the teacher **Background Information** for **Activity 3**, the voltage applied to a parallel circuit is equal to the voltage dissipated in each parallel branch of the circuit. This makes it convenient for calculating the power dissipated by appliances in a home; the voltage applied to a household circuit is 120 volts AC and, since homes are wired so that each appliance on a circuit (e.g., each light bulb, TV set, radio, etc.) is the only device in its particular branch, the voltage dissipated by each device also is 120 volts AC. Therefore, if the power rating of an appliance is not known it can be calculated using one of the above equations if either the current or resistance for the appliance is known (or can be measured).

To calculate the load limit of a household circuit in watts it is necessary only to multiply 120 volts by the rating in amperes of the circuit breaker or fuse which projects the circuit. In most homes each circuit is protected by either a 15 amp or 20 amp circuit breaker or fuse. Using the equation $P = IV$, the load limit for a circuit having a 15 amp fuse would be 120 volts × 15 amps = 1800 watts; for 20 amp protection the power would be 120 volts × 20 amps = 2400 watts.

The relationship between the load limit of a household circuit and the power ratings of the appliances connected to the circuit is very simple:

$$P_{\text{limit}} = P_1 + P_2 + P_3 + \ldots$$

When the sum of the power ratings of the individual appliances connected to the circuit ($P_1 + P_2 + P_3 + \ldots$) exceeds the load limit (P_{limit}) the circuit breaker automatically shuts the circuit off to prevent overheating of the wires in the home.

For the HFE dwelling the power of the appliances operating at any time may not exceed the specified 2400-watt load limit.

Students who do the **Stretching Exercise** for **Activity 4** may discover that the load center in a home has some large circuit breakers which are rated at as much as 50 amps. Such circuit breakers control 240-volt AC devices such as kitchen ranges, hot water heaters or other high power devices. For your information, the power company provides a home with 240 volts and either 100 or 200 amp (60 amps for some old homes) capability for current. The 240 volts is applied in entirety to devices such as kitchen ranges and is divided into two 120-volt components which, in turn, are applied to the home's normal circuits. The combined current which can be carried simultaneously by all of the circuits in the home is determined by the 100- or 200-amp rating of the load center. The 240-volt aspect of household electricity will not be involved in the HFE dwelling.

© It's About Time

Active-ating the Physics InfoMall

What Do You Think? asks about fuses and circuit breakers. Both of these can be searched for on the InfoMall. Suggested search words are "fuse*" or "circuit breaker*" where the asterisks have been included to allow for the singular and plural forms of the words.

For You To Do has a calculation and an equation for power. The InfoMall has many different avenues for exploring equations and calculations. You can go to the Equations Dictionary by clicking the button in the lower right-hand corner of the entrance screen (the button with equations all over it). This will give you a directory to choose from. When you have found the equation you can also ask for examples. Or you can browse through any of the textbooks, since most of them will have this equation. If you want homework-style problems, visit the Problems Place. You can choose from several collections of problems, including *Schaum's 3000 Solved Problems in Physics*. This would allow you to hand out problems and their solutions if you wish.

Physics To Go Part 7 mentions horsepower. A search using "horsepower" quickly provides several definitions of "horsepower" along with the historical context and conversion factors.

Planning for the Activity

Time Requirements

Allow about 10 - 20 minutes for the demonstration.

Materials Needed

For the class:

- 120-volt circuit for demonstrating overload (See **Advance Preparation and Setup**)
- fuses, 3 amp (5)
- fuses, 8 amp (5)

© It's About Time

Advance Preparation and Setup

Arrange a tabletop array of 120-volt AC light bulbs in porcelain sockets, wired in parallel with lamp cord. The number of bulbs and their chosen (equal) wattage's must be sufficient to blow the fuse or circuit breaker that controls the 120-volt AC circuit used as the energy source. Preferably, the bulbs will be the only load on the circuit during the demonstration, and the fuse or breaker will be visible to the students. The latter can be accomplished by rigging a subpanel for a load center with a fuse and/or breaker and a cord and plug so that it can be plugged into a classroom outlet. It must be fused at less than the panel that feeds the classroom circuit. It would be best to demonstrate both a fuse and circuit breaker.

If you have a 15-A fuse (or circuit breaker), with 120-volt circuit, and 60-W light bulbs, and 100-W light bulbs, calculate the total wattage, for each type of bulb, and make sure you have enough sockets. ($P = IV$; $P = 15\ \text{A} \times 120\ \text{V} = 1800\ \text{W}$. Therefore, you would need 18 – 100-W light bulbs, or 30 – 60-W light bulbs. As this is a large number of light bulbs, try using a smaller fuse, or use light bulbs, and a 1000-W hair dryer, and a 800-W kettle. Have the bulbs put a load of about 300, plug in the kettle and then plug in and turn on the hair dryer. Do this in different order, to see that it doesn't matter what the order is, it blows the fuse every time it is over the fuse rating.

If you have the time and equipment, place a voltmeter and an ammeter in the circuit to show the changes in the current, and the fact that the voltage stays the same.

Teaching Notes

Review from **Activity 2** that household circuits are parallel circuits and review the definitions of the coulomb and the ampere from the **PTG** in **Activity 2**. Do the demonstration, the **FYTD**, first, and then define the watt and the volt. Tread lightly when defining these terms, because prior background about work and energy cannot be assumed; you may wish to describe a joule as the work done when lifting a "quarter pounder" (the burger) one meter against gravity. The desired outcome is for students to have the idea that voltage is a measure of energy per unit of charge and that wattage is a measure of energy per unit time. Put it all together in the equation $P = IV$ using dimensional representation of the variables in the equation, and then set students to answering the questions in the **FYTD**.

To keep students aware that these activities contribute to knowledge needed for the **Chapter Challenge** be sure to link this activity to the load limit of the HFE electrical system. The power limit in the HFE system is 2400 watts. Be sure that students calculate the maximum current which the 120-volt system can accommodate at any one time.

To ensure the safety of the teacher and the students, do not leave the homemade circuit unattended. The household circuit of 15 A is more than enough to kill. As an added precaution, a GFCI plug should be used. If, as a teacher, you are uncomfortable with building your own circuit, ask for the electrician from your school district office to come out and build the circuit.

Students should come to understand that the current draw in any circuit is due to the resistance in the circuit. In keeping with Ohm's Law ($V = IR$), voltage is constant, and the resistance is a property of the substance (heating coils of a kettle for example). Therefore, by placing a 1200-W toaster with a resistance of 10 Ω, the amperage in a 120-volt circuit is 10 A. Show this by placing an ammeter in your circuit, and turning on the toaster. (Be sure to bring some bread and butter, to keep the students' attention.)

NOTES

© It's About Time

Activity 4 Load Limit

GOALS

In this activity you will:

- Define: watt, insulator, conductor.
- Apply $P = IV$.
- Measure the power limit of a 120-V household circuit.
- Differentiate between a fuse and a circuit breaker.
- Understand the need for circuit breakers and fuses in a home.

What Do You Think?

Everybody has at one time "blown a fuse" or "tripped" a circuit breaker.

- **What is a fuse or circuit breaker?**
- **Exactly what conditions do you think make a fuse "blow" or a circuit breaker "trip"?**

Record your ideas about these questions in your *Active Physics* log. Be prepared to discuss your responses with your small group and the class.

Activity Overview

This is a short demonstration activity, which should only take about 10 to 20 minutes. You may need to find out the size of the circuit breaker in your classroom from your building custodian. Make sure that the circuit you make in the classroom has a smaller amperage than the demonstration circuit.

Student Objectives

Students will:

- Define: watt, insulator, conductor.
- Apply $P = IV$.
- Measure the power limit of a 120-volt household circuit.
- Differentiate between a fuse and a circuit breaker.
- Understand the need for circuit breakers and fuses in a home.

ANSWERS FOR THE TEACHER ONLY

What Do You Think?

A fuse or a circuit breaker both operate on the principle that if the current surpasses the current allowed by the fuse, the fuse burns, or the breaker trips, opening the circuit. This is to prevent the overloading of a circuit, which would cause an increase in the heat energy in the circuit, and may lead to a fire.

For example, a 15-amp circuit breaker will trip when there is more than 15 amps on one circuit. This can happen if you have a toaster drawing 7 amps, and a kettle drawing 6 amps, (for a total of 13 amps), and someone plugs in the coffee maker which will draw 8 amps, adding up to a total of 21 amps. The circuit will trip, causing you to go to the breaker box, and turn the breaker back on. With a fuse, the fuse will burn out, and would have to be replaced.

Chapter 7

© It's About Time

For You To Do

1. Your teacher will do this part of the activity as a demonstration for the entire class. You will intentionally exceed the load limit of the circuits that supply electricity to your classroom. Observe the number of light bulbs needed to reach the load limit and the manufacturer's ratings of the bulbs.
   - a) Calculate the load limit of the circuit, in watts. This will apply to the load limit of the circuits for your Homes For Everyone (HFE) dwelling.
2. Knowing that the energy per unit of electric charge supplied to the circuit is 120 V (volts), use the answer to the previous calculation and the equation:

   Power = current × voltage ($P = IV$)

   to calculate the current, in amperes, that must have been flowing in the circuit when the load limit was reached.
   - a) Show your calculation in your log. Assume a voltage of 120 V.
   - b) How much current, in amps, must be flowing through the filament of a 60-W light bulb when it is operating in a 120-V household circuit? A 100-W light bulb? Show your calculations in your log.
   - c) Fuses and circuit breakers are rated in amperes, usually 15 or 20 A for most household circuits. In your log use the equation $P = IV$ to show how you can calculate the load limit, in watts, of a 120-V household circuit protected by a 15-A circuit breaker.

ANSWERS

For You To Do

1. a) The load limit of your circuit will depend on the fuse. For this part of the activity, count the number of bulbs (in W) and other appliances you have on the circuit when the circuit blows.

2. a) $P = IV$, and the $I = P/V$.

   b) $P = IV$, and the $I = P/V$, therefore, $I = 60$ W/120 V therefore, $I = 0.50$ A;

   $P = IV$, and the $I = P/V$, therefore, $I = 100$ W/120 V therefore, $I = 0.83$ A;

   c) $P = IV$, therefore, $P = 15 \text{ A} \times 120 \text{ V} = 1800$ W.

© It's About Time

PHYSICS TALK

Did you ever wonder why the lights go out in some rooms but not others when a fuse blows? Why does turning on the hair dryer blow a fuse but turning on the radio does not?

In the appliance package that you will be creating, there is a load limit of 2400 W. You are not restricted to a load limit in your own home. If you have the money, the electric company will set up more circuits allowing you to use more electricity. The company can then collect more money from you. There may actually be a limit. One person could not be allowed to use all the electricity from a massive power plant—at least, we hope not.

There are some restrictions on each circuit in your home. A fuse or circuit breaker protects each circuit. The protection is needed because too much current can heat or burn the wires in the walls of your home. An electrical fire can be extremely dangerous.

You probably think about the fuse or circuit breaker only when it turns off the electricity to a part of your home. The circuit is shut down when too much current is required. In most household circuits, the maximum current is usually 15 or 20 A (amperes). Each appliance in your home requires a certain amount of current to run properly. The power rating of the appliance, the voltage of the circuit, and the current required are related by the following equation:

$$P = VI$$

where P is the power in watts (W),

I is the current in amperes (A), and

V is the voltage in volts (V).

This equation can be rearranged to calculate the value of any of the terms.

$$I = \frac{P}{V} \qquad V = \frac{P}{I}$$

→

Chapter 7

© It's About Time

The voltage of almost all of the circuits in your home is 120 V. If the fuse or circuit breaker can handle a maximum of 15 A, then the power limit is (120 V) × (15 A) or 1800 W. If a toaster is 1200 W and a hair dryer is 1000 W, they cannot both be on any one line in your house since they total 2200 W, which is more than the 1800-W limit. If you want to run both appliances at the same time, you must use different circuits, each with its own 1800-W limit.

A second way of viewing the circuit is to look at the current requirements of each appliance. Since the appliances are in parallel, the total current will be the sum of all of the individual currents. In the example above, the current of the 1200-W toaster can be found using the power equation.

$$P = VI$$

$$1200\ W = 120\ V \times I$$

$$I = 10\ \text{A}$$

Similarly, the hair dryer requires about 8 A.

The total current is about 18 A. This is more than the 15-A fuse can tolerate.

Sample Problem 1

A 12-V starter battery in a car supplies 48 A of current to the starter. What is the power output of the battery?

Strategy: You are asked to find the power, so you use the power equation that is specific to electrical circuits. Power is defined as change in energy or work per unit time. However, in electrical circuits it can be found by multiplying the voltage (1 V = 1 J/C) times the current (1 A = 1 C/s). You notice that this results in joules per second, which are watts, the unit for power (1 J/s = 1 W).

Givens:

$V = 12\ \text{V}$

$I = 48\ \text{A}$

© It's About Time

Solution:

$$P = VI$$
$$= (12\text{ V})\ (48\text{ A})$$
$$= 576\text{ W (or about 580 W)}$$

Sample Problem 2

A 75-W study lamp is plugged into the 120-V household outlet in your room. What current does the outlet supply to the light bulb?

Strategy: Again, use the power equation but rearrange the equation to solve for current.

Givens:

$P = 75\text{ W}$

$V = 120\text{ V}$

Solution:

$$I = \frac{P}{V}$$
$$= \frac{75\text{ W}}{120\text{ V}}$$
$$= 0.63\text{ A}$$

Light bulbs do not draw a lot of current.

Sample Problem 3

A light bulb operating at 120 V draws 0.5 A. Determine the bulb's:

a) resistance

b) power

Strategy: Ohm's Law can be used to determine the resistance of the light bulb. The power can be determined using $P = VI$.

→

Chapter 7

© It's About Time

Givens:

$V = 120\text{ V}$

$I = 0.5\text{ A}$

Solution:

a) $V = IR$

Solving for R,

$$R = \frac{V}{I}$$

$$= \frac{120\text{ V}}{0.50\text{ A}}$$

$$= 240\ \Omega$$

b) $P = VI$

$$= (120\text{ V})(0.50\text{ A})$$

$$= 60\text{ W}$$

The power could also be calculated using either of the following:

$$P = I^2R$$

or $$P = \frac{V^2}{R} \quad (\text{Since } I = \frac{V}{R},\ I^2 = \frac{V^2}{R^2})$$

These relationships are derived using Ohm's Law.

Reflecting on the Activity and the Challenge

The load limit of the electrical system for the universal dwelling is set at 2400 W, as outlined in the chapter **Scenario**. It is also established by the design of the windmill power plant that 120 V will be applied to circuits within the dwelling. In this activity you learned what load limit means, and how to relate it to current and voltage. If the people in the dwelling try to run appliances that require more than 2400 W, the fuse will blow. In your home, you can always choose a different line to run the

© It's About Time

extra appliances. With only one generator, this is not an option in the universal dwelling. This will have direct application soon when you begin selecting appliances to be used in the dwelling.

Physics To Go

1. Explain in detail what load limit means, and include maximum current, in amperes, as part of your explanation.
2. Find out about the power rating, in watts, of at least six electrical appliances. You can do this at home, at a store that sells appliances, or by studying a catalog. Some appliances have the watt rating stamped somewhere on the device itself, but for others you may have to check the instruction book for the appliance or find the power rating on the original package. Also, your local power company probably will provide a free list of appliances and their power ratings on request. Bring your list to class.

   If the appliance lists the current in amps, you can assume a voltage of 120 V and calculate the power (in watts) by using the equation $P = IV$.

⚠ Make sure small appliances are unplugged before handling them.

3. List three appliances you would include in the HFE appliance package that will be part of the chapter **Assessment**. For each appliance, list the power demand. For each appliance, describe how it will contribute to the well-being of the people living in the dwelling.
4. An electric hair dryer has a power rating of 1200 W and is designed to be used on a 120-V household circuit. How much current flows in the hair dryer when it is in use?
5. A 120-V circuit for the kitchen of a home is protected by a 20-A circuit breaker. What combinations of the appliances listed below can be used on the circuit at the same time without the circuit breaker shutting off the circuit?
   - 1000-W toaster
   - 1200-W frying pan
   - 300-W blender
   - 600-W coffeemaker

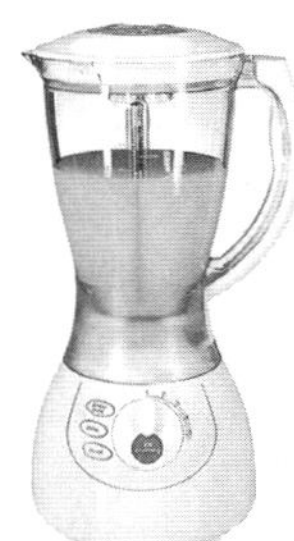

ANSWERS

Physics To Go

1. Load limit refers to the maximum amount of current that flows through the circuit. If it exceeds this maximum current, the fuse or circuit breaker will blow, stopping the flow of current. This is to protect the circuit from over-heating. (Refer to teacher background for more information.)
2. Students' answers will vary. However, they should have several appliances with a similar calculation to the following:

   $P = IV$ using a toaster-oven rated at 1500 W. therefore, $I = P/V$; $I = 1500\text{ W}/120\text{ V} = 12.5\text{ A}$)
3. Students' answers will vary. Students should take into consideration the time of day the appliance will be used, and what other appliances may be needed at the same time. For example, having a coffee perk rated at 600 W would be better to use than the drip coffee maker rated at 1100 W, if you were also making toast at the same time (toaster rated at 1100 W) with two 100-W light bulbs on at the same time.
4. $P = IV$, therefore $I = P/V$; therefore $I = 1200\text{ W}/120\text{ V} = 10\text{ A}$)
5. First, find the total power rating for the circuit. $P = IV$; $= 20\text{ A} \times 120\text{ V} = 2400\text{ W}$. Therefore, the total of the appliances must not exceed 2400 W: toaster and frying pan = 2200 W; toaster and blender and coffee maker = 1900 W; frying pan, blender and coffee maker = 2100 W; etc.

© It's About Time

ANSWERS

Physics To Go

(continued)

6. Total power: $P = IV$;
$P = 15\text{ A} \times 120\text{ V} = 1800\text{ W}$. Therefore, the number of light bulbs = 1800 W/60 W = 30 bulbs. For fluorescent bulbs 1800 W/22 W = 81 light bulbs (actually 81.8 light bulbs, but how would that .8 of a light bulb plug in?)

7. If one HP = 746 W, then the vacuum will have a power rating of 0.8 HP × 746 W = 597 W. The current that the vacuum draws is

$$I = \frac{P}{V} = \frac{597\text{ W}}{120\text{ V}} = 5\text{ A}.$$

8. $P = IV$, therefore
$P = 6\text{ A} \times 120\text{ V}$
$= 720\text{ W}$.

9. $P = VI$

$$I = \frac{P}{V}$$

$$= \frac{1500\text{ W}}{120\text{ V}}$$

10. a) $P = V \cdot I$
$= (120\text{ V})\ (8.0\text{ A})$
$= 960\text{ W}$

b) Since the appliances would be connected in parallel, they would have the same voltage drop across them. Being identical, they would have the same resistance and therefore the same current. Two toasters each drawing 8.0-A connected in parallel would result in a 16.0-A combination. This would be small enough not to trip a 20-A circuit breaker but too large for a 15-A one. So you are working with a 15-A breaker or fuse.

11. a) $I = \dfrac{P}{V} = \dfrac{3\text{W}}{120\text{V}} = 0.025\text{A}$

Electricity for Everyone

6. How many 60-W incandescent light bulbs can be operated at the same time on a 120-V, 15-A circuit in a home? How many energy-efficient 22-W fluorescent bulbs can operate on a similar circuit?

7. Some electrical appliances are rated in horsepower (HP).
1 HP = 746 W
What amount of current flows in a 0.8 HP vacuum cleaner operating on a 120-V circuit?

8. Some electrical appliances are rated in amps. What is the power in watts of a 6-A appliance designed to operate on a 120-V circuit?

9. A 1500-W hair dryer is plugged into the outlet in your bathroom. How much current does this hair dryer draw?

10. When you turn on the toaster in the kitchen, it draws 8.0 A of current from the line.
a) Find the power output of the toaster.
b) You plug another toaster in the same outlet and the circuit breaker snaps off. What do you think the maximum current is for the type of breaker you are using?

11. A 3-W clock operates at 120 V.
a) How much current does the clock draw?
b) Determine the clock's resistance.
c) If the maximum current that can be drawn from the outlet is 15 A, how many clocks would it take to blow the fuse? (Assume all clocks could be plugged into the outlet.)

12. An iron has a resistance of 13.1 Ω. Could two identical irons operate on the same fuse? (Assume $V = 120$ V and $I_{max} = 15$ A.)

13. The load limit for a particular fuse is 2400 W. Which combination of the following devices would blow the fuse: mini-refrigerator ($P = 300$ W), microwave oven ($R = 19\ \Omega$), hair dryer ($I = 12$ A), coffeemaker ($R = 9.2\ \Omega$).

Active Physics 470

ANSWERS

Physics To Go

(continued)

b) $R = \dfrac{V}{I} = \dfrac{120\text{ V}}{0.025\text{ A}}$
$= 4800\ \Omega$

c) Number of clocks = I_{max}/I_{clock} = (15 A)/(0.025 A) = 600 clocks
The 601st clock would blow the fuse!

12. The current drawn by each iron:

I = V\R = 120 V\13.09 Ω = 9.17 A

Both irons would require 18.34 A, which is greater than the fuse can deliver.

© It's About Time

Stretching Exercises

Find out about the electrical system of your home or the home of a friend or acquaintance. **With the approval of the owner or manager, and with adult supervision**, locate the load center, also called the main distribution panel, for the electrical system. Open the panel door and observe whether the system uses circuit breakers or fuses. How many are there, and what is the ampere rating shown on each fuse or circuit breaker? You may find some larger fuses or breakers that control large, 240-V appliances such as a kitchen range (electric stove); if so, how many are there, what are their ampere ratings and, if you can, determine what they control.

Warning: The inside of a load center is a dangerous area. Do not touch anything. Doing so could cause injury or death. Always have a qualified person help you.

In some load centers there is a list of what rooms or electrical devices are controlled by each fuse or breaker, but often the list is missing or incomplete.

With the approval of the owner or manager, and with adult supervision, you can develop a list that indicates what each fuse or breaker controls. To do so, shut off one circuit and go through the entire house to find the lights and outlets that are "dead" (check outlets with a lamp that you can carry around easily). Those items that are "off" are controlled by that fuse or breaker. List them. Then repeat the same process.

Do not touch any exposed electrical connections or wiring harnesses. Do not attempt to look into or insert anything into any wiring entry points on the panel. Do not reset any circuit breakers or attempt to change any fuses.

Report your findings to your teacher in the form of a list or diagram of the house showing what is controlled by each fuse or circuit breaker.

ANSWERS

Stretching Exercise

This is an excellent exercise, to help the student label the circuits in their home. One of the things the students may find, however, is that some of the circuits may be overloaded, and may have to be changed. Ensure that the students will be doing this with permission and supervision.

ANSWERS

Physics To Go

(continued)

13. Assuming the fuse provides a voltage of 120 V, the maximum current which can be drawn is 20 A ($P = VI$). As long as the combination requires less than 20 A and 2400 W, the fuse will be okay. If either is exceeded, the fuse will blow. Determine the required current and power for each device.

    refrigerator:
    $P = 300$ W
    $P = VI$, solving for I, $I = P/V$
    $= (300 \text{ W}) / (120 \text{ V}) = 2.5$ A

    microwave:
    $I = V/R = (120 \text{ V}) / (19.2 \text{ W})$
    $= 6.25$ A
    $P = VI = (120 \text{ V})(6.25 \text{ A}) =$
    750 W

    hair dryer:
    $I = 12.5$ A
    $P = VI = (120 \text{ V})(12.5 \text{ A}) =$
    1500 W

    Coffee Maker:
    $P = VI = (120 \text{ V})(13 \text{ A}) =$
    1560 W

    Some combinations that would blow the fuse:

    - All four devices; $P = 4110$ W, $I = 34.25$ A
    - Coffee Maker and Hair Dryer; $P = 3060$ W, $I = 25.5$ A
    - Refrigerator, Microwave, and Hair Dryer; $P = 2550$ W, $I = 21.25$ A

Chapter 7

© It's About Time

ACTIVITY 5
Who's in Control?

Background Information

Principles introduced in **Activity** 5 include:

- using switches to control electric circuits.
- calculation of electric energy and its cost.

Switches. The "trick" involved with predicting the effect of placing a switch in a parallel circuit is to determine how the total current divides among the branches within the circuit. Refer to the circuit diagram in **Activity** 5, **For You To Do** in the student text. Visualize the current as a flow of electrons leaving the battery or generator and entering wire 1. The entire current will continue into wire 2 because there is no alternate path. A switch used to replace either wire 1 or wire 2 would control current flow to all bulbs.

At the far (left) end of wire 2 the current flow will divide into two parts, one part travelling through wire 11, bulb C and wire 12. Replacing either wire 11 or wire 12 would control only bulb C. The part of the current flow which leaves wire 2 and enters wire 3 is on its way to divide again at the next branch point. A switch replacing wire 3 would control both bulbs B and A.

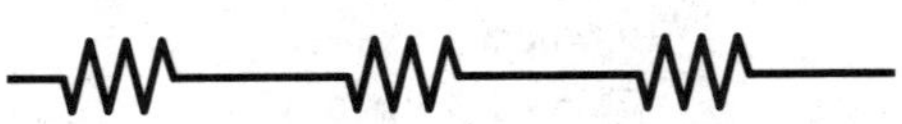

The current that flows through wire 3 divides into two parts, one part travelling through wire 13, bulb B and wire 14. A switch replacing either wire 13 or wire 14 would control only bulb B. The part of the current which leaves wire 3 to enter wires 4 and 5, bulb A, and wires 7 and 8 is on its way to recombine with the current flowing through bulb B. A switch replacing wire 4, 5, 7 or 8 would control only bulb A.

The currents flowing in wires 7 and 14 combine to flow into wire 8. A switch replacing wire 8 would control bulbs A and B.

The currents flowing through wires 8 and 12 combine to flow into wires 9 and 10 as the same amount of current that left the battery or generator to enter wires 1 and 2. A switch replacing either wire 9 or 10 would control all bulbs.

The kind of switch intended for **Activity** 5 is a "single pole single throw" switch which simply "makes" or "breaks" a circuit. Special switches often used in homes include 3-way and 4-way switches used, for example, to control one light bulb for two or more switching locations. Details about how special switches work can be found in manuals available in the electrical departments of most home building centers.

Electric energy/cost. Please first read the **Physics To Go** and **Physics Talk** sections of **Activity** 5 in the student text. The origin of the equation

Energy = VIt presented in **Physics Talk** is:

If, as defined, Power = (Energy)/(time)

Then, multiplying both sides of the above definition by time:

$$\textbf{Energy} = \textbf{(Power)} \times \textbf{(time)}$$

Substituting appropriate electrical terms for power in the above:

$$\textbf{Electric energy} = (IV)t$$

Energy, regardless of its form, should be expressed in joules. To test the above equation, units are substituted in the right side of the equation for I, V and t:

$$= \textbf{(coulomb/second)} \times \textbf{(joules/coulomb)} \times \textbf{(second)}$$

$$= \textbf{joules}$$

Unfortunately, the electric industry does not use joules as the unit of measurement for the energy it sells to customers. The industry uses kilowatt-hours (kWh). To understand this unit of measurement it is best to return to the definition of the watt as a unit of power:

Definition: 1 watt = 1 joule/second

Multiplying both sides of the definition by (second):

1 joule = 1 watt-second

Therefore, 1 watt-second is a unit of energy equivalent to 1 joule.

© It's About Time

The electric energy industry deals with large amounts of energy, and the watt-second is too small for their purposes. Therefore, the industry scales power upward from watts to kilowatts (1000 watts = 1 kilowatt) and scales time upward from seconds to hours (3600 seconds = 1 hour) resulting in:

1 kWh = (1000 watts) × (3600 seconds) = 3,600,000 watt-seconds = 3600,000 joules

For **Physics To Go**, students are asked to identify the power ratings of several appliances and estimate the amount of time each is used during one month. If time estimates are made in hours and power ratings of appliances are in watts, the monthly energy use in kWh will result if the power ratings in watts are divided by 1000 to convert to kilowatts, and then the kilowatt ratings are multiplied by time in hours. This will lead to consideration of the cost of operating appliances based on the power company's cost per kilowatt hour in **Activity 6**.

Active-ating the Physics InfoMall

Again in this activity, we use circuits and make predictions. The InfoMall was searched for information related to both of these in earlier activities.

For You To Read discusses circuit diagrams. A quick search for "circuit diagram*" produces many hits with several graphics. In this case, the graphics were not searched for, only the words. But on the InfoMall, figures and graphics appear in blue and are hyperlinks to the graphic. If you do this search, simply scan the page for blue text, click on it, and the graphic will appear. These can also be copied and pasted into your own handouts. Finding the circuit diagram you want is unlikely, however, so these may serve best as examples of how to draw diagrams.

Planning for the Activity

Time Requirements

This activity will take approximately 20 minutes, provided the groups have all the materials they need at their station. Allow about 10 to 20 minutes to get around to all groups before they are allowed to turn on their circuit.

© It's About Time

Materials Needed

For the class:

- 1100 Watt hair dryer to blow fuses
- source of cold water
- source of hot water

For each group:

- Alligator clip leads (14)
- manual electric generator: a hand-operated DC generator capable of a maximum output of about 300 milliamps at 6 volts is recommended for this and several other activities in the remaining part of this and the next chapter. The Genecon manufactured by Nakamura Scientific Co., Ltd. is recommended. It is available from several United States science supply houses. Generators are considered essential for student use in this and the next chapter, and batteries or power supplies should not be substituted for them.
- calculator
- light bulb, miniature screwbase (flashlight bulbs): Just about any low voltage bulb will do; a No. 13 (3.6 volt, .30 amp) is a good choice.
- bulb base for miniature screwbase bulbs: any kind (4)
- switch, spst (3)

Advance Preparation and Setup

If time allows, you may want to demonstrate how a bi-metallic strip works as the heart of a thermostat, and demonstrate other automatic switching devices such as a mercury switch or a photocell. Discuss the conditions to which each responds.

For this activity, have the materials available in small packages or in separate bins. The actual design of the circuit will depend on the materials available. To save time, precut all wires, and label (1, 2, 3,...). The students would not have to do any cutting, and the circuits will all look the same and this will speed up the checking of the circuits. An example might be to cut wire numbers 1, 5, 6, 10, 11, 12, 13, 14 all the same length, using one color, and wires 2, 3, 4, 7, 8, 9 all the same length, with a different color. Then instruct the students to get 8 of one color, and 6 of the other color.

Teaching Notes

Begin with a demonstration of automatic switching devices (see above recommendation). Ask students to offer explanations of the conditions which control the devices and what the devices must be doing to affect the electric circuits in which they are placed—how do they accomplish on/off actions? Then ask students to complete **What Do You Think?** to identify more ways that automatic switching is accomplished.

Before proceeding to **For You To Do**, plan to save some time at the end of the class period to discuss the **Reflecting on the Activity and the Challenge** and to be sure that the students understand how to complete **Physics To Go**.

Make sure the students are reminded of the dangers of electricity, and the precautions that must be taken. Under no circumstances should the students be plugging their circuit into the wall outlets. This circuit is designed for low current, and plugging into a wall outlet could cause serious injury or death. There is also a fire hazard, as the wires will not be able to handle the load.

SAFETY PRECAUTION: Students must not proceed until the teacher has checked the circuits.

Automatic switching devices may seem complicated, but once dissected, they are very simple. Show the students a number of different switching devices, and even allow the students to take apart some discarded switches. Go to an electronics store, or an electrician, for old switches. You may want to have an electrician or someone from the local power company, come into class to demonstrate the hazards of electricity, and how to handle it safely.

NOTES

© It's About Time

NOTES

© It's About Time

Activity Overview

The students will be investigating how switches work, and how to set up switches to operate a parallel circuit.

Student Objectives

Students will:

- Explain how a variety of automatic electrical switches work.
- Select switches and control devices to meet particular needs.
- Insert a switch in a parallel circuit to control a particular lamp.

ANSWERS FOR THE TEACHER ONLY

What Do You Think?

The number of switching devices the students list will be numerous. Some of them might include: street lights, timers for homes, timers on stoves, timers for traffic lights, thermostats in furnaces and air conditioners, signal lights in cars, automatic switches which turn stoves, on or off, thermocouples, security systems which operate on infrared, or movement, voice activated microphones, etc. The switching on or off will often utilize an electromagnet, or photocell.

With an electromagnetic switch, movement of the electromagnet, creates an electric current. This is attached to a relay. When an electric current runs through a relay, the current causes the relay switch to open or close (depending on the desired effect). This, then, activates the switch in another circuit to turn on your lights, or stove, or appliance.

Electricity for Everyone

Activity 5 Who's In Control?

GOALS

In this activity you will:

- Explain how a variety of automatic electrical switches work.
- Select switches and control devices to meet particular needs.
- Insert a switch in a parallel circuit to control a particular lamp.

What Do You Think?

Many electrical switches are operated manually (by hand), and many others are automatic, turning appliances on and off in response to a variety of conditions.

- **List as many different kinds of automatic switching devices as you can.**
- **What are the conditions that cause the on/off action of the switch?**

Record your ideas about these questions in your *Active Physics* log. Be prepared to discuss your responses with your small group and the class.

Active Physics 472

© It's About Time

For You To Do

1. Assemble the circuit as shown in the diagram. (Each number corresponds to a different wire.)

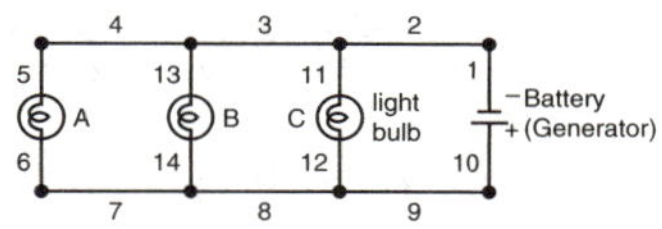

a) Draw a diagram of the circuit. Label the bulbs: A, B, C. (The bulb on the left should be A, the bulb in the middle B, and the bulb on the right C.)

Have your teacher approve your circuit before proceeding.

2. Compare the circuit you assembled with the one in the following diagram that shows the circuit you used in **Activity 2**. This time you used additional wires, 14 wires in total. It is sometimes easier to place switches into circuits if you use a few additional wires in the circuit. Crank the generator to be sure that all bulbs operate in this circuit.

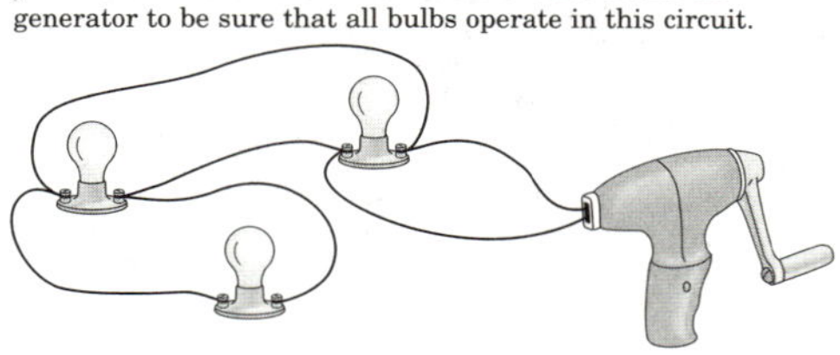

a) Identify the additional new wires on your circuit diagram.

b) What type of circuit (series or parallel) is represented in both diagrams?

3. Predict which wire could be replaced with a switch if you wished to turn all three bulbs on and off.

a) Record your prediction in your log.

b) Replace that wire with a switch. Does it work as predicted?

Answers

For You To Do

1. a) Student activity.

2. a) Students' answers will vary. You could refer to all the "inside" (1, 11, 13, 5, 6, 14, 12, 10) as new wires.

   b) Parallel circuit.

3. a) Students' prediction.

   b) A switch placed in wire 1, 2, 9, 10 would open the circuit to all three bulbs. See diagram.

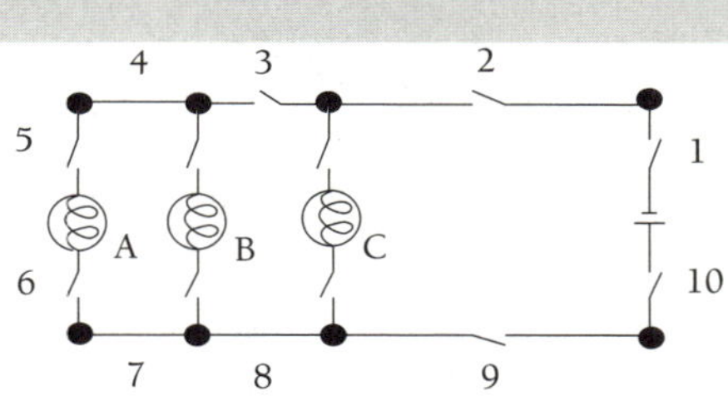

© It's About Time

ANSWERS

For You To Do
(continued)

4. a) Students' prediction.

   b) A switch placed in wire 5, 6, would open the circuit to bulb A.

   c) Student's diagram.

5. i) a) Students' prediction.

   b) A switch placed in wire 13, 14, would open the circuit to bulb B.

   c) Students' diagram.

   ii) a) Students' prediction.

   b) A switch placed in wire 11, 12, would open the circuit to bulb C.

   c) Students' diagram. See diagram.

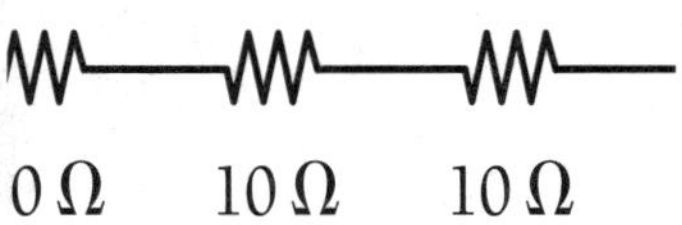

4. Predict which wire could be replaced with a switch if you wished to turn only bulb A on and off?
   a) Record your prediction in your log.
   b) Replace that wire with a switch. Does it work as predicted?
   c) Mark the location of the switch on the circuit diagram in your log by writing "Switch A" and drawing an arrow from the word "Switch A" to the place where the switch should be placed.

5. Repeat **Step 4** for bulb B and then for bulb C.
   a) Remember to record your predictions.
   b) Replace the wire you chose with a switch. Are your predictions beginning to improve?
   c) Draw two additional diagrams to show the location of switch B and switch C.

FOR YOU TO READ

Switches

Regardless of how an electrical switch may be activated, most switches work in the same basic way. When a switch is "on," a good conductor of electricity, usually copper, is provided as the path for current flow through the switch. Then, the circuit containing the switch is said to be "closed," and the current flows. When a switch is turned "off," the conducting path through the switch is replaced by an air gap. Since air has very high resistance, the current flow through the switch is interrupted, and the circuit is said to be "open."

Circuit Diagrams

An electrical circuit can be represented by a simple line diagram, called a schematic diagram, or wiring diagram. You may have seen wiring diagrams for cars, stereo systems, or homes. They may appear complicated, but if you know what the different symbols represent, they are relatively easy to follow.

The wire that carries the electric current is represented by a straight line. If a wire crosses over another wire, a line crossing a straight line is shown. If the wires join, a heavy black dot is shown. A lamp or light bulb is shown by a loop in the center of a circle. A battery or a generator is represented by a number of unequal lines. A switch is a line shown at an angle to the wire. There are many other symbols that are used by electricians.

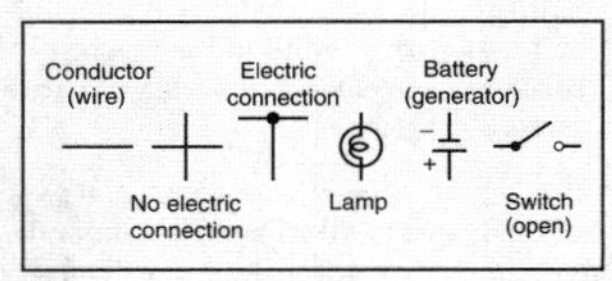

© It's About Time

PHYSICS TALK

Multiple resistors, like the light bulbs in **Activity 2**, can be set up in series or parallel. The properties of the series and parallel circuits are quite different. Light bulbs in series will all go out when one bulb is removed. Light bulbs in parallel will remain on when one bulb is removed. In this activity, you learned how switches could control which resistors in a parallel circuit will get current.

Computing the voltage, current, resistance, and power of appliances in your circuit require you to understand Ohm's Law ($V = IR$) and to calculate power ($P = VI$). Series and parallel circuits distribute the current differently and require different analysis.

We will begin with a few general organizing principles for circuits and then show how you can use these principles to determine the voltage, current, power, and resistance for each appliance in a circuit and for the circuit as a whole.

The current, measured in amperes (A), in an electrical circuit is a measure of the flow of charge. One ampere is defined as one coulomb per second. If charge is flowing in the circuit, then the current must remain constant. In a series circuit, the current that goes through the first light bulb must then go through the second light bulb and then through the third light bulb. The equation that describes this is:

$$I_{total} = I_1 = I_2 = I_3$$

In a parallel circuit, the current splits at certain junctions and then joins together at other junctions. The current entering any junction must equal the current leaving any junction if charge is to be conserved. In the part of the circuit shown on the following page, you can see that 1 A and 8 A enter the junction and 2 A, 3 A, and 4 A leave the junction.

→

© It's About Time

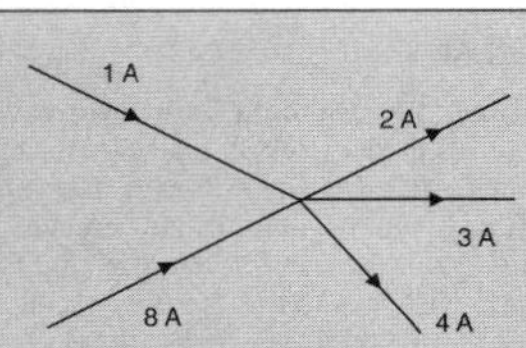

The equation that describes this is:

$$I_{total} = I_1 + I_2 + I_3$$

The voltage (also called the potential difference), measured in volts, in an electrical circuit is a measure of the energy per charge. One volt is defined as one joule per coulomb. The battery provides a certain amount of energy to every coulomb of charge. The energy must be conserved. The energy in the circuit is distributed to the resistors (appliances) in the circuit.

In a series circuit, all of the charges go through all of the resistors. The energy that the charge begins with must be distributed to the resistors. If a battery provides 6 V, 2 V may be transferred to the first resistor, 3 V to the second resistor and 1 V to the third resistor. The equation that describes this is:

$$V_{total} = V_1 + V_2 + V_3$$

In a parallel circuit, the charges that go through one resistor do not go through any other resistor. All of the energy of that charge must be transferred to that one resistor. If a battery provides 6 V, then 6 V must be transferred to each resistor. In the parallel circuit shown in the diagram on the next page, there are 8 A of current. The charges in these 8 A have been provided with 6 V by the battery (6 V = 6 J/C). The 8 A of current splits and 2 A go through the first resistor, 1 A through the second resistor and 5 A through the third resistor. The three currents are shown in three different colors in the sketch on the next page. The 2 A of "red" current drops all of its

© It's About Time

6 V in the first resistor. The 1 A of "blue" current drops all of its 6 V in the second resistor. The 5 A of "green" current drops all of its 6 V in the third resistor.

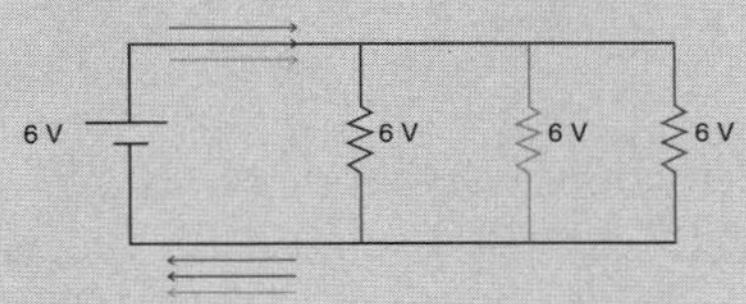

The equation that describes this is:

$$V_{total} = V_1 = V_2 = V_3$$

Having used the conservation of charge and the conservation of energy to determine how currents and voltages are distributed in series and parallel circuits you can now use Ohm's Law to derive equations for the total resistance of multiple resistors.

Series Circuit

Current: $I_{total} = I_1 = I_2 = I_3$

Voltage: $V_{total} = V_1 + V_2 + V_3$

Resistance: $R_{total} = R_1 + R_2 + R_3$

Ohm's Law: Ohm's Law can be applied to the entire circuit or to any resistor in the circuit.

$V_{total} = I_{total}R_{total}$ $V_1 = I_1R_1$ $V_2 = I_2R_2$ $V_3 = I_3R_3$

Power: Power can also be applied to the entire circuit or to any resistor in the circuit.

$P_{total} = V_{total}I_{total}$ $P_1 = V_1I_1$ $P_2 = V_2I_2$ $P_3 = V_3I_3$

→

Chapter 7

© It's About Time

Sample Problem 1

Given the following series circuit, find:

a) the total resistance
b) the total current
c) the current through each resistor
d) the voltage for each resistor
e) the total power
f) the power through each resistor

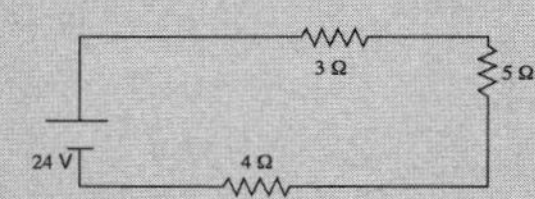

Strategy: This is a series circuit. All the current goes through each of the resistors. First calculate the total resistance. Then using Ohm's Law, calculate the total current. You can also use Ohm's Law to calculate the voltage for each resistor. Then, using the equation for power, calculate the total power and the power through each resistor.

Givens:

$V = 24\ \text{V}$

$R_1 = 3\ \Omega$

$R_2 = 5\ \Omega$

$R_3 = 4\ \Omega$

Solution:

a) Total resistance: $R_{\text{total}} = R_1 + R_2 + R_3$

$= 3\ \Omega + 5\ \Omega + 4\ \Omega$

$= 12\ \Omega$

b) Total current: $V_{total} = I_{total} R_{total}$

$24\ \text{V} = I_{total}(12\ \Omega)$

$I_{total} = 2\ \text{A}$

© It's About Time

c) If you know that the total current is 2 A, then you know that the current through each resistor is 2 A. It is helpful to place this information in the diagram.

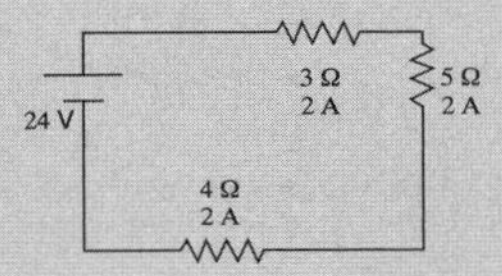

d) Once you know two of the four variables (*V*, *I*, *P*, *R*), you can find the other two variables. In this case, you know *I* and *R* and can find the voltage using Ohm's Law for each resistor.

$V_1 = I_1R_1$	$V_2 = I_2R_2$	$V_3 = I_3R_3$
$V_1 = (2\text{ A})(3\text{ W})$	$V_2 = (2\text{ A})(5\text{ W})$	$V_3 = (2\text{ A})(4\text{ W})$
$V_1 = 6\text{ V}$	$V_2 = 10\text{ V}$	$V_3 = 8\text{ V}$

Notice the sum of the voltage drops:

6 V + 10 V + 8 V = 24 V

This is the voltage supplied by the battery.

e) You can now find the total power.

$P_{total} = V_{total}I_{total}$

$P_{total} = (24\text{ V})(2\text{ A})$

$P_{total} = 48\text{ W}$

f)

$P_1 = V_1I_1$	$P_2 = V_2I_2$	$P_3 = V_3I_3$
$P_1 = (6\text{ V})(2\text{ A})$	$P_1 = (10\text{ V})(2\text{ A})$	$P_1 = (8\text{ V})(2\text{ A})$
$P_1 = 12\text{ W}$	$P_1 = 20\text{ W}$	$P_1 = 16\text{ W}$

Notice the sum of the power in the resistor:

12 W + 20 W + 16 W = 48 W

That is the power supplied by the battery.

→

Chapter 7

© It's About Time

Parallel Circuits

Current: $I_{total} = I_1 = I_2 = I_3$

Voltage: $V_{total} = V_1 + V_2 + V_3$

Resistance: $\dfrac{1}{R_{total}} = \dfrac{1}{R_1} + \dfrac{1}{R_2} + \dfrac{1}{R_3}$

Ohm's Law: Ohm's Law can be applied to the entire circuit or to any resistor in the circuit.

$V_{total} = I_{total}R_{total} \quad V_1 = I_1R_1 \quad V_2 = I_2R_2 \quad V_3 = I_3R_3$

Power: Power can also be applied to the entire circuit or to any resistor in the circuit.

$P_{total} = V_{total}I_{total} \quad P_1 = V_1I_1 \quad P_2 = V_2I_2 \quad P_3 = V_3I_3$

Sample Problem 2

Given the following parallel circuit, find:

a) the current through each resistor
b) the total current
c) the total power
d) the power in each resistor
e) the total resistance

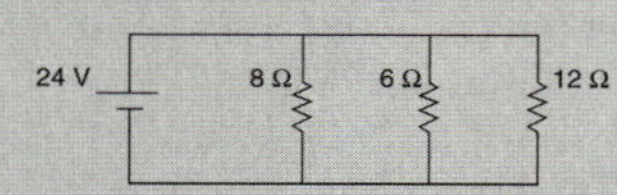

Strategy: This is a parallel circuit. The current follows different paths to each resistor. In a parallel circuit, the voltage drops across each resistor are equal. You can put this information in the diagram immediately.

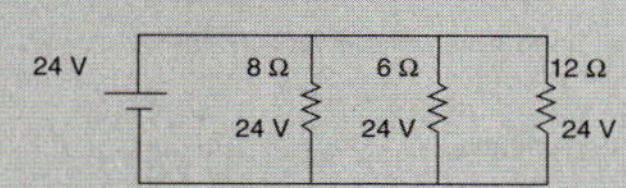

© It's About Time

Once you know two of the four variables (V, I, P, R), you can find the other two variables. In this case, you know V and R. You can find the current using Ohm's Law for each resistor. You can find the total current by adding the currents through each resistor.

Use the power equation to calculate the power once you know the current.

a) $V = IR$

$I_1 = \frac{V_1}{R_1}$ $\quad I_2 = \frac{V_2}{R_2}$ $\quad I_3 = \frac{V_3}{R_3}$

$I_1 = \frac{24\ \text{V}}{8\ \Omega}$ $\quad I_2 = \frac{24\ \text{V}}{6\ \Omega}$ $\quad I_3 = \frac{24\ \text{V}}{12\ \Omega}$

$I_1 = 3\ \text{A}$ $\quad I_2 = 4\ \text{A}$ $\quad I_3 = 2\ \text{A}$

b) If the resistors have currents of 3 A, 4 A and 2 A, then the total current must be equal to

3 A + 4 A + 2 A = 9 A

c) $P_{total} = V_{total}I_{total}$

$P_{total} = (24\ \text{V})(9\ \text{A})$

$P_{total} = 216\ \text{W}$

d) $P_1 = V_1I_1$ $\quad P_2 = V_2I_2$ $\quad P_3 = V_3I_3$

$P_1 = (24\ \text{V})(3\ \text{A})$ $\quad P_1 = (24\ \text{V})(4\ \text{A})$ $\quad P_1 = (24\ \text{V})(2\ \text{A})$

$P_1 = 72\ \text{W}$ $\quad P_1 = 96\ \text{W}$ $\quad P_1 = 48\ \text{W}$

Notice the sum of the power in the resistors:

72 W + 96 W + 48 W = 216 W

That is the power supplied by the battery.

e) You can find the total resistance of the circuit easily by using Ohm's Law for the entire circuit.

$V_{total} = I_{total}R_{total}$

$R_{total} = \frac{V_{total}}{I_{total}}$

→

Chapter 7

© It's About Time

$$= \frac{24\ \text{V}}{9\ \text{A}}$$

$$= 2.67\ \Omega \text{ (or about } 3\ \Omega)$$

You can also find the total resistance by adding the individual resistors in parallel. Notice that in this equation, you are dealing with fractions.

$$\frac{1}{R_{total}} = \frac{1}{R_1} + \frac{1}{R_2} + \frac{1}{R_3}$$

$$= \frac{1}{8\ \Omega} + \frac{1}{6\ \Omega} + \frac{1}{12\ \Omega}$$

Using the least common denominator of 24:

$$\frac{1}{R_{total}} = \frac{3}{24\ \Omega} + \frac{4}{24\ \Omega} + \frac{2}{24\ \Omega}$$

$$= \frac{9}{24\ \Omega}$$

$$R_{total} = \frac{24\ \Omega}{9}$$

$$= 2.67\ \Omega$$

This agrees with the simpler method above, as it must.

Reflecting on the Activity and the Challenge

Part of the problem you are facing with the electrical system for the Homes For Everyone (HFE) dwellings is to assure that the people who live in them will not exceed the 2400 W power limit of the system by having too many appliances in use at any one time. Depending on what you choose to include in the HFE appliance package, it may be necessary to use switching devices—automatic, manual, or both—to assure that the people who live in the homes will stay within the power limit as they use their appliances.

Perhaps you could also use automatic switches as "fail safe" devices to prevent accidentally using up too much electrical energy by, for example, forgetting to shut off lights that are not in use.

© It's About Time

Physics To Go

1. In your log, describe several possibilities for using switching devices to address the power limit problem in your universal dwelling. Write your ideas in your log.

2. Electric switches are available which act as timers to turn appliances on and off at chosen times or for chosen intervals. Identify one or more ways a timer switch would be useful in an HFE dwelling.

3. Look at the wiring diagrams shown. Copy each into your log. Position and draw a switch in each circuit which would allow you to turn two lights on, and leave one light off.

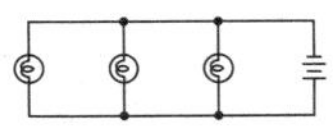

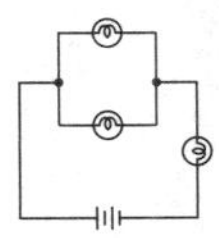

4. A 12-Ω, 14-Ω, and 4-Ω resistor are connected in series to a 12-V power supply. Find:
   a) the total resistance of the circuit
   b) the current flowing through each resistor
   c) the voltage drop across each resistor

5. Three resistors of 8 Ω, 6 Ω, and 4 Ω are connected in series to a battery of six 1.5-V dry cells, which are also connected in series.
   a) Draw a circuit diagram for this situation.
   b) Calculate the total voltage provided by the battery.
   c) Calculate the total resistance.
   d) Find the total current.
   e) What is the voltage drop across each individual resistor?

6. The following chart represents a set of three resistors arranged in series. Draw a diagram of the circuit. Then use the facts that you know about series circuits to fill in the empty spaces.

	V	I	R
Total	120 V		
R_1			10 Ω
R_2			5 Ω
R_3			25 Ω

ANSWERS

Physics To Go

1. Some of the devices that could be used are timers, that automatically shut off small appliances or lights. For example, a switch could be installed in bathroom lights to automatically shut off after fifteen minutes (typically the length of time needed for a shower). Other examples may include switches that work toward warning that the power level limit is close. Another practical way can be listing appliances available and their power ratings, so that people in the home will know what is on and what can be turned on without blowing the breaker. Some other devices could use motion detectors that automatically turn on when someone enters a room, and turn off after a time limit or, no motion or "clap-on/clap-off".

2. Answers will vary. See answer for #1.

3. See diagrams to the right and on next page.

ANSWERS

Physics To Go
(continued)

4. a) $R_{total} = R_1 + R_2 + R_3$
$= 12 + 14 + 4$
$= 30$

b) $V = \frac{I}{R}$ total

$I = \frac{I}{R}$ total

$= \frac{12\text{ V}}{30\ \Omega}$

$= 0.4\text{ A}$

c) $V_1 = I_{total}R_1$
$= (0.4\text{ A})(12\ \Omega)$
$= 4.8\text{ V}$

$V_2 = I_{total}R_2$
$= (0.4\text{ A})(14)$
$= 5.6\text{ V}$

$V_3 = I_{total}R_3$
$= (0.4\text{ A})(4)$
$= 1.6\text{ V}$

(continued on next page)

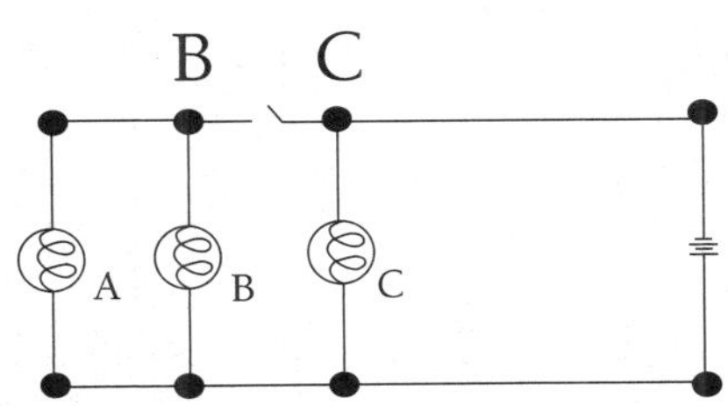

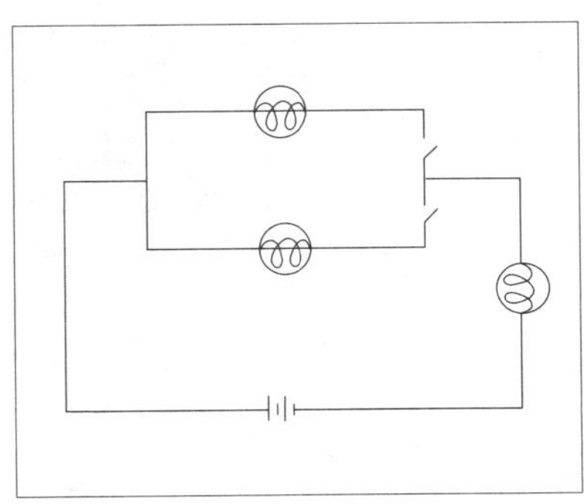

© It's About Time

ANSWERS

Physics To Go
(continued)

As a check to the solution, we can apply the law of conservation of energy:

$$V_{total} = V_1 + V_2 + V_3$$
$$= 4.8\text{ V} + 5.6\text{ V} + 1.6\text{ V}$$
$$= 12\text{ V}$$

5. a)

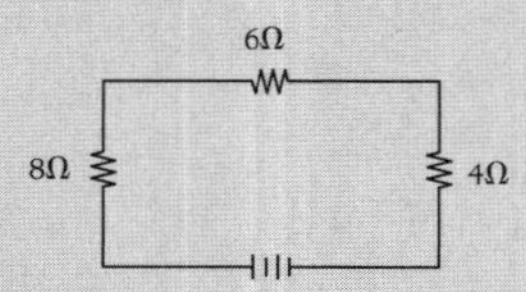

b) A series connection of cells results in a voltage that is the arithmetic sum of the individual voltages. In this case we have six identical cells so the total voltage is 6 times the individual voltage:

$$V_{total} = 6(V_{individual})$$
$$= 6\ (1.5\text{ V})$$
$$= 9\text{ V}$$

c) $R_{total} = R_1 + R_2 + R_3$

$$= 8 + 6 + 4$$
$$= 18$$

d) $V = IR$

$$I = \frac{V}{R_{total}}$$

e) Knowing that the current will be the same in each resistor, we can find the individual voltage drops with Ohm's Law and the individual resistances.

$$V_1 = I_{total}R_1$$
$$= (0.5\text{ A})(8\ \Omega)$$
$$= 4.0\text{ V}$$
$$V_2 = I_{total}R_2$$
$$= (0.5\text{ A})(6\ \Omega)$$
$$= 3.0\text{ V}$$
$$V_3 = I_{total}R_3$$
$$= (0.5\text{ A})(4\ \Omega)$$
$$= 2.0\text{ V}$$

6. The diagram produced by the students will be similar to the one in **Question 5**.

© It's About Time

7. A 6-Ω, 3-Ω, and 18-Ω resistor are connected in parallel to an 18-V power supply. Find:
   a) the total resistance of the circuit
   b) the total current through the circuit
   c) the current flowing through each resistor

8. You are given three 10-Ω resistors by your teacher. You are told to arrange them in the following ways. Sketch a diagram for each arrangement. What values will you have for total resistance in each case?
   a) All three resistors in series.
   b) All three resistors in parallel.
   c) One resistor in series and two in parallel.
   d) Two resistors in series with one in parallel.

9. The following chart represents a set of three resistors arranged in parallel. Draw a diagram of the circuit. Then use the facts that you know about parallel circuits to fill in the empty spaces.

	V	I	R
Total	24 V		
R_1			18 Ω
R_2			12 Ω
R_3			36 Ω

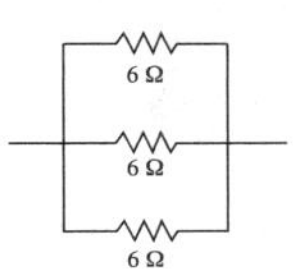

10. Which circuit segment below has the same total resistance as the circuit segment shown in the diagram to the left?

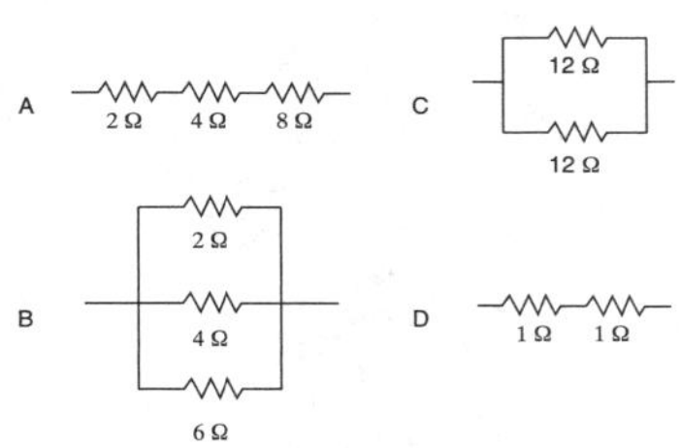

	V	I	R
Total	120 V	3 A	40 Ω
R_1	30 V	3 A	10 Ω
R_2	15 V	3 A	5 Ω
R_3	75 V	3 A	25 Ω

© It's About Time

ANSWERS

Physics To Go

(continued)

7. a) $\frac{1}{R_{total}} = \frac{1}{R_1} + \frac{1}{R_2} + \frac{1}{R_3}$

$= \frac{1}{6\ \Omega} + \frac{1}{3\ \Omega} + \frac{1}{18\ \Omega}$

$= \frac{3}{18\ \Omega} + \frac{6}{18\ \Omega} + \frac{1}{18\ \Omega}$

$\frac{1}{R_{total}} = \frac{10}{18\Omega}$

$\frac{1}{R_{total}} = 1.8\ \Omega$

b) $V_{total} = \frac{I_{total}}{R_{total}}$

$I_{total} = V_{total} = R_{total}$

$= \frac{18\ V}{1.8\ \Omega}$

$= 10\ A$

c) $I_1 = \frac{V}{R_1}$

$= \frac{18\ V}{6\ \Omega}$

$= 3\ A$

$I_2 = \frac{V}{R_2}$

$= \frac{18\ V}{3\ \Omega}$

$= 6\ A$

$I_3 = \frac{V}{R_3}$

$= \frac{18\ V}{18\ \Omega}$

$= 1\ A$

8. a) Add the resistances arithmetically.

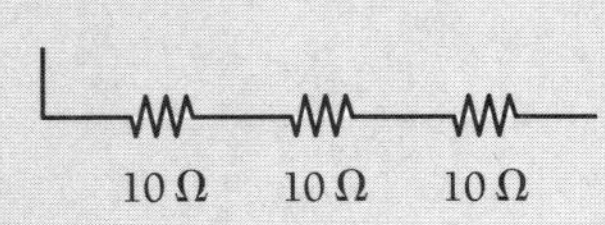

$R_T = R_1 + R_2 + R_3$

$= 10\ \Omega + 10\ \Omega + 10\ \Omega$

$= 30\ \Omega$

Chapter 7

ANSWERS

Physics To Go
(continued)

b) Use the parallel resistance formula.

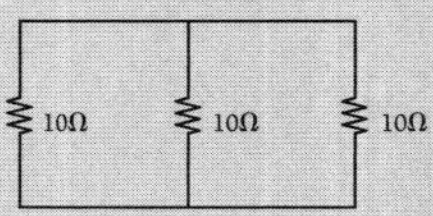

$$\frac{1}{R_T}=\frac{1}{R_1}+\frac{1}{R_2}+\frac{1}{R_3}$$

$$\frac{1}{R_T}=\frac{1}{10\ \Omega}+\frac{1}{10\ \Omega}+\frac{1}{10\ \Omega}$$

$$\frac{1}{R_T}=\frac{3}{10\ \Omega}$$

$$R_T=\frac{10\ \Omega}{3}=3.3\ \Omega$$

c) Start with the two in parallel and replace them with an equivalent resistor.

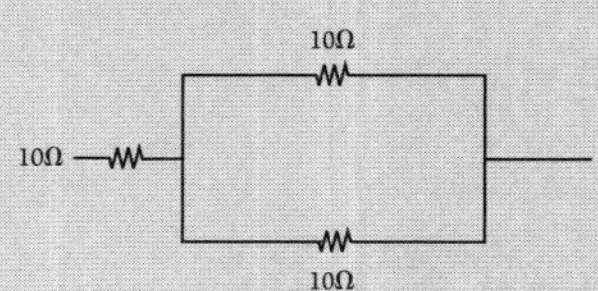

$$\frac{1}{R_{equivalent}}=\frac{1}{R_2}+\frac{1}{R_3}$$

$$\frac{1}{R_{equivalent}}=\frac{1}{10\ \Omega}+\frac{1}{10\ \Omega}$$

$$=\frac{2}{10\ \Omega}$$

$$R_{equivalent}=5\ \Omega$$

Add the value of the first resistor and the equivalent of the two parallel resistors.

$$R_T=R_1+R_2$$

$$=10\ \Omega+5\ \Omega$$

$$=15\ \Omega$$

d) Consider the two series resistors first.

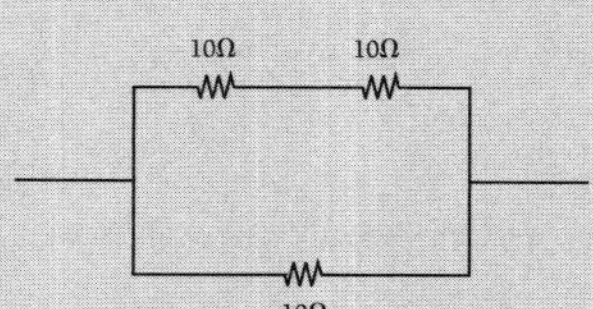

$$R_T=R_1+R_2$$

$$=10\ \Omega+10\ \Omega$$

$$=20\ \Omega$$

© It's About Time

ANSWERS

Physics To Go
(continued)

Add this to resistor three using the parallel resistance formula.

$$\frac{1}{R_T} = \frac{1}{R_{equivalent}} + \frac{1}{R_3}$$

$$\frac{1}{R_T} = \frac{1}{20\ \Omega} + \frac{1}{10\ \Omega}$$

$$= \frac{3}{20\ \Omega}$$

$$R_T = 0.15\Omega$$

9. Students should use the facts about parallel circuits and the method of finding total resistance, as well as Ohm's Law to fill in the spaces on the chart. This is similar to working on a crossword puzzle. Having students work in teams on this, talking about the problem is a very effective pedagogical tool. Most students enjoy this and will ask for more problems like this!

	V	I	R
Total	24 V	4.0 A	6 Ω
R_1	24 V	1.3 A	18 Ω
R_2	24 V	2.0 A	12 Ω
R_3	24 V	0.67 A	36 Ω

10. D

© It's About Time

11. 1 A in the direction P to W.

12. 1 A

13. 3 A

11. The diagram to the right shows the current in three branches of a direct-current electric circuit. What is the value and the direction of the current between junction P and point W?

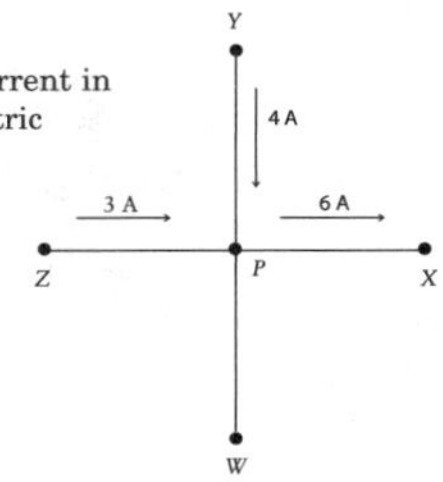

12. What is the current in the circuit represented in the diagram below?

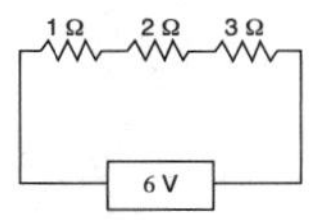

13. What is the current in the ammeter A in the diagram below?

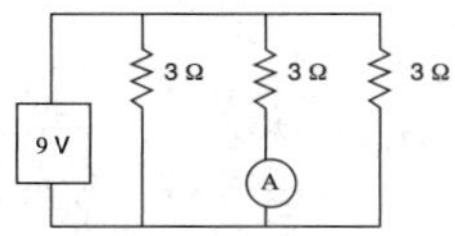

Stretching Exercises

Shop, in a store or catalog, for electrical switching devices controlled by a variety of conditions such as light/dark, high/low temperature, motion, etc. The devices may be either "built-in" to appliances (example: waterbed thermostat) or separate (example: a clock timer designed to control appliances plugged into it). Find as many different kinds of switching devices as you can, and note which ones may be useful for the HFE electrical system, and for what purpose. In your log, write a brief report on your findings.

Stretching Exercises

Student reports. Expect a variety of answers. The students will be looking for devices that will be useful in their HFE home. Look for energy-saving devices, timers, low wattage lights with timers, etc.

© It's About Time

Circuit Diagrams

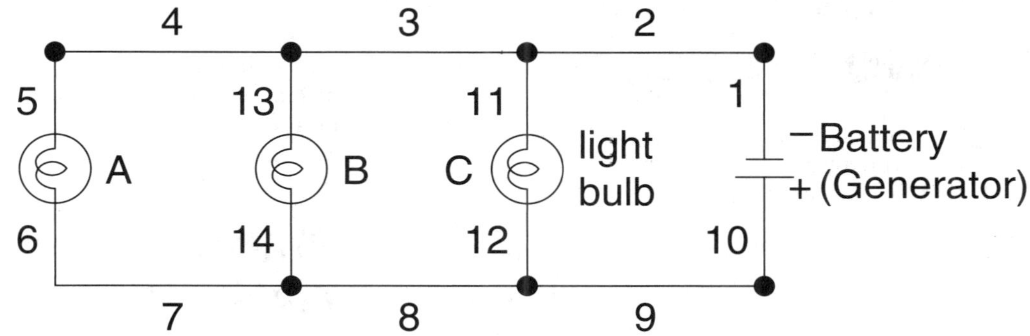

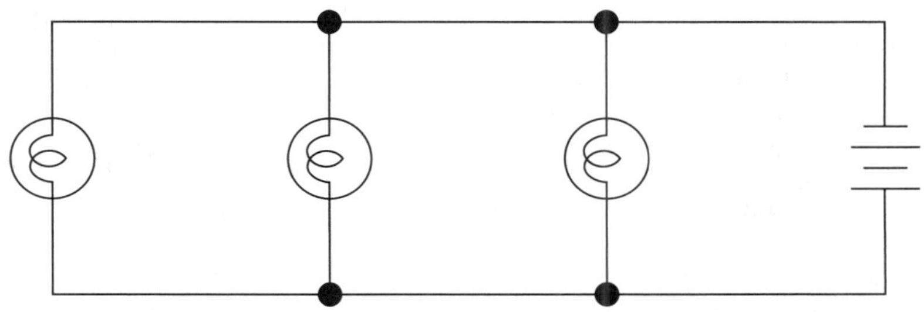

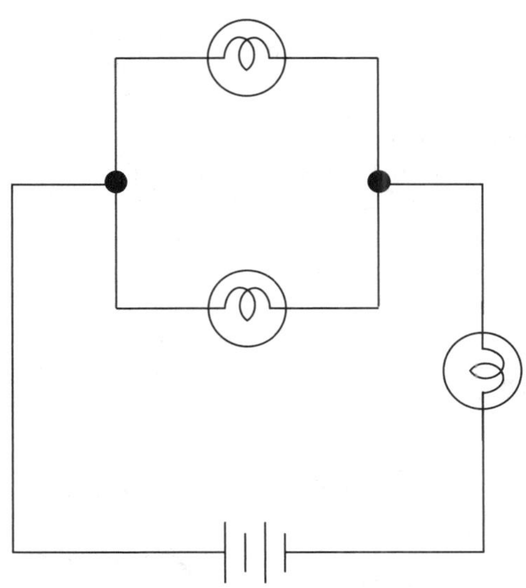

© It's About Time

For use with *Electricity For Everyone*, Chapter 7, Activity 5: Who's In Control

ACTIVITY 6
Cold Shower

Background Information

Activity 6 combines principles of heating water with principles of electrical energy introduced in **Activity 5** of this chapter.

You may wish to review the following:

Heat lost = Heat gained in a transfer of energy between two objects.

$$\text{Heat} = mc\Delta T$$

Specific heat of water = 4180 (joule)/(kilogram)(degree Celsius)

The purpose of this activity is to use an electric heater to transfer electrical energy to water in the form of heat, accounting for the electrical energy input and the heat received by the water.

The electrical energy delivered to the electric heater [a resistor having a known power rating (in watts) which is plugged into a 120-volt outlet for a measured amount of time in seconds] is:

Energy (joules) = Power (watts) × time (seconds)

The heat received by the water is:

Energy (joules) = mass (kilograms) × specific heat (joule)/(kilogram)(degree Celsius) × temperature change (degree Celsius)

In principle, the electrical energy expended in the heater should equal the heat received by the water. However, less than 100 % efficiency can be expected for the energy transfer due to heat loss from the system to the surrounding environment.

Active-ating the Physics InfoMall

For You To Do Step 1 makes use of a calorimeter. A search for "calorimeter" provides many examples of calorimeters, together with diagrams. If you need a calorimeter, the Catalog Corner has *Vernier, Central Scientific*, and *PASCO* catalogs to choose from. This is a CD-ROM, however, so the catalogs are not exactly new. They can give you a great starting point, though.

Physics Talk discusses specific heat and resistors. Both of these topics are discussed at various level of detail in many places on the InfoMall. These are very good topics for which to prepare your own handouts, with examples and graphics included.

Physics To Go Step 1 mentions conduction, convection, and radiation. While these terms may be familiar to your students, it might not be a bad idea to have some extra information ready, and you can easily find it on the InfoMall.

Planning for the Activity

Time Requirements

Allow about 10 - 15 minutes to set up the apparatus, and about 20 minutes to do the activity and record their results. If there is time, have the students run the activity with different quantities of water, in order to get a set of data to graph.

Materials Needed

For the class:

- electric calorimeter

For each group:

- calculator
- graduated cylinder, 100 ml
- thermometer or temperature probe
- resistive immersion heater of known wattage: Many are available, designed for labs at high cost, but the little portable heaters for making coffee in a hotel room do the job and are less expensive. A 200-watt, 120-volt AC one works fine when used with at least 300 mL of water.
- insulated calorimeter container: You can get fancy, but a Styrofoam® cup works about as well as anything. Just be sure the heater doesn't contact the wall of the cup. Use a ringstand and clamp to suspend the coil in the water without touching the cup.
- stirring rod
- stopwatch

Advance Preparation and Setup

Have students complete the **What Do You Think?**. Students may not know what specific heat capacity is. Briefly explain that SHC is referring to the amount of heat energy that is required to raise

© It's About Time

the temperature one degree Celsius of a specific quantity of water. The SHC of water is 4180 J/kg°C. This means 1 kg of water requires 4180 J of energy to raise the temperature 1°C. Emphasize to the students that it requires more energy to heat water than many other materials. Therefore, hot water for HFE dwellings presents a problem. Pose the For You To Do activity as an exploration of the viability of using the HFE electrical system for providing hot water for the universal dwellings. Plan to save some time at the end of class to assist students with the principles and calculations involved in the **Physics To Go**, and the **Reflecting on the Activity and the Challenge**.

Teaching Notes

Expect that students will need help with the energy and efficiency calculations involved in this activity.

Work through the **Physics Talk** with the students to establish that, since power is energy per unit time, energy can be thought of, and calculated, as power multiplied by time ($E = Pt$). Using the dimensions of the quantities, establish that watt-seconds are the equivalent of joules.

You will need to explain (Q) = $mc\Delta T$, where Q is the heat energy in joules, m is the mass in kilograms, c is the specific heat capacity of water (4180 J/kg°C), and ΔT is the change in temperature. This formula will help the students understand the quantity of heat energy required to heat water.

Be sure the students understand that, in principle, the energy lost by the hot resistor can be set equal to the energy gained by the cold water. Therefore, by calculating the heat energy lost by the resistor, ($E = I^2Rt$), and equating it to heat gained by the water, we can determine the amount of power required to heat a known quantity of water.

Again emphasize the need for safe practices when handling electric devices near water. Ensure that the students have properly assembled the calorimeters.

Reiterate the idea that there is not a 100% transfer of energy from the resistor to the water. They will calculate the efficiency of the transfer, and should be able to account for the "lost" energy!

© It's About Time

Activity Overview

In this activity, the students will be measuring the effect of a resistor heating a quantity of water. From this data they will be able to determine the efficiency of the resistor heating water, and the energy required. After they have determined efficiency, they will be able to apply it to their own lives, and then to their HFE home.

Student Objectives

Students will:

- Calculate the heat gained by a sample of water.
- Calculate the electrical energy dissipated by a resistor.
- Recognize and quantify an ideal mechanical/electrical equivalent of heat.
- Calculate the efficiency of a transformation of electrical energy to heat.
- Explore the power ratings and energy consumption levels of a variety of electrical appliances.

Activity 6 Cold Shower

GOALS

In this activity you will:

- Calculate the heat gained by a sample of water.
- Calculate the electrical energy dissipated by a resistor.
- Recognize and quantify an ideal mechanical/electrical equivalent of heat.
- Calculate the efficiency of a transformation of electrical energy to heat.
- Explore the power ratings and energy consumption levels of a variety of electrical appliances.

What Do You Think?

The entire daily energy output of a Homes For Everyone (HFE) generator would not be enough to heat water for an average American family for a day.

- **If an electrical heating coil (a type of resistor) were submerged in a container of water, and if a current were to flow through the coil to make it hot, what factors would affect the temperature increase of the water? Identify as many factors as you can, and predict the effect of each on the water temperature.**

Record your ideas about this question in your *Active Physics* log. Be prepared to discuss your responses with your small group and the class.

ANSWERS FOR THE TEACHER ONLY

What Do You Think?

Answers will vary. Amount of water, as you increase the mass, you increase the time of heating; current in the wire, heating increases as you increase the current ($E = I^2Rt$); resistance in the wire, as you increase the resistance, you increase the energy ($E = I^2Rt$); time, as you increase the time of heating, the temperature will increase; the type of wire or heating coil changes the resistance; the initial temperature, the lower the temperature, the longer it will take to heat up to a particular temperature.

© It's About Time

For You To Do

1. Assemble and use an electric calorimeter according to the directions given by your teacher. Add a measured amount of cold tap water to the calorimeter.
   a) Record the mass of the water. (You will need to find the mass of an empty container as well as the mass of the container and the measured amount of water.)
   b) Measure and record the temperature of the water.
   c) Record the watt rating of the resistor that will be used to heat the water.

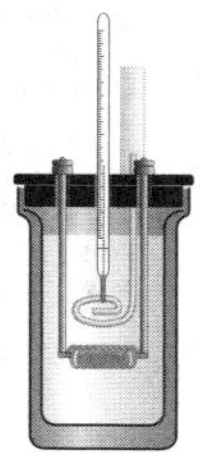

Always make sure the coil (or heater) is completely submerged in the water.

Promptly dry up any spilled water.

Do not try to stir the water with the thermometer. If the thermometer should break, immediately notify your instructor.

2. Mark the time at which you begin sending electric current through the resistor. Keep the electric heater operating for the amount of time recommended by your teacher. When you stop the current, note the time, stir the water, and measure the final, maximum temperature of the water.

Mass of water (grams):
Cold water temperature (°C):
Resistor power (watts):
Heating time (seconds):
Final water temperature (°C):

   a) In your log, record the final water temperature (°C) and the heating time (seconds).

Chapter 7

ANSWERS

For You To Do

1. a) Student observation and collected data.
   b) Student observation and collected data.
   c) Student observation and collected data.
2. a) Student observation and collected data.

© It's About Time

ANSWERS

For You To Do
(continued)

3. a) Student data and calculation. The calculation should look like this (contrived data):

heat gained = $mc\Delta T$
= 0.045 kg × 4.180 J/kg°C × 35°C
= 65,835 J of energy.

4. a) Student data and calculation. The calculation should look like this (contrived data):

energy = Pt
= 1500 W × 90 s
= 135,000 J

5. a) Since the input energy (from the heater) of 135,000 J is greater than the output energy (heating of the water) of 65,835 J, we must conclude that some of the energy is "lost."

b) In fact, the energy is not lost (First Law of Thermodynamics), but converted into a useless form of energy, for the purposes of this experiment, that of heat to the environment.

3. The heat energy gained by an object can be found using the equation:

   Heat energy = mass of object × specific heat of material × temperature change

   Use the equation to calculate the heat energy gained by the water in the calorimeter. The specific heat of water is 4180 J/kg°C.

   a) Show your calculation in your log.

4. Power rating, in watts, is expressed as the amount of energy that something consumes per unit of time. Mathematically, power is expressed as

$$\text{Power} = \frac{\text{energy}}{\text{time}}$$

   Then it is true that: Energy = power × time

   Energy can be expressed in kiloWatt-hours (kWh) if power is in kiloWatts and time is in hours, OR energy can be expressed in joules, if power is expressed in watts, and time in seconds. Use the equation to calculate the electrical energy, in joules, consumed by the resistor in the calorimeter.

   a) Show your calculation in your log.

5. The energy calculated from the temperature change of the water should be equal to the energy calculated from the power rating of the appliance and the time.

© It's About Time

Activity 6 Cold Shower

a) Compare these two values for the heat energy. If the values are not the same, what must you conclude?

b) Where could the "lost" energy have gone?

6. If all of the electrical energy did not heat up the water, then you know your system for heating water is less than 100% efficient. Calculate the efficiency of your water heating device. (Hint: If the appliance were 100% efficient, all of the electrical energy would have heated the water; if the appliance were 50% efficient, half of the electrical energy would have heated the water.)

a) Record your calculations in your log. How efficient was the calorimeter?

PHYSICS TALK

What is the world's greatest invention? A comedian once replied: "The thermos. When you put a cold drink in it, it stays cold. When you put a hot drink in it, it stays hot. How does the thermos know?"

A cup of hot water left on a table cools down. A cup of cold water left on the same table warms up. How can you change the temperature or keep the temperature constant?

Thermal energy is a form of energy. Energy in one form must come from energy in another form. Energy must be conserved. If you wish to heat up water, you must supply it with a source of energy like a flame, an electrical heater, or a hot metal.

If a system is isolated from outside sources of energy, one part of the system may warm up and another part may cool down, but the total energy must remain constant.

In this activity, you calculated the electrical energy that was used to heat the water. Energy can be calculated using the following equation:

→

Answers

For You To Do

(continued)

6. a) Student data and calculation. The calculation should look like this (contrived data):

From the previous questions:

efficiency = input − output/input

× 100%

$$= \frac{135{,}000\ J - 65{,}835\ J}{135{,}000\ J} \times 100\%$$

$$= 51\%$$

Chapter 7

© It's About Time

$$E = VIt$$

where E is the energy in joules (J),

V is the voltage in volts (V),

I is the current in amperes (A), and

t is the time in seconds (s).

Since $P = VI$, the equation may also be written as:

$$E = Pt$$

Sample Problem 1

A 12-V starter battery in a car supplies 48 A of current to the starter. If the starter draws energy for 15 s, how much energy does the starter use?

Strategy: You can use the equation to calculate energy.

Givens:

$V = 12$ V

$I = 48$ A

$t = 15$ s

Solution:

$E = VIt$

$= (12 \text{ V})(48 \text{ A})(15 \text{ s})$

$= 8640$ J (or about 8600 J)

Recall that volts are joules per coulomb and amperes are coulombs per second. When you multiply these together you get joules per second, the power unit. Multiplying this by seconds gives you joules, units of energy.

Heat energy is measured in units of joules (J). The heat energy gained by an object can be found using the following equation:

Heat energy = mass of object × **specific heat** of material × temperature change

Physics Words

specific heat: the amount of energy required to raise the temperature of 1 kg of a material by 1°C.

© It's About Time

Where heat energy is expressed in joules (J),

mass is in kilograms (kg),

specific heat is in joules per kilogram degree Celsius (J/kg°C), and temperature is in degrees Celsius (°C).

You may also see this equation written as follows:

$$E_H = mc\Delta T$$

The symbol for specific heat is *c*. The symbol Δ represents a change in something, in this case, temperature.

To calculate the heat energy gained by a material, you must know the specific heat of the material. Different materials require different amounts of energy to raise the temperature of a given mass of the substance. Different substances have different capacities to hold heat. Water holds heat very well, whereas silver does not hold heat well. Water is said to have a higher specific heat than silver. The specific heat of a material is the amount of energy required to raise 1 kg of the material by 1°C.

The specific heat of water is 4180 J/kg°C.

Sample Problem 2

You apply 8640 J of energy to 100.0 g of water. How much would it change the temperature of the water?

Strategy: The amount of temperature change depends on several factors. One is the specific heat of the material with which you are working. In this case, the material is water. The other factor is the amount of material you have.

Givens:

$E = 8640$ J

$m = 100.0$ g $= 0.10$ kg

$c = 4180$ J/kg°C

→

Chapter 7

© It's About Time

ANSWERS

Physics To Go

1. Student answers will vary. Some of the ways in which the heat energy escaped will be through the insulation, through the top, heating the thermometer, residual heat left in the resistor, etc.

2. If you stand beside the hot water heater, you can feel the heat. Therefore, some of the heat is lost through the insulation. Some of the heat is lost up the chimney, from the bottom where the element or burner is, and through the pipes of the water which has been heated. Some of this could be prevented by increasing the insulation, more efficient burners, modifying the chimney to trap the heat (without the carbon monoxide), use cleaner fuel, insulating the pipes.

Solution:

$$E = mc\Delta T$$

$$\Delta T = \frac{E}{mc}$$

$$\Delta T = \frac{8640 \text{ J}}{(0.10 \text{ kg})(4180 \text{ J/kg°C})}$$

$$= 20.6°\text{C (or about } 21°\text{C)}$$

Reflecting on the Activity and the Challenge

In this activity your knowledge of how to calculate electric power consumption was extended to include how to calculate electric energy consumption. You also learned that heating water electrically requires a lot of energy and can be quite inefficient. All of this knowledge applies directly to the selections you will make for electrical appliances to be used in the universal dwelling.

Physics To Go

1. The calorimeter did not allow the water to trap 100% of the energy delivered to it by the resistor. Some of the heat energy probably escaped from the water. Identify and explain ways in which you think heat energy may have escaped from the water, reducing the efficiency of the calorimeter.

2. The calorimeter used in this activity can be thought of as a scaled-down, crude version of a household hot water heater. The efficiencies of hot water heaters used in homes range from about 80% for older models to as much as 92% for new, energy-efficient models. Identify and explain ways in which you think heat escapes from household hot water heaters, and how some of the heat loss could be prevented.

© It's About Time

3. From what you have learned so far, discuss the possibilities for providing electrically heated water for Homes For Everyone (HFE). Is a standard water heater of the kind used in American homes desirable, or possible, for HFE? What other electrical options exist for accomplishing part or all of the task of heating water for HFE?

4. For most Americans, the second biggest energy user in the home, next to the heating/air conditioning system, is the water heater. A family of four that heats water electrically (some use gas or oil to heat water) typically spends about \$35 per month using a 4500-W heater to keep a 160-L (40 gallon) tank of water hot at all times. The water is raised from an average inlet temperature of 10°C (50°F) to a temperature of 60°C (140°F), and the average family uses about 250 L (60 gallons) of hot water per day for bathing and washing clothes and dishes.

   In the above description, explain what each of the following numbers represents: 35, 4500, 160, 40, 10, 60, 50, 140, 250, 60.

5. Make a list with three columns in your log. The first column will be Name of Appliance, the second column will be Power (Watts), the third column will be Form of Energy Delivered.

   Choose 5 appliances from the list of Home Electrical Appliances beginning on page 496 that have a wide range of power. Use information from the list of appliances to fill in the first two columns. In the third column, write the form of energy (heat, light, motion, sound, etc.) that you think each appliance is designed to deliver.

   What pattern, if any, do you think exists between the power rating of an appliance and the form of energy it is designed to provide? Explain your answer.

6. Make a new list with six columns similar to the one shown on the next page.

   Choose 5-10 appliances from the list. Record the name and power rating of that appliance. Record the approximate time that this appliance is used in one day. Calculate the time that this appliance is used in one month (assuming that a month has 30 days). Calculate the electrical energy that the appliance consumes.

ANSWERS

Physics To Go

(continued)

3. An electric water heater is rated at 4500 W. The amount of water used will determine how long the hot water heater must run. There should be a more efficient method by which you can heat the water. Some other ways may be using a gas heater if it is available, or other fuel source such as heating on a wood stove. Students may come up with many different answers.

4. 35 – cost of running a hot water heater; 4500 – the rating of power, or the energy used in joules per second; 160 – liters of water; 40 – gallons of water; 10 – degrees Celsius, temperature of the water at the inlet to house; 60 – degrees Celsius, temperature of the hot water; 50 – degrees Fahrenheit, temperature of the water at the inlet to house; 140 – degrees Fahrenheit, temperature of the hot water; 250 – liters of hot water used by the average household; 60 – gallons of hot water used by the average household.

Chapter 7

5. Students' answers will vary. The students may notice that if an appliance is designed for heating, it generally uses more energy.

6. Students' answers will vary. One example row should look like the following:

Appliance	Power (W)	# hours/day	Energy/mo. (Wh)	Energy/mo. (kWh)
MW oven	1200	0.75	27,000	27

© It's About Time

ANSWERS

Physics To Go
(continued)

7. $E = VIt$

$= Pt$

$= (1500\text{ W})(300\text{ s})$

$= 450{,}000\text{ J}$

8. $E = mc\Delta T$

$= (0.2\text{ kg})(387\text{ J/kg°C})(150\text{ °C} - 20\text{ °C})$

$= 10{,}062\text{ J}$

9. $E = mc\Delta T$

$\Delta T \frac{E}{mc}$

$= \frac{42\text{ J}}{(0.02\text{ kg})(140\text{ J/kg°C})}$

$= 15\text{°C}$

Appliance	Power (Watts)	#hours/ day (est.)	#hours/ month	Energy/month (Wh/month)	Energy/month (kWh/month)

The energy in watt-hours (Wh) is found by multiplying the power times the time in hours in one month.

Energy = power × time

Calculate the energy in kiloWatt-hours (kWh) by dividing the watt-hours by 1000 since there are 1000 W in 1 kW.

7. You use a 1500-W hair dryer for 5 minutes every morning to dry your hair.
   a) How much electrical energy are you changing to heat every day?
   b) If you could transfer all of that energy to heating up water for coffee, how much water could you heat from 20°C to 90°C?

8. Copper wire has a specific heat of 387 J/kg°C. How many joules of energy would it take to heat 200 g of this copper wire from 20°C to 150°C?

9. If 42 J of energy are supplied to a 20-g sample of mercury (c = 140 J/kg °C), what is the temperature change of the sample?

© It's About Time

Stretching Exercises

1. Find out about EnergyGuide® labels. Recently, the United States government established a federal law that requires EnergyGuide labels to be displayed on major appliances such as water heaters, refrigerators, freezers, dishwashers, clothes washers, air conditioners, furnaces, and heat pumps. The bright yellow EnergyGuide label allows consumers to compare the energy costs and efficiencies of appliances. Visit a store where appliances are sold and record the information given on the EnergyGuide labels of competing brands of one kind of appliance, such as water heaters. Prepare a short report on which appliance you would purchase, and why.
2. Research ways to reduce the amount of electrical energy needed to provide hot water for your own home or an HFE dwelling. Some possibilities may include (a) using solar energy and/or a "tempering tank" to heat the water partially, followed by "finishing off" the heating electrically, and (b) tankless "instant" water heaters which use electricity to heat water, but only when hot water is needed. Prepare a report on your findings.

Chapter 7

ANSWERS

Stretching Exercises

1. Student reports. In the reports look for the understanding of the EnergyGuide label. Have the students understand that one appliance may be rated lower than the other, but need to run it longer to do the same job. Also, some appliances will have a higher rating, (e.g., convection ovens), but don't necessarily need to be preheated to do the same job, and can do the same baking in less time with a lower temperature. The students need to understand the criteria for testing. For example, the freezer is less efficient when half empty, and will require more energy. Some appliances will not operate as efficiently in certain locations. For example, putting a refrigerator next to the furnace will cause the refrigerator to run longer, thereby cutting into the efficiency.
2. Students' answers will vary.

© It's About Time

Home Electrical Appliances

Average Power and Average Monthly Energy Use for a Family of Four

Family Data

	Power (watts)	Energy/mo. (kWh/mo.)
Big Appliances		
Air Conditioner		
(Room)	1360	
(Central)	3540	
Clothes Washer	512	
Clothes Dryer	5000	
Dehumidifier	645	
Dishwasher	1200	
Freezer	400	
Humidifier	177	
Pool Filter	1000	
Kitchen Range	12,400	
Refrigerator	795	
Space Heater	1500	
Waterbed	350	
Water Heater	4500	
Small Refrigerator	300	
Lights & Minor Appliances (combined)		
Kitchen		
Baby Food Warmer	165	
Blender	300	
Broiler (portable)	1200	
Can Opener	100	
Coffee Maker		
Drip	1100	
Percolator	600	
Corn Popper		
Oil-type	575	
Hot Air-type	1400	
Deep Fryer	1500	
Food Processor	370	
Frying Pan	1200	
Garbage Disposal	445	
Sandwich Grill	1200	
Hot Plate	1200	
Microwave Oven	750	
Mixer	150	

© It's About Time

Home Electrical Appliances

	Power (watts)	Energy/mo. (kWh/mo.)
Roaster	1400	
Rotisserie	1400	
Slow Cooker	200	
Toaster	1100	
Toaster Oven	1500	
Trash Compactor	400	
Waffle Iron	1200	
Entertainment		
Computer	60	
Radio	70	
Television	90	
Stereo	125	
VCR	50	
Personal Care		
Air Cleaner	50	
Curling Iron	40	
Hair Dryer	1200	
Hair Rollers	350	
Heat Blanket	170	
Heat Lamp	250	
Heat Pad	60	
Iron	1100	
Lighted Mirror	20	
Shaver	15	
Sun Lamp	300	
Toothbrush	1	
Miscellaneous		
Auto Engine Heater	850	
Clock	3	
Drill (1/4")	250	
Fan (attic)	375	
Fan (window)	200	
Heat Tape	240	
Sewing Machine	75	
Skill Saw	1000	
Vacuum Cleaner	650	
Water Pump (well)	335	

Please note: Average values of power are shown. The power of a particular appliance may vary considerably from the value in the table. Energy use will vary with family size (a four-member family is assumed for the tabled values), personal preferences and habits, climate, and season. Similar information in greater detail is available free upon request from most electric utilities.

© It's About Time

Home Electrical Appliances

Average Power and
Average Monthly Energy Use for a Family of Four

Family Data		
	Power (watts)	Energy/mo. (kWh/mo.)
Big Appliances		
Air Conditioner		
(Room)	1360	
(Central)	3540	
Clothes Washer	512	
Clothes Dryer	5000	
Dehumidifier	645	
Dishwasher	1200	
Freezer	400	
Humidifier	177	
Pool Filter	1000	
Kitchen Range	12,400	
Refrigerator	795	
Space Heater	1500	
Waterbed	350	
Water Heater	4500	
Small Refrigerator	300	
Lights & Minor Appliances (combined)		
Kitchen		
Baby Food Warmer	165	
Blender	300	
Broiler (portable)	1200	
Can Opener	100	
Coffee Maker		
Drip	1100	
Percolator	600	
Corn Popper		
Oil-type	575	
Hot Air-type	1400	
Deep Fryer	1500	
Food Processor	370	
Frying Pan	1200	
Garbage Disposal	445	
Sandwich Grill	1200	
Hot Plate	1200	
Microwave Oven	750	
Mixer	150	
Roaster	1400	
Rotisserie	1400	

© It's About Time

For use with *Electricity For Everyone*, Chapter 7, Activity 6: Cold Shower

Home Electrical Appliances

	Power (watts)	Energy/mo. (kWh/mo.)
Slow Cooker	200	
Toaster	1100	
Toaster Oven	1500	
Trash Compactor	400	
Waffle Iron	1200	
Entertainment		
Computer	60	
Radio	70	
Television	90	
Stereo	125	
VCR	50	
Personal Care		
Air Cleaner	50	
Curling Iron	40	
Hair Dryers	1200	
Hair Rollers	350	
Heat Blanket	170	
Heat Lamp	250	
Heat Pad	60	
Iron	1100	
Lighted Mirror	20	
Shaver	15	
Sun Lamp	300	
Toothbrush	1	
Miscellaneous		
Auto Engine Heater	850	
Clock	3	
Drill (1/4")	250	
Fan (attic)	375	
Fan (window)	200	
Heat Tape	240	
Sewing Machine	75	
Skill Saw	1000	
Vacuum Cleaner	650	
Water Pump (well)	335	

Please note: Average values of power are shown. The power of a particular appliance may vary considerably from the value in the table. Energy use will vary with family size (a four-member family is assumed for the tabled values), personal preferences and habits, climate, and season. Similar information in greater detail is available free upon request from most electric utility companies.

For use with *Electricity For Everyone*, Chapter 7, Activity 6: Cold Shower

© It's About Time

ACTIVITY 7
Pay Up

Background Information

Activity 7 deals with the comparison of monthly electric bills and the cost of operating various electric appliances.

Several factors may account for differences in electric bills for different households, including but not limited to: family size, appliances used, different energy rates for different power companies.

As a direct extension of **Activity** 5, the cost of operating various electric appliances is calculated based on power ratings, time of use and cost per unit of energy. Students have gathered data on monthly energy used for several appliances in units of kWh/month. The monthly energy used per appliance multiplied by the billing rate in $/kWh gives the monthly cost of operating each appliance.

Active-ating the Physics InfoMall

This activity is pretty well self-contained. For fun, you can search the InfoMall for "electric bill". There are a small handful of hits for this search, and most of them are examples you may wish to use.

Of course, the CD-ROM is useful for this activity, but by now you know what you need to look up, and how to do it. If you need to know what a kilowatt-hour is, the InfoMall will provide the answer. If you want more problems for **Physics To Go**, simply look in the Problems Place.

Planning for the Activity

Time Requirements

This activity should take about 40 minutes to complete, including the Physics To Go.

Materials Needed

For each group:

- calculator
- graduated cylinder, 100 ml
- resistive immersion heater of known wattage: Many are available, designed for labs at high cost, but the little portable heaters for making coffee in a hotel room do the job and are less expensive. A 200-watt, 120-volt AC one works fine when used with at least 300 mL of water.
- sample household electric bills from students or others
- stirring rod
- thermometer or temperature probe

Advance Preparation and Setup

Contact your local electric company for power bills. Ask for several different ones, and both residential and commercial. Alternatively, ask the students to bring in their monthly bills, from different times of the year if possible (make sure they first ask for permission from their parents, and black out any personal information). Collect them, then distribute them to the groups, providing a variety of bill types for each group.

Teaching Notes

Have students complete the **What Do You Think?** and then discuss their responses to the question. Be sure that it is clear to students that the two factors which determine the cost of operating any electrical appliance are (in addition to the rate established by the power company) the power rating of the appliance and the amount of time for which the appliance is used during a billing period.

Being sure to protect the anonymity of individuals, proceed to the analysis of the students' household electric bills in **For You To Do**.

© It's About Time

NOTES

© It's About Time

Activity Overview

Students will be analyzing electric bills from trends, cost per kWh, etc. From this they should get a better idea on how power is used, and whether it is used wisely. Taking this knowledge, they will apply it to their own HFE home, and hopefully, they will become more aware of how they can reduce the use of electricity in their own lives.

Student Objectives

Students will:

- Analyze household electric bills, relating energy consumed, billing rate, and cost.
- Compare the costs of operating a variety of electrical appliances in terms of power ratings, amount of time each appliance is used, and billing rate.
- Appreciate energy consumption differences across cultures.

ANSWERS FOR THE TEACHER ONLY

What Do You Think?

Many factors are involved, and the students will provide answers based on the area in which they live. Some of the factors which affect the rate at which the power company charges for use of the electricity include: major industry close by; distance between populated centers; distance from power source; cost of the source; number of customers on the power grid; primarily urban or rural customers; how new or how old the grid is; etc. The two things which affect the amount of money charged is the power rating of the appliance, and how long it is operated in one month.

Electricity for Everyone

Activity 7 Pay Up

GOALS

In this activity you will:

- Analyze household electric bills, relating energy consumed, billing rate, and cost.
- Compare the costs of operating a variety of electrical appliances in terms of power ratings, amount of time each appliance is used, and billing rate.
- Appreciate energy consumption differences across cultures.

What Do You Think?

Eggs are priced by the dozen, and electricity is priced by the kiloWatt-hour.

- **What factors determine the amount of money the electric company charges you for the use of an appliance?**

Record your ideas about this question in your *Active Physics* log. Be prepared to discuss your responses with your small group and the class.

Active Physics 498

© It's About Time

For You To Do

1. Look at the copies of the electric bills provided. If possible, also obtain a copy of the monthly electric bill for your home or that of an acquaintance. Compare the electric bills provided below, or the ones from the homes of individuals in your group.

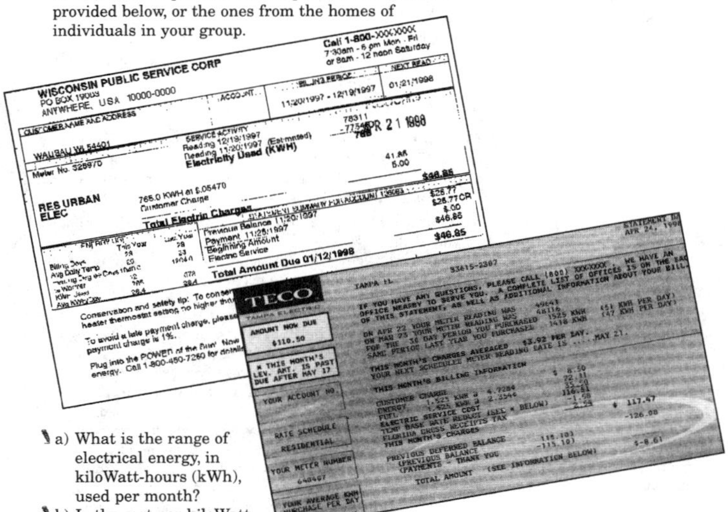

a) What is the range of electrical energy, in kiloWatt-hours (kWh), used per month?

b) Is the cost per kiloWatt-hour the same on all bills, or are there differences?

c) What factors might account for differences among electric bills within your group? Identify as many factors as you can and explain what you think is the effect of each factor on the bill.

d) What is the average monthly amount of electrical energy used in the homes represented in your group, and how does it compare to the 90 kWh per month available to Homes For Everyone (HFE) dwellings?

ANSWERS

For You To Do

1. a) This will vary with the type of bill.

   b) The price will probably be different depending on whether it is commercial, residential, urban, rural, etc.

   c) As mentioned in the **What Do You Think?**.

   d) The average monthly bill would probably be many times larger than the 90 kWh per month available to the HFE homes.

© It's About Time

2. In a previous activity, you estimated the time that an appliance is used, and the amount of energy required to use it for one month. Now you will calculate the expense of that energy. Combine the lists from the group members and calculate the expense of running the electrical appliances by using the following equation:

   Cost = Energy × Price per unit of energy

   The price for electricity can be taken from the electric bills that were provided.

   a) Record your calculations in your log.

3. Compare the cost of electricity in your town to other places in the United States. The "low" for electrical costs is approximately $0.03 per kWh. The "high" for electrical costs is approximately $0.16 per kWh.

   a) What are the high and low costs of the electrical bills that you examined?

   b) How might your consumption of electricity change if you had to pay for your own electrical bills?

FOR YOU TO READ

Determining Energy Consumption

The electrical energy consumed by appliances can be determined in different ways. If the power rating of an appliance is known, and if the power remains at a steady value while the appliance is in use (example: light bulb), the energy, in kWh, can be calculated by multiplying the power of the appliance, in kiloWatts, by the time, in hours, for which the appliance is used (Energy = Power × Time).

Determining the energy consumed by some appliances, however, is tricky, because the appliances vary in power while they are in use. For example, a refrigerator may cycle on and off under the control of a thermostat. Therefore, it is not operating during all of the time it is "plugged in" and the calculation described above would lead to a misleadingly high value for the heater's energy consumption. For such appliances whose power varies throughout time, a kiloWatt-hour meter is used to measure the total energy consumed, with variations in power throughout time taken into account. The same kind of meter, a kiloWatt-hour meter, is used by the power company to measure the total electrical energy used in your home. The meter is mounted somewhere at your home so that it can be read by the power company's "meter reader" person.

Another way to determine the energy used by electrical appliances is to use the experience of power companies or corporations that sell electrical supplies. Extensive lists of appliances, their power ratings, and each appliance's average energy use per month by a typical family are available free from such sources.

ANSWERS

For You To Do *(continued)*

2. a) Student data. Sample calculation: (microwave oven used 0.75 h/day)

   Cost = Energy × price per unit of energy

   = 27 kWh × $0.07/kWh

   = $1.89

3. a) Student response from data from electric bills.

   b) Student response, but students will probably say their consumption will go down (hopefully!).

© It's About Time

Activity 7 Pay Up

PHYSICS TALK

The cost of operating an electrical appliance can be calculated using the following equation:

Cost = Energy × Price per unit of energy

where energy is in kiloWatt-hours (1000 Wh = 1 kWh).

Sample Problem

An electric coffeemaker uses an average of 22.5 kWh of energy each month. If the family is charged $0.12/kWh for electricity, what is the average monthly cost of operating the coffeemaker?

Strategy: Use the equation for cost.

Givens:

Energy = 22.5 kWh

Price = $0.12/kWh

Solution:

Cost = Energy × Price per unit of energy

= 22.5 kWh x $0.12/kWh

= $2.70

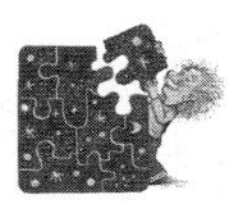

Reflecting on the Activity and the Challenge

In this activity you learned that the average American family's energy use exceeds the maximum available for our HFE dwellings. Imagine how your own standard of living would change if you were restricted to 90 kWh per month.

The **Chapter Challenge** requires that you need to be thinking about two things at the same time as you select appliances for the universal dwelling: the power consumed by each appliance and the amount of time for which each appliance is used. Both need to be taken into account to stay within the power and energy limits of the HFE electrical system.

© It's About Time

ANSWERS

Physics To Go

1. a) 30 days × 0.5 hours /day = 15 hours total

   Therefore, $E = Pt$
   $= 1200 \text{ W} \times 15 \text{ h}$
   $= 18{,}000 \text{ Wh}$

   b) In kilowatt hours, divide the Wh by 1000

   $= 18{,}000 \text{ Wh}/1000 \text{ Wh/kWh}$
   $= 18 \text{ kWh}$

2. The cost for the month is 18 kWh × \$0.15 /kWh = \$2.70/mo. The cost per year is multiplied by 12 = \$2.70/mo. × 12 mo. = \$32.40.

3. 1.350 kW × 24 h = 32.4 kWh. The cost for one day then is 32.4 kWh × \$0.12 /kWh = \$3.89/ day. For the 122 days, the cost of the heater is 122 days × \$3.89 /day = \$475.00.

4. Student activity, but the students should be listing low wattage appliances, and low wattage bulbs, etc.

5. a) Student response: Sample data = 10 hours.

   b) $E = Pt$
   $= 20 \text{ W} \times 10 \text{ h}$
   $= 200 \text{ Wh or } 0.200 \text{ kWh}$

   c) Student response: sample cost of \$2.00

   d) Cost per kWh = \$2.00 / 0.200 kWh = \$10.00 per 4 batteries or \$2.50 per battery.

   e) If you were to use the CD player for the same amount of time using the electricity from the utilities, the cost would be = 0.200 kWh × \$0.10 /kWh = \$0.02 for using the CD player for 10 hours.

Physics To Go

1. A 1200-W hair dryer is used by several members of a family for a total of 30 minutes per day during a 30-day month. How much electrical energy is consumed by the hair dryer during the month? Give your answer in:
   a) watt-hours
   b) kiloWatt-hours

2. If the power company charges \$0.15 per kWh for electrical energy, what is the cost of using the hair dryer in **Question 1** during the month? What is the cost for a year?

3. Not enough heat from the furnace reaches one bedroom in a home. The homeowner uses a portable 1350-W electric heater 24 hours per day to keep the bedroom warm during four cold winter months. At \$0.12 per kiloWatt-hour, how much does it cost to operate the heater for the four months? (Assume two 30-day and two 31-day months.)

4. Prepare your personal list of electrical appliances to recommend to your group to be included in the HFE appliance package. Remember that you will have to justify why you chose a certain appliance. Be prepared to contribute your ideas to your group's decision-making process when completing the challenge.

5. A portable CD player is rated at approximately 20 W and uses 4 AA batteries.
   a) Estimate the number of hours that you can listen to the music on a CD player before the batteries need replacing.
   b) Calculate the energy requirements of the CD player.
   c) Estimate the cost of 4 AA batteries.
   d) Calculate the cost per kiloWatt-hour of a battery.
   e) Compare battery costs with the cost of electricity from the utilities (use approximately \$0.10 per kiloWatt-hour).

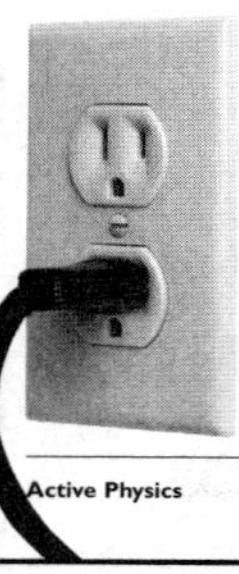

© It's About Time

Sample Electric Bill, New York

30-1285-0615-0010-7 OCT 16, 1998 718-409-7100

79 Service to: JOHN HORTON
at: 26 SMITH ST 2FL

BILLING SUMMARY AS OF 09/18/98

Amount due last bill...	$186.10
Payments through 09/16	NONE
Balance remaining......	$186.10
CURRENT ELEC CHARGES...	$122.11
CURRENT GAS CHARGES....	$36.22
TOTAL AMOUNT NOW DUE...	$344.43

BILLING DETAIL

ELECTRIC USE - RATE EL1 RESIDENTIAL

09/17/98 reading (Actual).......	50435
08/18/98 reading (Actual).......	-49688
Total KWH used in 30 days......	747

CHARGES FOR ELECTRICITY USED

	KWH	COST/KWH	
Basic service charge: (does not include usage)			$7.42
First	250.0 @	13.6720¢	34.18
Next	497.0 @	14.6640¢	72.88
Fuel adjustment @		.3922¢	2.93
Sales tax @ 4.000%			4.70

AVERAGE DAILY ELECTRIC USE

▮ = Actual, ▯ = Estimated
KWH ▥ = Other Customers This Period

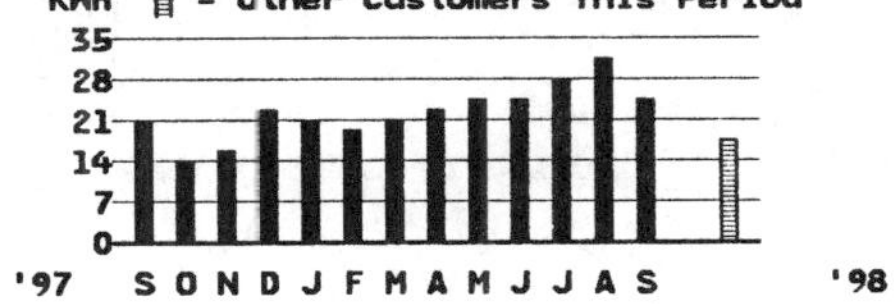

DAILY AVERAGES

	KWH USED
THIS PERIOD:	24.90
SAME PERIOD LAST YEAR:	21.09

CURRENT ELECTRIC CHARGES $122.11

© It's About Time

For use with *Electricity For Everyone*, Chapter 7, Activity 7: Pay Up

Sample Electric Bill, Connecticut

ELECTRIC USE - RATE EL1 RESIDENTIAL

05/07/98 reading (Estimated)....	8190
04/08/98 reading (Estimated)....	-8168
Meter reading difference	22
Meter multiplier	x10
Total KWH used in 29 days......	220

CHARGES FOR ELECTRICITY USED

	KWH	COST/KWH	
Basic service charge: (does not include usage)			$7.18
First	220.0 @	13.6909¢	30.12
Fuel adjustment @		.6909¢	1.52
Sales tax @ 4.000%			1.55

CURRENT ELECTRIC CHARGES	**$40.37**

AVERAGE DAILY ELECTRIC USE

█ = Actual, [] = Estimated

KWH [] = Other Customers This Period

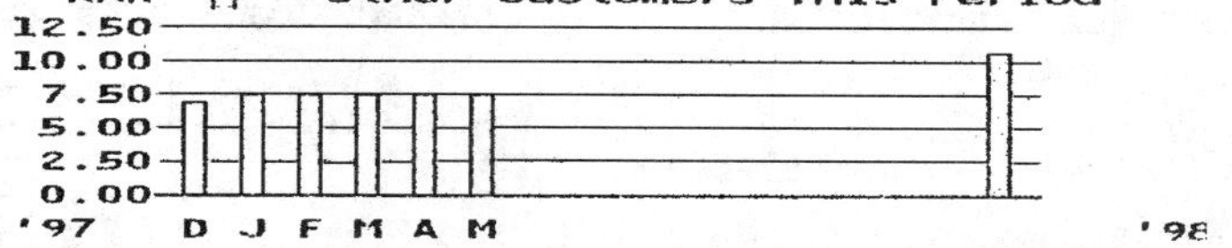

BILLING SUMMARY AS OF 05/08/98

Amount due last bill...	$39.95
Payments through 05/06	$39.95
Balance remaining......	NONE
CURRENT ELEC CHARGES...	**$40.37**
TOTAL AMOUNT NOW DUE...	**$40.37**

© It's About Time

For use with *Electricity For Everyone*, Chapter 7, Activity 7: Pay Up

NOTES

© It's About Time

ACTIVITY 8
More for Your Money

Background Information

The efficiency of electric appliances is introduced in **Activity 8**.

Three different heating appliances—a microwave oven, a hot plate and a heating coil—are used, one at a time, to effect equal temperature changes in standard samples (equal masses) of water in standard containers. The time to accomplish the temperature change for each heating device is recorded.

The electrical energy used by each appliance in joules is calculated by multiplying its power rating in watts by the heating time in seconds.

Since each appliance is used to accomplish the same heating task, the efficiencies of the appliances can be compared in an informal way on the basis of the energy used by each appliance; students perhaps will intuitively agree that the most efficient appliance is the one which used the least amount of energy to accomplish the task.

For your information and in case you wish to formalize treatment of efficiency for the appliances used for the particular task encountered in this activity:

Eff. (%) = [Heat gained by water (joules)/Energy used by appliance (joules)] × 100%

Active-ating the Physics InfoMall

Perhaps the most interesting use of the InfoMall in this activity is exploring how microwave ovens work. Simply search for "microwave oven" and you will quickly find the answer.

Since this activity is about efficiency, you may note that one of the hits from this search is "Calibration and efficiency of microwave ovens," from *The Physics Teacher*, vol. 28, issue 8, 1990. (When you find this article, the citation on the InfoMall is incorrect—it lists it as vol. 25.) You may find this article interesting and you should check it out.

Planning for the Activity

Time Requirements

Allow at least 40 minutes for this activity. Depending on the appliances chosen, the activity may take a longer or shorter amount of time. Try this experiment, and evaluate what is the best quantity of water to use would be in order to get the best results in the shortest amount of time.

Materials Needed

For the class:

- microwave oven, known wattage on high setting
- cold tap water
- source of hot water

For each group:

- Pyrex® beaker, 250 mL.
- Pyrex® beaker, 1-L capacity (2)
- electric hot plate, known wattage

Advance Preparation and Setup

Whether you do the **FYTD** as a demonstration or with students working in groups will depend upon the number of microwave ovens, hot plates and immersion coils available, and the power limit of the circuit(s) in your classroom. You will need to know the power ratings of the appliances used, and you may wish to measure them in advance. Also, you will need to know if the microwave oven power is steady as the oven operates or if the power cycles vary as it operates.

Because it may be difficult to get to get enough microwaves and hot plates for all groups, the teacher may want to do the microwave and hot plate portions as a demonstration, and then allow the students to use the immersion coils.

Use a small quantity of water for the immersion heater activity, such as 1 cup (250 mL). Try this experiment before the students do it, to ensure that the results are sufficiently differentiated. The time for the microwave oven should be about two minutes to heat the water almost to boiling. The

© It's About Time

time for the hot plate, may be anywhere between 3 or 4 minutes up to 10 minutes to get the same temperature change.

Teaching Notes

Have the students complete the **What Do You Think?**. This discussion could lead to the idea of a "payback" period. In other words, if it costs more for the energy-efficient bulb, how long does it have to be on at a smaller wattage, in order to make up the difference? This depends on the cost of power, the difference between the inefficient bulb and the efficient bulb, and the length that each will last. It would have to be stated also, that an energy-efficient bulb will probably last longer, because it will not burn as hot as the other bulb.

When doing this activity, one also must remember that, although the microwave heats more efficiently, and faster, the hot plate will keep the water hotter for longer periods of time. This is necessary in countries where boiling the water is necessary to kill microbes and other organisms which may cause disease. This gives the students another criterion with which to make their decision.

SAFETY PRECAUTION: Safety concerns are that of operating any electrical appliance, as well as handling hot containers. Remind the students to be careful, and use oven mitts or tongs when handling the glassware. If the glass breaks, or if there are any cracks in the glass, remind the students to let the teacher know immediately.

© It's About Time

NOTES

© It's About Time

Activity 8 More for Your Money

GOALS

In this activity you will:

- Measure and compare the energy consumed by appliances.
- Measure and compare the efficiencies of appliances.
- Choose an appropriate appliance for efficiency, based on tests performed.

What Do You Think?

Some hot water heaters and furnaces for homes are more than 90% efficient.

- **If high-efficiency appliances cost more, are they worth the added cost?**

Record your ideas about this question in your *Active Physics* log. Be prepared to discuss your responses with your small group and the class.

For You To Do

1. Place one liter of cold tap water in a beaker made of Pyrex® glass. Make sure the outside of the beaker is dry so it does not slip from your grasp. Measure the temperature of the water.

a) Record the temperature of the water in your log.

b) Record the quantity of water in milliliters and grams. (1 mL of water has a mass of 1 g.)

Activity Overview

Students will be evaluating the efficiencies of different types of appliances that are designed to do the same job. This will be the culmination of the chapter, and therefore, allow them to choose the most energy efficient appliances for their HFE home.

Student Objectives

Students will:

- Measure and compare the energy consumed by appliances.
- Measure and compare the efficiencies of appliances.
- Choose an appropriate appliance for efficiency, based on tests performed.

ANSWERS FOR THE TEACHER ONLY

What Do You Think?

In the long term they are worth the extra cost. Energy-efficient appliances also save on the environment because the power companies do not have to produce as much electricity, and therefore, the companies do not have to use nonrenewable resources and pollute as much to keep up with the demand.

ANSWERS

For You To Do

1. a) Student observation and data.

b) Student observation and data.

© It's About Time

ANSWERS

For You To Do
(continued)

2\. a) Student activity.

b) The longer you wait to take the temperature, the more heat energy will leave the water and be transferred to the surroundings.

3\. a) Student observation and data.

4\. a) Student observation and data.

2. Place the beaker of water in a microwave oven of known power, in watts. Mark the time at which the oven is turned on at its highest power level. After two minutes, stop the time measurement. Carefully check that the beaker is not too hot to grasp and remove the beaker from the oven, stir the water, and check the water temperature, all as quickly as possible.

microwave oven power = ________ watts
time required to heat water = ________ seconds
final water temperature = ________ °C

a) Why is it important to complete the temperature measurement as quickly as possible? Record your answer in your log.

3. Prepare an identical Pyrex beaker containing the same amount of cold tap water, preferably at the same original temperature as the water used above.

a) Record the mass and temperature of the water in your log.

4. Have ready a hot plate that has not been turned on (a "cold" hot plate) and that has a known power rating, in watts. Also have ready a clock or stopwatch capable of measuring time, in seconds, for an interval of several minutes. Place the beaker of cold water on the hot plate, and mark the time at which the hot plate is turned on at its highest setting. Gently stir the water while it is heating and monitor the temperature of the water. When the temperature of the water has increased to the value of the water from the microwave, mark the time and shut the hot plate off.

a) Record the power of the hot plate, the time required to heat the water, and its final temperature in your log.

© It's About Time

5. Repeat the process a third time with the coil.
   a) Record all your observations in your log.
6. Energy (in joules) = Power (in watts) × Time (in seconds)
   Calculate the energy used by each appliance to cause equal temperature increases in equal amounts of water.
   a) Show your calculations in your log.
   b) Which appliance is the "winner"? Choose a way to express a comparison of the performance of the three appliances.
   c) Was the method used to compare the three appliances fair? How could the fairness be improved?
   d) The beaker that served as the container for the water also was heated by the three appliances. Did it affect the outcome of the comparison? Might another kind of container be more or less effective to use with either appliance; for example, might using a metal pan as the container on a hot plate make the hot plate perform more efficiently?

Reflecting on the Activity and the Challenge

You know that the electrical appliances used in the universal dwelling cannot exceed 90 kWh per month of energy consumption. In this activity you discovered that some appliances are more efficient than others. That is, a highly efficient appliance can accomplish a task while using less electrical energy than a low-efficiency appliance for the same task. Obviously you will want to make a careful selection for each kind of appliance based on efficiency so that Homes For Everyone (HFE) inhabitants can have the greatest possible benefit from the electrical system.

ANSWERS

For You To Do
(continued)

5. a) Student observation and data.

6. a) Students will have their own data. Here is a sample data and calculations:

   Microwave: $E = Pt$
   $= 750$ W 120 s
   $= 90$ kWs

   Hot plate: $E = Pt$
   $= 1200$ W 600 s
   $= 720$ kWs
   Immersion

   coil: $E = Pt$
   $= 1000$ W 180 s
   $= 180$ kWs

   b) The winning appliance will be the appliance which has the lowest energy rating. In this case it is the microwave at 72 kWs.

   c) Probably not, as the microwave oven does not lose a lot of residual heat, compared to the hot plate. The immersion coil will not lose a lot of the heat to the surroundings either. But that is what makes these more efficient.

   d) Using different utensils will change the results somewhat, because there is a more efficient heat exchange with some metals than with glass. Glass, as the students found out before, is not a good conductor of heat.

© It's About Time

ANSWERS

Physics To Go

1. Again, some utensils will be better for different purposes. For example, using a small pot on a large burner, will waste more heat than if it were on a small burner. Students will have more variations of this same theme.

2. The microwave oven is obviously the more efficient. However, in some third world countries, the need for clean water is as valuable as energy itself. Therefore, there is a need for looking at all alternatives when choosing appliances for the HFE home.

3. To calculate the actual efficiencies of the appliances, you would start with the same quantities of water, and heat them up for the same amount of time and measure the change in temperature. Calculate he heat energy, using $Q = mc\Delta T$, and then measure the efficiency.

4. a) $E = Pt$
   $= 1500 \text{ J/s} \times 180 \text{ s}$
   $= 270{,}000 \text{ J}$

   b) $E = Pt$
   $= 1200 \text{ J/s} \times 240 \text{ s}$
   $= 288{,}000 \text{ J}$

5. a) The 1500-W hair dryer is more efficient.

6. a) $P = IV$ (where I is the current = 10 A, and V is the voltage = 120 V)
   $= 10 \text{ A} \times 120 \text{ V}$
   $= 1200 \text{ W}$

   b) $E = Pt$
   $= 1200 \text{ J/s} \times 120 \text{ s}$
   $= 144{,}000 \text{ J}$

   c) Energy for one day = 144,000 J
   Energy for one year = 144,000 J/day × 52 weeks × 5 days /week = 37,440,000 J/year.
   One other way to look at it is 37,440,000 Ws.

   Convert this to kWh:

   37,440,000 /1000 = 37,440 kWs;
   37,440 kWs / 3600 s/h = 10.4 kWh.

Physics To Go

1. Are some cooking utensils (pots, pans, etc.) better than others for certain purposes? Write what you think about the effectiveness of different cooking utensils, and what you could do to find out about their comparative effects on efficiency.
2. Does either the hot plate or the microwave oven seem to be a good choice to include in the HFE appliance package? Why, or why not?
3. You probably have concluded that the most efficient appliance of the three tested is the one that used the least energy, but do you know the actual efficiency of the appliances? Explain how you could calculate the efficiencies, and try it.
4. Calculate the energy used, in joules, by each of the following:
   a) a 1500-W hair dryer operating for 3 minutes
   b) a 1200-W hair dryer operating for 4 minutes
5. If both situations described in **Question 4** result in the same dryness of hair, which hair dryer is more efficient?
6. A 10-A electric pencil sharpener is used 2 minutes every working day.
   a) Calculate the power of the electric pencil sharpener.
   b) How much energy is used by the sharpener in one day?
   c) How much energy could be saved in one year (assume 5 working days per week all year) by using a manual pencil sharpener?

© It's About Time

PHYSICS AT WORK

The Schultzes

LIVING CLEAN AND FREE

Bob-o Schultze and his family live "off-the-grid," miles from the nearest electrical power line. Their 2,400 square foot home in northern California includes two computers, a dishwasher, a microwave oven, a washing machine, and a satellite television. It is entirely powered by clean, free, renewable energy.

"Living off-the-grid does not require any major lifestyle changes," explains Mr. Schultze who has been living without traditional energy for over 30 years. "It takes no more technical sophistication or work than the basic upkeep of any house or car."

The Schultzes use three natural power sources: water, wind, and Sun. There is a hydroelectric facility in a creek on the property, a wind machine behind the house, and photovoltaic facilities on tracks following the Sun. All three feed into one large industrial, deep-cycle, 6-V battery and there is never an electric bill. "We are fortunate to have a great site, but every site has at least one source of natural, renewable energy and most have two. The Sun is always an option and if conditions allow for a wind machine, which must be 20-30 feet higher than anything within 500 feet, the power of the wind can also be harnessed."

"I was inspired by the Grateful Dead," explains Bob-o whose name was created from Robert by children many years ago and has stuck. "I moved to a very rural area—30 miles from any electrical power—and I wanted my music. I started using regular D batteries, which drained very quickly, then a car battery, which drained almost as quickly, and then I developed a hydroelectric plant."

All of the appliances in the Schultze's home are the most energy-efficient available. "Compact fluorescent light uses only 25% of the power of standard lighting—and simply turning things off when they are not being used can save 500 watt-hours a day," says Bob-o, emphasizing that energy should be conserved regardless of its source. "We use nine to ten kiloWatt-hours per day, and the average for a family of four is about thirty. As more people become aware of the finite nature of traditional energy sources, the use of renewable energy will increase," claims Bob-o. The Schultzes are just a little ahead of many people in this regard.

507

Active Physical Science

Chapter 7

© It's About Time

Chapter 7 Assessment

You learned a great deal about electricity and electrical terms in this chapter. Read the Scenario from the beginning of the chapter once again. Do you now understand all the terms used in the description of the wind-generator system?

Now that you have completed the activities in this chapter, you are ready to complete the **Chapter Challenge**. You will be asked to do the following:

- **List the appliances to be included in the HFE "appliance package." Your list must be as comprehensive as possible, and it must be clear how each appliance will enhance the health or well-being of the people who live in the dwelling.**
- **Develop an outline of the training manual for HFE volunteers. Your manual must explain the difference between 2400 W and 3 kWh. It must also give clear examples of how use of the appliances in the package can be scheduled to stay within the power and energy limits of the electrical system on both a daily and a long-term basis.**

Review the criteria which you and your class established for how your appliance package and training manual will be graded. Your work will be judged on the basis of 100 points. If necessary, discuss the criteria once again, add details to the criteria if it would be helpful, and agree on the finalized point allocation.

Physics You Learned

Simple circuits

Generators

Series circuits

Parallel circuits

Power

$P=IV$ (watts)

Energy = Pt (kiloWatt-hours; joules)

Load limits

Fuses

Switches

Utility bills

Costs for electricity

Electrical efficiency

© It's About Time

Alternative Chapter Assessment

Select the best response for each statement or question.

1. Which property of a material describes the amount of heat required to increase the temperature of 1 kilogram of the material by 1° Celsius?

   A. Heat conductivity.

   B. Insulating effect.

   C. Specific heat.

   D. Heat of vaporization.

2. A light bulb transforms electrical energy into:

   A. heat only.

   B. mechanical energy.

   C. light only.

   D. heat and light.

3. A generator transforms:

   A. electrical energy into mechanical energy.

   B. heat into mechanical energy.

   C. mechanical energy into electrical energy.

   D. mechanical energy into heat.

4. The output energy of a generator:

   A. is much greater than the input energy.

   B. is slightly greater than the input energy.

   C. is equal to the input energy.

   D. is less than the input energy.

5. The ohm is a unit of electrical:

   A. current.

   B. energy per unit of charge.

   C. resistance.

   D. power.

© It's About Time

6. The volt is a unit of electrical:

   A. current.

   B. energy per unit of charge.

   C. resistance.

   D. power.

7. The ampere, or amp, is a unit of electrical:

   A. current.

   B. energy per unit of charge.

   C. resistance.

   D. power.

8. The watt is a unit of electrical:

   A. current.

   B. energy per unit of charge.

   C. resistance.

   D. power.

9. The equation $P = IV$ is used to calculate electrical:

   A. current.

   B. energy per unit of charge.

   C. resistance.

   D. power.

10. Electrical energy can be calculated by multiplying:

    A. current × voltage.

    B. power × resistance.

    C. current × power.

    D. power × time.

© It's About Time

11. The kilowatt-hour is a unit of electrical:

    A. energy.

    B. voltage.

    C. power.

    D. current.

12. Both a joule and a watt-second are units of:

    A. energy.

    B. voltage.

    C. power.

    D. current.

13. The cost of operating a 1000-watt heater for 24 hours when electricity costs 10 cents per kWh equals:

    A. $0.10

    B. $2.40

    C. $240

    D. $2,400

14. Which of the items listed below require the most energy for an average American family?

    A. food mixers.

    B. water heaters.

    C. televisions sets.

    D. lights.

15. One ampere, or amp, equals:

    A. one joule per coulomb.

    B. one ohm per volt.

    C. one joule per second.

    D. one coulomb per second.

16. One volt equals:

    A. one joule per coulomb.

    B. one ohm per volt.

    C. one joule per second.

    D. one coulomb per second.

© It's About Time

17. One watt equals:
    A. one joule per coulomb.
    B. one ohm per volt.
    C. one joule per second.
    D. one coulomb per second.

18. One kilowatt-hour equals:
    A. 1 joule.
    B. 1000 joules.
    C. 3600 joules.
    D. 3,600,000 joules.

19. The maximum current for a 120-volt circuit rated at 2400 watts is:
    A. 20 amps.
    B. 0.05 amps
    C. 288 amps.
    D. 288,000 amps.

20. If the current flow in a 120-volt vacuum cleaner is 5 amps, the power of the vacuum cleaner is:
    A. 0.042 watts.
    B. 24 watts.
    C. 600 watts.
    D. 600 kilowatts.

© It's About Time

Alternative Chapter Assessment Answers

1. C
2. D
3. C
4. D
5. C
6. B
7. A
8. D
9. D
10. D
11. A
12. A
13. B
14. B
15. D
16. A
17. C
18. D
19. A
20. C

© It's About Time

NOTES

© It's About Time

Chapter 8

TOYS FOR UNDERSTANDING

Chapter 8- Toys for Understanding

National Science Education Standards

Chapter Challenge

Homes for Everyone (HFE) identified a need to educate children about electricity and how it is generated. HFE wants to approach this with toys. Students are challenged to prepare a kit, with materials and instructions, that children use to build a toy with a motor and/or generator. This toy will serve as a tool to illustrate how the electric motors in home appliances work or how electricity can be produced from an energy source such as wind, moving water, or some external force.

Chapter Summary

To gain understanding of the science concepts of energy conversions necessary to meet this challenge, students are engaged in activities to learn about electricity and magnetism. These experiences engage students in the following content from the National Science Education Standards.

Content Standards

Unifying Concepts

- Evidence, models and explanations
- Form and function

Science as Inquiry

- Identify questions and concepts that guide scientific investigations
- Use technology and mathematics to improve investigations
- Formulate and revise scientific explanations and models using logic and evidence

Science and Technology

- Abilities of technological design
- Understanding about science and technology

History and Nature of Science

- Historical perspectives

Science in Personal and Social Perspectives

- Natural resources
- Science and technology in local, national, and global challenges

Physical Science

- Motion and force
- Conservation of energy and increase in disorder
- Interactions of energy and matter

© It's About Time

Key Physics Concepts and Skills

Activity Summaries	Physics Principles
Activity 1: The Electricity and Magnetism Connection Students investigate the relationship between electricity and magnetism by testing the effect of a magnetic field on current-bearing wire and on a compass.	• Electricity • Magnetism • Magnetic fields
Activity 2: Electromagnets Using the hand generator to construct an electromagnet is the first step in this continued investigation of the relationship between electricity and magnetism. Students test the strength and find polarity of electromagnets made with different core materials.	• Electromagnets • Solenoids
Activity 3: Detect and Induce Currents Students construct a galvanometer as they learn that a compass can detect the presence of a magnetic field. They use the galvanometer to create a current similar to the process used by Faraday and Henry, manually alternating the motion of a magnet.	• Galvanometers • Induced currents
Activity 4: AC & DC Currents The use of human energy to produce electricity is replaced in this activity by a rotating coil motor. While using this motor, students test and describe the voltage in this induced current. They learn the difference between how AC and DC currents are generated. Students also learn to sketch output wave forms.	• Energy conversion • AC and DC currents • Electrical waves
Activity 5: Building an Electric Motor Students construct, operate, and explain the workings of a DC motor. This enables them to measure and express the efficiency of energy transfers. They also read to learn more about the discoveries that led to the generators and motors we use today to obtain useable power and electricity.	• Electricity • Magnetism • Energy transfer
Activity 6: Building a Motor/Generator Toy In this final activity, students apply what they have learned about the workings of an electromagnetic motor and how both AC and DC currents are generated. They must use given materials to design, construct, and demonstrate the physics of a motor or generator.	• Energy conversion • Electricity • Magnetism • AC and DC currents • Energy conversion • Energy flow and power • Electrical efficiency

Chapter 8

© It's About Time

Equipment List For Chapter 8

PART	ITEM	QTY	ACTIVITY	TO SERVE
AC-6002-H3	Alligator Clip Leads	2	3, 5, 6	group
BS-1596-H3	Battery, D-Cell	1	5, 6	group
CP-6081-H3	Compass	1	1, 2	group
CS-6047-H3	Cup, Styrofoam	6	5	group
GH-0001-H3	Galvanometer, 0-500 ma.	1	4	group
GH-0002-H3	Tangent Galvanometer	1	1	class
GH-7704-H3	DC Generator, Hand-Operated W/Wire Leads	1	2, 3, 5	group
LS-6959-H3	Miniature Bulb, Light	3	3	group
LS-6960-H3	Bulb Base for Miniature Screwbase	1	3	group
MS-1886-H3	Magnet,Ceramic Ring,Refrigerator	2	5	group
MS-0090-H3	Magnet, Small Bar	1	6	group
MS-0354-H3	Magnet, Bar, Large	1	3	group
NS-7613-H3	Nail, 16 d.	1	2	group
PA-7614-H3	Clips, Paper, Small	10	2	group
PS-7620-H3	Safety Pins, Large	2	5	group
RB-6108-H3	Rubber Band, #64	1	5	group
SS-6078-H3	Sandpaper, Fine Square	6	2, 3, 5	group
SS-6113-H3	Drinking Straw, Each	10	2	group
TS-2662-H3	Tape, Masking, 3/4 × 60 yds.	1	3, 5	group
TS-5000-H3	Test Tube, Plastic, 2 × 10 cm	1	5	group
TS-7601-H3	Tube, Cardboard, 1 × 4	1	3	group
WS-1148-H3	#24 Magnet Wire	60FT	2, 3, 5	group
WS-7612-H3	Wire, Insulated Copper For Tangent Galvanometer	18 INCH	1	group
ZZ-0066-H3	66 Quart Clear Plastic Storage Container	1		group
	TEACHER DEMONSTRATION ITEMS			
AC-7600-H3	AC/DC Demonstration Generator	1	4	class
BS-7602-H3	DC Source, High Current	1	1	class
CT-0006-H3	5' Copper Tube For Cow Magnet	1	4	class
HS-7617-H3	1 ¾" Buret Clamp (Use To Support Copper Tube)	1	4	class
MS-0061-H3	Magnet, Cow	1	4	class
	THINGS NEEDED NOT SUPPLIED			
	All materials needed have been supplied			

© It's About Time

Organizer for Materials Available in Teacher's Edition

Activity in Student Text	Additional Material	Alternative / Optional Activities
ACTIVITY 1: The Electricity and Magnetism Connection p. 512		
ACTIVITY 2: Electromagnets p. 517		
ACTIVITY 3: Detect and Induce Currents p. 521		Activity 3 A: Twin Coil Swings p. 386
ACTIVITY 4: AC & DC Currents p. 526	Current Induced by AC and DC Generators pp. 398-399	Activity 4 A: Falling Magnet p. 397
ACTIVITY 5: Building an Electric Motor p. 533		
ACTIVITY 6: Building a Motor/Generator Toy p. 539		

© It's About Time

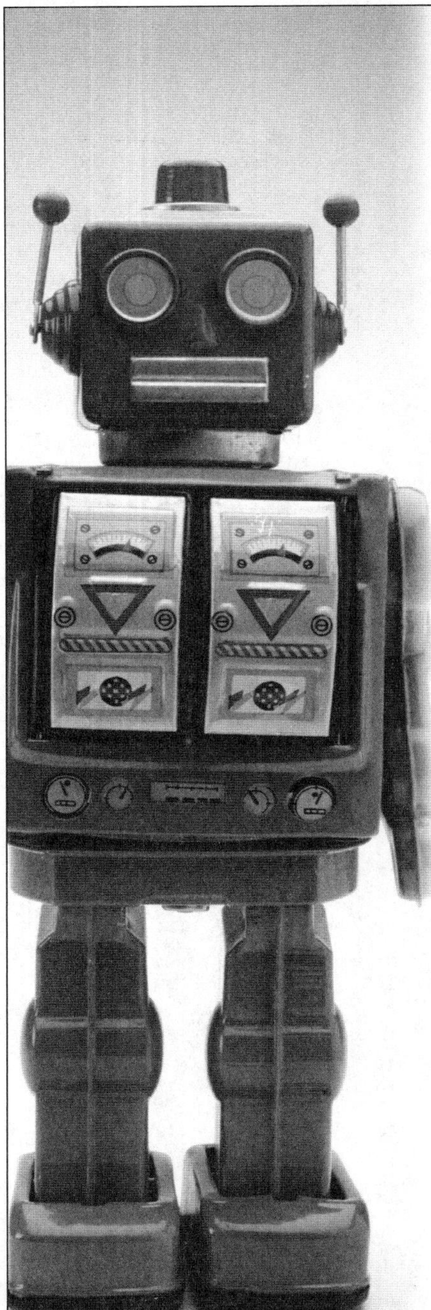

Scenario

In this *Active Physics* chapter, you will try to help educate children through the use of toys. With your input, the Homes for Everyone (HFE) organization has developed an appliance package that will allow families living in the "universal dwelling" to enjoy a healthy and comfortable lifestyle. The HFE organization would now like to teach the children living in these homes, and elsewhere, more about electricity and the generation of electricity. They hope that this may encourage interest in children to use electricity wisely, as well as encourage development of alternative sources for electrical energy by future generations.

The HFE organization will work with a toy company to provide kits and instructions for children to make toy electric motors and generators. These toys should illustrate how electric motors and generators work and capture the interest of the children.

In an effort to help others, people often make changes or introduce new products without considering the personal and cultural impact on those whom they are trying to assist. If you ever become involved in a self-help community group, such as HFE, it would be important for you to work together with the people you are assisting to assess their needs, both personal and cultural. Although that is not possible given your limited time in class, you should recognize the need for collaborative teamwork in evaluating the impact of any new product on an established community.

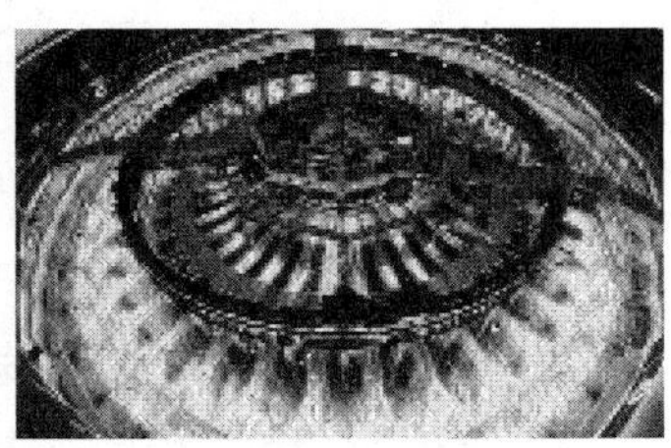

© It's About Time

Chapter and Challenge Overview

In this chapter, the students are applying some of the principles they have learned about electricity and applying it to the building of a toy. Although the toy is designed for children, it is important for the students to focus on the learning that is taking place. A generator can be a motor, and vice versa. The students need to realize that the chapter is designed to get them to apply a particular concept (electromagnetism) to a real life situation. It is hoped that the students will gain a greater respect for physics and be able to apply their knowledge in developing their overall appreciation of science.

A specific design for a DC motor which could also serve as a generator is not presented until the final activity of this chapter. Earlier, perhaps right away, you may wish to have students search for alternate designs—references on science projects would be good sources. Also, you may wish to make a rule on whether motor/generator kits available from hobby stores or toy kits will be allowed; greater benefit may result from having students build the device from "junk."

An understanding of the principles of an electric motor or generator must be brought forward to the students. However, quantitative analysis is not necessary at this stage. Some aspects of the quantitative forces may be used as an alternative project for enrichment.

© It's About Time

Challenge

Your task is to prepare a kit of materials and instructions that the toy company will manufacture. Children will use these kits to make a motor or generator, or a combination electric motor/generator. It will serve both as a toy and to illustrate how the electric motors in home appliances work or how electricity can be produced from an energy source such as wind, moving water, a falling weight, or some other external source.

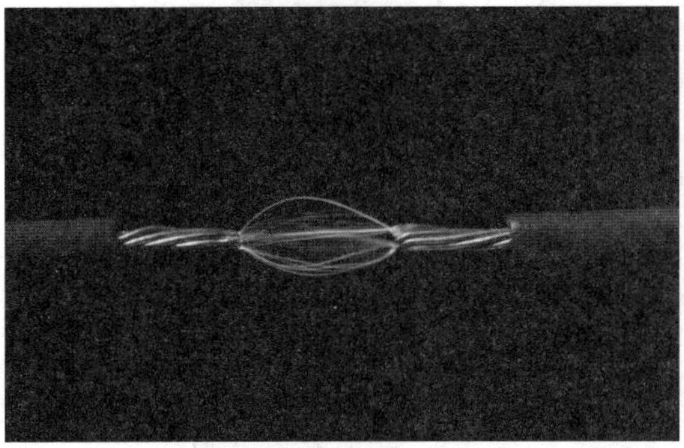

Criteria

Your work will be judged by the following criteria:

- **(30%) The motor/generator is made from inexpensive, common materials, and the working parts are exposed but with due consideration for safety.**
- **(40%) The instructions for the children clearly explain how to assemble and operate the motor/generator device, and explain how and why it works in terms of basic principles of physics.**
  - **(30%) If used as a motor, the device will operate using a maximum of four 1.5-V (volt) batteries (D cells), and will power a toy (such as a car, boat, crane, etc.) that will be fascinating to children.**

  OR

  - **(30%) If used as a generator, the device will demonstrate the production of electricity from an energy source such as wind, moving water, a falling weight, or some other external source and be fascinating to children.**

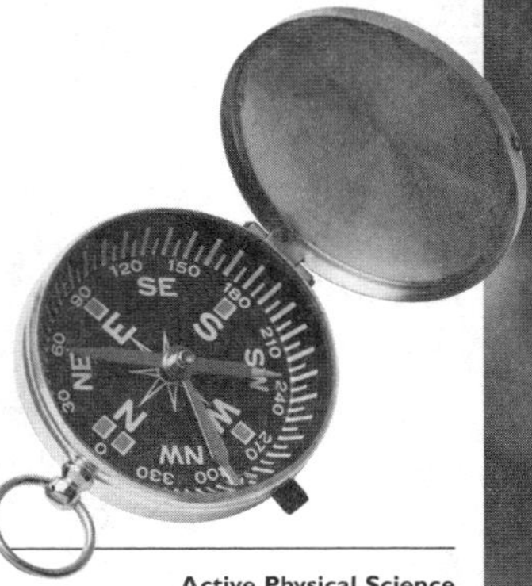

© It's About Time

Assessment Rubric for Challenge

The following is a possible guideline for evaluation of the challenge.

Place a check mark in the appropriate box. If you would rather, you could mark it holistically, where all the statements taken together earn, for example, a mark of 30%.

You may change the criteria, by adding or deleting any of the points, determined by a class decision.

Descriptor	Yes	No
Construction of motor/generator (30%)		
made from simple, inexpensive materials		
easy to assemble (or already assembled)		
working parts are exposed		
safe device		
creative and imaginative		
Instructions (40%)		
well-written		
clear and concise		
explanation of how it works is clear		
simple explanation of the physics involved		
includes safe operating instructions		
clearly stated as motor or generator (or both)		
provides evidence of testing with children		
clear understanding that there is no "creation " of electricity, but transformation		
A) Motor – toy (30%)		
will operate on maximum of four 1.5-v batteries		
will operate a toy		
is fascinating to children		
is durable in the hands of children		
toy shows character and imagination		
OR B) Generator (30%)		
clear understanding as to difference between motor and generator		
demonstrates the production of electricity from an energy source		
is fascinating to children		
is durable in the hands of children		
generator shows character and imagination in the transformation of energy		

For use with *Toys For Understanding*, Chapter 8

© It's About Time

NOTES

© It's About Time

What is in the Physics InfoMall CDROM for Chapter 8?

Although this chapter is called *Toys for Understanding*, this chapter is not really about toys, but about motors and generators. And we will want to use the Physics InfoMall CD-ROM to gather information about these.

After entering the InfoMall, click on "Functions" from the menu bar and select "Compound Search". This will bring up a window which will allow you to specify the search you wish to do. (Caution: this information is generated on a Macintosh, and may vary slightly from what you will find on a PC. The same CD-ROM works on both platforms, and the database is the same. The interface is a little different.) Click the "Search under databases" button and select all databases, (it is sometimes faster to de-select the Keyword Kiosk, since it often provides information you may not want) then click "Apply". Enter "dc generator" in the uppermost box followed by clicking "OK". The search will take less than a minute, and will return all occurrences of the words "dc generator". The first hit on the list of search results is pretty good. You should check them out — they are on the InfoMall in their entirety. And the text is all live, meaning you can print it, copy it, and even paste it into your own handouts. This first hit even has reasonably good graphics

Because this entire chapter deals with electricity, it is a good idea to find out what problems students have with learning about electricity and electrical circuits. Perform a search using "student" AND "difficult*" AND "electric*" on the entire CD-ROM. This search initially provides "Too Many Hits" so it must be limited. One method is to limit it with Terms Must Appear in the Same Paragraph. This finds several references on the difficulties students have with concepts related to electricity. Knowing these difficulties will allow the teacher to expect them, and have ideas on how to overcome them. A similar search can be done using "misconcept*" in place of "difficult*". The results are not identical, so both searches should be tried.

Another idea for a search comes from realizing that the word "student" in the previous searches is almost redundant; we should not expect to find "difficulties" or "misconceptions" in the same paragraph with "electricity" very often unless we are talking about students. In addition, getting rid of the word "student" also gets rid of the "Too Many Hits" problem, so we can search entire articles again. If you perform the searches without "student" you will find many additional references, including a passage from the "Current Electricity" chapter of *A Guide to Introductory Physics Teaching* found in the Book Basement: "Research has been showing that the most basic concepts underpinning simple direct current (d.c.) circuits offer very serious difficulties to many students and that certain misconceptions are widely prevalent [Arons (1982); Cohen, Eylon, and Ganiel (1983); Fredette and Clement (1981)]. As in the case of static electricity, the learning problems are aggravated by the remoteness of the underlying phenomena from direct sense perception. The observable effects are not easily linked to abstractions such as "electrical charge," "current," and "energy." Since students are aware that batteries "run down" and that one "uses" household electricity, they believe that "something is used up" in electric circuits, and, to many of them, the most reasonable thing to be "used up" is "electricity" itself.

Another hit from this search is "LAB MANUAL: Recipes for Science: Misconceptions." You should check this out.

We searched above for "dc generator." What if we search now for "motor"? You might expect many, many hits. And you might expect that many of those hits are useless. Try it anyway. The very first hit is from *Household Physics* from the Textbook Trove, and is a great source of information, including a discussion of the theory of electric motors. We should not be surprised that this book shows up on our search, since this *Active Physical Science* book is about the home. Look through the rest of the hits. Some even have nice graphics and homework problems you can use.

© It's About Time

ACTIVITY 1
The Electricity and Magnetism Connection

Background Information

The physics phenomenon involved in **Activity 1** is the magnetic field surrounding a long, straight wire carrying a direct current.

Magnetic field near a current-carrying straight wire. Before providing **Background Information** about the magnetic field near a current-carrying straight wire, two definitions are offered which apply to magnetic fields in general:

Definition: The direction of a magnetic field at a particular location is the direction that the north-seeking pole of a magnetic compass would point if the compass were placed at the location.

Definition: The strength of a magnetic field, in Tesla, at a particular location is the force, in newtons, that a long, straight wire carrying a current of one ampere would experience per meter of wire length when the wire is placed at the location so that the length of the wire is oriented perpendicular to the direction of the magnetic field.

1 Tesla = 1 newton/(ampere)(meter) = 1N/(amp)m

Magnetic field strength is a vector quantity since it has both direction and magnitude.

As will be indicated by the compass in **For You To Do, Activity 1**, the magnetic field near a long, straight wire carrying a direct current has a circular shape which is concentric on the wire as shown below:

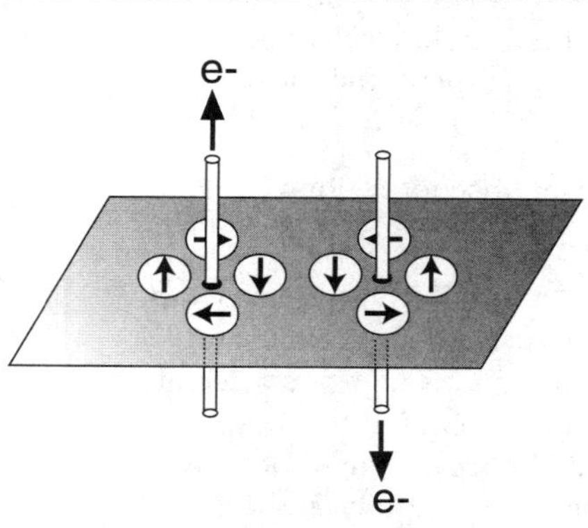

The direction of the magnetic field depends on the direction of the current flow; if the flow of electrons is reversed, so is the direction of the magnetic field around the wire.

SAFETY PRECAUTION: Some physics textbooks use the flow of positive charges (generally called conventional current) as the direction of current flow; this text uses the flow of [negative] electrons which travel around a circuit from the negative terminal of the energy source to the positive terminal.

A convenient rule for determining the direction of the magnetic field surrounding a straight wire in relation to the direction of the current in the wire is shown below as a left-hand rule:

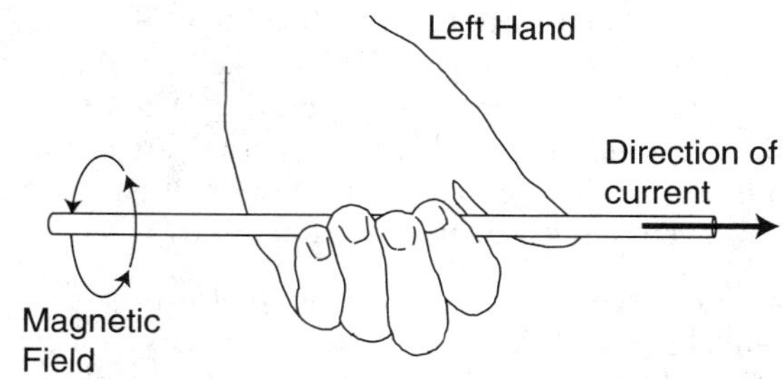

Left-hand rule for the direction of the magnetic field surrounding a straight current-carrying wire. Use the left hand to grasp the wire with the extended thumb pointing in the direction of the current in the wire. The fingers point in the direction in which the magnetic field surrounds the wire.

The strength of a magnetic field near a straight current-carrying wire. The strength of the magnetic field near a straight wire varies directly with the amount of current flowing in the wire and inversely with the distance from the wire. The equation is:

$$B = \mu_0 I/2\pi d$$

where B is the field strength in Tesla, μ_0 is the permeability of air or vacuum which has the value $4\pi \times 10^{-7}$ newton/(amp)^2, I is the current in amps, and d is the distance from the wire in meters.

Active-ating the Physics InfoMall

One of the topics in this activity is the magnetic field. While students often have difficulties learning about fields, this is one area that was not studied widely in time for inclusion on the Physics InfoMall. Some studies have been done more recently, so there is some information to be found, but little of it is on the CD-ROM. Still, you may wish to try a few searches to see what information does exist in this database.

A good search for this activity (with results that we will want in later activities) is "current" AND "compass". You should limit this search to Within

© It's About Time

Paragraph. Some of the results are great, including this passage, this time taken from the Electromagnetism chapter of *A Guide to Introductory Physics Teaching:* found in the Book Basement: "Very few students will conduct a meaningful investigation without guidance, however. The homework assignment should guide them Socratically into: (1) investigating the compass deflection both above and below the current-carrying wire; (2) investigating the effect of reversing the connection to the battery terminals; (3) ascertaining the pattern of the effect all the way around the wire—not just above and below; (4) qualitatively noting the effect of changing the distance between the wire and the compass needle; (5) qualitatively studying the strength of the effect on the compass needle (held at fixed distance from the wire) when additional bulbs are inserted in the circuit either in series or in parallel with an initial single bulb; (6) studying the effect of introducing an additional battery in series with the first; (7) forming, from synthesis of the observations, the right-hand rule mnemonic for the direction of magnetic field around the current-carrying wire".

What is this "right-hand rule" just mentioned? To find out, conduct a search on the list of hits from the previous search. To do this, simply change the Search Category to "Search Hits" and have the engine look for "hand rule" (after all, we don't want to leave out the left-handed versions). A quick look through these hits provides plenty of useful information, again including graphics. Don't forget to reset the search options before your next search!

Planning for the Activity

Time Requirements

If the students have to make the apparatus, allow about twenty minutes to set up the lab. It should then take about 20 - 30 minutes to complete the lab activity.

Depending on available equipment, this may be done either as a demonstration or in groups.

© It's About Time

Materials Needed

For the class:

- DC power supply or automobile-type battery capable of producing high current (5 amps or more)

For each group:

- heavy-gauge (approx. #12) insulated copper wire vertically arranged to penetrate a horizontal surface such as a piece of cardboard
- magnetic compass, 1 or more
- tangent galvanometer

Advance Preparation and Setup

Try this activity in advance to be sure the field produced by the current deflects the compass to indicate a field concentric on the wire. Parallel currents in a bundle of wire (formed by looping one long wire) may be used to enhance the effect.

Adjust the current at each station in advance, and emphasize to students that they should not change the amount of current. Show students how to safely change the direction of the current.

Teaching Notes

Present the **WDYT** question to the students. The student responses may be explored in class discussion and noted on the board, or individually recorded by students and a discussion at the end of the lesson.

If there are not enough stations for each group, have each group come to the power supply **(FYTD #5)**, while the other groups work on mapping the magnetic field **(FYTD #2-4)** of the bar magnet.

Again, a warning regarding the power supply is in order. The settings must not be changed. Warn them of the danger of any electric source, and that electricity is not something to take for granted. If there are any students with serious heart problems, they should probably not handle the equipment.

Students sometimes will refer to the poles of a magnet, in the same sense as the poles of an electrical charge. Whereas, the electric charges of positive and negative can be isolated, the north and south poles can never be isolated. As you break a bar magnet in half, you end up with two N-S magnets, not a magnet with a north pole at both

Activity Overview

In this activity, the students will be observing the effects of a current on a compass. This will lay the foundation for developing an understanding of the relationship between electricity and magnetism.

Student Objectives

Students will:

- Describe electric motor effect and generator effect in terms of energy transformations.
- Use a magnetic compass to map a magnetic field.
- Describe the magnetic field near a long, straight current-carrying wire.

ANSWERS FOR THE TEACHER ONLY

What Do You Think?

Answers will vary. Some answers may include: no electricity; no TVs; video games; hair dryers; etc. Look for thoughtful answers where the students may refer to life before electricity was commonplace, and some may look at it as though they were camping or living at a cabin without power.

Without motors, some students may reflect that they would have to return to the steam engine, or even earlier, where horsepower, and muscle power, provided all the advantages that are enjoyed by people today. Some may refer to the advantages found in the kitchen (blenders, mix masters, coffee grinders, etc.).

Toys for Understanding

Activity 1 The Electricity and Magnetism Connection

GOALS

In this activity you will:

- Describe electric motor effect and generator effect in terms of energy transformations.
- Use a magnetic compass to map a magnetic field.
- Describe the magnetic field near a long, straight current-carrying wire.

What Do You Think?

Generators produce electricity. Motors use electricity.

- **What is the significance of motors and generators to your standard of living? That is, how would your life be different if you had no motors or generators?**

Write your answer to these questions in your *Active Physics* log. Be prepared to discuss your ideas with your small group and other members of your class.

For You To Do

1. Set up the equipment as shown in the diagram, or as directed by your teacher.

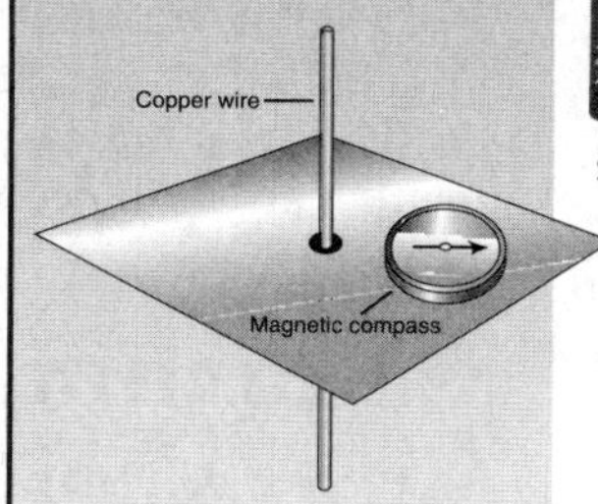

2. The needle of a compass is a balanced magnet. It can be used as a magnetic field detector. If any magnet is present, the compass will respond. It usually aligns itself with Earth's magnetic field. With no current flowing in the wire, verify that the compass always points in the same direction, north, no matter where it is placed on the horizontal surface.

Active Physics 512

ANSWERS

For You To Do

1. Student activity.

2. a) Student data. The compass should point in the same direction regardless of where on the horizontal surface it is placed. (Note: If some students don't see these results, it may be that they are close to a large power supply in the walls or floors of your school. This can be a teaching moment, where you can have the students try to locate the supplies.)

© It's About Time

a) Sketch the compass direction at different places on the horizontal surface in your log.

3. Bring another type of magnet, such as a bar magnet, into the area near the compass needle.

a) Describe your observations in your log.

b) What happens to the dependable north-pointing property of a compass when the compass is placed in a region where magnetic effects, in addition to Earth's magnetic field, exist?

4. You will now make a map of the magnetic field of the bar magnet. Place the magnet on another piece of paper and trace its position. Place the compass at one location and note the direction it points. Remove the compass.

a) Put a small arrow at the location from which you removed the compass to signify the way in which it pointed.

b) Place the compass at a second location about at the tip of the first arrow. Remove the compass and place another small arrow in this location to signify the way in which the compass pointed.

c) Repeat the process at an additional 20 locations to get a map of the magnetic field of a bar magnet. Tape the piece of paper of the map in your log.

Do not adjust the power supply settings provided by your teacher.

5. Return the compass to the horizontal surface surrounding the wire. Observe the orientation of the compass. Send a current through the wire. The direction of the flow of electrons which make up the current in the wire is from the negative terminal of the power supply to the positive terminal. Move the compass to different locations on the horizontal surface, observing the direction in which the compass points at each location. Make observations on all sides of the wire, and at different distances from the wire.

a) Record how the compass was oriented when the bar magnet was removed.

b) Describe any pattern that you observe about how the compass behaves when it is near the current-carrying wire. Use a sketch and words to describe your observations in your log.

c) From your observations, what effect does the electric current appear to have on the wire?

ANSWERS

For You To Do

(continued)

3. a) The students should note that the compass needle will point either toward the magnet or away depending on whether the north or south pole of the magnet is pointed toward the compass.

b) At that particular spot, the magnetic field of the magnet will be larger than the magnetic field of Earth. The students can see this measurable effect by moving the magnet farther way, until it no longer has the effect of "pulling" the compass pointer.

4. a) Student data.

b) Student data.

c) The students should note a pattern like the diagram below:

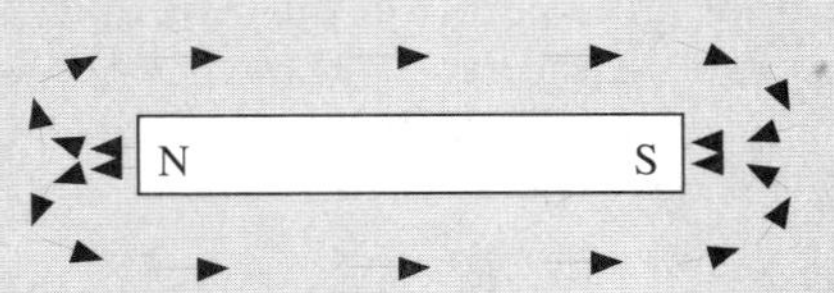

5. a) Student data.

b) Student data. Students should have a sketch showing the compass pointing in a circular path around the current-carrying wire.

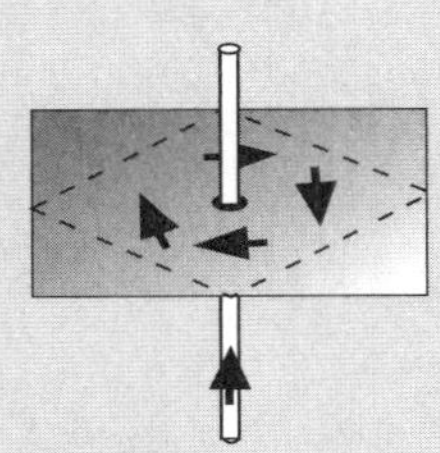

c) The electric current appears to create a magnetic field. The compass indicates the direction of the magnetic field.

© It's About Time

ANSWERS

For You To Do
(continued)

6. a) The students will note that the direction of the compass will be opposite to the previous question. Diagram will be opposite.

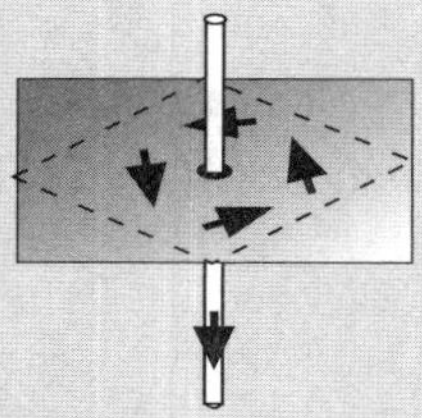

b) Students' answers will vary. Most will state that if the current is going up, the magnetic field will be directed clockwise. If the current is going down, the magnetic field is directed counter-clockwise. If they use the left-hand rule, they should say that the thumb points in the direction of the current and the fingers point in the direction of the magnetic field.

Toys for Understanding

6. Reverse the direction of the current in the wire by exchanging the contacts of the power supply. Repeat your observations.
   a) Describe the results.
   b) Make up a rule for remembering the relationship between the direction of the current in a wire and the direction of the magnetism near the wire (i.e. when the current is up, the magnetic field . . .). Anyone told your rule should be able to use it with success. Write your rule in your log. Include a sketch. (Hint: One of the rules that physicists use makes use of your thumb and fingers.)

Reflecting on the Activity and the Challenge

This activity has provided you with knowledge about a critical link between electricity and magnetism, which is deeply involved in your challenge to make a working electric motor or generator. The response of the compass needle to a nearby electric current showed that an electric current itself has a magnetic effect which can cause a magnet, in this case a compass needle, to experience force. You have a way to go to understand and be able to be "in control" of electric motors and generators, but you've started along the path to being in control.

Physics To Go

1. If 100 compasses were available to be placed on the horizontal surface to surround the current-carrying wire in this activity, describe the pattern of directions in which the 100 compasses would point in each of the following situations:
   a) no current is flowing in the wire
   b) a weak current is flowing in the wire
   c) a strong current is flowing in the wire
2. If a vertical wire carrying a strong current penetrated the floor of a room, and if you were using a compass to "navigate" in the room by always walking in the direction indicated by the north-seeking pole of the compass needle, describe the "walk" you would take.

Active Physics 514

ANSWERS

Physics To Go

1. a) All the compasses will be pointing toward the north.

   b) The students would notice that the compasses that are closest to the wire, will be pointing in a circular path surrounding the wire, in agreement with the left-hand rule. As the compasses get further and further away, they would then point toward the north.

   c) The students should notice that all of the compasses should be pointing in a circular path.

2. The path you would take would be in a circular path around the current-carrying wire. If the direction of the current is up, then the path would be clockwise. If the current is down, then the path would be counter-clockwise.

© It's About Time

3. Use the rule which you made up for remembering the relationship between the direction of the current flowing in a wire and the direction of the magnetic field near the wire to make a sketch showing the direction of the magnetic field near a wire which has a current flowing:
   a) downward
   b) horizontally

4. Physicists remember the orientation of the magnetic field of a current by placing their left thumb in the direction of the electric current and noting whether the fingers of their left hand curve clockwise or counterclockwise. Copy the following diagrams into your log. Use this rule to sketch the direction of the magnetic field in each case.

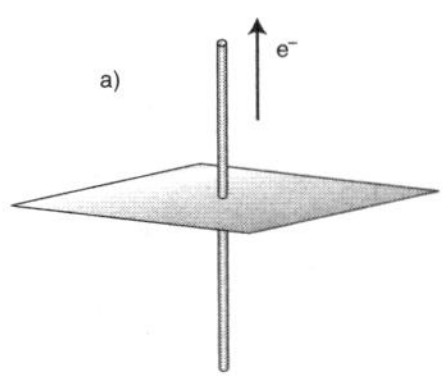

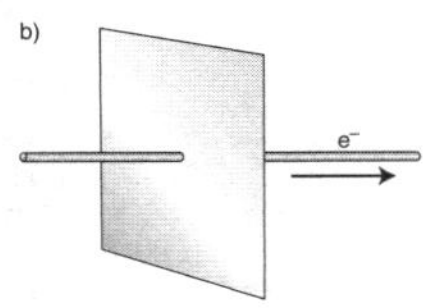

5. Imagine that a second vertical wire is placed in the apparatus used in this activity, but not touching the first wire. There is room to place a magnetic compass between the wires without touching either wire. If a compass were placed between the wires, describe in what direction the compass would point if the wires carry equal currents:
   a) which are in opposite directions
   b) which are in the same direction

ANSWERS

Physics To Go

(continued)

3.a)

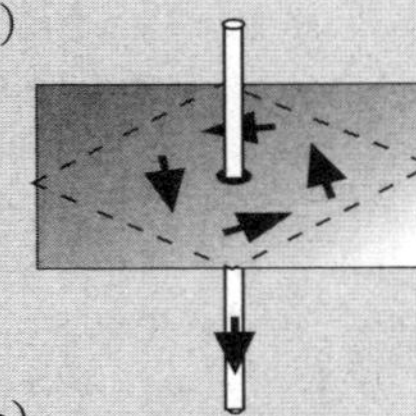

b)

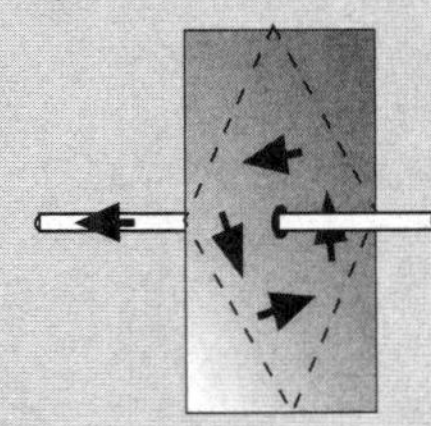

4.a)

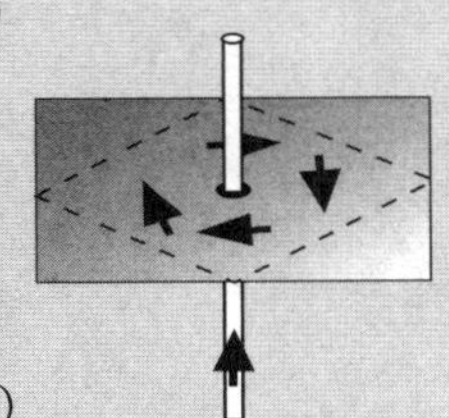

b)

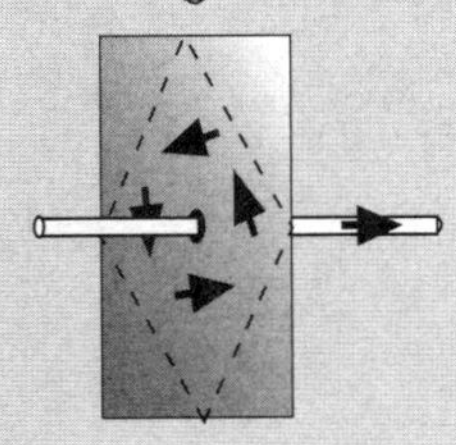

5.a) In this diagram, the compass will appear to go back and forth between the north and south poles.

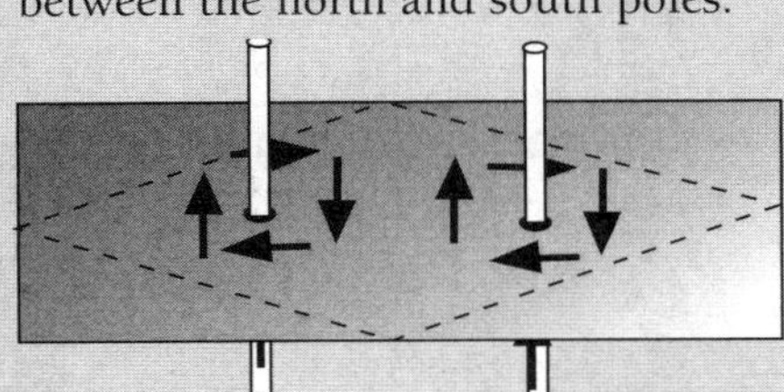

(answer to 5.a)

b) In this diagram, the compass will point in one direction (top of page).

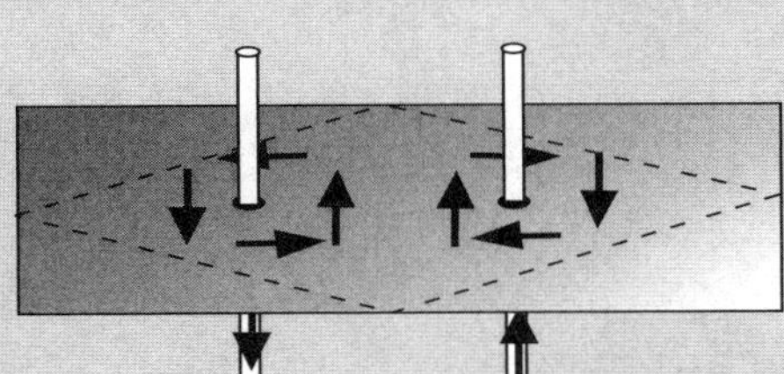

Chapter 8

© It's About Time

ANSWERS

Physics To Go
(continued)

6. The compass will point to the left-hand side of the page. If the students don't see this, make a large loop of wire, and have them use the first left-hand rule to see the direction of the compass. They should note that wherever you put the compass inside the loop, it will point in the same direction. The compass will point in the same direction outside of the loop also. This will help them understand the next left-hand rule in the following activity.

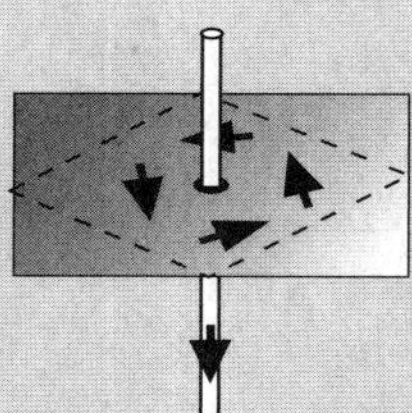

Toys for Understanding

6. A hollow, transparent plastic tube is placed on a horizontal surface as shown in the diagram. A wire carrying a current is wound once around the tube to form a circular loop in the wire. In what direction would a compass placed inside the tube point? (Plastic does not affect a compass; only the current in the wire loop will affect the compass.)

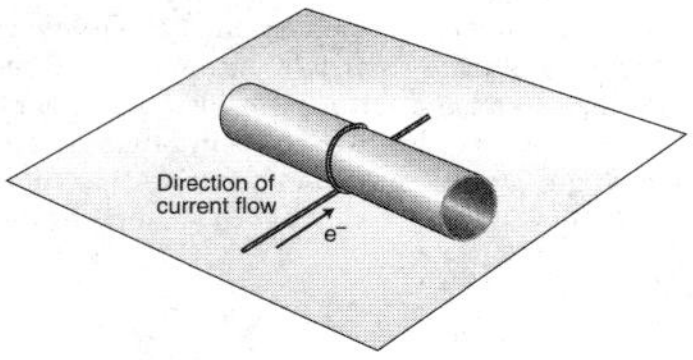

Stretching Exercises

Use a compass to search for magnetic effects and magnetic "stuff." As you know, a compass needle usually aligns in a north-south direction (or nearly so, depending on where you live). If a compass needle does not align north-south, a magnetic effect in addition to that of the Earth is the cause, and the needle is responding to both the Earth's magnetism and some other source of magnetism. Use a compass as a probe for magnetic effects. Try to find magnetic effects in a variety of places and near a variety of things where you suspect magnetism may be present. Try inside a car, bus, or subway. The structural steel in some buildings is magnetized and may cause a compass to give a "wrong" reading. Try near the speaker of a radio, stereo, or TV. Try near electric motors, both operating and not operating.

Do not bring a known strong magnet close to a compass, because the magnet may change the magnetic alignment of the compass needle, destroying the effectiveness of the compass.

Make a list of the magnetic objects and effects that you find in your search.

516

ANSWERS

Stretching Exercises

Students will enjoy this activity, because they will find that there are many different sources of magnetism, even within their own home. Most electrical appliances will cause the compass to move, but may not show an accurate reading due to AC current.

© It's About Time

NOTES

© It's About Time

ACTIVITY 2
Electromagnets

Background Information

The physics phenomenon involved in **Activity 2** is the magnetic field of a solenoid.

A wire wound into the shape of a solenoid as shown in **For You To Do, Activity 2**, and carrying a direct current behaves similar to a bar, magnet with the ends having north and south poles. Considering short segments of the wire of the solenoid as straight pieces of wire, the direction of the magnetic field of each segment can be found using the left-hand rule introduced in the teacher **Background Information** for **Activity 1**. The combined effect of the magnetic fields of segments summed over the entire solenoid results in a magnetic field for the solenoid as shown below:

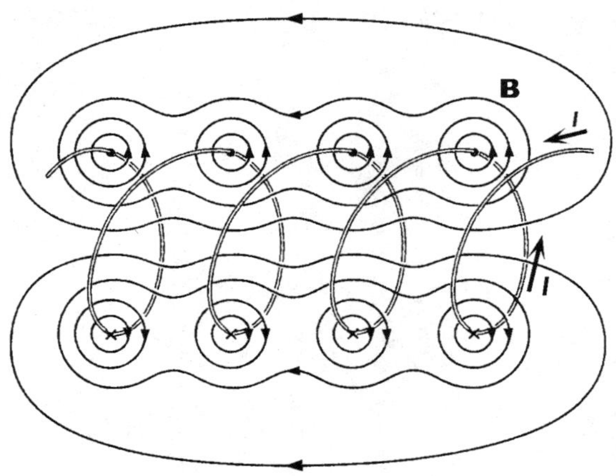

Left-hand rule for the polarity of a solenoid. A convenient rule for determining the magnetic polarity of a solenoid in relation to the direction of the current in the solenoid is to use the left hand to grasp the solenoid with the fingers pointing in the direction in which the current circles the windings. The extended left thumb points to the north pole of the solenoid.

The strength of the magnetic field near the center of a solenoid. The strength near the center (inside) a solenoid varies directly with the number of turns of wire, directly with the current flowing in the wire, and inversely with the length of the solenoid. The equation is:

$$B = \mu_0 NI/L$$

where B is the magnetic field strength in Tesla, μ_0 is the permeability of air or vacuum (if the core is air or vacuum) in newtons/amp^2, N is the number of windings, I is the current in amps, and L is the length of the solenoid in meters.

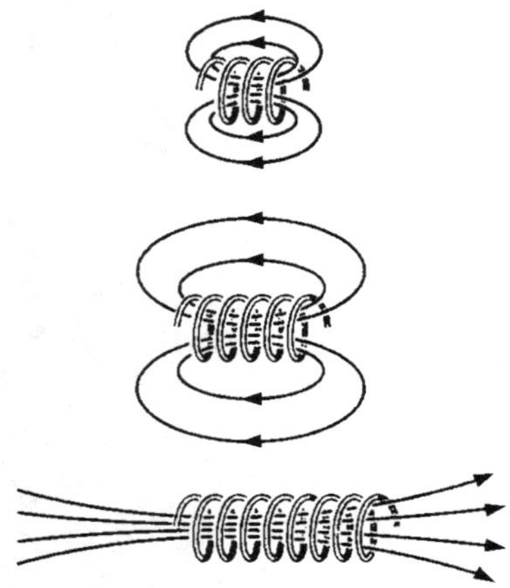

The effect of core material on the strength of a solenoid. If a ferromagnetic material such as iron (viz., a material having high tendency to become magnetized) is used as the core of a solenoid the magnetism of the core greatly adds to the strength of the magnetism. In such a case the constant μ_0 in the above equation is replaced by the magnetic permeability, μ, of the core material which commonly increases the magnetic effect by a factor of thousands.

Iron can be treated so that when used as a core material for a solenoid its magnetism "turns on and off" in concert with the electric current in the solenoid.

Notice in the diagram of the solenoid's field that inside the core of the solenoid the field lines are close together and point toward the right; outside the solenoid the field lines are farther apart and point toward the left. If a current-carrying wire were used as a probe to sample the magnetic field strength it would show that the field is stronger inside the solenoid than outside. This allows development of an alternate model for expressing the strength of a magnetic field in terms of the number of field lines which penetrate a unit of area oriented perpendicular to the direction of the field; the greater the "density" of lines penetrating a unit of area, the greater the magnetic field strength.

In the teacher **Background Information** for **Activity 1** the strength of a magnetic field was defined

© It's About Time

in terms of the force on a current-carrying wire oriented perpendicular to the direction of the field:

1 Tesla = 1 newton/ampere-meter

The magnetic field line model provides an alternate definition of field strength:

1 Tesla = 1 Weber per square meter

= 1 Wb/m^2

where the Weber (Wb) represents the number of magnetic field lines (and is called the "magnetic flux") and m^2 is the area penetrated by the flux [in more advanced treatments, the area is expressed as a vector; this will not be done here].

Symbolically, the alternate definition is expressed as:

$B = \phi/A$

where B is the magnetic field strength in Tesla, ϕ is the magnetic flux (number of field lines) in Weber, and A is the area in m^2. Therefore, the strength of a magnetic field may be expressed in two equivalent ways:

1 Tesla = 1 newton/ampere-meter
= 1 Weber/square meter

(Better to visualize magnetic fields penetrating small areas, physicists sometime prefer to express magnetic field strength in yet another unit, the "gauss" 1 $Weber/m^2$ = 10,000 gauss.)

Active-ating the Physics InfoMall

A search for "electromagnet" will not be disappointing for this activity. Try it for yourself.

In **For You To Do, Step 3,** students are asked to make a prediction. The importance of the prediction should not be overlooked; indeed, predictions force students to examine their understanding of a phenomena and actively engage thought. If you were to search the InfoMall to find more about the importance of predictions in learning, you would find that you need to limit your search. Sadly, not much information exists on the InfoMall regarding predictions about electromagnets. A search for "prediction*" AND "inertia". Try it. The prediction itself is important. Too often when we observe something new, we think "that's what I thought would happen" and learn nothing. By consciously recognizing our preconceptions, we have a chance to change the misconceptions.

The **Stretching Exercise** asks about magnetic levitation. A search for "magnetic levitation" produces a wonderful hit from "Large-scale applications of superconductivity," *Physics Today*, vol. 30, issue 7, 1977.

Planning for the Activity

Time Requirements

The class should be conducted in 40 - 50 minutes. If there is time, you may have the students do several different solenoids, to get enough data to see the relationship to the number of windings, and the relative strength of the electromagnet. Allow another class period to do the **Inquiry Investigation.**

Materials Needed

Prepared solenoids with taps for varying the number of turns and a variety of core materials are available from science suppliers and may be used; however, greater impact may result from students winding their own solenoids. For the "homemade" version shown, you will need:

For each group:

- #24 magnet wire (60 ft)
- drinking straw (10)
- nail, 16 d
- paper clips (10)
- hand generator (or power supply with a variable current)
- sandpaper for removing insulation from the wire ends (6)
- magnetic compass

Advance Preparation and Setup

Demonstrate the magnetic field of a bar magnet and a solenoid, noting similarities. Apparatus for demonstrating the fields on an overhead projector are available from several science supply houses, and employ iron filings, or transparent compasses to visualize the fields.

© It's About Time

Teaching Notes

Begin the lesson pursuing the students' understanding of electromagnetism. As an extension of **What Do You Think?**, demonstrate the fields of a bar magnet and solenoid (see **Materials Needed**). For the solenoid, it would be helpful to use a diagram to show that the field can be inferred by considering short wire segments of the solenoid as straight conductors, each having a surrounding circular field which contributes to the total field of the solenoid (see **Background Information**). Be sure to include in the demonstration the obvious, but important, characteristic that the electromagnet can be turned on and off, while the bar magnet's magnetism can't.

Ask students which factors may be changed to affect the strength of the electromagnet or the bar magnet, and use this to establish the purpose of the **For You To Do** activity.

Here you may wish to establish, as a matter of definition, that the north pole of an electromagnet (or any magnet) is that pole which repels the north-seeking pole of a compass needle [this, of course, dictates that Earth's north magnetic pole is a south pole! – be prepared to deal with this if you bring this up, because it is likely that a student will come up with it].

Also, you may wish to share the second left-hand rule for remembering which is the north pole of a solenoid. This will be useful for students when in the next activity they wind a coil to make a galvanometer, and when later they wind armatures for motor/generators.

SAFETY PRECAUTION: Remind students they are working with electricity, and caution must be taken. Anytime there is electricity running through a resistor wire, there can be a buildup of heat if the current is on for extended periods of time. Using a generator will help because students will not want to crank it for long periods of time. However, if you are using a power supply, make sure that the students turn it off after use, and not leave it on for long periods of time, or while they are writing up the lab activity.

© It's About Time

NOTES

© It's About Time

NOTES

© It's About Time

Activity 2 Electromagnets

GOALS

In this activity you will:

- Describe and explain the magnetic field of a current-carrying solenoid.
- Compare the field of a solenoid to the field of a bar magnet.
- Identify the variables of an electromagnet and explain the effects of each variable.

What Do You Think?

Large electromagnets are used to pick up cars in junkyards.

- **How does an electromagnet work?**
- **How could it be made stronger?**

Write your answer to these questions in your *Active Physics* log. Be prepared to discuss your ideas with your small group and other members of your class.

For You To Do

1. Wind 50 turns of wire on a drinking straw to form a solenoid as shown in the diagram on the next page. Use sandpaper to carefully clean the insulation from a short section of the wire ends to allow electrical connection of the solenoid to the generator.

517

Active Physical Science

ANSWERS

For You To Do

1. Student data.

Activity Overview

Students will be building their own electromagnet, using simple materials. They will discover the variables which give an electromagnet the force necessary to pick up large objects. There may be a need for some discussion about permanent and temporary magnets.

Student Objectives

Students will:

- Describe and explain the magnetic field of a current-carrying solenoid.
- Compare the field of a solenoid to the field of a bar magnet.
- Identify the variables of an electromagnet and explain the effects of each variable.

ANSWERS FOR THE TEACHER ONLY

What Do You Think?

Electromagnets are the result of a changing electrical current in a wire surrounding a soft iron core. The greater the current flowing through the wire, or the greater the number of coils around the core, the greater the magnetic force will be.

Chapter 8

© It's About Time

ANSWERS

For You To Do

(continued)

2. a) Student data. The compass needle will orient either toward or away from the solenoid.

   b) If the compass needle points toward the solenoid, that will be the "south" pole (north-seeking pole). If the compass needle points away from the solenoid, that will be the "north" pole.

3. a) Change the wires to the different terminals of the generator.

   b) Students may also predict that reversing the direction of the generator will have the same effect.

4. a) Student data.

5. a) Students should notice that the solenoid causes the compass needle to point in the same direction as before without the nail inside the straw. There may be a visible increase in the strength of the electromagnet now.

6. a) Student data.

   b) There is an increase in the magnetic field, due to the iron core. The evidence is that the number of paper clips will be larger than before.

7. a) Student data.

   b) As the students increase the speed of the generator, there will be an increase in the number of paper clips picked up. If you have a variable current power source, you may be able to show the effect is linear by graphing the number of paper clips picked up versus the current in the power source. (Depending on your power source, the effect may be hard to see. Try this before the students attempt it.)

8. a) Student data.

   b) Again, there should be a noticeable increase in the number of paper clips picked up as you increase the number of turns. Continue to use the nail as the core, and increase the number of turns by at least 20 turns. Again, the results may be graphed.

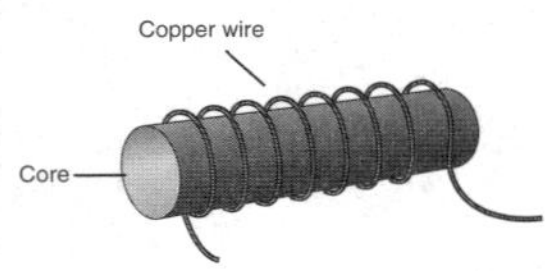

2. Carefully connect the wires from the generator to the wire ends of the solenoid. Bring one end of the solenoid near the magnetic compass and crank the generator to send a current through the solenoid. Observe any effect on the compass needle. Try several orientations of the solenoid to produce effects on the compass needle.
   a) Record your observations in your log.
   b) How can you tell the "polarity" of an electromagnet; that is, how can you tell which end of an electromagnet behaves as a north-seeking pole?

3. Predict what you can do to change the polarity of an electromagnet.
   a) Write your answer in your log.
   b) Test your prediction.

4. Use the solenoid wound on the drinking straw as an electromagnet to pick up paper clips.
   a) Record your observations in your log.

5. Carefully, slip a nail into the drinking straw to serve as a new core. Again, test the effect on a compass needle.
   a) Record your observations in your log.

6. Use the solenoid wound on the nail to pick up paper clips.
   a) Record your observations in your log.
   b) What evidence did you find that the choice of core material for an electromagnet makes a difference?

7. Predict what will happen when you increase the current running through the coiled wire solenoid. This can be done by increasing the speed at which you crank the generator.
   a) Write your answer in your log.
   b) Test your prediction by measuring how many paper clips can be picked up.

8. Predict what will happen when you increase the number of turns of wire forming the solenoid.
   a) Write your answer in your log.
   b) Test your prediction by measuring how many paper clips can be picked up.

© It's About Time

Reflecting on the Activity and the Challenge

An electromagnet, often constructed in the shape of a solenoid, and having an iron core, is the basic moving part of many electric motors. In this activity you learned how the amount of current and the number of turns of wire affect the strength of an electromagnet. You will be able to apply this knowledge to affect the speed and strength with which an electric motor of your own design rotates.

Physics To Go

1. Explain the differences between permanent magnets and electromagnets.
2. The diagram below shows an electromagnet with a compass at each end. Copy the diagram and indicate the direction in which the compass needles will be pointing when a current is generated.

3. Which of the following will pick up more paper clips when an electric current is sent through the wire:
   a) A coil of wire with 20 turns, or a coil of wire with 50 turns?
   b) Wire wound around a cardboard core, or wire wound around a steel core?
4. Explain conditions necessary for two electromagnets to attract or repel one another, as do permanent magnets when they are brought near one another.
5. Explain what you think would happen if, when making an electromagnet, half of the turns of wire on the core were made in one direction, and half in the opposite direction.

Answers

Physics To Go

1. Temporary magnets are ferromagnetic material, which have been treated in order to turn on and off with the current. An electromagnet is also considered a temporary magnet, but as will be shown in this activity, is much stronger when there is a soft iron core. Permanent magnets are ferromagnetic material (iron, nickel, cobalt, and gadolinium) which retain their magnetic properties even when the current is removed.

2. The right side of the picture will be the north pole (compass needle will be pointing away), and the south pole (the compass needle will be pointing toward the solenoid) will be on the left side of the picture.

3. a) 50 turns.

   b) Steel core.

4. In order for the electromagnets to repel each other, the wire would have to be wound opposite to each other:

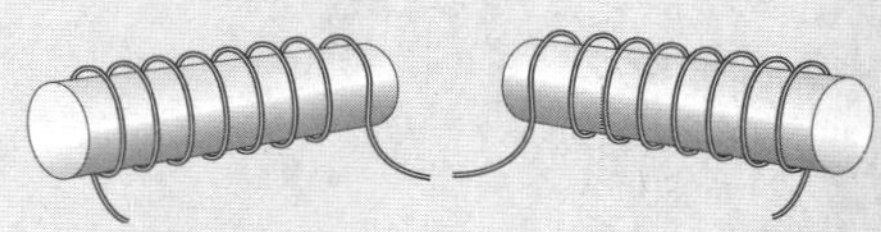

5. The currents would be in opposite directions. This would create magnetic fields in opposite directions. The strength of the electromagnet would be close to zero.

© It's About Time

Stretching Exercises

1. Find out how both permanent magnets and electromagnets are used. Do some library research to learn how electromagnets are used to lift steel in junkyards, make buzzers, or serve as part of electrical switching devices called "relays." For other possibilities, find out how magnetism is used in microphones and speakers within sound systems, or how "super-strong" permanent magnets made possible the small, high-quality, headset speakers for today's portable radio, tape and CD players. Prepare a brief report on your findings.

2. Do some research to find out about "magnetic levitation." "Maglev" involves using super-conducting electromagnets to levitate, or suspend objects such as subway trains in air, thereby reducing friction and the "bumpiness" of the ride.

   a) What possibilities do "maglev" trains, cars, or other transportation devices have for the future?
   b) What advantages would such devices have?
   c) What problems need to be solved? Prepare a brief report on your research.

3. Identify as many variables as you can that you think will affect the behavior of an electromagnet, and design an experiment to test the effect of each variable. Identify each variable, and describe what you would do to test its effects. After your teacher approves your procedures, do the experiments. Report your findings.

ANSWERS

Stretching Exercises

1. Student report.

2. a) Students answers will vary. Some students may report that it will reduce the amount of pollution in the atmosphere. Remind the students that the electricity to run the trains must come from some source (usually coal-fired, gas-fired or nuclear power). It might be possible to reduce the number of car and train accidents, the trains may run on time, without worrying about trains having to stop for other trains.

   b) Other advantages might be that cars would all move at the same speed, reducing the need for police; reduced need for fossil fuels; reduced noise pollution in cities; etc.

   c) The technology has not moved very quickly, and there are some problems with getting the trains to run consistently. Also, there is a great quantity of energy needed to run the trains.

© It's About Time

ACTIVITY 3
Detect and Induce Currents

Background Information

The physics phenomenon involved in **Activity 3** is electromagnetic induction.

As discussed for generators in the student text for **Chapter 8, Activity 4**, which it is suggested you read before proceeding here, an electromagnetic force, or voltage, is produced in a conductor when relative motion between the conductor and a magnetic field happens in a way that causes the conductor to "cut" magnetic flux (or, in other words, cut magnetic field lines) as shown below.

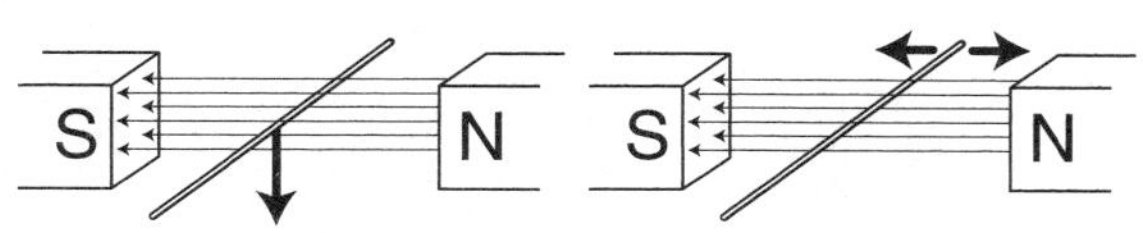

In the above example, thrusting the wire downward causes the field lines in the gap between the magnet poles to be cut by the conductor. This causes electrons in the wire to surge to the far end of the wire, giving that end of the wire an excess negative charge. The near end of the wire then has a deficiency of electrons and is positively charged. The voltage developed across the two ends of the wire is:

$$V = -\Delta\phi/\Delta t$$

where V is the voltage induced, or caused to arise, between the ends of the wire, and $\Delta\phi/\Delta t$ is the number of field lines cut by the wire per unit of time. The negative sign indicates that the current which would flow if the wire were to be part of a complete circuit would produce a magnetic field, and a force, which would oppose the original thrusting action which caused the induced current.

The induced voltage, and current if the wire is part of a circuit, happens only while lines are being cut by the conductor, and the amount of induced voltage at any instant depends directly on the rate at which the magnetic field lines are being changed (cut) at that instant. Also, the direction of the induced voltage depends on the direction of the cutting action. If in the above diagram the wire had been thrust upward, the electrons would have moved in the opposite direction. Finally, if a bundle of individual wires were used above instead of a single wire, the effect of induction would be multiplied by the number of wires.

There are many ways to stage situations where a conductor cuts magnetic flux to "induce" or "generate" electricity. One way is to rotate a loop, or coil, of wire in a magnetic field, as in a generator. Another is used by students in this activity: plunging a bar magnet in and out of a solenoid. In both cases, the principle involved is the one described above; only the geometric configuration of the conductors and fields are different.

Active-ating the Physics InfoMall

Our search for "electromagnet" in **Activity 2** also produced hits useful for this activity, including information on galvanometers. However, you should perform additional searches for "galvanometer," too.

This activity also provides a great chance to search the calendar cart for information on Hans Christian Oersted. Of course, you should also search the entire InfoMall for information; what were the circumstances of his discovery?

For the **Stretching Exercise**, you can do a search of the InfoMall, or you can recall that the Textbook Trove's *Household Physics* was useful for much of this *Active Physical Science* book. Indeed, there is an entire chapter on House Wiring.

Planning for the Activity

Time Requirements

Allow at least one class period (40 – 50 minutes). If time allows, students may wish to change the number of variables in order to see different results (number of windings on the solenoid, on the galvanometer, etc.).

© It's About Time

Materials Needed

For each group:

- #24 magnet wire
- alligator clip leads
- bulb base for miniature screwbase
- hand generator (or power supply with a variable current)
- miniature bulb, light
- sandpaper, fine square (6)
- tape, masking (3/4" × 60 yds)
- tube, cardboard (1 × 4)
- magnet, bar large

Teaching Notes

Have students complete **What Do You Think?** In discussing **What Do You Think?**, mention that science, contrary to common notion, does not always proceed in deliberate ways. Sometimes accidental (serendipitous) discoveries provide breakthroughs, and Oersted's discovery involves one such occasion in the history of science.

SAFETY PRECAUTION: Regarding heating of the solenoid and the galvanometer, as in the previous activity.

© It's About Time

Activity 3 Detect and Induce Currents

GOALS

In this activity you will:

- Explain how a simple galvanometer works.
- Induce current using a magnet and coil.
- Describe alternating current.
- Recognize the relativity of motion.

What Do You Think?

In 1820, the Danish physicist Hans Christian Oersted placed a long, straight, horizontal wire on top of a magnetic compass. Both the compass and the wire were resting on a horizontal surface, and both the length of the wire and the compass needle were oriented north-south. Next, Oersted sent a current through the wire, and happened upon one of the greatest discoveries in physics.

- **What do you think Oersted saw?**

Write your answer to this question in your *Active Physics* log. Be prepared to discuss your ideas with your small group and other members of your class.

For You To Do

1. Wrap 10 turns of wire to form a coil that surrounds a magnetic compass. Wrap the wire on a diameter corresponding to the north-south markings of the compass scale, as shown in the diagram. Hold the turns of wire in place with tape, or use the method recommended by your teacher. Use sandpaper to carefully remove the insulation from a short section of the wire ends to allow electrical connection.

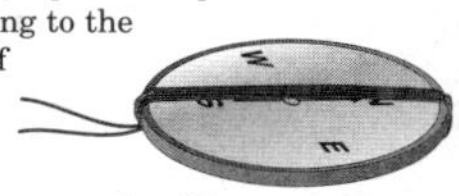

2. In **Step 1**, you constructed a galvanometer, a device to detect and measure small currents. Carefully connect a hand generator, a light bulb, and the galvanometer, as shown on the next page (in a series circuit). Rest the galvanometer so that the compass is horizontal, with the needle balanced, pointing north, and free to rotate. Also, turn the galvanometer, if necessary, so that the compass needle is aligned parallel to the turns of wire which pass over the top of the compass.

ANSWERS

For You To Do

1. Student activity.
2. Student activity.

Activity Overview

The students will build a galvanometer, test to see that it works, then show that moving a magnet through a solenoid, will produce a current. Any moving current will produce a magnetic field, and the magnetic field produced by the generator will cause the compass needle to move (depending on the direction of the current). You may want to remind the students of the left-hand rules to determine in which direction the field is, and in which direction the current is moving.

Student Objectives

Students will:

- Explain how a simple galvanometer works.
- Induce current using a magnet and coil.
- Describe alternating current.
- Recognize the relativity of motion.

ANSWERS FOR THE TEACHER ONLY

What Do You Think?

Students' answers will vary. Oersted saw that when there was a current flowing through the wire, the compass needle oriented itself in a particular direction.

Chapter 8

© It's About Time

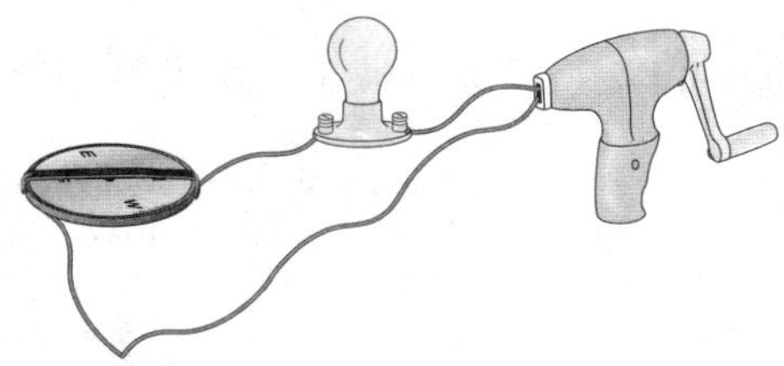

3. Crank the generator to establish a current in the circuit. Think of the compass needle as a meter such as the one in the speedometer of a car. The amount it moves corresponds to the amount of current. The glow of the light bulb verifies that current is flowing.
   a) Does the compass-needle galvanometer also indicate that current is flowing? How? In your log, use words and a sketch to indicate your answer.

4. The amount of current flowing in the circuit can be varied by changing the speed at which the generator is cranked, and the amount of current is indicated by the brightness of the light bulb. Vary the speed at which you crank the generator, and observe the galvanometer.
   a) How does the galvanometer indicate changes in the amount of current? Use words and sketches to indicate your answer.

5. Change the direction in which you crank the generator.
   a) What evidence does the galvanometer provide that changing the direction in which the generator is cranked has the effect of changing the direction of current flow in the circuit? Use words and sketches to give your answer.

6. Carefully connect each wire end of a galvanometer to a wire end of a solenoid wound on a hollow core of non-magnetic

ANSWERS

For You To Do *(continued)*

3. a) Due to the magnetic field created by the moving current, the needle of the compass moves. The more it moves, the greater the current.

4. a) The galvanometer indicates changes in the current by deflecting more away from equilibrium, when there is no current flowing.

5. a) The compass needle should be deflected in the other direction.

6. Student activity.

© It's About Time

Activity 3 Detect and Induce Currents

material, such as a cardboard tube. Orient the galvanometer so that it is ready to detect current flow.

7. Hold a bar magnet in one hand and the solenoid steady in the other hand. Rapidly plunge one end of the bar magnet into the hollow core of the solenoid, and then stop the motion of the magnet, bringing the end of the magnet to rest inside the solenoid. Another person should hold the galvanometer in a steady position so that it will not be disturbed if the solenoid is moved. Observe the galvanometer during the sequence. You may need to practice this a few times.
   a) Write your observations in your log.

8. Remove the magnet from the solenoid with a quick motion, and observe the galvanometer during the action.
   a) Record your observations.
   b) A current is produced! How does the direction of the current caused, or induced, when the end of the magnet is entering the solenoid, compare to the direction of the current when the magnet is leaving the solenoid?
   c) How can you detect the direction of the current in each case?

ANSWERS

For You To Do *(continued)*

7. a) Students should notice a deflection of the needle.

8. a) Again, the students should notice a deflection of the needle, only in the opposite direction.

   b) When going in, the current is one direction and when coming out, the current is going in the other direction.

   c) The direction of the current is observed by the deflection of the compass needle, first in one direction and then in the other.

© It's About Time

ANSWERS

For You To Do
(continued)

9. a) Students will see that the current will still be induced, but in the opposite direction of the first case.

   b) As you increase the speed of the movement, the current will be increased, and the compass needle will show a greater deflection.

   c) The magnet must be moving to induce a current.

   d) There will be no difference, as it is the relative movement that causes the induction, not the movement of one or the other.

   e) This will in effect create a greater relative movement, and therefore have the effect of increasing the induced current.

9. Modify and repeat **Steps 7** and **8** to answer the following questions:
   a) What, if anything, about the created or induced current changes if the opposite end of the bar magnet is plunged in and out of the solenoid?
   b) How does the induced current change if the speed at which the magnet is moved in and out of the solenoid is changed?
   c) What is the amount of induced current when the magnet is not moving (stopped)?
   d) What is the effect on the induced current of holding the magnet stationary and moving the solenoid back and forth over either end of the magnet?
   e) What is the effect of moving both the magnet and the solenoid?

Reflecting on the Activity and the Challenge

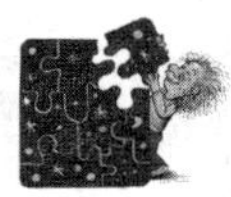

In this activity you discovered that you can produce electricity. A current is created or induced when a magnet is moved in and out of a solenoid. The current flows back and forth, changing direction with each reversal of the motion of the magnet. Such a current is called an alternating current, and you may recognize that name as the kind of current that flows in household circuits. It is frequently referred to by its abbreviated form, "AC." It is the type of current that is used to run electric motors in home appliances. Part of your **Chapter Challenge** is to explain to the children how a motor operates in terms of basic principles of physics or to show how electricity can be produced from an external energy source. This activity will help you with that part of the Challenge.

Physics To Go

1. An electric motor takes electricity and converts it into movement. The movement can be a fan, a washing machine, or a CD player. The galvanometer may be thought of as a crude electric motor. Discuss that statement, using forms of energy as part of your discussion.

ANSWERS

Physics To Go

1. The electric current creates a magnetic field, which induces a current in a coil of wires. This coil of wire (galvanometer) creates a magnetic field, which causes the needle to move (like poles repel, unlike poles attract). In terms of energy, the magnetic field causes a force in the wire, which moves the electrons (electrical kinetic energy). These moving electrons cause a force to act on the magnet (compass needle), thereby converting electrical energy to the kinetic energy of the needle.

© It's About Time

2. Explain how the galvanometer works to detect the amount and direction of an electric current.
3. How could the galvanometer be made more sensitive, so that it could detect very weak currents?
4. An electric generator takes motion and turns it into electricity. The electricity can then be used for many purposes. The solenoid and the bar magnet, as used in this activity, could be thought of as a crude electric generator. Explain the truth of that statement, referring to specific forms of energy in your explanation.
5. If the activity were to be repeated so that you would be able to see only the galvanometer and not the solenoid, the magnet, and the person moving the equipment, would you be able to tell from only the response of the galvanometer what was being moved, the magnet or the solenoid or both? Explain your answer.
6. Part of the **Chapter Challenge** is to explain how the motor and generator toy works.
   a) Write a paragraph explaining how a motor works.
   b) Write a paragraph explaining how a generator works.
7. In generating electricity in this activity, you moved the magnet or the coil. How can you use each of the following resources to move the magnet?
   a) wind
   b) water
   c) steam

Stretching Exercise

Find out about the 120-V (volt) AC used in home circuits. If household current alternates, at what rate does it surge back and forth? Write down any information about AC that you can find and bring it to class.

ANSWERS

Stretching Exercise

AC (alternating current) is named as such due to the "back and forth" motion of the electrons in the wire. Most household circuits operate on 60 Hz (60 cycles per second). European homes operate on 50 Hz, but with a 240-v circuit. If you try to use an appliance (i.e., hair dryer, shaver, radio, etc.) designed for Europe in North America, you can cause serious damage to the appliance, and possibly to yourselves or the circuits in your home.

ANSWERS

Physics To Go
(continued)

2. The galvanometer needle is deflected one way with the current moving in one direction. When the current is moving in the opposite direction the galvanometer will be deflected the other way.
3. The greater number of coils around the galvanometer will cause a greater deflection with the same current, so increasing the number of coils, will give a greater deflection for a smaller current.
4. As you move your hand in and out of the solenoid (kinetic energy) electrical energy is produced in the wire of the solenoid.
5. It is only the relative movement of the magnet and the solenoid that creates the electric current. You would not be able to tell which was being moved.
6. a) Answers will vary. Look for the changing electrical current to produce a force on the magnet inside the coils of wire (solenoid) which cause a rotor or axle to turn.
   b) Answers will vary. As you move a magnet through a coil of wire, an electrical current is produced.
7. a) Wind is the kinetic energy required to move the blades of the generator. This will cause the magnet to move through the coils of wire. This, then, generates the electrical current.
   b) The gravitational potential energy or the kinetic energy of moving water will turn the axle of the (turbine) generator, which moves a magnet through the coils of wires. Again, this produces an electrical current.
   c) Using steam drives a turbine which creates the electrical current as the magnet is turned through the coils of the solenoid.

© It's About Time

Activity 3 A: Demonstration

Twin Coil Swings

FOR YOU TO DO

The following is a demonstration to illustrate the change from mechanical energy to electrical energy. This will be the demonstration: (You may ask some students to design this and report this to the class as an alternative project. Alternately, you may have this set up as a curiosity as the students enter the classroom. Then ask for volunteers to come up and see what they can do to make one side move, by touching only the other side.)

Demonstrate that the twin coil swings, moving one to cause movement in the other due to induced current. Ask students to identify which is acting as a generator, and which as a motor. Switch to move the opposite coil – reversible?

Ask for, and try, changes in the system that will change the swings from moving the "same way" to "opposite ways". Try moving one of the magnets instead of a swing to start up the system; can you "pump" the swing if you hit the proper rhythm (the natural period of the swings)?

Insert a galvanometer to detect the induced current. What is the "E & M connection"? Expect that students will regard this system as a curiosity and want to know what makes it work. Do not offer a full explanation, but suggest that soon students will have a basis for understanding its principles of operation. If possible, leave the apparatus set up in the classroom so that students can return to it to develop explanations of it as their knowledge increases. For now, use the device to establish the meanings of the generator effect (mechanical to electrical energy) and the motor effect (electrical energy to mechanical energy).

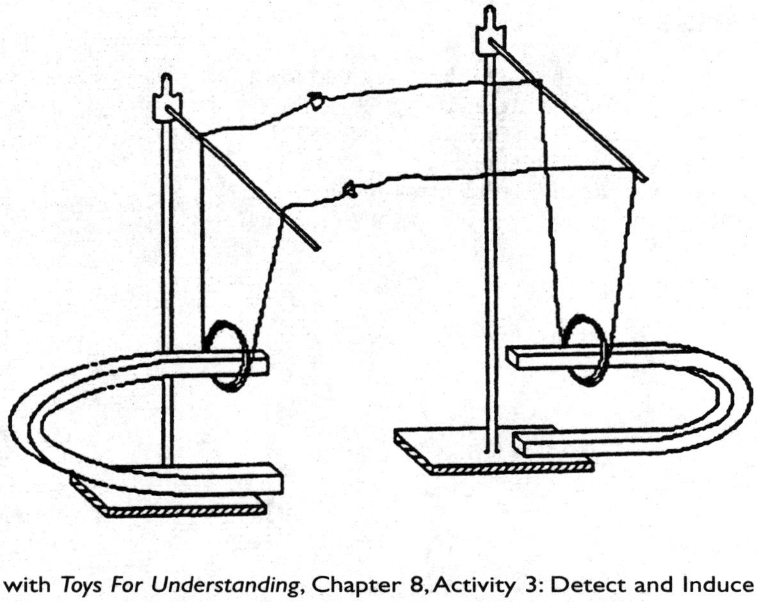

© It's About Time

For use with *Toys For Understanding*, Chapter 8, Activity 3: Detect and Induce Currents

NOTES

Chapter 8

© It's About Time

ACTIVITY 4
AC & DC Currents

Background Information

The principles of electromagnetic induction introduced in **Activity 3** are applied to AC and DC generators in **Activity 4**. It is suggested that you read the **Background Information** for **Activity 3** if you have not already done so.

It is also suggested that you read the detailed treatment of AC and DC generators presented in the **For You To Read** section of **Activity 4** before conducting the activity with students.

The hertz (Hz) as a unit of frequency. The hertz is introduced in the **Stretching Exercise** of **Activity 4** as the unit frequency; here, it is applied to AC electricity. A common misconception about this unit of measurement is shown when a frequency of, for example, 60 Hz is expressed as 60 cycles per second, or 60 vibrations per second, or 60 [fill in the noun] per second. While it is true that terms such as "cycles", "vibrations" or other descriptive nouns may enhance communication, it is essential to recognize that, by definition:

1 hertz = (second)-1 = 1/second

Therefore, the mathematically appropriate way to express frequency in equations is, for example, 60 Hz = 60/s; descriptive words such as "cycles" or "vibrations" are not included because they are not included in the formal definition of the hertz as a unit of measurement. Carrying such descriptive terms in the numerator of expressions of frequency within calculations lead to trouble with dimensional analysis of units because the terms do not cancel; within calculations involving equations it is best to express Hz as reciprocal seconds.

Electric power plants in the United States are required to maintain 60 Hz as the precise frequency of AC voltage distributed on the power grid. The reason for maintaining a dependable frequency is that many devices such as clocks and motors are designed to operate in synchronization with the frequency, or a multiple thereof. It should also be recognized that one complete AC cycle of 1/60 second duration contains two pulses of current, one in each direction. Therefore, the "pulse frequency" of 60 Hz electricity is 120 Hz; sometimes it can be heard being emitted from electrical devices as a "hum" corresponding to a pitch of 120 Hz.

Active-ating the Physics InfoMall

We already did a search earlier for "DC generator" so now we should add to this a search for "AC generator." Again, the results are pleasing. Combine this search with the DC search, and much of this activity is covered. You can also find graphs of the generator outputs in these searches.

For You To Do Step 4 begins working with graphs. Graphs are a tool that can be difficult for some students. Search for "difficult*" AND "graph*" in the Same Paragraph. The first hit is "Student difficulties in connecting graphs and physics: Example from kinematics," from the *American Journal of Physics*, vol. 55, issue 6, 1987. The second hit is "Student difficulties with graphical representations of negative values of velocity," from *The Physics Teacher*, vol. 27, issue 4, 1989. While these and many of the other hits are from kinematics, the findings related to graphs can still be enlightening. You should read these.

Planning for the Activity

Time Requirements

Allow about one 40 – 50 minute class period for this activity.

Materials Needed

For the class (demonstration):

- 1 3/4" buret clamp (use to support copper tube)
- 5' copper tube for cow magnet
- AC/DC demonstration generator
- magnet, cow

For each group:

- galvanometer, 0-500 ma.

Advance Preparation and Setup

It is assumed that limitations on equipment will require that this activity be performed as a teacher demonstration. Large, low voltage, demonstration AC and DC generators are available from science suppliers which have exposed parts, and many are convertible, serving four-ways (AC, DC generators and motors). Such a device, or the equivalent, is needed for this activity.

© It's About Time

Be familiar with the progression of **For You To Do.** The students may or may not be able to fully grasp the concept of a sinusoidal curve. They can better appreciate it when they have seen the dimming of the bulb, and the changing of the galvanometer.

Teaching Notes

Review with the students the result of passing a wire through a permanent magnet. They need to understand that only when the wire is moving perpendicular to the magnetic field lines will a current be produced. Ask them questions such as: What if the wire is moving at an angle? Therefore, only the perpendicular vector portion of the wire will produce the current. Have the students come up with the fact that the magnitude of the current will be less than when passed through the magnetic field lines at 90°. This will help them understand that the current is not simply on and off, but variable.

It may be difficult for all the class to see the galvanometer and the changes, so it is important to have the setup of the apparatus so that all will be able to see. If a physics software simulation program is available, you might allow the students some time to use the software during another class.

SAFETY PRECAUTION: Safety issues are the same as they are with any electronic equipment. Do not let the students operate the equipment without proper supervision.

The students may have to be reminded of the Law of Conservation of Energy. Energy cannot be created nor destroyed, only converted from one form to another. Some students may believe that the only energy that has to be added to the system of the generators of power plants is just enough to overcome the effects of friction of the mechanical parts. If this were the case, you would be creating energy from nothing which is impossible. The energy input must be greater than the energy output in order to account for the "loss" of energy due to friction.

© It's About Time

Activity Overview

This activity expands on the information presented in **Activity 3**.In the previous activities, we saw how the movement of a wire in a magnetic field produced a current in that wire. In this activity, the students will explore step by step, how an AC generator works. It involves the rotation of a coil of wire in a magnetic field to produce a continuous current that can be used for an electrical device.

Student Objectives

Students will:

- Describe the induced voltage and current when a coil is rotated in a magnetic field.
- Compare AC and DC generators in terms of commutators and outputs.
- Sketch sinusoidal output wave forms.

ANSWERS FOR THE TEACHER ONLY

What Do You Think?

Many different answers will come from the students. Look for thoughtful responses. Some answers might include: heat, geothermal, mechanical, nuclear, light, wind, water, gas-fired hydroelectric, etc.

Activity 4 AC & DC Currents

GOALS

In this activity you will:

- Describe the induced voltage and current when a coil is rotated in a magnetic field.
- Compare AC and DC generators in terms of commutators and outputs.
- Sketch sinusoidal output wave forms.

What Do You Think?

In the last activity, you used human energy to produce motion to generate electricity.

- **What other kinds of energy can generate electricity?**

Write your answer to this question in your *Active Physics* log. Be prepared to discuss your ideas with your small group and other members of your class.

For You To Do

AC Generator

1. Your teacher will explain and demonstrate a hand-operated, alternating current (AC) generator. During the demonstration, make the observations necessary to gain the information needed to answer these questions:

a) When the AC generator is used to light a bulb, describe the brightness of the bulb when the generator is cranked slowly, and then rapidly. Write your observations in your log.

b) When the AC generator is connected to a galvanometer, describe the action of the galvanometer needle when the generator is cranked slowly, and then rapidly.

ANSWERS

For You To Do

1. a) The brightness of the bulb appears to increase as the speed is increased.

   b) When the generator is cranked slowly, there is only a small deflection of the galvanometer needle. As the speed increases the needle's deflection increases. The needle also goes back and forth or positive and negative as the generator is cranked.

© It's About Time

2. It is easier to understand the creation of a current if you think of a set of invisible threads to signify the magnetic field of the permanent magnets. The very thin threads fill the space and connect the north pole of one magnet with the south pole of the other magnet. If the wire of the generator is imagined to be a very thin, sharp knife, the question you must ask is whether the knife (the wire) can "cut" the threads (the magnetic field lines). If the wire moves in such a way that it can cut the field lines, then a current is generated. If the wire moves in such a way that it does not cut the field lines, then no current is generated.

a) Look at the diagrams of the magnetic fields shown. In which case, I, II, or III will a current be generated?

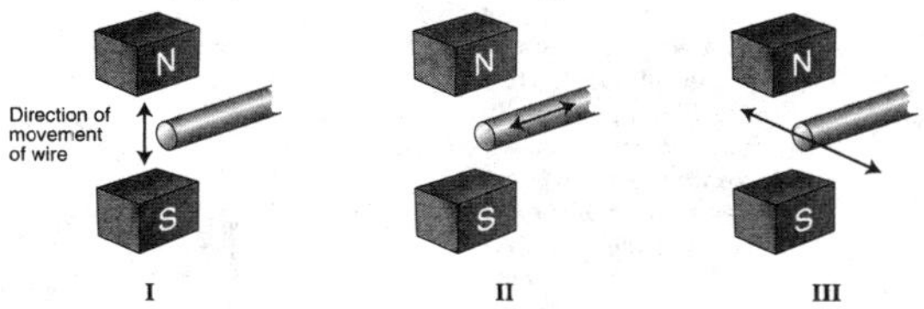

3. The following diagram shows the position of the rotating coil of an AC generator at instants separated by one-fourth of a rotation of the coil. Build a small model of the rectangular coil so that you can move the model to help you understand the drawings. The coil model can be constructed by carefully bending a coat hanger into the shape of the rectangular coil. Rest the coil between two pieces of paper—label the left paper N for the north pole of a magnet; label the right paper S for the south pole of a magnet.

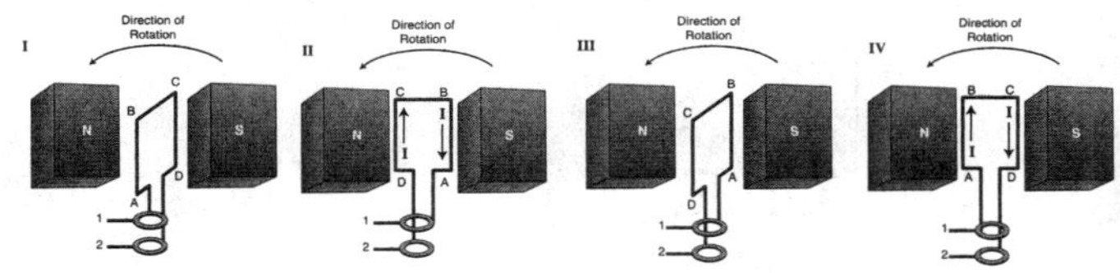

ANSWERS

For You To Do *(continued)*

2. a) Case III will generate a current.

3. Student activity.

© It's About Time

Chapter 8

4. For the purpose of analyzing the rotating coil figure, the four sides of the rectangular coil of the AC generator will be referred to as sides AB, BC, CD, and DA. Side DA is "broken" to allow extension of the coil to the rings. The "brushes," labeled 1 and 2, make sliding contact with the rings to provide a path for the induced current to travel to an external circuit (not shown) connected to the brushes. The magnetic field has a left-to-right direction (from the north pole to the south pole) in the space between the magnets in the rotating coil figure. It is assumed that the coil has a constant speed of rotation.

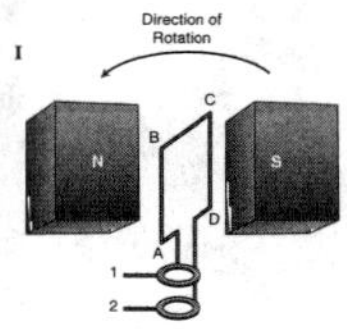

a) When the generator coil is in position I shown in the rotating coil, is a current being generated? A current is produced if the wire cuts the magnetic field lines. Record your answer and the reason for your answer in your log.

b) Use a graph similar to the one shown below. Plot a point at the origin of the graph, indicating the amount of induced current is zero at the instant corresponding to the beginning of one rotation of the coil.

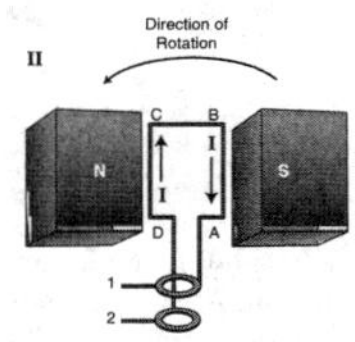

c) One-fourth turn later, at the instant when the rotating coil is in position II, is a current being generated? Record your answer and the reason for your answer in your log.

d) On your graph, plot a point directly above the $\frac{1}{4}$-turn mark at a height equal to the top of the vertical axis to represent maximum current flow in one direction.

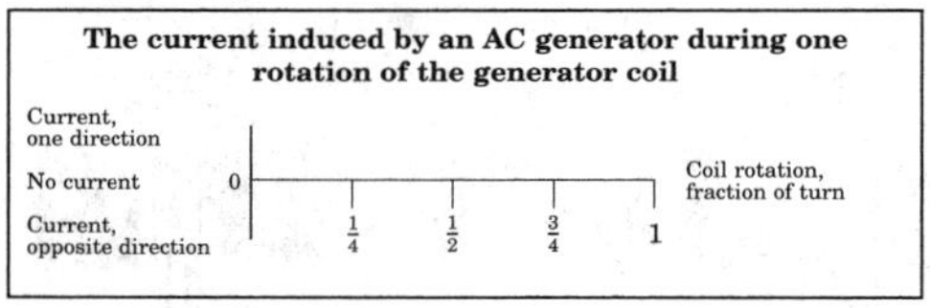

ANSWERS

For You To Do *(continued)*

4. a) There is no current as the wire is moving parallel to the magnetic field.

b) Student activity.

c) Current is generated and should be at a maximum.

d) Student activity.

© It's About Time

e) One-half turn into the rotation of the coil, at the instant shown in the rotating coil position III, the current again is zero because all sides of the coil are moving parallel to the magnetic field. Plot a point at the $\frac{1}{2}$ mark on the horizontal axis to show that no current is being induced at that instant.

f) At the instant at which $\frac{3}{4}$ of the rotation of the coil has been completed, shown by the rotating coil in position IV, the induced current again is maximum because coil sides AB and CD again are moving across the magnetic field at maximum rate. However, this is not exactly the same situation as shown in the rotating coil position II; it is a different situation in one important way: the direction of the induced current has reversed. Follow the directions of the arrows which represent the direction of the current flow in the coil to notice that, at this instant, the current would flow to an external circuit out of brush 2 and would return through brush 1. On your graph, plot a point below the $\frac{3}{4}$-turn mark at a distance as far below the horizontal axis as the bottom end of the vertical axis. This point will represent maximum current in the opposite, or "alternate," direction of the current shown earlier at $\frac{3}{4}$-turn.

III

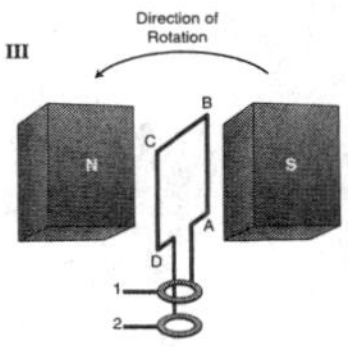

IV

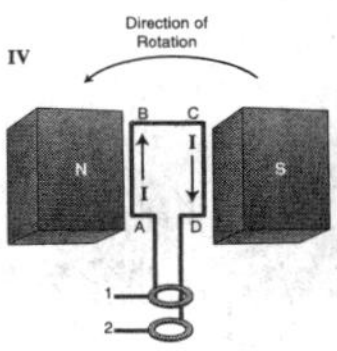

g) The rotating coil in position I is used again to show the instant at which one full rotation of the generator coil has been completed. Again, all sides of the coil are moving parallel to the magnetic field, and no current is being induced. Plot a point on the horizontal axis at the 1-turn mark to show that the current at this instant is zero.

5. You have plotted only 5 points to represent the current induced during one complete cycle of an AC generator.

a) Where would the points that would represent the amount of induced current at each instant during one complete rotation of the generator coil be plotted?

b) What is the overall shape of the graph? Should the graph be smooth, or have sharp edges? Sketch it to connect the points plotted on your graph.

c) What would the graph look like for additional rotations of the generator coil, if the same speed and resistance in the external circuit were maintained?

ANSWERS

For You To Do *(continued)*

e) Student activity.

f) Students will note that the current should be flowing in the opposite direction, but at a maximum.

g) Student activity.

5. a) The points should be plotted along a line that would eventually be a sinusoidal curve.

b) Students may have to be shown that the points do in fact represent a smooth curve rather than a straight line of best fit.

c) The graph will have a consistent up and down wave-like motion.

© It's About Time

DC Generator

6. Your teacher will explain and demonstrate a hand-operated, direct current (DC) generator. During the demonstration, make the observations needed to answer these questions:
   - a) When the DC generator is used to light a bulb, describe the brightness of the bulb when the generator is cranked slowly, and rapidly. Write your observations in your log.
   - b) When the DC generator is connected to a galvanometer, describe the action of the galvanometer needle when the generator is cranked slowly, and rapidly.

7. The diagram shows important parts of a DC generator. As in **Step 3**, build a model of the generator to help you analyze how it works.
8. Use a graph similar to the one shown below. Complete the graph using the same pattern of analysis applied to the AC generator.
   - a) At the instant shown in the DC generator diagram, the induced current is maximum. The instant corresponds to the rotating coil II. Plot a point on the graph directly above the $\frac{1}{4}$-turn mark at a height equal to the top of the vertical axis to represent maximum current flow at that instant.
   - b) At the instant $\frac{1}{4}$-turn earlier than the instant shown in the DC generator figure, corresponding to the zero mark of rotation, the current would have been zero because all sides of the coil would have been moving parallel to the direction of the magnetic field. Therefore, plot a point at the origin of the graph.
   - c) Similarly, the induced current again would be zero at the instant $\frac{1}{4}$-turn later than the instant shown in the DC generator figure; therefore, plot a point on the horizontal axis at the $\frac{1}{2}$-turn mark.

The current induced by a DC generator during one rotation of the generator coil

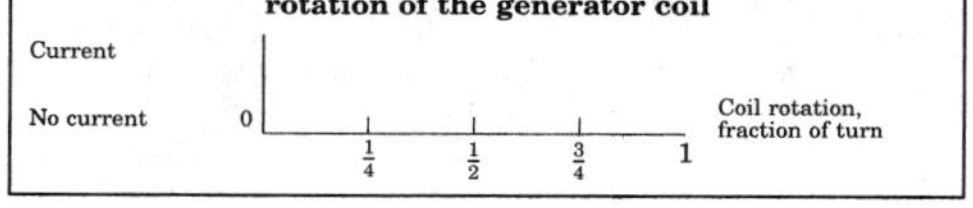

ANSWERS

For You To Do *(continued)*

6. a) Again, the students should note that the light dims and brightens regularly, with the brightness increasing as the cranking increases.

   b) The galvanometer needle will go from the equilibrium position to one side or the other. It will not go back and forth across the equilibrium as it did for the AC generator. Increasing the speed of the cranking will increase the amount of the deflection.

7. Student activity.

8. a) Student activity.

   b) Student activity.

   c) Student activity.

© It's About Time

9. Notice the arrangement used to transfer current from the generator to the external circuit for the DC generator. It is different from the arrangement used for the AC generator. The DC generator has a "split-ring commutator" for transferring the current to the external circuit. Notice that if the coil shown in the DC generator figure were rotated $\frac{1}{4}$-turn in either direction, the "brush" ends that extend from the coil to make rubbing contact with each half of the split ring would reverse, or switch, the connection to the external circuit. Further, notice that the connection to the external circuit would be reversed at the same instant that the induced current in the coil reverses due to the change in direction in which the sides of the coil move through the magnetic field. The outcome is that while the current induced in the coil alternates, or changes direction each $\frac{1}{2}$-rotation, the current delivered to the external circuit always flows in the same direction. Current that always flows in one direction is called direct current, or DC.

   a) Plot a point on the graph at a point directly above the $\frac{3}{4}$-turn mark at the same height as the point plotted earlier for the $\frac{1}{4}$-turn mark.

   b) As done for the AC generator, find out how to connect the points plotted on this graph to represent the amount of current delivered always in the same direction to the external circuit during the entire cycle.

Reflecting on the Activity and the Challenge

It is time to begin preparing for the **Chapter Challenge**. Now that you know how a generator works, you should begin to think about toys that might generate electricity. You should also think about how you could assemble "junk" into a toy generator, or do some research on homemade generators and motors.

ANSWERS

For You To Do *(continued)*

9. a) Student activity.

   b) Again, the students should notice that the points should give a smooth curved line. However, the shape of the graph will differ in that the points will only be on the positive side of equilibrium in the shape of a letter "m". The pattern will appear to be a series of bumps rather than a sinusoidal curve.

© It's About Time

Chapter 8

ANSWERS

Physics To Go

1. a) An electric generator is used to convert mechanical energy (cranking the generator) to electrical energy.

   b) An electric motor is used to convert electrical energy to mechanical energy.

2. Direct current travels in one direction only, whereas alternating current appears to travel back and forth.

AC:

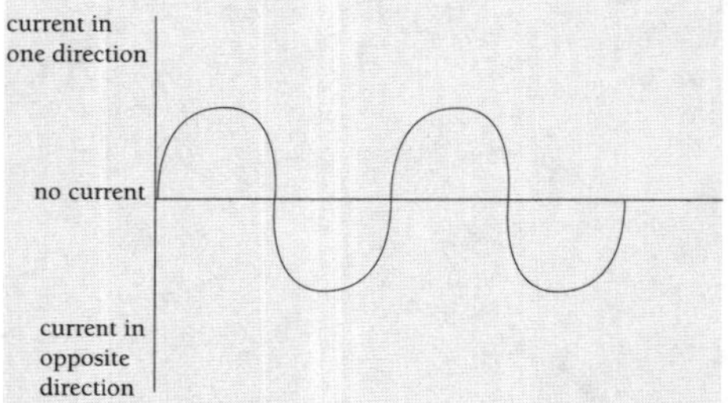

DC:

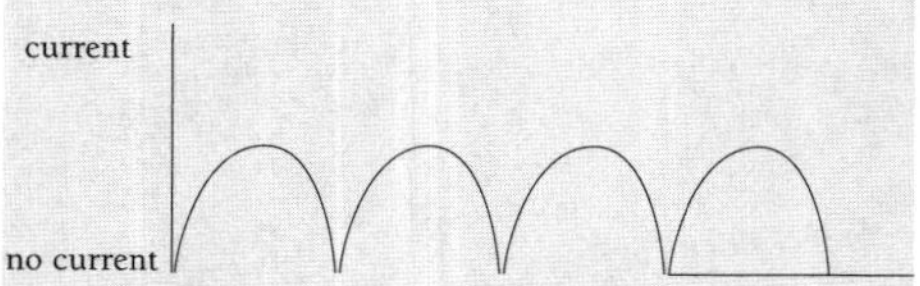

3. A current is produced in a wire only when the wire cuts through the magnetic field lines at a 90° angle. The wire must be moving in order for a current to be produced.

Toys for Understanding

Physics To Go

1. What is the purpose of:
   a) An electric generator?
   b) An electric motor?
2. How does a direct current differ from an alternating current? Use graphs to illustrate your answer.
3. In an electric generator, a wire is placed in a magnetic field. Under what conditions is a current generated?

Stretching Exercises

1. What is the meaning of "hertz," abbreviated "Hz," often seen as a unit of measurement associated with electricity or stereo sound components such as amplifiers and speakers?
2. What does it mean to say that household electricity has a frequency of 60 Hz?
3. Have you ever heard 60 Hz AC being emitted from a fluorescent light or a transformer?
4. Look at a catalog or visit a store where sound equipment is sold, and check out the "frequency response" of speakers—what does it mean?
5. Heinrich Hertz was a 19th-century German physicist. Find out about the unit of measurement named after him, and write a brief report on what you find.

532

ANSWERS

Stretching Exercises

1. Hertz (Hz) is the unit of measurement for frequency.
2. In electrical appliances, it is referring to the AC circuit, which in North America is on a 60 Hz frequency. It means that the circuit is changing directions 60 times per second.
3. The sound might be referred to as a low hum. It can be heard by high voltage power lines, as well as in-ground transformers.
4. In speakers the frequency is referring to the frequency of sound. 20 Hz is generally accepted as the lowest sound heard, and depending on the individual, the highest frequency is about 12,000 to 16,000 Hz. In sound Hz is referring to the number of vibrations per second. The greater the number of vibrations, the higher the frequency.
5. Student report.

© It's About Time

Activity 4 A: Demonstration

Falling Magnet

FOR YOU TO DO

Equipment (teacher demonstration):

- rigid copper pipe, 3/4-inch diameter, 6-foot length (approx.), sold at plumbing supply stores in 10 foot lengths.
- cow magnet (or a cylindrical magnet having a diameter less than the inside diameter of the copper pipe).
- clamps and supports to hold the copper pipe in a vertical orientation with some space between the ends of the pipe and the classroom floor and ceiling.
- stepladder.

As an extension of **What Do You Think?**, ascend the stepladder and announce that you are going to drop the magnet through the copper pipe. Ask students if a current should be induced in the pipe, reminding them of the effect of relative motion between a magnet and a conductor. Also ask what effect, if any, there should be on the free fall of the magnet, neglecting any small amount of friction if the magnet hits the wall of the pipe during its fall. Then drop the magnet through the pipe. Students will observe that the fall of the magnet is profoundly retarded.

Ask students why the fall of the magnet was retarded. As a hint, suggest that they think back to using the hand generator under "load" and "no load" conditions — does an induced current "fight back" against the action that causes the induced current [Lenz's Law]? How does the energy output of a generator compare to the energy input? If the magnet was inducing a current in the copper pipe during its fall, how does that explain the retarded fall of the magnet?

© It's About Time

For use with *Toys For Understanding*, Chapter 8, Activity 4: AC & DC Currents

AC Generator

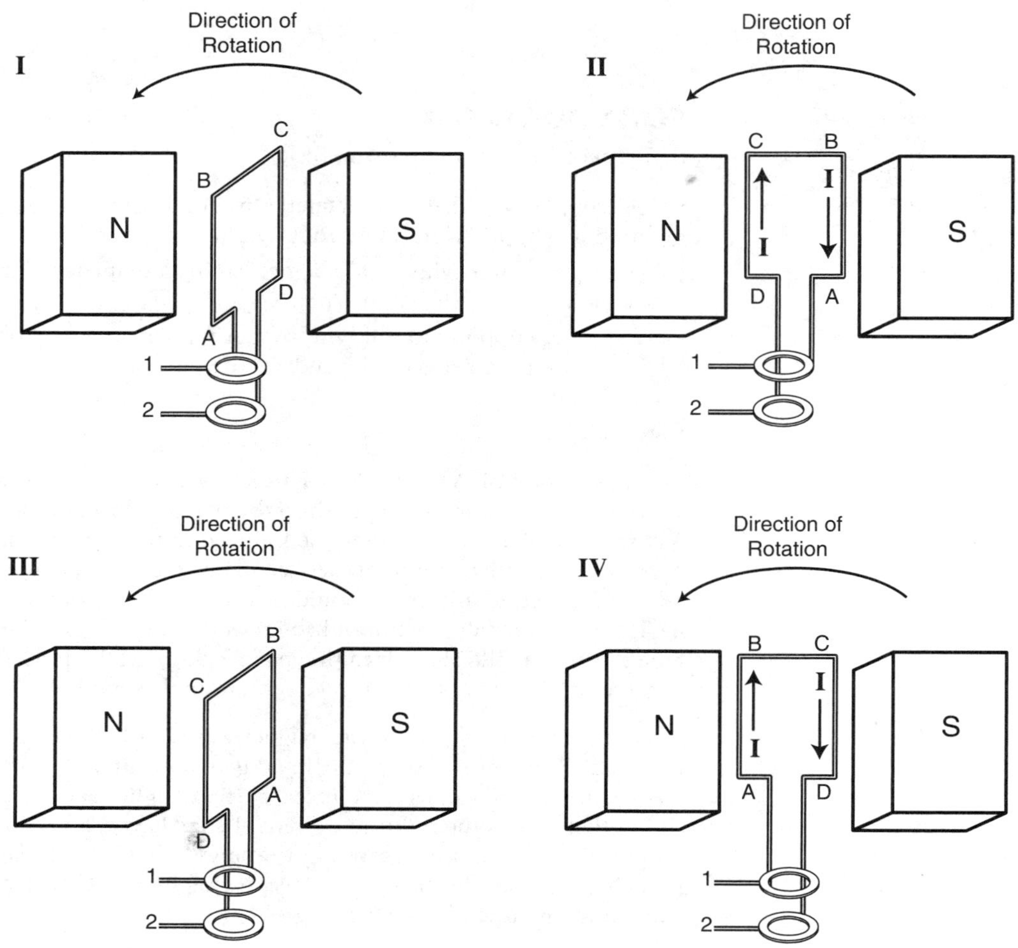

The current induced by an AC generator during one rotation of the generator coil

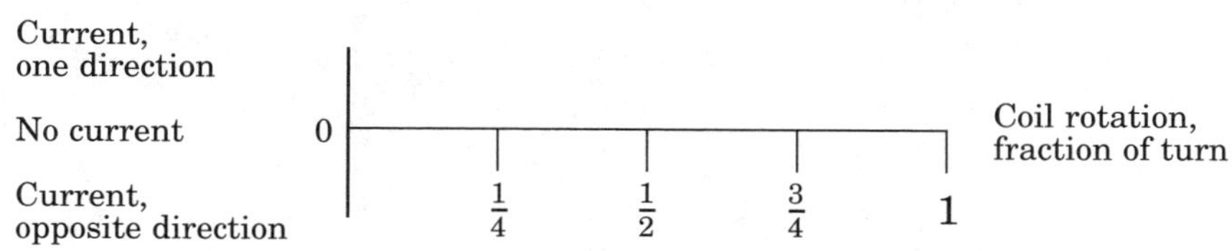

© It's About Time

For use with *Toys For Understanding*, Chapter 8, Activity 4: AC & DC Currents

DC Generator

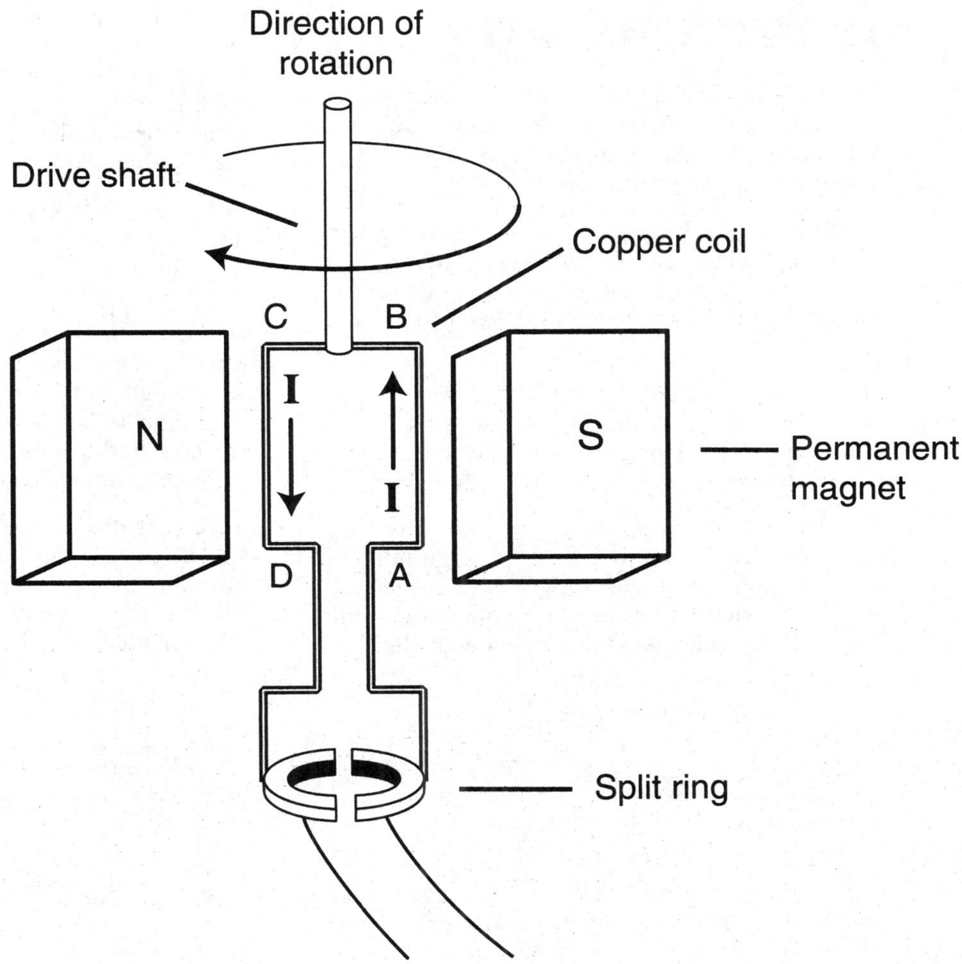

The current induced by a DC generator during one rotation of the generator coil

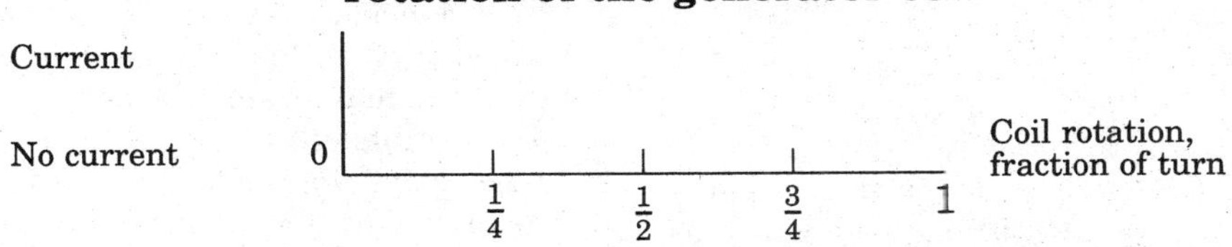

© It's About Time

For use with *Toys For Understanding*, Chapter 8, Activity 4: AC & DC Currents

ACTIVITY 5
Building an Electric Motor

Background Information

Principles of the electric motor are introduced in **Activities Five** and **Six**. The **Background Information** presented here will serve for both activities and will be limited to the DC motor.

A DC motor can be thought of as a DC generator "running backwards," and vice versa. Indeed, Gramme's discovery described in the student text for **Activity 5** shows that this is true in fact as well as principle.

How an electric motor converts electrical energy into mechanical energy is understood by considering what happens when an electric current travels in a region in which a magnetic field exists.

The diagram below shows a permanent "horseshoe" magnet which has a strong magnetic field in the region between the poles of the magnet. Magnet field lines extend from the north pole to the south pole of the magnet, as shown by the arrowheads on the field lines.

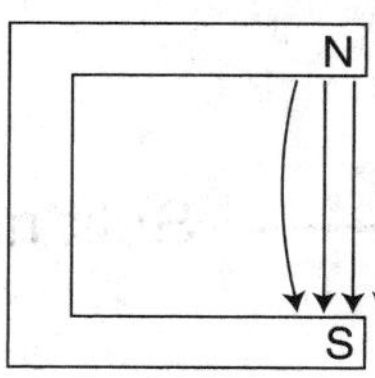

The smallest concentric circle in the diagram below represents a segment of wire viewed from one end. The "x" inside the smallest circle indicates that an electric current is flowing in the wire and has a direction away from the reader, into the page. The outer circles, with arrowheads, represent a magnetic field, caused by the current, surrounding the wire in the counterclockwise direction.

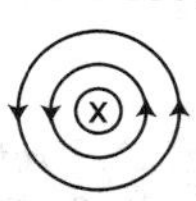

Below is shown a model which explains what happens when a current-carrying wire which has its associated circular magnetic field is placed in the magnetic field created by the permanent magnet.

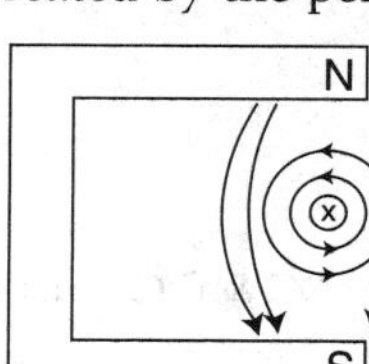

Directly to the left of the wire the magnetic field caused by the current has the same direction (downward) as the field established by the permanent magnet. Directly to the right of the wire the opposite is true; the field lines are in opposite directions. Lenz's Law suggests that forces arise to prevent changes in magnetic field lines, and, in this case, motion of the wire to the right would tend to preserve uniformity of the magnetic field; an electromagnetic force, *F*, would move the wire to the right in an attempt to move the high density field to the left of the wire to "fill in" the low density field to the right of the wire in the same way that "nature abhors a vacuum".

The amount of force on the wire in the simple case when the current passes perpendicularly through the field created by the permanent magnetic is:

$$F = BIL$$

where F is the force on the wire in Newtons, B is the strength of the magnetic field of the permanent magnet in Tesla, I is the current in the wire in amperes, and L is the length in meters along which the wire is embedded in the field.

A convenient "left-hand rule" for predicting the direction of the force in a current-carrying wire travelling through a magnetic field in a direction which is perpendicular to the direction of the magnetic field lines is shown in the diagram below:

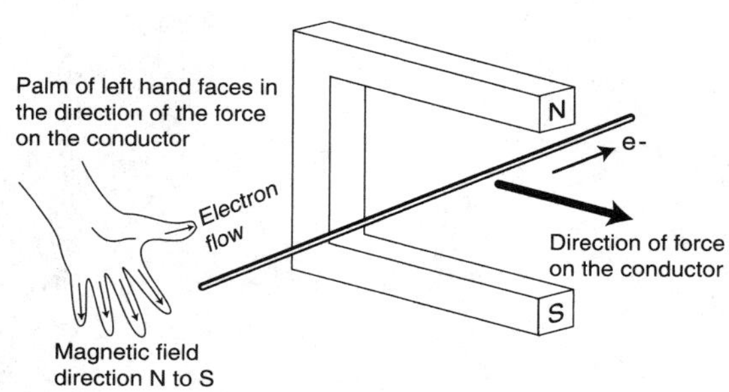

Left-hand motor rule

If the wire did indeed move to the right in the above position, it soon would exit the magnetic field, and the force would disappear. Therefore, electric motors are designed to employ a coil which rotates in a magnetic field. The above described motor effect can be applied in a "thought experiment" in which a current is fed into the coil of a DC generator to cause the coil to rotate. It is suggested that you satisfy yourself that motors and generators are basically the same in construction, the difference being that electrical energy is the output for the generator and is the input for the motor.

© It's About Time

Planning for the Activity

Time Requirements

Allow at least one 40-minute period. Students may want to explore different arrangements and configurations in order to speed up and slow down the motor, etc.

Materials Needed

For each group:

- #24 magnet wire (60 ft)
- alligator clip leads (2)
- battery, D-cell
- cup, Styrofoam® (6)
- hand generator (or power supply with a variable current)
- magnet, ceramic ring, refrigerator (2)
- rubber band #64
- safety pins, large (2)
- sandpaper, fine square (6)
- tape, masking (3/4" × 60 yds)
- test tube, plastic (2 × 10 cm)

Advance Preparation and Setup

Make a demonstration motor for the students to see that it is possible to build it, and show that it does work. Try to build it exactly the same way as it is in the book, and anticipate the different problems that the students might encounter.

Teaching Notes

Take the students back to the diagram of the DC generator in **Activity 4**, and use the diagram to develop an explanation for what happens when a DC current is fed to the armature through the split ring commutator/brushes. Essentially, the motor is running backwards. There may be a need to describe to the students what is happening with regard to the creation of a magnetic field around the wire, which interacts with the magnet of the generator. This creates the force by which the rotor is turned, thus creating the motor effect.

An alert student might observe that the design for the Basic Motor does not include split rings to reverse the current in the coil each 1/2 turn. That is true, and the motor, in principle, should not work. However, the motor does work, but only because lack of symmetry of the coil causes the coil to jump during each rotation, breaking and reestablishing contact to allow the coil to keep rotating. Indeed, students who do a very careful job of shaping the coil may find that their motor will not function, and the solution is to bend the coil a bit to cause some nonsymmetry.

If a student does not bring up the lack of provision for reversing the input current each 1/2 turn, you should bring it up for discussion at the end of the activity — it will be important to deal with it, because students will need to provide a split ring system for the motor/generator which they will design and build for the next activity.

SAFETY PRECAUTION: If students will be operating the hand generator, it is important to remind them of the danger of electrical shock, and the safety procedures required.

© It's About Time

NOTES

© It's About Time

Activity 5 Building an Electric Motor

GOALS

In this activity you will:

- Construct, operate, and explain a DC motor.
- Appreciate accidental discovery in physics.
- Measure and express the efficiency of an energy transfer.

What Do You Think?

You plug a mixer into the wall and turn a switch and the mixer spins and spins—a motor is operating.

- **How do you think the electricity makes the motor turn?**

Write your answer to this question in your *Active Physics* log. Be prepared to discuss your ideas with your small group and other members of your class.

For You To Do

1. Study the diagram on the following page closely. Carefully assemble the materials, as shown in the diagram, to build a basic electric motor. Follow any additional directions provided by your teacher.

Activity Overview

The students will be building a simple electric motor. The more powerful the magnet, the more dramatic are the results.

Student Objectives

Students will:

- Construct, operate, and explain a DC motor.
- Appreciate accidental discovery in physics.
- Measure and express the efficiency of an energy transfer.

ANSWERS FOR THE TEACHER ONLY

What Do You Think?

Students' answers will vary, but some will note that it has to do with the opposite of the generator effect. Some may even say something similar to what Faraday said, that if electricity can be made from moving a wire through a magnetic field, then why can't movement be created by a moving current in a magnetic field?

ANSWERS

For You To Do

1. Student activity.

Chapter 8

© It's About Time

ANSWERS

For You To Do
(continued)

2. a) Student data: changing polarity of the battery, more than one magnet, another magnet positioned at different locations around the wire.

3. a) Student observation. The orientation of the magnet needs to be opposite to the orientation of the other magnet in order for there to be magnetic field lines between the two magnets.

4. a) Again, the magnets must create magnetic field lines in order for there to be a motor effect.

5. a) Students' answers will vary. Changing the speed will involve an increase in the force that moves the wire. The force is influenced by the current, the magnetic field strength, and the length of the wire in the magnetic field.

6. a) Student data. They will probably discover that the faster you crank the generator the faster the motor will turn. Some other observations might be that at certain times the motor does not move. At this point, the motor is said to be in a dead zone. This is when the wire is in between the north and south poles so that the armature has an equal force acting on both sides, opposing motion.

7. a) Student answers will vary. Examples could include: washer, dryer, blender, grinder, food processor, can opener, electric razor, electric drill, electric fan.

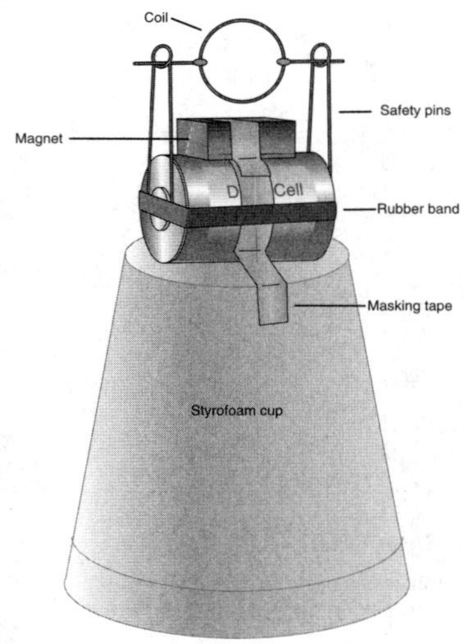

2. When your motor is operating successfully, find as many ways as you can to make the motor change its direction of rotation.
   a) Describe each way you tried and identify the ways that were successful.

3. Hold another magnet with your fingers and bring it near the coil from above, facing the original magnet, as the motor is operating.
   a) Describe what happens. Does the orientation of the second magnet make a difference?

4. Replace the single magnet with a pair of attracting magnets on top of the battery.
   a) What is the effect?

5. Think of other ways to change the speed of the motor. With the approval of your teacher, try out your methods.
   a) Describe ways to change the speed of the motor.

6. Use a hand generator as the energy source instead of the battery. You can disconnect the battery without removing it from the structure by placing an insulating material, such as a piece of cardboard, between the safety pin and the battery to open the circuit at either end of the battery. Then clip the wires from the generator to the safety pins to deliver current from the generator to the motor.
   a) Discuss what you find out.

7. Your motor turns! Chemical energy in the battery was converted to electrical energy in the circuit. The electrical energy was then converted to mechanical energy in the motor.
   a) List at least three appliances or devices where the motor spins.

© It's About Time

8. The spin of the motor occurs because the current-carrying wire has a force applied to it. You know if something moves, a force must be applied. As you observed, when the battery connection was broken, the motor stopped turning. You know from a previous activity that a current-carrying wire creates a magnetic field. Pause for a bit to remind yourself of the behavior of magnets. Take a bar magnet and place its north pole near a compass. The compass is a tiny bar magnet that can easily turn.
   a) Draw a sketch to show the orientation of the compass.

9. Shift the compass to the south pole of the bar magnet.
   a) Draw a sketch to show the orientation of the compass.

10. The north pole of the bar magnet repelled the north pole of the compass. The south pole of the bar magnet attracted the north pole of the compass. This attraction and repulsion is the result of a force on the compass. You can now investigate the force between the poles of a magnet and the magnetic field of a current-carrying wire.

8. a)–9. a)

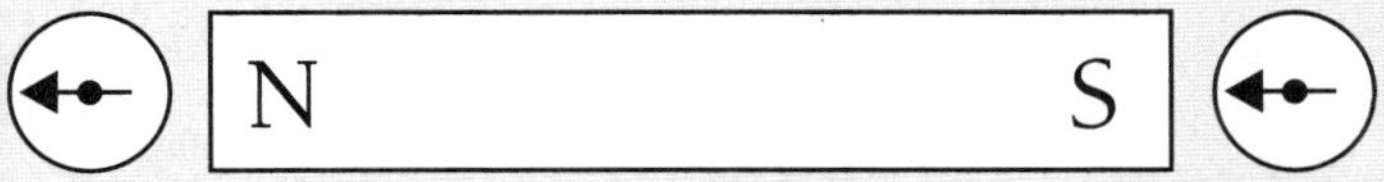

© It's About Time

ANSWERS

For You To Do
(continued)

11. a) The compasses would attract one another.

b) The compasses would repel one another.

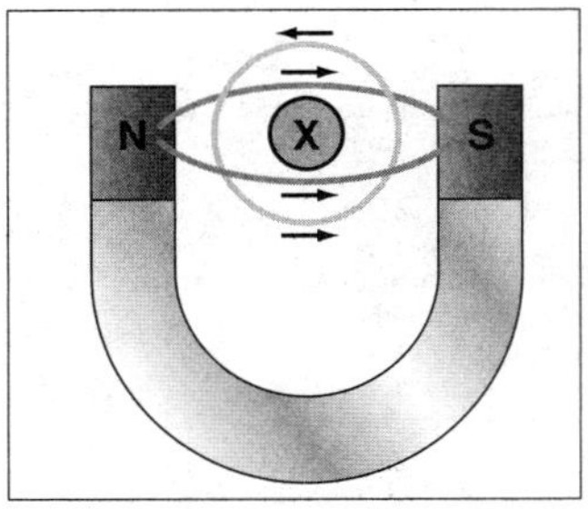

The magnetic field lines are drawn for a horseshoe magnet. The direction of the magnetic field lines is identical to a direction that a compass would point. The compass would point away from the north pole and toward the south pole. The magnetic field of the current-carrying wire is circular, as you investigated in an earlier activity. Compare the direction of this magnetic field to the direction of the magnetic field of the horseshoe magnet.

11. Think of the magnetic field lines above as small compasses.
   a) Write down whether the compasses above the wire attract one another or repel one another?
   b) Write down whether the compasses below the wire attract one another or repel one another?

12. This attraction/repulsion causes the wire to jump. There is a force on the wire. This force on the current-carrying wire is the basis for the electric motor that you built in this activity. The use of the loop of the wire allows the wire to rotate instead of jumping in the way a single wire would.

13. It is the moving electrons in the wire that create the current. In some TV sets, there is an electron beam that shoots the electrons from the back of the TV to the front. There are horseshoe magnets of a sort in the television. The moving electrons experience a force. The electrons' path is affected by the magnetic field. By varying the strength of the magnetic field, the electron beam can hit all parts of the screen and you receive a TV image.

© It's About Time

FOR YOU TO READ

The history of science is filled with discoveries that have led to leaps of progress in knowledge and applications. This is certainly true of physics and, in particular, electricity and magnetism. These discoveries "favor" the prepared mind. Oersted's discovery in 1820 of the magnetic field surrounding a current-carrying wire already has been mentioned. Similarly, Michael Faraday

Michael Faraday

discovered electromagnetic induction in 1831. Faraday was seeking a way to induce electricity using currents and magnets; he noticed that a brief induced current happened in one circuit when a nearby circuit was switched on and off. (How would that cause induction? Can you explain it?) Both Oersted and Faraday are credited for taking advantage of the events that happened before their eyes, and pursuing them.

About one-half century after Faraday's discovery of electromagnetic induction, which immediately led to development of the generator, another event occurred. In 1873, a Belgian engineer, Zénobe Gramme, was setting up DC generators to be demonstrated at an exposition (a forerunner of a "world's fair") in Vienna, Austria. Steam engines were to be used to power the generators, and the electrical output of the generators would be demonstrated. While one DC generator was operating, Gramme connected it to another generator that was not operating. The shaft of the inactive generator began rotation—it was acting as an electric motor! Although Michael Faraday had shown as early as 1821 that rotary motion could be produced using currents and magnets, a "motor effect," nothing useful resulted from it. Gramme's discovery, however, immediately showed that electric motors could be useful. In fact, the electric motor was demonstrated at the very Vienna exposition where Gramme's discovery was made. A fake waterfall was set up to drive a DC generator using a paddle wheel arrangement, and the electrical output of the generator was fed to a "motor" (a generator running "backwards"). The motor was shown to be capable of doing useful work.

© It's About Time

ANSWERS

Physics To Go

1. Students' answers will vary. Some advantages: increase the field strength of the magnet in order to increase the output from the motor; maintain a constant magnetic field in order to have more control of the speed. Some disadvantages: reduction in the efficiency of the engine reduction in the energy output; need for cooling of the motor as the current running through the electromagnet will heat up.
2. Student design.
3. Student response. Many of the different kinds of materials might be children's construction toys such as Mechano™, wheels, gears, elastic bands, pulleys, string, bearings, strips of wood, plastic wheels, gears etc., popsicle sticks, toothpicks, straws, etc.
4. An electric motor operates when an electric current is moving through a magnetic field.
As a moving current produces a magnetic field around the wire, it is either attracted to or repelled from the magnets in the motor, which forces the wires to move. This turns the armature.
A generator works on a similar principle in that a moving wire is passed through a magnetic field and a current is induced in the wire.

Reflecting on the Activity and the Challenge

Decision time about the **Chapter Challenge** is approaching for your group. In this activity you built a very basic, working electric motor. This is an important part of the **Chapter Challenge** However, knowing how to build an electric motor is only part of the challenge. Your toy must be fascinating to children. You must also be able to explain how it works.

Physics To Go

1. Some electric motors use electromagnets instead of permanent magnets to create the magnetic field in which the coil rotates. In such motors, of course, part of the electrical energy fed to the motor is used to create and maintain the magnetic field. Similarly, electromagnets instead of permanent magnets are used in some generators; part of the electrical energy produced by the generator is used to energize the magnetic field in which the generator coil is caused to turn. What advantages and disadvantages would result from using electromagnets instead of permanent magnets in either a motor or generator?
2. Design three possible toys that use a motor or a generator or both. One of these may be what you will use for your project.
3. The motor/generator you submit for the **Chapter Challenge** must be built from inexpensive, common materials. Make a list of possible materials you could use to construct an electric motor.
4. In the grading criteria for the **Chapter Challenge**, marks are assigned for clearly explaining how and why your motor/generator works in terms of basic principles of physics. Explain how an electric motor or generator operates.

© It's About Time

NOTES

© It's About Time

ACTIVITY 6
Building a Motor/ Generator Toy

Background Information

Background Information for this activity was provided in the background for **Activity 5**.

Active-ating the Physics InfoMall

Ideas for searches include "electric motor" and "Faraday." Most other useful information has already been found!

Planning for the Activity

Time Requirements

Allow at least 4 class periods, to collect, design, and build their toys. One class can be used for different groups to try the toys to see if they work and if they are fun. At least one more class will be needed to write instructions on how to build the motor/generator, and how it works in the toy.

Materials Needed

For each group:

- alligator clip leads (2)
- battery, D-cell
- magnet, small bar

Advance Preparation and Setup

You will need to decide how you will acquire the items needed for groups who wish to pursue alternate designs.

Teaching Notes

Set a time limit and communicate this to the class. You will have to set aside some time for testing the motor/generators, and for incorporating the motor in a toy. As these will be going to the HFE volunteers, the students will also need some time to prepare the instructions and explain the device to HFE inhabitants.

For assessment, you will need to decide how each group's products will be judged against the assessment criteria. Peer judging can be very effective, as long as the understanding is that they are fair. Have the students arrive at a consensus on exactly how the assessment will be done.

SAFETY PRECAUTION: Safety issues must be addressed. Working with electricity, can be dangerous (although the students are working with small voltages and amperages). There are some sharp objects involved, and students should be cautioned.

© It's About Time

NOTES

© It's About Time

NOTES

© It's About Time

Activity 6 Building a Motor/Generator Toy

GOALS

In this activity you will:

- Design, construct and operate a motor/generator.

What Do You Think?

You may have heard the following expression used before: "The difference between men and boys is the cost of their toys."

- **What characteristics make an item a toy?**

Write your answer to this question in your *Active Physics* log. Be prepared to discuss your ideas with your small group and other members of your class.

For You To Do

1. Confer within your group and between your group and your teacher about whether you will pursue, as a basis for the motor/generator kit for the **Assessment**, the motor design presented in this activity, an alternate design, or both. Whatever

Activity Overview

This activity is the culmination of **Chapter 8** and the the three chapters on the home (**Chapters** 6, 7, 8). Students will be using their knowledge of motors and generators, to build a toy. This must be a functional toy with the criteria as set out in the **Assessment** on page 544 and should be a fun activity for the students.

Student Objectives

Students will:

- Design, construct and operate a motor/generator.

ANSWERS FOR THE TEACHER ONLY

What Do You Think?

Students' answers will vary. Many of the responses will probably include characteristics such as fun, easy to work, durable, inexpensive, not predictable, etc.

© It's About Time

design(s) your group chooses to pursue, you are encouraged to be creative. Most designs can be improved in some way or another by substituting materials or making other changes. There is no single "best way" to go about designing the motor/generator and making it function within a toy or to produce electrical energy from another form of energy. The best way for your group is the way that the group can get the job done.

a) When you have decided on a design, submit your design to your teacher for approval.

2. In your group decide how you will make the motor/generator fascinating to children. You may wish to use some of the ideas you generated in answering the **What Do You Think?** question.

a) Record your ideas in your log.

b) Describe and make a sketch of your final design, and submit it to your teacher for approval.

3. Use the design for a DC motor as shown in the diagram as a basis to begin your construction. It can be adapted, as required, for the **Chapter Assessment**, to power a toy. Also as required, the motor could be adapted to be driven "backwards" by an external energy source to function as a DC generator.

[The motor design shown was adapted from the following public domain work: Educational Development Center, Inc., *Batteries and Bulbs II* (New York: McGraw-Hill, 1971), pp. 85-88.]

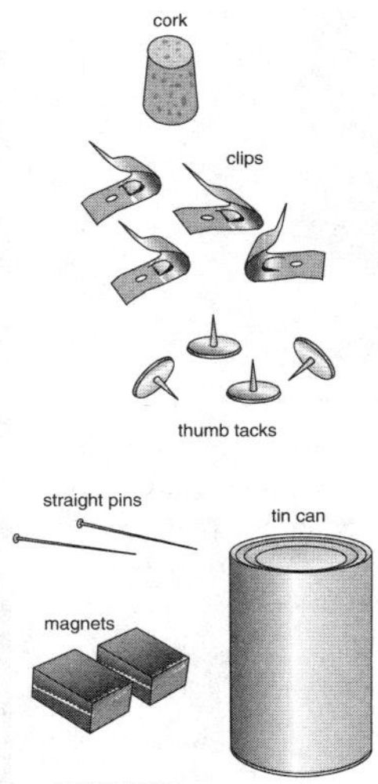

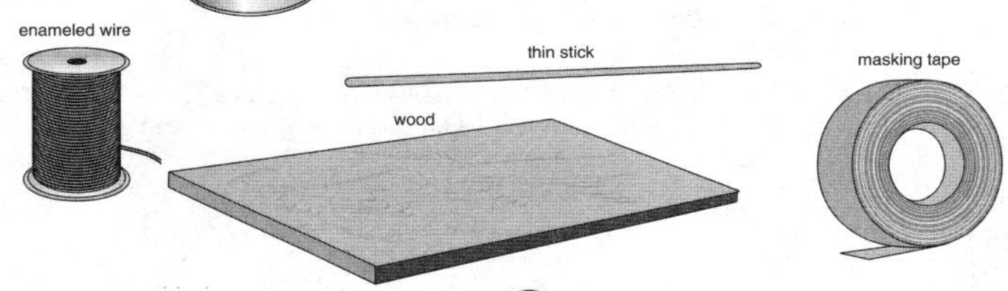

ANSWERS

For You To Do

1. a) Student design of the project.

2. a) Student responses.

 b) Student design and sketch.

3. Student activity.

© It's About Time

Activity 6 Building a Motor/Generator Toy

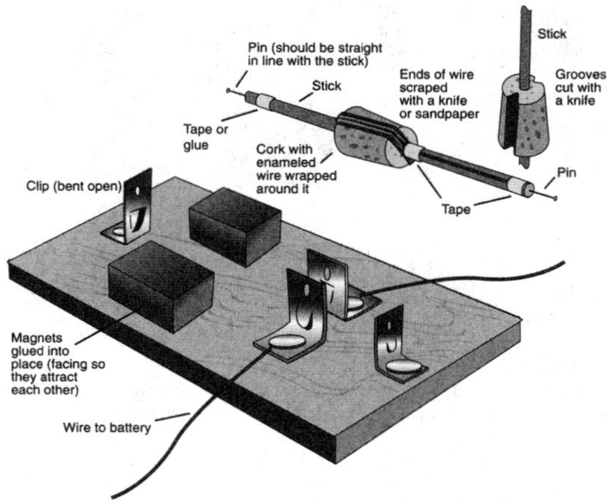

Reflecting on the Activity and the Challenge

You are now well on your way to completing the **Chapter Challenge**. You have decided on the design for your motor/generator and the toy it will power.

Physics To Go

Your assignment is to prepare to meet the criteria of the **Assessment** of the **Chapter Challenge**.

ANSWERS

Physics To Go

Build the motor/generator.

© It's About Time

PHYSICS AT WORK

Uriah Gilmore

HEADED FOR THE STARS

Uriah Gilmore loved to take electric appliances apart when he was growing up. "I couldn't always get them back together," he admits, "but I was so curious I couldn't help myself. I just had to see how they worked." Fortunately, Uriah's parents supported his curiosity.

Uriah and his fellow teammates from Cleveland, Ohio's East Technical High School recently won first place at the National High School Robotics Tournament at Epcot Center in Orlando, Florida for building a robot. "We were counseled along the way by engineers from NASA," he enthusiastically explains. "We called our robot Froggy and painted it green," Uriah continues, "and we used noisemakers so it even sounded like a frog." During the final contest "Froggy" was put in a pit with two other robots and had to place balls of a certain color in a specified area. The robot who got the most balls in won the contest.

"In my sophomore year the school I was attending closed and I went to East Technical High School which was the best thing that happened to me." He entered the engineering program and became a member of the engineering team—a team that is more popular than any sports team in his school.

Uriah attended Morehouse College on a NASA scholarship. "But," he states, "it's not enough to be a good student. You also have to be involved with your school and your community." Uriah once led a march on the Cleveland, Ohio, City Hall to protest a law which threatened to fire certain teachers, including one who inspired Uriah and was responsible for the revitalization of East Technical High School.

"My ultimate goal is to travel in space and explore the galaxy," he states. A shorter term goal is to be as involved in college as he has been in high school.

© It's About Time

543

Active Physical Science

Chapter 8

© It's About Time

Toys for Understanding

Chapter 8 Assessment

Your task is to prepare a kit of materials and instructions that a toy company will manufacture. Children will use these kits to make a motor or generator, or a combination electric motor/generator. It will serve both as a toy and to illustrate how the electric motors in home appliances work or how electricity can be produced from an energy source such as wind, moving water, a falling weight, or some other external source.

Review and remind yourself of the grading criteria that you and your classmates agreed on at the beginning of the chapter. The following was a suggested set of criteria:

- **(30%) The motor/generator is made from inexpensive, common materials, and the working parts are exposed but with due consideration for safety.**
- **(40%) The instructions for the children clearly explain how to assemble and operate the motor/generator device, and explain how and why it works in terms of basic principles of physics.**
- **(30%) If used as a motor, the device will operate using a maximum of four 1.5-volt batteries (D cells), and will power a toy (such as a car, boat, crane, etc.) that will be fascinating to children.**

OR

- **(30%) If used as a generator, the device will demonstrate the production of electricity from an energy source such as wind, moving water, a falling weight, or some other external source and be fascinating to children.**

Physics You Learned

Motors

Generators

Galvanometers

Magnetic field from a current

Solenoids

Electromagnets

Induced currents

AC and DC generators

Active Physics 544

© It's About Time

Alternative Chapter Assessment

Select the best response for each statement or question.

1. Normally, the magnetism of Earth causes a magnetic compass to point in the direction:

   a) North

   b) South

   c) East

   d) West

2. When a magnetic compass is placed near a wire carring an electric current, the direction in which the compass points is influenced by:

   a) only Earth's magnetism.

   b) only the wire's magnetism.

   c) neither Earth's magnetism nor the wire's magnetism.

   d) both Earth's magnetism and the wire's magnetism.

3. A generator transforms:

   a) electrical energy into mechanical energy.

   b) heat into mechanical energy.

   c) mechanical energy into electrical energy.

   d) mechanical energy into heat.

4. The output energy of a generator:

   a) is much greater than the input energy.

   b) is slightly greater than the input energy.

   c) is equal to the input energy.

   d) is less than the input energy.

5. The shape of the magnetic field caused by an electric current flowing in a straight wire is:

   a) along a straight line.

   b) square.

   c) circular.

   d) impossible to predict.

© It's About Time

6. Which factor or factors listed below affects the magnetic strength of a solenoid which is carrying an electric current?

   1.The amount of the current flowing in the solenoid.

   2. The number of turns of wire of the solenoid.

   3. The length along which the turns of wire are wound on the solenoid.

   a) Factor (1) only.

   b) Factor (2) only

   c) Factor (3) only

   d) Factors (1), (2) and (3).

7. If both the current and number of turns of wire per unit of a length of a solenoid are doubled, the magnetic strength of the solenoid should:

   a) remain unchanged.

   b) decrease by a factor of two.

   c) increase by a factor of two.

   d) increase by a factor of four.

8. Which choice of core material for a solenoid will result in the strongest electromagnetic if all other factors remain equal?

   a) wood.

   b) plastic.

   c) aluminum.

   d) iron.

9. If the number of turns of wire on a solenoid is doubled along a constant length of core material, and if the amount of current flowing in the solenoid is reduced to one-half the original amount, the magnetic strength of the electromagnetic will:

   a) remain unchanged.

   b) decrease by a factor of two.

   c) increase by a factor of two.

   d) increase by a factor of four.

10. A galvanometer is meant to be used to:

    a) detect electric currents.

    b) detect magnetic fields.

    c) measure frequency.

    d) measure the efficiency of a generator.

© It's About Time

11. The hertz is a unit of:

    a) energy.

    b) voltage.

    c) frequency.

    d) current.

12. Household electricity in the United States is:

    a) DC.

    b) direct.

    c) inverse.

    d) AC.

13. The number of complete cycles per second for household electricity in the United states is:

    a) 12

    b) 60

    c) 120

    d) 240

14. Split rings and brushes are used in a DC motor to:

    a) increase speed.

    b) increase efficiency.

    c) increase magnetic field strength.

    d) reverse current.

15. Using a conductor to cut through a magnetic field can result in:

    a) increased resistance.

    b) induced current.

    c) power loss.

    d) reduction in temperature.

16. When an electric generator is set in motion to produce electricity, the amount of energy needed to keep the generator going is:

    a) only the amount needed to overcome mechanical friction in the generator.

    b) greater than or equal to the energy output of the generator.

    c) impossible to predict.

    d) less than or equal to the energy output of the generator.

© It's About Time

17. When an electric motor is set in motion to produce electricity, the amount of energy needed to keep the motor going is:

a) only the amount needed to overcome mechanical friction in the motor.

b) greater than or equal to the energy output of the motor.

c) impossible to predict.

d) less than or equal to the energy output of the motor.

18. Which of the conditions listed below cause electric current to flow in a coil of wire when a magnet is near the coil of wire?

1. Only the magnet is moved.

2. Only the coil of wire is moved.

3. Both the magnet and the coil of wire are moved.

a) Condition (1) only.

b) Condition (2) only.

c) Condition (3) only.

d) All of the conditions.

19. If a direct current is fed into the coil of a generator which normally is used to produce DC electricity:

a) nothing will happen.

b) electricity will be generated as usual.

c) the generator will function as an electric motor.

d) it is not possible to predict what will happen.

20. An AC generator:

a) always produces AC having a frequency of 60 Hz.

b) produces direct current.

c) always produces 120-V AC.

d) produces one AC cycle during each turn of the coil.

© It's About Time

Alternative Chapter Assessment Answers

1. a
2. d
3. c
4. d
5. c
6. d
7. d
8. d
9. a
10. a
11. c
12. d
13. b
14. d
15. b
16. b
17. b
18. d
19. c
20. d

© It's About Time

NOTES

© It's About Time

Chapter 9

ATOMS ON DISPLAY

Chapter 9- Atoms on Display

National Science Education Standards

Chapter Challenge

Creating a science-museum exhibit provides the scenario for this chapter. Students are challenged to design an exhibit that gives visitors to the museum an understanding of the atom. Their exhibit must capture the visitor's attention within 30 seconds. Students will also be expected to submit a building plan and cost estimates.

Chapter Summary

To meet this challenge students will work on activities to learn about static electricity, Coulomb's Law, the quantization of electric charge, and the existence and relative size of the nucleus. They will study the Bohr and Schrödinger models of the atom and will be introduced to the wave/particle duality of both light and electrons. A simulation will provide students with an understanding of the stability of the nucleus, which further leads to a description of nuclear fission and fusion. They also will use Einstein's famous equation, $E = mc^2$, to calculate binding energies of nuclei. These experiences engage students in the following content from the *National Science Education Standards*.

Content Standards

Unifying Concepts

- System, order, and organization
- Evidence, models, and explanations
- Constancy, change, and measurement

Science as Inquiry

- Abilities necessary to do science inquiry
- Understanding about scientific inquiry

Science and Technology

- Abilities of technological design
- Understanding about science and technology

History and Nature of Science

- Historical perspectives

Science in Personal and Social Perspectives

- Personal and community health
- Natural and human induced hazards
- Science and technology in local, national, and global challenges

Physical Science

- Structure of atoms
- Motion and forces
- Conservation of energy
- Interactions of energy and matter

© It's About Time

Key Physics Concepts and Skills

Activity Summaries	Physics Principles
Activity 1: Static Electricity and Coulomb's Law Students begin by confronting what they presently know about atoms by commenting on a popular (though flawed) representation of the atom. They then investigate static electricity and Coulomb's Law to prepare them to understand the forces holding the atom together.	• Electrostatic forces • Coulomb's Law • Conservation of charge
Activity 2: Tiny and Indivisible In a simulation of Millikan's oil-drop experiment, students use inquiry to find the number of coins enclosed in a film canister. They then learn how related techniques were used to determine that electric charge is quantized.	• Charge is quantized • Millikan oil-drop experiment • Charge on an Electron
Activity 3: How Big Is Small? Using indirect measurements, students are able to accurately determine the size of a penny. They compare their statistical approach with direct measurement. Finally, they compare their experiment with Rutherford's experiment and are able to recognize the size of the nucleus in relation to the atom and the evidence we have for that knowledge.	• Indirect measurements • Size of the nucleus • Rutherford's scattering experiment
Activity 4: Hydrogen Spectra and Bohr's Model Students investigate the spectral lines from three gases using a diffraction grating and measure the wavelengths of light. They then learn about the Bohr model of the atom and are able to calculate the wavelengths of light as electrons jump from one orbit to another.	• Light spectra • Models of the atom • Conservation of energy
Activity 5: Extending and Amending the Model The wave and particle nature of light is explored by investigations of two-slit interference and the photoelectric effect. Electrons are restricted to a one-dimensional box so that their wave characteristics can be studied. A new interpretation of the Bohr orbits as standing waves of electrons is introduced, with a non-mathematical introduction of the Schrodinger equation.	• Diffraction of light • Interference of light • Wavelength of light • Photoelectric effect • Probability functions • Electron wavelength
Activity 6: Inside the Nucleus With a huge Coulomb repulsion pushing protons apart, the need for a strong attractive force in the nucleus is investigated. Students are then introduced to Feynman diagrams as a means of understanding how forces are transmitted.With a huge Coulomb repulsion pushing protons apart, the need for a strong attractive force in the nucleus is investigated. Students are then introduced to Feynman diagrams as a means of understanding how forces are transmitted.	• Neutrons • Isotopes • Nuclear forces • Feynman diagrams
Activity 7: Radioactive Decay and the Nucleus Students investigate the statistical properties of randomly tossing sugar cubes. This is then related to the radioactive decay of nuclei and the use of half-life as a clock. Students then complete nuclear equations for alpha, beta, and gamma decays.	• Radioactive decay • Half-life • Alpha, beta, gamma emission • Decay series of unranium
Activity 8: Holding the Nucleus Together Students calculate the mass defect and binding energies of nuclei by using Einstein's famous equation $E = mc^2$.	• $E = mc^2$ • Kinetic energy • Mass defect • Binding energy
Activity 9: Breaking Up Is Hard to Do After graphing the binding energy per nucleon curve for all of the elements, students explore nuclear fission and fusion. Chain reactions in fission reactions are also studied.	• Binding energy • Binding energy per nucleon graph • Fission • Fusion

© It's About Time

Equipment List For Chapter 9

PART	ITEM	QTY	ACTIVITY	TO SERVE
BP-0001	Pith Ball Electroscope	1	1	group
BP-1000	1000 mL Plastic Beaker	1	7	group
CM-1108	Calculator, Basic	1	3, 4, 5, 6, 7, 8	group
DH-9725	Diffraction Grating, 5 cm Square	1	5	group
FI-0007	Plastic Film Storage Containers W/Lids	10	2	group
FS-0001	Plastic Friction Rod W/Silk Pad	1	1	group
GF-0012	Glass Friction Rod	1	1	group
GS-7706	Graph Paper, Pad of 50 Sheets	1	7, 9	group
IC-1000	Interlocking Gram Cubes-Substituted for sugar cubes	1	7	group
LH-6534	Laser Pointer	1	5	group
MS-1425	Marker, Felt-Tip	1	7	group
RS-2826	Ruler, cm	1	3, 5	group
SP-1226	Cardboard Spectroscope	1	4	group
TF-0001	Fork, Tuning	1	5	group
TS-6059	Spool of Thread-To Restring Pith Ball Electroscope	1	1	group
WL-0404	Wool Pad	1	1	group
ZZ-0027	27 Qt Clear Plastic Container	1	1	group
TEACHER DEMONSTRATION ITEMS				
EL-4181	Electronic Scale	1	2	class
PS-6200	Spectrum Tube Power Supply	1	4	class
PS-6201	Hydrogen Spectrum Tube	1	4	class
HS-0001	Helium Spectrum Tube	1	4	class
NS-0001	Neon Spectrum Tube	1	4	class
THINGS NEEDED NOT SUPPLIED				
	Can of Aluminum Spray Paint To Spray Pith Balls	1	1	class
	UV Light Source		1	class
	Scotch Tape	1 Roll	1	class
	Pennies	100	2	class
	Nickels	100	2	class
	Piece of Paper	1	3, 5	class
	Pencil	1	3	class
	2 Pieces of Cardboard, Different Size Slits	1	5	group

© It's About Time

Organizer for Materials Available in Teacher's Edition

Activity in Student Text	Additional Material	Alternative / Optional Activities
ACTIVITY 1: Static Electricity and Coulomb's Law, p. 548	Conservation of Charge, p. 441	
ACTIVITY 2: Tiny and Indivisible p. 556	Millikan's Oil-Drop Experiment p. 449	
ACTIVITY 3: How Big Is Small? p. 561	Pencil-Drop Activity Card, p. 459 Rutherford's Scattering Experiment, p. 460	
ACTIVITY 4: Hydrogen Spectra and Bohr's Model p. 567	Lyman, Balmer, and Paschen Series p. 472	
ACTIVITY 5: Extending and Amending the Model p. 576	Diffraction of Light, p. 476	
ACTIVITY 6: Inside the Nucleus p. 587	Feynman Diagrams, p. 490	
ACTIVITY 7: Radioactive Decay and the Nucleus p. 595	Data Table for Activity 7, p. 498 Radioactive Decay of C-14, p. 499 Radioactive Decay of I-131, p. 500 Chart of Radioactive Decay of Uranium, p. 505	
ACTIVITY 8: Holding the Nucleus Together p. 605		
ACTIVITY 9: Breaking Up Is Hard to Do p. 612	Binding Energies per Nucleon, p. 520 Graph of Binding Energies, p. 522	

© It's About Time

Scenario

High-school students in Oregon walk into a giant mouth and down a cavernous esophagus into a spherical and moist room filled with liquids and loud sounds. An elementary student in California melds a picture of his or her face with that of a friend to see how pretty or ugly a composite face would be. Middle-school students in New York try to improve their "major-league" batting skills or their "professional" golf swing using computer video analysis. All of these students are visiting their science museums.

Science museums ensure that learning and fun are indistinguishable. They feature hands-on exhibits that stimulate, educate, and entertain. They provide exciting experiences that help visitors develop an understanding of science.

The visitors at a science museum can come and go as they please. Research has shown that there are only 30 seconds available at an exhibit to capture people's interest or they walk right by. Once they stop at the exhibit, you must get them involved. Often this involvement includes some kind of interaction with the exhibit. Some of the exhibits are targeted for a specific age group. Other exhibits are for a broad audience.

If you have never been to a science museum, it's worth a visit. If you are not able to get to a city that has a museum, you may want to visit one of the many virtual science museums on the Internet.

© It's About Time

NOTES

© It's About Time

Challenge

Your *Active Physics* class has been asked to develop an exhibit that will provide visitors to a science museum with an understanding of the atom.

The exhibit must:

- **include distinct features of the structure of the atom**
- **communicate the size and scale of the parts of the atom**
- **provide information on how the atom is held together**
- **capture the visitor's attention within 30 seconds**
- **include written matter that will further illuminate the concepts**
- **have a model, a T-shirt, a poster, a booklet, or a toy that can be sold at the museum store**
- **have a building plan and cost estimate for the exhibit**
- **include safety features**
- **be interactive—visitors should not merely read**

Criteria

You will be presenting your museum-exhibit plan to your class. As a class, decide how your work will be evaluated. Some required criteria are listed in the **Challenge**. Are there other criteria that you think are worth including?

How should each of the criteria be weighted? Should the building plan and cost estimate be worth as many points as the item for the museum store? Should the "30-second" criterion be worth more than anything else? Since the museum exhibit is supposed to educate the visitor, how much should the content be worth? Work with your class to agree on how many points should be assigned for each criterion. The total points should add up to 100.

Once you decide on the point allocation, you will have to decide how you can judge the assigned point value. For instance, assume that you chose the 30-second criterion to be worth 15 points. How will your class decide on whether your exhibit gets the full 15 points or only 10 points? It is worth knowing how each criterion will be judged so that you can ensure success and a good grade.

You will probably want to assign some strict guidelines and also leave room for some extra points. Learning how to judge the quality of your own work is a skill that all businesses expect to see in their professionals.

© It's About Time

Chapter and Challenge Overview

Science museums throughout the world must combine both science content and entertainment to be effective. Museum visitors have the option of walking by an exhibit if they are not captivated by what is presented. Research involving museum visitors demonstrate that the museum visitor will give a display 30 seconds of attention before moving along. For those visitors who do remain, the science content must be correct, interactive, engaging, and informative. The creation of a successful museum exhibit involves the collaboration of scientists, educators, artists, graphic designers, carpenters, electricians, and a host of others. This is a complex task and one that will challenge all students in your *Active Physical Science* classroom.

Students will learn about the structure of the atom in this chapter in order to communicate that knowledge in a successful museum exhibit. No one exhibit will be able to encompass all of the physics content presented in the chapter and student groups will have to choose which physics concepts will be their focus. All exhibits will have to meet certain criteria that museum directors are well aware of. All exhibits will have to capture the visitor's attention within the first 30 seconds. All will have to be accompanied by a building plan and cost estimates. All will be required to include safety features as well as be interactive. Finally, students will be able to further test their creativity and entrepreneurship by inventing a T-shirt, toy, poster, or similar artifact that can be on sale at the museum shop.

The chapter begins with a study of static electricity that leads to a description of Coulomb's Law. It is the electrostatic attraction of the positive nucleus and the negative electrons that holds atoms together and creates the stability of matter. In an effort to show the symmetry of the physics laws, Newton's Law of Gravitational Forces is compared with Coulomb's Law of Electrostatic Forces. Students then simulate the Millikan oil-drop experiment in order to learn about the quantization of electric charge. The experimental evidence for the existence of the nucleus and the relative size of the nucleus emerges from a simulation of the Rutherford scattering experiment.

Electron structure in the atom emerges from a look at atomic structure that then leads to a Bohr model of the atom, followed by a Schrödinger model of the atom. This more complex model requires introducing the students to the wave/particle duality of both light and electrons.

The final sections of the chapter move students to an investigation of the nucleus. The attraction of all the protons in the nucleus requires the introduction of a nuclear force. In another comparison of forces, Feynman diagrams are introduced to compare and contrast the electrostatic force and the nuclear force. The stability of the nucleus in radioactive decay is simulated through the toss of sugar cubes. The exploration of the stability of the nucleus then leads to descriptions of both nuclear fission and nuclear fusion. The chapter concludes with Einstein's famous equation $E = mc^2$ and its use in calculating binding energies of nuclei.

© It's About Time

ACTIVITY 1
Static Electricity and Coulomb's Law

Background Information

It is suggested that you read the **For You To Read** section for **Activity 1** in the student text before proceeding in this section.

Understanding the electrostatic force between charges is crucial for explaining how atoms are held together. It is also the basis of chemical bonding. There are two types of charges, positive and negative. Like charges attract and unlike charges repel.

In the everyday interaction of objects, it is the electrons – the negative charges – that get exchanged between objects. An excess of electrons accounts for a negative charge while a deficiency of electrons accounts for a positive charge.

There are few conserved quantities in nature. The conservation of electric charge is one of these few. The total charge in a system remains the same. The charges can be exchanged, but the total charge before any interaction must equal the total charge after the interaction.

Coulomb's Law describes the force between charges. The force is proportional to the charges and inversely proportional to the square of the distance between the charges. Charles Augustus Coulomb experimentally determined the equation for electric forces in 1784.

$$F = \frac{kq_1q_2}{d^2}$$

where F is the force in newtons (N)

q_1 and q_2 are charges in coulombs (C)

d is the distance between the centers of the charges in meters

k is Coulomb's constant and always equal to $9 \times 10^9\ \text{Nm}^2/\text{C}^2$.

The similarity between Coulomb's Law and Newton's Law of Gravitation hints at an underlying symmetry of nature. The investigation of symmetries like this (often on a more subtle mathematical level) drives much physics research.

Planning for the Activity

Time Requirements

Allow one class period for the activity and one class period for the calculations and use of Coulomb's Law.

Materials Needed

For the class:

- 1 roll Scotch® tape (1 roll)
- can of aluminum spray paint
- UV light source

For each group:

- 27-quart clear plastic container
- glass friction rod
- plastic friction rod w/silk pad
- pith ball electroscope
- spool of thread
- wool pad

Advance Preparation and Setup

You may wish to attach the string to the Styrofoam ball and suspend the ball before students arrive for class.

Teaching Notes

While other brands of type could be used, the Scotch brand gives the best and most consistent results.

The charging of the rubber rod and the glass rod requires lots of friction. If the experiments are done in a dry environment, they work well. If there is lots of humidity, it may be difficult to hold the charge.

You may want to have students try to create charges with other materials. Plastic strips in bags from dry cleaners work very well. A plastic comb brushed in hair produces a charge. Students can compare this to the rubber rod and ball and discover its charge.

The calculations involving Coulomb's Law should be approached in two ways. There is the direct substitution into the equation so that the force can be found given the charges and the distance between them. Remind students that the distance in

© It's About Time

the denominator of the fraction is squared. You may decide when reviewing problems to write the number twice in the denominator rather than using the "square" symbol.

The second approach is perhaps more important. It provides practice to the student in comparing forces when the charges or the distance between charges change. This requires the students to understand the equation and to see how changes in one variable affect the force. When you approach this type of analysis with the class, you should first have the students focus only on changes in one charge (e.g., if one charge is tripled, the force is tripled; if one charge is halved, the force is halved). The symmetry of the equation informs us that it does not matter which charge is changed. You can then explore what happens when both charges change (e.g., if you triple one charge and quadruple the other charge, the force is 3×4 or 12 times as large; if you triple one charge and halve the other charge, the force is $3 \times 1/2$ or 3/2 times as large).

After students are comfortable with changes in charges, you can then move to a discussion of how the force changes with changes in distance and no changes in charge. The students should first identify whether the distance change will result in a larger force (a smaller distance) or a smaller force (a larger distance). They can then find out how many times larger or smaller. They compare the new distance with the original distance (e.g., if the distance is tripled, the force will be smaller by a factor of $3 \times 3 = 9$).

After success with changing the charges or changing the distance, you can then challenge your students with changes in both the charges and the distances.

NOTES

© It's About Time

Activity Overview

Students begin by confronting what they presently know about atoms by commenting on a popular (though flawed) representation of the atom. They then investigate static electricity and Coulomb's Law to prepare them to understand the forces holding the atom together.

Student Objectives

Students will:

- Charge a ball with a positive charge.
- Charge a ball with a negative charge.
- Observe that like charges repel and unlike charges attract.
- Calculate the electrical force using Coulomb's law.
- Apply the organizing principle of conservation of charge.
- Recognize the similarities and differences between Coulomb's law and Newton's Law of Gravitation.
- Explain how Coulomb was able to measure Coulomb's constant using a torsion balance.

ANSWERS FOR THE TEACHER ONLY

What Do You Think

In *Atoms on Display,* this section is intended to elicit students' prior understandings. Students are supposed to share their answers. The role of the teacher is to listen and clarify. The teacher is not supposed to correct or modify answers. Most certainly, the teacher is not supposed help the students reach closure on what is the "right" answer to the question.

The popular diagram of the atom has a number of problems that the students will become aware of while learning the content of the chapter. The scale of the nucleus to the atom is exaggerated. The electrons are not little balls. The atom has no color. The electrons do not revolve about the nucleus in fixed elliptical orbits.

Atoms on Display

Activity 1 Static Electricity and Coulomb's Law

GOALS

In this activity you will:

- Charge a ball with a positive charge.
- Charge a ball with a negative charge.
- Observe that like charges repel and unlike charges attract.
- Calculate the electrical force using Coulomb's Law.
- Apply the organizing principle of conservation of charge.
- Recognize the similarities and differences between Coulomb's Law and Newton's Law of Gravitation.
- Explain how Coulomb was able to measure Coulomb's constant using a torsion balance.

What Do You Think?

"Everyone" has heard of atoms, but no one has ever seen an atom. Look at the sketch below of an atom that you often see in advertisements and some science books.

- **How would you describe what is shown in the sketch of an atom?**
- **Are there any problems with this depiction of the atom? Explain your answer.**

Record your ideas about these questions in your *Active Physics* log. Be prepared to discuss your responses with your small group and your class.

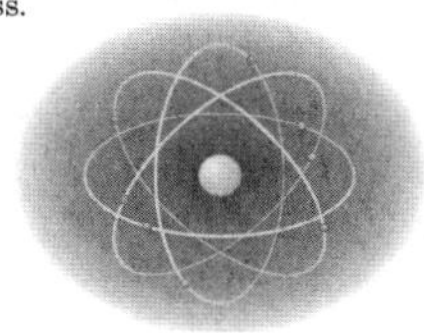

Active Physics 548

© It's About Time

For You To Do

1. Have you ever seen a tremendous lightning storm? Bolts of lightning ignite the sky as they streak toward the Earth. A tiny "lightning storm" also takes place when you get an electric shock. Think back to the last time you got a shock.
   a) Were you inside or outside?
   b) Was it winter or summer?
   c) What did you touch to get the shock?
   d) Include in your *Active Physics* log any additional details.

2. The study of lightning, shocks, and "static cling" can reveal important physics. Cut two strips of Scotch® Magic™ tape about 12 cm long. Bend one end of each strip under to form a tab. Place one strip sticky side down on a table and label the tab B, for "bottom." Place the other strip sticky side down on top of the first strip and label the tab T, for "top." With one hand, peel off the top strip using the tab. Then pick up the bottom strip with the other hand. Hold both strips apart, allowing them to hang down. Slowly bring the hanging strips toward each other, but do not let them touch.

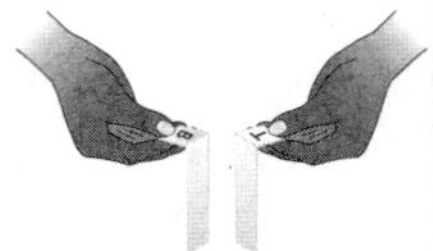

   a) Record your observations.
   b) If the strips accelerated (moved) toward or away from each other, Newton's Second Law tells you that there must be a force. (Newton's First Law tells you that an object at rest will tend to stay at rest. Newton's Second Law says that if an object accelerates, it's being acted on by a force.) Was there a force present?
   c) If a force was present, was it attractive or repulsive? Explain.

3. Make a second set of strips as in **Step 2**.
   a) Predict what you think will happen if the two top strips are picked up, one from each set, and brought toward each other. Record your prediction in your *Active Physics* log.

   Pick up the two top strips by the tabs, allowing both strips to hang down. Slowly bring them toward each other.
   b) Record your observations.
   c) Was the force attractive or repulsive? Explain.
   d) Predict what you think will happen if the two bottom strips of tape are picked up and brought toward each other. Record your prediction.

ANSWERS

For You To Do

1. a) Students probably received electric shocks when they were inside.
   b) It was probably in the winter (when the air is dry).
   c) They may have touched a doorknob, a light switch, or another person.
   d) Answers will vary.
2. a) The two strips will attract each other.
   b) Yes, they accelerated towards each other. There was a force.
   c) The force was attractive.
3. a) Encourage students to record their predictions.
   b) The two strips repel each other.
   c) The force was repulsive.
   d) Students record their prediction.

© It's About Time

ANSWERS

For You To Do
(continued)

3. e) The two strips also repel each other.
   f) The force was repulsive.
4. a) The rod first attracts the ball. It then touches the ball and the ball is repelled. The ball continues to be repelled as the rod is brought closer.
5. a) The charged ball is attracted to the glass rod.
6. a) Some of the electrons (negative charges) were transferred to the ball.
   b) Assuming that the wool was initially neutral, if some of the electrons (negative charges) left the wool, the wool would have become positively charged.
   c) The rubber rod can be rubbed in the wool and then used to charge the ball by sharing the negative charge. The rubber rod can be rubbed in the wool again and the wool can be quickly brought near the negative ball to see if it attracts the negative ball.
7. a) If the glass rod transferred electrons *to* the silk, then the glass rod becomes positively charged.
   b) If the silk gained those electrons, it would have become negative.
   c) Immediately after rubbing the glass rod in the silk, we could place the silk next to a negatively charged ball and see if it is repelled.

Pick up the two bottom strips by the tabs, allowing both strips to hang down. Slowly bring these toward each other.

e) Record your observations.

f) Was the force attractive or repulsive? Explain.

4. Charge up a rubber rod by vigorously rubbing it with a piece of wool. Use lots of friction to get the rubber rod charged up. Touch the rod to a small Styrofoam ball coated with a conducting paint. Observe the interaction of the rod and the ball.

   a) In your *Active Physics* log describe the interaction you observe.

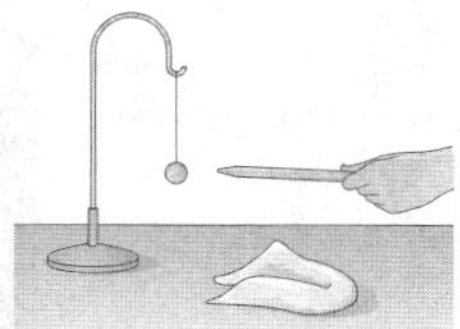

5. Charge up a glass rod by rubbing it with a piece of silk. Bring it toward the charged ball but try not to let them touch. Observe the interaction of the glass rod and the ball.

   a) In your *Active Physics* log describe the interaction you observe.

6. Assume that the rubber rod gained electrons (negative charges) and became negatively charged when rubbed with the wool.

   a) What do you think could have happened to some of the excess electrons on the rubber rod when it touched the Styrofoam ball? Use your observations to justify your answer.

   b) If the rubber rod got the excess electrons from the wool, what would be the wool's charge?

   c) Explain how you could check to see if your answer to **(b)** is correct.

7. Assume that the glass rod transferred some electrons to the silk.

   a) What would be the charge of the glass rod? Use your observations to justify your answer.

   b) If the silk gained electrons, what would be the silk's charge?

   c) Explain how you could verify your answer to **(b)**.

8. Get rid of any excess electrons on the Styrofoam ball by touching it. (This is made more effective if you moisten your finger. The slight moisture helps remove the charge.)

© It's About Time

Removing electrons to make an object neutral is called grounding. Adding electrons to make an object neutral is also called grounding. The "ground" is able to supply or remove electrons to make an object neutral. In this situation you are connected to the "ground."

9. Negatively charge the rubber rod. Bring it close to the neutral Styrofoam ball, but do not touch the rod to the ball.
   a) Describe your observations.
   b) Like charges repel, and unlike charges attract. How would you explain the interaction between the charged rubber rod and the uncharged ball? (Hint: The electrons in the ball are able to move about the ball so that some are able to move closer to the rubber rod or further from the rubber rod.) Use a diagram to help you explain your observations.

Physics Words

Coulomb's Law: the relationship among electrical force, charges, and the distance between the charges.

FOR YOU TO READ

Coulomb's Law

In physics, scientists often search for an equation that can provide a clearer, more precise description of what they observe. Charles Augustin de Coulomb experimentally determined the equation for electric forces in 1784.

$$F = \frac{kq_1q_2}{d^2}$$

where F is the force in newtons (N),

q_1 and q_2 are charges in coulombs (C),

d is the distance between the centers of the charges in meters (m), and

k is Coulomb's constant, always equal to $9 \times 10^9\ \text{Nm}^2/\text{C}^2$.

Sample Problem

Two small charged spheres are placed 0.2 m apart. The first sphere has a charge of $+3.0 \times 10^{-6}$ C and the second sphere has a charge of -4.0×10^{-6} C. Calculate the force between them.

Strategy: You can use the equation for electric forces to calculate the force between the spheres.

Givens:

$q_1 = +3.0 \times 10^{-6}$ C

$q_2 = -4.0 \times 10^{-6}$ C

$d = 0.2$ m

Solution:

$$F = \frac{kq_1q_2}{d^2}$$

$$F = \frac{(9 \times 10^9\ \text{Nm}^2/\text{C}^2)(+3.0 \times 10^{-6}\ \text{C})(-4.0 \times 10^{-6}\ \text{C})}{0.2\ \text{m}^2}$$

$F = -2.7$ N

The negative sign indicates that the force is attractive.

→

ANSWERS

For You To Do

(continued)

8. Students ground the ball by touching it.

9. a) The negatively charged rubber rod will attract the neutral ball.

   b)

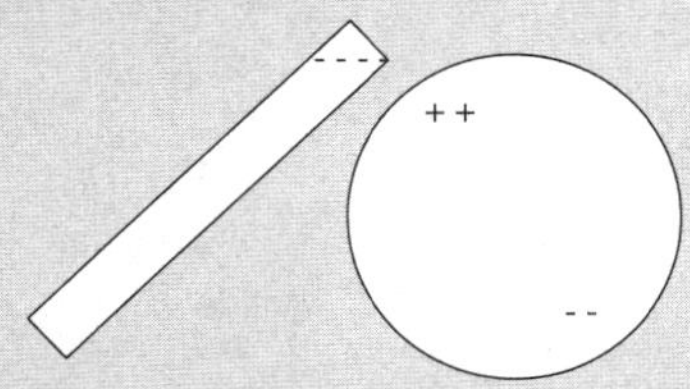

The negative rod repels the electrons in the ball to the far end away from the rod. The near side of the ball becomes positive. The near side of the ball and the rod attract each other.

© It's About Time

Static Charges and Forces

In this activity, you noticed that a force was present between the pieces of tape. It's an invisible force. You couldn't see the force, but you saw evidence that it was there because it moved the tape. It's tough to believe in invisible things, but since this force had (and consistently does have) visible and measurable effects, it would be tougher not to believe in it.

You also observed that the force could be attractive or repulsive. One explanation might be that there are two types of charges and that the interaction of the charges creates a force. If the two pieces of tape have the same charges, the force is repulsive. If the two pieces of tape have the opposite charges, the force is attractive. An alternative explanation is that there is one type of charge. If the two pieces of tape have an excess amount of charge, there is repulsion. If the two pieces of tape have a deficiency of charge (they've each lost some charge), there is repulsion. If one piece of tape has a deficiency and the other piece of tape has an excess of the charge, there is attraction.

You then found evidence for forces that you could attribute to excess charges. The rubber rod removed excess charges from the wool. These excess (negative) charges were shared with the Styrofoam ball. The rod and the ball now had the same charge and you observed a force of repulsion that pushed the ball and rod apart from each other. When the glass rod, which had a deficiency of (negative) charges, was placed near the (negatively) charged ball, there was a force of attraction.

In the 1700s Benjamin Franklin performed experiments similar to those you did in this activity. He defined which charge would be called positive and which would be called negative. A rubber rod rubbed in wool gains a negative charge. A glass rod rubbed in silk becomes positively charged. The glass rod did not gain positive charges but became positive by losing negative charges. Before Benjamin Franklin, investigators supposed that, since there was both attraction and repulsion, there were two kinds of electricity.

Conservation of Charge

A rubber rod gains electrons, while a piece of wool loses electrons. A glass rod loses electrons, while a piece of silk gains electrons. These situations are good examples of one of the major organizing principles of physics—the conservation of charge. In any system, the total amount of charge must remain constant. If negative charge is removed from the wool, that negative charge must go somewhere. For example, if 15 bits of negative charge are removed from the wool, then 15 bits of negative charge are transferred to the rubber rod. There are only a few quantities discovered that nature conserves—charge is one of them.

© It's About Time

Activity 1 Static Electricity and Coulomb's Law

Perhaps without being aware of it, you used the concept of conservation of charge in your analysis of what you observed.

A few simple problems illustrate this conservation law. When two identical spheres touch, they will end up with identical charges. (If they did not end up with identical charges, then there would be a way to distinguish them. They would not be identical.) One sphere with a charge of –10, touches a neutral, identical sphere. After touching and separating, each sphere has a charge of – 5. The charge was transferred, but the total charge remained the same.

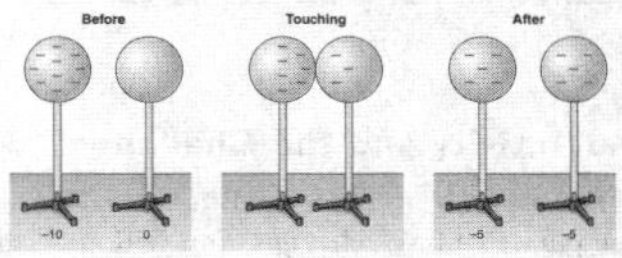

Keep in mind that neutral matter already contains enormous amounts of pluses and minuses, but in equal amounts. Thus the "excess" and "deficiency" are relative to a baseline of starting out neutral.

Coulomb's Law and Newton's Law of Gravitational Attraction

You found you could actually calculate the force of attraction or repulsion of charges by using **Coulomb's Law**:

$$F = \frac{kq_1q_2}{d^2}$$

Here's an incredible part of physics that you did not examine in this activity. Coulomb's Law for electrostatic attraction and repulsion is very similar to **Newton's Law of Gravitational Attraction**!

$$F = \frac{Gm_1m_2}{d^2}$$

where F is the force in newtons (N),

m_1 and m_2 are masses in kilograms (kg),

d is the distance between the centers of the masses in meters (m), and

G is the gravitational constant, always equal to 6.67×10^{-11} Nm²/kg².

Look at the similarities:

- Both laws show forces that decrease with the square of the distance.
- Both laws show forces that depend on the product of the masses or charges.
- Both laws have constants.

Look at the differences:

- Electric forces are attractive and repulsive; gravitational forces are only attractive.
- Charges come in two varieties, + and –. Mass comes in one variety, +.
- The electric force constant is quite large, while the gravitational force constant is quite small.

If you look at the gravitational and electrical forces between two electrons, the gravitational force is much, much smaller. It is so small, you don't even take it into account when describing the behavior of the charges.

The experimental techniques to find the value of k and G are quite similar. In Coulomb's experiment, two spheres were attached to the ends of a rod and the rod was suspended by a wire. These spheres were charged, and similarly charged spheres were brought near the ends of the rods. The repulsive force caused the wire to twist. The twist was a measure of the force, and Coulomb was able to verify his law.

→

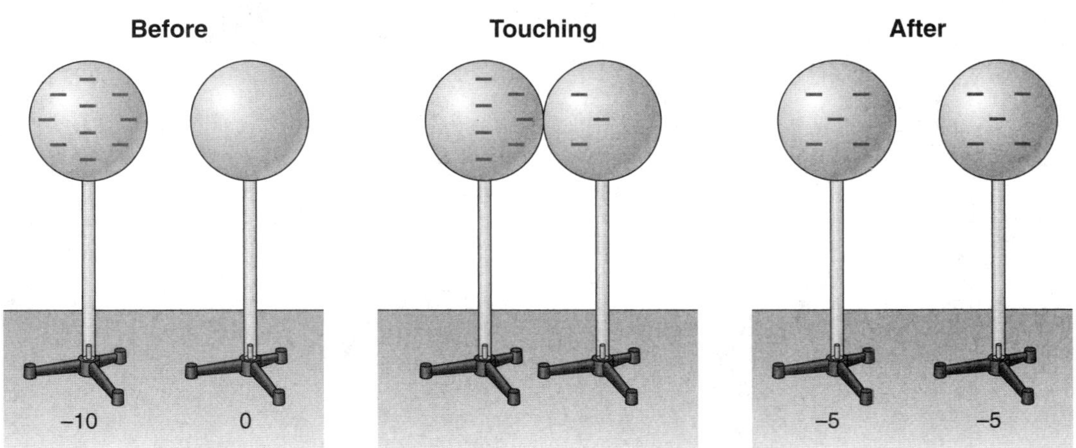

© It's About Time

Chapter 9

ANSWERS

Physics To Go

1. a) The material will become negative.

   b) The rod will become positive.

2. The rubber rod picked up electrons from the wool. The excess electrons on the rod made it negative. The deficiency of electrons on the wool made it positive.

3. The total charge before contact is – 1 + – 3 = – 4. The total charge after contact must also be – 4. Since the spheres are identical, they will each get a charge of – 2.

4. Total charge = + 1 + – 5 = – 4. Each sphere will have a charge of – 2 after touching.

Cavendish's experiment was similar, but the attraction between the pairs of spheres was due to their gravitational attraction. This tiny force was measured by the twist in the wire.

This symmetry of what appears to be two unrelated forces provides a glimpse into a beauty of our world. Physicists remark on this beauty. It drives them to try to find out if there are other underlying understandings of the two forces because of that symmetry. This is what physicists are exploring when you hear about their work on unified theories.

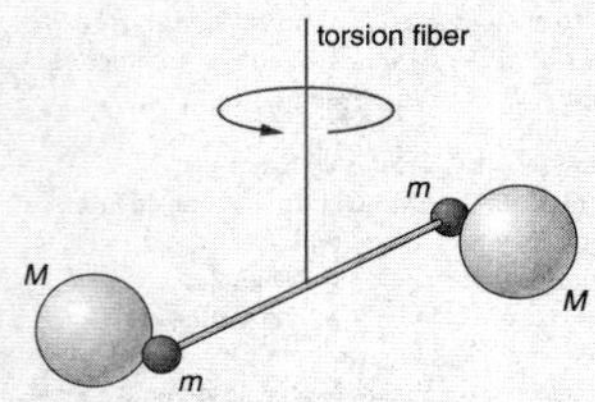

Reflecting on the Activity and the Challenge

You probably know that atoms are composed of electrons and other particles. These particles have charges. You will need to provide a description of the interaction between the charges when you provide a description of the atom. You may find a way to include in your exhibit the larger concepts of conservation of charge or the ability to actually calculate these forces of attraction and repulsion.

Physics To Go

1. Electrons are transferred from a rod to a piece of material.
   a) Which object will become negatively charged?
   b) Which object will become positively charged?
2. A rubber rod is negatively charged after being rubbed by wool. Explain how this happens.
3. Two identical spheres are mounted on insulated stands. The first sphere has a charge of –1. The second sphere has a charge of – 3. After the spheres touch, what will the charge on each be?
4. One of two identical metal spheres has a charge of +1 and

Physics Words

Newton's Law of Gravitational Attraction: the relationship among gravitational force, masses, and the distance between the masses.

© It's About Time

the other sphere has a charge of – 5. Compare the total charge on the spheres before and after contact.

5. Charge A is $+2.0 \times 10^{-6}$ C and charge B is $+3 \times 10^{-6}$ C. The charges are 3 m apart. What is the force between them? Is it attractive or repulsive?

6. Charge A is -4.0×10^{-6} C and charge B is $+2 \times 10^{-6}$ C. The charges are 5 m apart. What is the force between them? Is it attractive or repulsive?

7. When the air is dry and you walk on a wool carpet with your shoes, you may experience a shock when you touch a doorknob. Explain what is happening in terms of electric charge. (Hint: Your shoes are similar to the rubber rod.)

8. Compare and contrast Coulomb's Law and Newton's Law of Gravitational Attraction. Provide at least one similarity and one difference.

9. Coulomb's Law states that the electric force between two charged objects decreases as the square of the distance increases. Suppose the original force between two objects is 60 N, and the distance between them is tripled; the new force would be 3^2, or 9 times weaker. The new force would be 60 N ÷ 9 = 6.7 N or 7 N.

   Find the new forces if the original distance was:

   a) doubled
   b) quadrupled
   c) halved
   d) quartered

10. Sketch a graph that shows how the electrostatic force defined by Coulomb's Law varies with the distance.

11. A single electron has a charge of 1.6×10^{-19} C. Show why it takes 6.25×10^{18} electrons to equal 1 C.

12. How could you depict the invisible electrostatic force in a museum exhibit?

ANSWERS

Physics To Go *(continued)*

5. $F = kq_1q_2/d^2$

   $F = (9 \times 10^9 \text{ Nm}^2/\text{C}^2)(+2.0 \times 10^{-6} \text{ C})(+3 \times 10^{-6} \text{ C})/(3 \text{ m})^2$

   $F = +6 \times 10^{-3}$ N

   The force is repulsive (like charges repel).

6. $F = kq_1q_2/d^2$

   $F = (9 \times 10^9 \text{ Nm}^2/\text{C}^2)(-4.0 \times 10^{-6} \text{ C})(+2 \times 10^{-6} \text{ C})/(5 \text{ m})^2$

   $F = -2.9 \times 10^{-3}$ N

   The force is attractive (unlike charges attract).

7. The rubber soles of your shoes pick up electrons from the wool carpet. This makes your shoes negative. Since you are a good conductor of electric charge, the negative charge spreads throughout your body. When you touch the doorknob, you are grounded and the charge leaves your body at once producing a shock.

8. They are similar because:

   They both decrease with the square of the distance between the objects.

   They both have constants.

   They are different because:

   The Coulomb constant is much greater than the gravitational constant.

   Electrical forces are attractive and repulsive while gravitational forces are only attractive.

   Masses come in one variety while charges come in two varieties (+/–).

9. a) If distance is doubled, the force is 1/4: 60/4 = 15 N

   b) If distance is quadrupled, the force is 1/16: 60/16 = 3.75 N

   c) If the distance is halved, the force is 4 times as large: 60 × 4 = 240 N

ANSWERS

Physics To Go *(continued)*

   d) If the distance is quartered, the force is 16 times as large: 60 × 16 = 960 N

10.

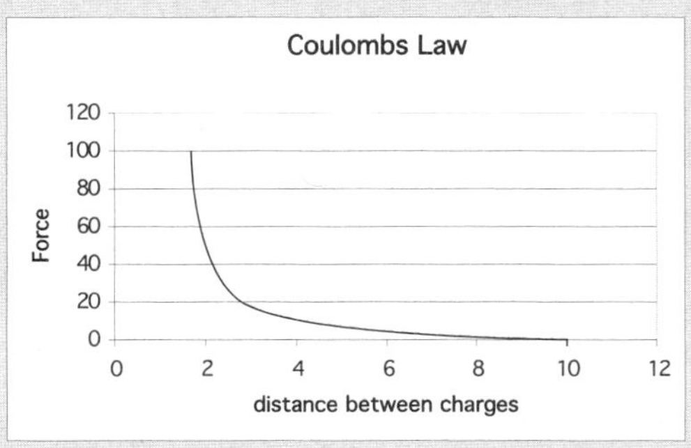

11. (1.6×10^{-19} C/electron) × (6.25×10^{18} electrons) = 1.0 C

12. Answers will vary. Encourage students to be creative.

© It's About Time

Chapter 9

NOTES

© It's About Time

ACTIVITY 2
Tiny and Indivisible

Background Information

It is suggested that you read the **For You To Read** section of the student text for **Activity 2** before proceeding in this section.

Electric charge is quantized. You can have 1 charge, 2 charges, 3 charges, etc., but you can never have fractional charges such as 1.4 charges or 2.7 charges. This is quite different from other quantities studied in physics. Mass does not seem to be quantized – any mass is permissible. Time does not seem to be quantized – any time is permissible. Charge, however, comes in units that are multiples of 1.6×10^{-19} C.

Millikan found out that charge was quantized experimentally by suspending a charged oil droplet within an electric field. The electric force up could be manipulated by varying the voltage between the plates of the capacitor. When the electric force was equal to the gravitational force, the oil drop would be at rest or would move up or down with a constant velocity. The gravitational force or weight of the oil droplet could be determined by watching it fall without an electrical force. When the droplet fell, it would acquire a terminal velocity that was dependent on its mass and properties of the air. The entire experiment had to be conducted through a microscope.

Quantum mechanics reveals that there are other quantities of the electron, in addition to charge, that are quantized. Energy levels in atoms are quantized, angular momentum is quantized and spin is quantized as well. The quantum numbers help define the electron configuration of atoms.

Planning for the Activity

Time Requirements

One class period is required to complete this activity.

Materials Needed

For the class:

- electronic scale
- nickels (100)
- pennies (100)

For each group:

- plastic film canisters w/lids (10)

Advance Preparation and Setup

You can get as many film canisters as required from a one–hour photo shop.

You will have to have the film canisters with pennies ready to distribute when the class begins. (You may want to have none of the canisters contain just a single penny for some groups.)

If a balance is not available, you will have to mass each canister and write it on a label for the film canister.

Teaching Notes

Students should be given the time to explore how they would go about determining the mass of the pennies before they begin their investigations. As you circulate about the room, you may want to ask students what would happen to their approach if none of the containers had only one penny.

Encourage the different lab groups to share their results. Ask the students if their confidence in their result is increased after comparisons with other lab groups. This is an example of how scientists work together and support one another's discoveries.

Students may want to open the film canisters to count the number of pennies. This should be discouraged. In the real Millikan oil-drop experiment, Millikan was never able to "check" to see if he got the charges correct. Students should take as much data as they like to reach their conclusion, but in real science you must convince yourself and others – there is no "answer key."

There are commercially available computer simulations of the oil-drop experiment. You may wish to show this to the students or have them work with it or, as the **Stretching Exercise** describes, have them create one of their own.

© It's About Time

Activity Overview

In a simulation of Millikan's oil-drop experiment, students use inquiry to find the number of coins enclosed in a film canister. They then learn how related techniques were used to determine that electric charge is quantized.

Student Objectives

Students will:

- Determine the number of hidden pennies in a container.
- Explain why the masses of containers with pennies can only take on certain values.
- Describe the Millikan oil-drop experiment
- Explain what we mean when we say that electric charge is quantized

Answers for the Teacher Only

What Do You Think?

It will be interesting to see how students interpret the idea of divisible. Is a single egg divisible? Is an Oreo cookie divisible if one person gets the top half and crème and the other gets the bottom half? Is a penny divisible?

Atoms on Display

Activity 2 Tiny and Indivisible

GOALS

In this activity you will:

- Determine the number of hidden pennies in a container.
- Explain why the masses of containers with pennies can only take on certain values.
- Describe the Millikan oil-drop experiment.
- Explain what we mean when we say that electric charge is quantized.

What Do You Think?

It's easy to share a box of popcorn with a friend at the movies. It's a bit tougher to share a slice of pizza equally, but it can be done with a knife.

- **Can you think of something that cannot be shared? Describe something that cannot be broken into two parts.**

Record your ideas about this question in your *Active Physics* log. Be prepared to discuss your responses with your small group and your class.

For You To Do

1. Your teacher will provide you with a set of film canisters or other containers that contain pennies. Do not open the containers. Your goal is to determine the mass of a single penny. You can find the mass of an empty container with a balance. You can also find the mass of each container, including the pennies, using a balance.
   a) Explain how you can find the mass of one penny using the measurements you are permitted to make. It's very possible that no container has only one penny. Write down your strategy.

556

Active Physics

Answers

For You To Do

1. You can find the mass of the empty container. You can then find the masses of many containers. If the containers contain pennies, then all the masses inside the containers should be a multiple of the mass of one penny. In math, it would be similar to finding the largest common denominator.

© It's About Time

2. Assume that each container has a mass of 5 g and each penny has a mass of 2 g.

   a) Make a list of possible masses of containers that have 1 penny, 2 pennies, 5 pennies, etc. Do this for at least ten containers.

   b) Given only the masses of the ten containers that you calculated in **(a)**, describe how you can find the mass of a single penny.

3. The container does not have a mass of 5 g, and each penny does not have a mass of 2 g. Using the balance and the containers of pennies, determine the mass of a single penny.

   a) Record the mass of a single penny.

   b) Compare your value of the mass of a penny with that of another group. How does your confidence in the value change as you compare your value with more and more groups? This is how the sharing and comparing of results of different teams of scientists make scientific progress.

4. Suppose you obtain a new set of containers with nickels in them. Assume that each container is 5 g and each nickel is 5 g.

   a) What are some possible masses you would expect for four of the containers?

   b) A lab group stated that they measured the mass of a container of nickels and it was 23 g. Your lab group thinks that they must have made an error. Explain to the first group in writing why you think that there is a problem.

   c) When your lab group measured the mass of the container, you also found it to be 23 g. You now have a problem—a mystery—a puzzle. It is this kind of puzzle that challenges and intrigues physicists. How could this be? It would be great to open up the container, but this may not be possible. Can you solve the puzzle? Suggest at least three different solutions to this puzzle.

ANSWERS

For You To Do *(continued)*

2. a)

Mass of container (g)	Number of pennies	Mass of pennies (g)	Total mass (g)
5	1	2	7
5	2	4	9
5	5	10	15
5	7	14	19
5	10	20	25
5	12	24	29
5	15	30	35
5	20	40	45
5	25	50	55
5	26	52	57

© It's About Time

ANSWERS

For You To Do

(continued)

b) By finding the masses of the pennies, you can see if all of the masses are divisible by the smallest mass of one container. In the chart on the previous page, the smallest mass of pennies is 2 g. All of the other masses are divisible by 2 g. Therefore, 2 g is probably the mass of a single penny. If the smallest container does not have one penny, but rather two pennies, we would see that all of the other masses are not divisible by that number. In the example, if the smallest container were 4 g, then we would see that this must be more than 1 penny since the other masses are not multiples of 4 g. We could divide 4 g by 2 and then arrive at the correct mass of one penny.

3. Students will conduct the experiment

FOR YOU TO READ

In 1910, Robert A. Millikan completed an experiment very similar to the activity you just completed. He did not measure containers of pennies. He measured the forces on charged oil drops. This allowed him to calculate the charge on each drop. He made thousands of measurements. He always found that the oil drop had 1 charge, 2 charges, 5 charges, 23 charges, and other whole numbers of charges. He never found 3.5 charges or 4.7 charges or 11.2 charges. The Millikan oil-drop experiment shows that you never seem to have a piece of a charge. The experiment has been conducted many times. Nobody has found fractional charges. Millikan concluded from his oil-drop experiment that there is a basic unit of charge.

Robert A. Millikan taken in 1923 at age 55

J.J. Thomson had "discovered" the **electron**, a tiny negatively charged particle, about ten years before Millikan's experiment. He did this by analyzing electron beams in a tube very similar to the tube where electrons travel in your television. It is these electrons that hit the screen and make the TV images.

Your penny lab was easy compared to Millikan's oil-drop experiment. The oil drops are so small that in Millikan's experiment he had to view them through a microscope. To find their mass required some ingenuity as well. Millikan sprayed the oil droplets between a positively charged plate and a negatively charged plate. If the oil drop had a negative charge, it would be repelled from the negative plate and attracted to the positive plate. If the positive plate were on top, the electric force would be pulling the drop up, while gravity would be pulling the drop down. If the two forces were equal, the drop would come to rest and remain suspended (or travel at a slow constant speed). By comparing the electrical force and the gravitational force, the charge on the oil drop (i.e., the charge of the electrons) could be found. Millikan won the Nobel Prize for this experiment.

The diagram shows the oil drop, the electrical plates, and the battery voltage that provides the charge on the plates.

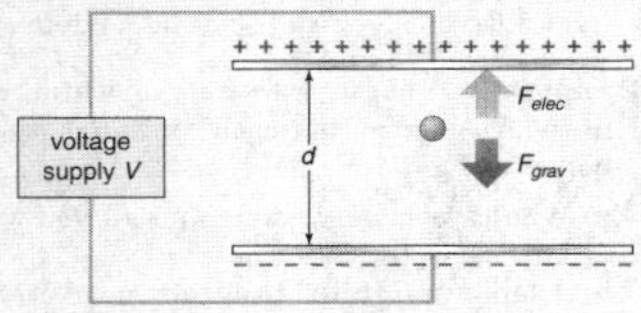

The weight of the oil drop (mg) can be found by observing the oil drop as it falls because of air resistance. You can calculate the electrical force from the voltage, charge, and distance between the plates. You can calculate the gravitational force from the mass and the acceleration due to gravity. Using these equations, you can find the charge on the oil drops. The voltage of the power supply and the distance between the plates can be measured:

$$F_{elec} = F_{grav}$$

$$\frac{qV}{d} = mg$$

$$q = \frac{mgd}{V}$$

4. a)

Mass of container (g)	Number of nickels	Mass of nickles (g)	Total mass (g)
5	2	10	15
5	5	25	30
5	7	35	40
5	12	60	65

b) If the container is 5 g, then your nickels must have a mass of 23 g – 5 g = 18 g. If nickels are 5 g each, then the mass can be 15 g or 20 g, but not 18 g.

c) Perhaps there are some other coins in there. Perhaps there is a piece of a nickel in there. Perhaps that container only weighs 3 g.

© It's About Time

From Millikan's and many additional experiments, the charge on an electron is determined to be 1.6×10^{-19} C (coulombs). You could expect to see twice this charge, three times this charge, nine times this charge, and any other whole multiple of this charge. If you never see a fractional part of this charge, you assume that charge is indivisible.

Current theories of physics state that a $\frac{1}{3}$ charge and a $\frac{2}{3}$ charge can exist. There is evidence for these fractional charges and the quarks (the tiniest components from which matter is made) associated with them. The Millikan oil-drop results led to the conclusion that quarks always join up to make a total charge of +1 or −1.

Reflecting on the Activity and the Challenge

Electrons are part of the atom. In your museum exhibit, you will have to include the charge on an electron. You may also find a way to include how the Millikan oil-drop experiment helped scientists find out about the charge and its indivisibility. You may choose to make a part of the exhibit deal with electrons and electric charge interesting by making it interactive.

Physics To Go

1. Two students are playing tug-of-war with a rope. How is this game similar to the two forces in the Millikan oil-drop experiment?
2. Describe how you could make Millikan's oil-drop experiment into an exciting interactive display.
3. A donut can be split into two pieces. Does the Millikan experiment prohibit the electric charge from being split into two pieces?
4. Assume that a container has a mass of 10 g and each penny has a mass of 3 g.
   a) Make a list of possible masses of five containers that have 1 penny, 2 pennies, 5 pennies, 10 pennies, and 12 pennies.
   b) List two masses you would not find for a container with pennies.
5. What is the net static electric charge on a metal sphere having an excess of +3 elementary charges (electrons)?
6. How many coulombs are equivalent to the charge of 100 electrons?

Physics Word

electron: a negatively charged particle with a charge of 1.6×10^{-19} coulombs and a mass of 9.1×10^{-31} kg.

ANSWERS

Physics To Go

1. In the Millikan oil-drop experiment, the suspended oil drop has a gravitational force pulling it down and an electrical force pulling it up. When the two forces are equal, the drop remains still or moves with a constant speed. The center of the rope is being pulled by the two students. If the forces are equal, the center of the rope remains at rest or moves with a constant velocity.
2. Answers will vary.
3. No, it does not prohibit it. Rather it shows that experimentally, a fractional charge has never been observed.

4. a)

Mass of container (g)	Number of pennies	Mass of pennies (g)	Total mass (g)
10	1	3	13
10	2	6	16
10	5	15	25
10	10	30	40
10	12	36	46

b) You could not find masses of 15 g or 26 g.

5. (3 electrons) × (1.6×10^{-19} C/electron) = 4.8×10^{-19} C.
6. (100 electrons) × (1.6×10^{-19} C/electron) = 1.6×10^{-17} C

© It's About Time

ANSWERS

Physics To Go

(continued)

7. For a charge to be possible it must be a multiple of (1.6×10^{-19} C).
   a) Not possible.
   b) possible (2 charges).
   c) Not possible.
   d) Not possible.
8. 4.8×10^{-19} C/(1.6×10^{-19} C/electron) = 3 electrons.
9. Student responses will vary.
10. Possible answers are suggested.
    a) +1 = (+2/3, + 2/3, −1/3)
    b) 0 = (+2/3, −1/3, −1/3)
    c) −1 = (−2/3, −2/3, +1/3)

7. Which electric charge is possible?
   a) 6.32×10^{-18} C
   b) 3.2×10^{-19} C
   c) 8.0×10^{-20} C
   d) 2.4×10^{-19} C
8. An oil drop has a charge of -4.8×10^{-19} C. How many excess electrons does the oil drop have?
9. J.J. Thomson, the discoverer of the electron, tried to describe the enormity of the discovery of this tiny particle: "Could anything at first sight seem more impractical than a body which is so small that its mass is an insignificant fraction of the mass of an atom of hydrogen, which itself is so small that a crowd of these atoms equal in number to the population of the whole world would be too small to have been detected by any means then known to science?" Create a quote of your own that captures the excitement of the discovery of the electron. Perhaps the quote could be displayed near the entrance to your proposed museum exhibit.
10. Quarks have charges $+\frac{1}{3}, -\frac{1}{3}, +\frac{2}{3}, -\frac{2}{3}$.
    a) Show how three quarks can combine to create a particle with a total charge of +1.
    b) Show how three quarks can combine to create a particle with a total charge of 0.
    c) Show how three quarks can combine to create a particle with a total charge of −1.

Stretching Exercise

The Millikan oil-drop experiment can be completed in a high school lab. It does require the use of a microscope and incredible patience to view the drops. Design a computer simulation or game that works like the Millikan oil-drop experiment. The simulation should have these features:

- The screen should look like the apparatus with a variable power supply and oil drops between the plates.
- The drop should be able to get a new charge.
- The drop should be able to move.
- The voltage should be able to be varied so that the net force on the drop is zero and the drop travels at constant velocity or is at rest.
- The velocity of the drop should be able to be measured to determine if it is traveling at a constant velocity.
- New drops should be able to be inserted between the plates.

© It's About Time

NOTES

© It's About Time

ACTIVITY 3
How Big Is Small?

Background Information

It is suggested that you read the **For You To Read** section of the student text for **Activity 3** before proceeding in this section.

The Rutherford scattering experiment is one of the hallmark achievements in physics. It not only yielded our understanding of the structure of the atom with a dense, positive nucleus at the center, but it provided a size for that nucleus. It created a gateway to our understanding of atomic structure that led to much of 20th century physics.

The idea of a nucleus being such a tiny part of the atom (1/100,000) and that the atom is mostly empty space is quite strange. The students may not be comfortable with this notion at first, but by trying to draw the atom, the students will get a good sense of the size relationship.

Some students may wonder what is between the nucleus and the electrons. There is nothing between. It is a vacuum. Matter is mostly empty space.

Planning for the Activity

Time Requirements

One class period is required to complete this activity.

Materials Needed

For the class:

- 1 pencil that can be dropped 100 times
- piece of paper

For each group:

- basic calculator
- ruler, cm

Advance Preparation and Setup

No advance preparation and setup are required for this activity unless you choose to reproduce the cards for the students. A template can be found in the Teacher's Edition at the end of this activity.

Teaching Notes

Students cannot perform the Rutherford scattering experiment, but this simulation is able to guide them to a good understanding of the techniques that Rutherford used.

Since this "penny" lab is a simulation, you must show a comparison between their experiment and Rutherford's. In their experiment, it is a "hit" or "miss." In Rutherford's experiment, the alpha particle was *deflected* from its path near the nucleus. The amount of deflection provided Rutherford's data. A copy of Rutherford's experiment is provided as a Blackline Master at the end of this activity.

You will also want to use the simulation to initiate a discussion of the nature of science. The concept of comparing the size of the penny to a value measured with a ruler and the human eye was not available to Rutherford. There was no "answer key" and no way to "check to see if he was right," other than to continue the experiment and take better data. Another aspect of the nature of science has to do with the collaboration of Rutherford with his two graduate students. Science is a "team" activity.

© It's About Time

Activity 3 How Big Is Small?

GOALS

In this activity you will:

- Measure the size of a penny using statistical techniques.
- Compare statistical measurements to direct measurements.
- Relate the Rutherford scattering experiment to the penny simulation.
- Describe the relative scale of the nucleus to the atom.

What Do You Think?

You may have heard that the salt you use on your food is made up of sodium chloride ions (NaCl).

- **If you could magnify a grain of salt so that the sodium ion was the size of a penny, where would you find the neighboring chloride ion? Where would you find the next sodium ion?**

Record your ideas about these questions in your *Active Physics* log. Be prepared to discuss your responses with your small group and your class.

For You To Do

1. Work with a partner. Use a ruler and a pencil to outline a square that is 10 cm × 10 cm on a card. Trace a penny as many times as you like within the square. Draw the circles so that they do not touch each other. Make the circles as close to the actual size of the penny as you can.

ANSWERS

For You To Do

1. Students construct card with outlines of pennies. A template for the card is also provided as a Blackline Master in this Teacher's Edition at the end of the activity.

Activity Overview

Using indirect measurements, students are able to accurately determine the size of a penny. They compare their statistical approach with direct measurement. Finally, they compare their experiment with Rutherford's experiment and are able to recognize the size of the nucleus in relation to the atom and the evidence we have for that knowledge.

Student Objectives

Students will:

- Measure the size of a penny using statistical techniques.
- Compare statistical measurements to direct measurements.
- Relate the Rutherford scattering experiment to the penny simulation.
- Describe the relative scale of the nucleus to the atom.

ANSWERS FOR THE TEACHER ONLY

What Do You Think?

If the sodium ion (the nucleus of sodium) were the size of a penny (approximately 1 cm) then the distance to the nearest chloride ion would be 1 kilometer away.

When listening to student responses to this question, encourage the students to place numbers on their answers rather than qualitative comparisons like near or very far.

© It's About Time

ANSWERS

For You To Do
(continued)

2. a) – b) Students record hit and miss data.

3. a) Students' calculations will vary. Let's assume 28 hits out of 100 drops.

   28 hits/100 drops = area of all pennies/100 cm^2

   Area of all pennies = 28 cm^2

4. a) Assuming 10 penny outlines on the card:

   Area of one penny = area of all pennies/10 = 28 cm^2/10 = 2.8 cm^2

5. a) Radius of a penny = 0.95 cm

   b) $A = \pi r^2 = (3.14)(0.95\ cm)^2 = 2.84\ cm^2$

   c) By lining up 10 pennies, you get to measure a larger distance and then divide by 10. The potential error in your number will now be only 1/10th as high as before.

6. a) Students respond based on their data.

   b) Students compare their data with other lab groups.

   c) The direct measurement with a ruler may seem more accurate, but if enough trials are done, then statistically, the dropping method will be more accurate.

   d) Aiming the pencil adds statistical basis.

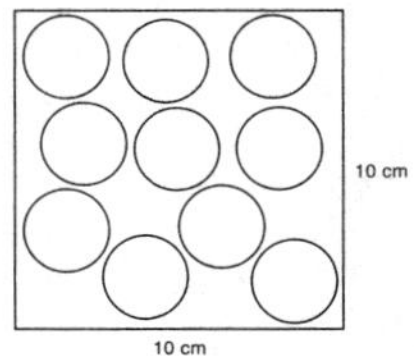

2. Drop a pencil onto the card so that the point hits within the square. Do not aim the pencil. It is actually better if you don't look. You want the drops to be as random as possible.
   a) Have your partner record whether you "hit" a circle, or "miss" a circle. If the pencil falls outside the square, ignore that drop.
   b) Make 50 "countable" drops. Then switch roles with your partner. Continue until 100 drops are recorded.

3. Find the area of all the pennies using the ratio:

$$\frac{\text{hits}}{\text{drops}} = \frac{\text{area of all pennies}}{\text{total area}}$$

   a) Show your calculations in your log.

4. Find the area of one penny by dividing the area of all the pennies by the number of penny outlines on your card.
   a) Record your calculation in your log.
   b) Compare your value with that of other lab groups. How close are the values?

5. You can also find the area of one penny using this equation: Area = πr^2, where $\pi = 3.14$ and r = the radius of the penny. To use this equation, you must find the radius of a penny.
   a) Measure the diameter of a one-penny circle on your card with a ruler. Record your measurement. (The radius is one-half the diameter.)
   b) Calculate the area of a penny and record your calculation in your log.
   c) If you wish to get a more accurate value of the diameter of a penny, you could line up 10 pennies before making the ruler measurement. Why would that give you a better result?

6. Compare the results you obtained using indirect measurement (dropping the pencil) and direct measurement (using a ruler).
   a) How close are the results you got using the indirect and direct methods?
   b) Compare your results with those of another lab group. How do your results compare?
   c) Which method is more accurate? Explain your answer.
   d) Why is it important not to aim the pencil in the indirect measurement?

© It's About Time

7. If possible, average the results of the indirect measurements and direct measurements from the entire class.
   a) How close are the results the class got using the indirect and direct methods?

8. Extend the experiment to an unknown object. Suppose a student conducts a similar experiment, but replaces the penny with a *single* unknown object. If that student got 50 hits out of 100 drops, then you would conclude that the unknown object's area was approximately 50% or one-half of the total area of the 10 cm × 10 cm square. You don't know from the data the shape or the location of the object. All of the following are possible because one half of the total area of each square is shaded.

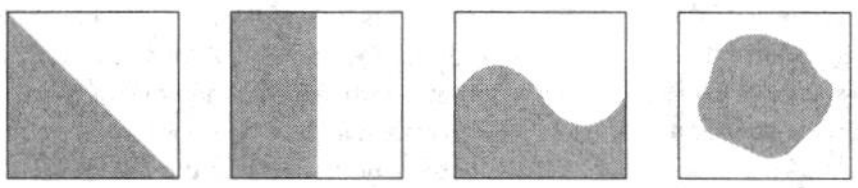

   a) What might the unknown object look like if it was reported to have 75 hits out of 100 drops? Draw the outer square and the unknown object's size within it.
   b) What might the unknown object look like if it was reported to have 25 hits out of 100 drops? Draw the square and the unknown object's size.
   c) What might the unknown object look like if it was reported to have one hit out of 100 drops? Draw the square and the unknown object's size.
   d) What might the unknown object look like if it was reported to have one hit out of 10,000 drops? Draw the square and the unknown object's size.

Answers

For You To Do
(continued)

7. The entire class's data should be very good for the indirect method.
8.

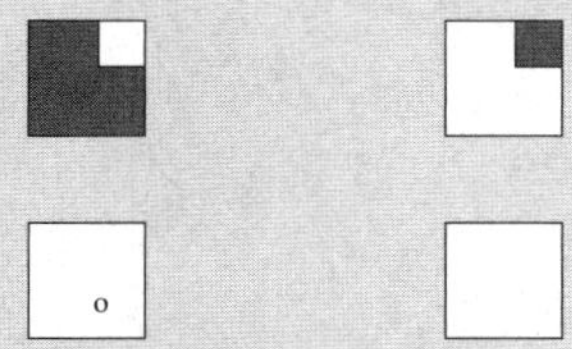

Too small to see

© It's About Time

Chapter 9

FOR YOU TO READ

Measuring the Size of the Nucleus

In this activity you used indirect measurements to find the area of a penny. Indirect measurement has been very useful in science. It continues to be useful. Finding the size of the penny without directly measuring it may seem like a good trick. However, you know that you can always verify the size by measuring a penny with a ruler. The **atom** cannot be measured with a ruler. Scientists must rely on indirect measurements.

A key scientific discovery—the discovery of the atomic **nucleus**—was made using a method similar to the one you used in this activity. Ernest Rutherford and his colleagues, Hans Geiger and Ernst Marsden, made the discovery. In the lab, the team bombarded a piece of thin gold foil with a beam of positively charged alpha particles. They originally thought that an atom had positive charges spread evenly through it, as shown in the diagram. If this idea were true, the alpha particles would go through the foil in nearly straight lines. The alpha-particle beam was their "dropping pencil." The foil was their card. Most of the particles did go straight through the gold foil. However, to the surprise of the team, Marsden saw that a very few of the alpha particles bounced back towards the source of the beam.

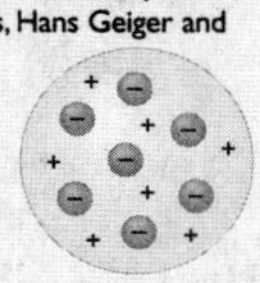

Rutherford thought deeply about these observations. He also thought about all of the ideas that scientists had about what was in atoms. He concluded that the positively charged alpha particles bounced back when they hit an area of concentrated positive charge. He also thought that the charge was in the center of the atom, the nucleus. Rutherford used his results to calculate the size of the nucleus of these atoms. He could have compared the number of "hits," particles that bounced back, with the total number of particles sent toward the foil. Then he could have used that ratio to determine the area of the foil where the atomic nuclei could be found. Rutherford's mathematics was a bit more complicated. In his situation the hit was not yes or no. It was an angle of deflection that could range from 0° to 180°. In future courses you may study those details. The result was that he calculated that the diameter of the atomic nucleus was 10^{-15} m. The diameter of the atom is 10^{-10} m. The nucleus is only $\frac{1}{100,000}$ the size of the atom.

RUTHERFORD'S EXPERIMENT

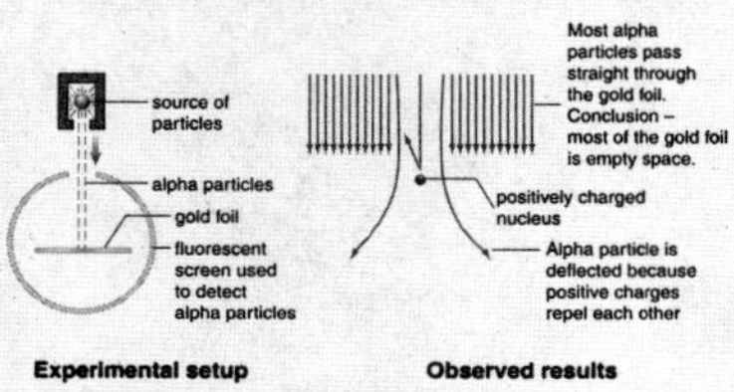

© It's About Time

It would be great if it were possible to check Rutherford's indirect measurement. A direct measurement of the size of a nucleus is impossible. Only the indirect measurement is available. It is used as evidence of the existence of the nucleus and its size.

Rutherford's nucleus provides a view of matter as mostly empty space. Electrons surround a tiny, tiny nucleus. Why then do solids appear so solid? How can the empty space of your fist hurt so much when it hits the empty space of a table? Electrons whizzing around the nuclei of the atoms in your fist are repelled by the electrons whizzing around the nuclei of the atoms in the table. The closer you try to bring your fist to the table, the stronger the force of repulsion between the electrons. The force of repulsion can be greater than the force of attraction holding your fist together and bones can break.

How can empty space exert such forces? Imagine a thin propeller blade. If the blade is still, it is surrounded by mostly empty space. When the blade rotates, it seems to fill that empty space and create a solid wall. The tiny electrons, in rapid motion, create a similar effect. It is as if the electrons are everywhere at once and the empty space appears solid.

Your sense of touch is the repulsion of electrons. It follows Coulomb's Law: $F = \frac{kq_1q_2}{d^2}$. It's amazing that you are able to feel this force when it is as tiny as a kiss or as large as striking a table with your fist. The next time you kiss someone, remember that you are experiencing the repulsion of electrons.

Reflecting on the Activity and the Challenge

By shooting particles at thin foils and seeing how the particles scatter, you can investigate the structure of matter. Rutherford's scattering experiment revealed that a tiny nucleus contains all of the positive charge of the atom. This also implies that most of the atom is empty space. In your museum exhibit, you will certainly want to help visitors to get a sense of the structure of the atom with its tiny nucleus. Is there a way in which you can get visitors to your exhibit to explore the size of the nucleus? You may also want to help visitors to understand how "mostly empty space" can make hard, solid objects.

Probability was important in this indirect method of measuring the size of a penny or the size of the nucleus. If you had aimed the pencil, the experiment would not have given good results. The use of probability and indirect measurements may be something you wish to include in your museum exhibit.

The museum exhibit must capture a visitor's attention within 30 seconds. How can you "grab" the visitor?

Physics Words

atom: the smallest particle of an element that has all the element's properties; it consists of a nucleus surrounded by electrons.

nucleus: the positively charged dense center of an atom containing neutrons and protons.

© It's About Time

ANSWERS

Physics To Go

1. a) Student results will vary

   b) The direct method: radius of a quarter = 1.2 cm

   $A = \pi r^2 = (3.14)(1.2\text{ cm})^2$
   $= 4.5\text{ cm}^2$

   c) Both methods are accurate. If we increase the number of drops, the indirect method will eventually become more accurate.

2. By aiming at the card, students will skew the result so that the quarter calculations emerge as much too large (if they aim at the quarter) or much too small (if they aim away from the quarter).

3. 10^{-10} is 10^5 or 100,000 times larger.

4. a) $A = \pi r^2 = (3.14)(4\text{ cm})^2$
   $= 50\text{ cm}^2$

   b) $A = \pi r^2 = (3.14)(7\text{ cm})^2$
   $= 154\text{ cm}^2$

   c) $A = \pi r^2 = (3.14)(10\text{ cm})^2$
   $= 314\text{ cm}^2$

5. Any statistical variations (a lucky hit) will be averaged out with more drops.

6. (23 hits/100 drops) = area of all coins/100 cm^2

   Area of all coins = 23 cm^2

   With 7 coins on the card:

   Area of one coin = area of all coins/7 = 23 cm^2/7 = 3.3 cm^2

7. Since the atom is 100,000 times (10^5) as large as the nucleus, then the atom would be 100,000 × 1 cm = 1000 m = 1 km and the next nucleus would be twice this distance or 2 km away.

8. As the positive alpha particles got closer to the positive nucleus, then the force would be larger and the deflection would be larger, as well.

9. Answers will vary.

Physics To Go

1. Determine the size of a quarter indirectly by repeating the penny experiment.
   a) Record your results.
   b) How close are the results you got using the direct and indirect method of measurement?
   c) Which method is more accurate? Explain your answer.

2. Repeat the quarter experiment, but this time aim at the card.
   a) Record your results.
   b) How does aiming change the results?

3. Which is greater, 10^{-15} m or 10^{-10} m? How many times greater?

4. Find the areas of circles with the following diameters:
   a) 4 cm
   b) 7 cm
   c) 10 cm

5. Why do you get better results when you drop the pencil 1000 times instead of 10 times?

6. You drop a pencil 100 times and get 23 "hits." There are seven coins on a card 10 cm × 10 cm. How large in area is each coin?

7. If the nucleus could be enlarged with a projector so that it was 1 cm in diameter, how far away would the next nucleus be? (Each nucleus is 10^{-15} m and each atom is 10^{-10} m.)

8. In the Rutherford scattering experiment, the alpha particles were deflected at different angles. It was not simply a hit or miss as it was in your penny simulation. How would you expect the deflection to be affected as the positive alpha particles came closer to a positive nucleus?

9. How might you show the proper scale of the nucleus and the atom in your museum exhibit?

© It's About Time

Blackline Master –
Pencil-Drop Activity Card

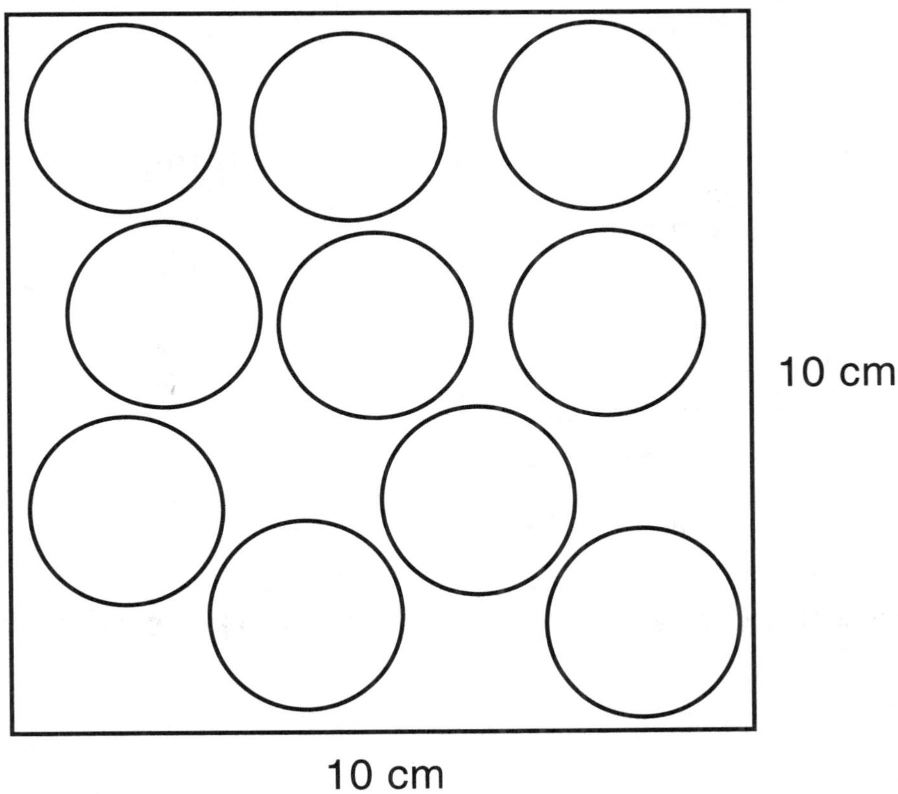

For use with *Atoms on Display*, Chapter 9, Activity 3: How Big Is Small?

© It's About Time

Blackline Master – Rutherford's Experiment

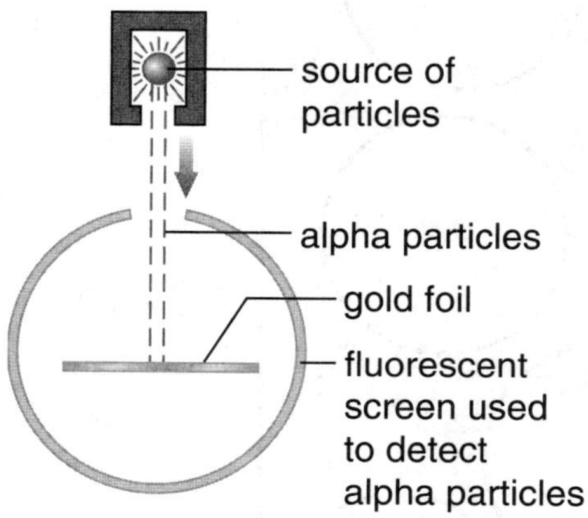

Experimental setup

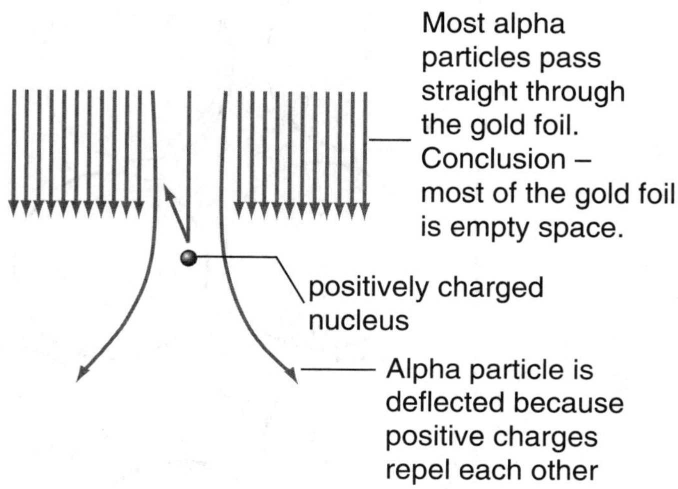

Observed results

© It's About Time

For use with *Atoms on Display*, Chapter 9, Activity 3: How Big Is Small?

NOTES

© It's About Time

ACTIVITY 4
Hydrogen Spectra and Bohr's Model

Background Information

It is suggested that you read the **For You To Read** section of the student text for **Activity 4** before proceeding in this section.

Bohr's analysis of the spectra of hydrogen provides a new understanding of the structure of the atom. In Bohr's model, the electron is restricted to certain orbits about the nucleus. When the electron jumps from one orbit to another, light of a specific wavelength is emitted. When the transitions end with the $n = 2$ energy level ($n = 3$ to $n = 2$, $n = 4$ to $n = 2$), then the wavelengths of light are in the visible spectrum.

The wavelengths of light that correspond to transitions to the $n = 1$ level are of higher energy and correspond to wavelengths in the ultraviolet region. The wavelengths of light that correspond to transitions to the $n = 3$ level are of lower energy and correspond to wavelengths in the infrared region.

The idea that electrons are restricted to certain wavelengths makes no sense from a classical perspective. The question of how an electron can make a transition from one orbit to another without ever being between orbits also makes no sense. It is a purely quantum perspective. Although there appears to be no reason why the orbits should be restricted, or how the electron makes the transitions, the values that correspond to the transitions give excellent predictions of the emitted wavelengths of light. In fact, not only was it able to account for wavelengths of light already known, the theory also predicted the wavelengths of light previously unknown. These wavelengths of light were then found experimentally.

The Bohr atom and its corresponding calculations of the wavelengths of light is a valuable example of the nature of science.

Planning for the Activity

Time Requirements

You will need two class periods to complete this activity.

Materials Needed

For the class:

- spectrum tubes of neon, hydrogen and helium
- power supplies for the spectrum tubes

For each group:

- basic calculator
- cardboard spectroscope

Advance Preparation and Setup

You should be able to darken the room to provide more brilliant spectra with the grating.

Teaching Notes

The more diffraction grating spectrometers you have, the more fun the lab will be. If you only have access to one portable, plastic spectrometer for each group, consider giving out diffraction gratings to other members of the group. The diffraction gratings will allow students to see the colors and to make a sketch of the different colors. The student with the spectrometer will then be able to announce the specific wavelengths that correspond to the colors.

The theory corresponding to the Bohr atom is a fascinating example of the nature of science in the 20th century. You should ensure that students take the time required in the **For You To Read** section. You may want to have a class discussion corresponding to the question of whether we should believe in things that go along with our common sense or things that produce excellent predictions.

The last paragraphs of the **For You To Read** section provide a brief glimpse of a philosophy of science in trying to summarize what makes a "good theory." The ability of a theory to predict something that nobody has any notion of and then to discover the existence of this property is a hallmark achievement of science.

Students will be uncomfortable with electrons that can disappear from one orbit and appear in another orbit without ever being between these two orbits. This is part of the fascination of quantum mechanics – one of the greatest achievements of intellectual thought in the past century. There are many popular books and articles describing both the history, philosophy, and physics of quantum mechanics. You should encourage your students to expand their knowledge through further reading.

© It's About Time

Activity 4 Hydrogen Spectra and Bohr's Model

GOALS

In this activity you will:

- Observe the colors of light given off by tubes of hydrogen, helium and neon.
- Record the wavelengths of the light given off by the gases as measured by the grating spectrometer.
- Calculate the energy levels of the Bohr hydrogen atom.
- Describe how the electrons jump from one orbit to another and give off light of a specific wavelength.
- Match the theoretical, calculated values for the wavelengths of light with the observed values of the wavelength.
- Cite evidence for why we think that the Sun is composed of hydrogen and helium gases.
- Calculate the wavelengths of light emitted from transitions of electrons in the Bohr atom.

What Do You Think?

It is said that there are no two snowflakes alike.

- **How do you distinguish one person from another?**

Record your ideas about this question in your *Active Physics* log. Be prepared to discuss your responses with your small group and your class.

For You To Do

1. You are going to use your observation of colors to shed light on the structure of the atom. Your teacher will set up tubes of hydrogen, helium, and neon gases and connect each to a high-voltage power supply. Observe the light of each tube with the naked eye. Then observe the same light with a spectrometer.

a) Record your observations in your *Active Physics* log.

b) The spectrometer has a scale and values within it that are measurements of the wavelengths of light. Record the wavelengths that correspond to each color of light that you are observing. (The wavelengths are measured in nanometers (nm). The prefix *nano* means 10^{-9}.)

ANSWERS

For You To Do

1. a) – b) The different tubes of gas will have different wavelengths of light.

Hydrogen: 434 nm, 486 nm, 656 nm

Helium: 471 nm, 588 nm, 668 nm

Neon: 586 nm, 630 nm, 643 nm (and many additional colors)

Activity Overview

Students investigate the spectral lines from three gases using a diffraction grating and measure the wavelengths of light. They then learn about the Bohr model of the atom and are able to calculate the wavelengths of light as electrons jump from one orbit to another.

Student Objectives

Students will:

- Observe the colors of light given off by tubes of hydrogen, helium, and neon.
- Record the wavelengths of the light given off by the gases as measured by the grating spectrometer.
- Calculate the energy levels of the Bohr hydrogen atom.
- Describe how an electron jumps from one orbit to another and gives off light of a specific wavelength.
- Match the theoretical, calculated values for the wavelengths of light with the observed values of the wavelength.
- Cite evidence for why we think that the Sun is composed of hydrogen and helium gases.
- Calculate the wavelengths of light emitted from transitions of electrons in the Bohr atom.

ANSWERS FOR THE TEACHER ONLY

What Do You Think?

We distinguish people by their height, weight, facial features, the way they walk and similar characteristics. We could also use fingerprints and DNA if we want to be more confident of the identification.

© It's About Time

Chapter 9

ANSWERS

For You To Do
(continued)

2. a) Students should be able to identify the gas from the wavelengths of light.
3. a) Three of the wavelengths are identical to hydrogen.
   b) Three of the wavelengths are identical to helium.
   c) You can conclude that the Sun is composed of these two gases: hydrogen and helium.

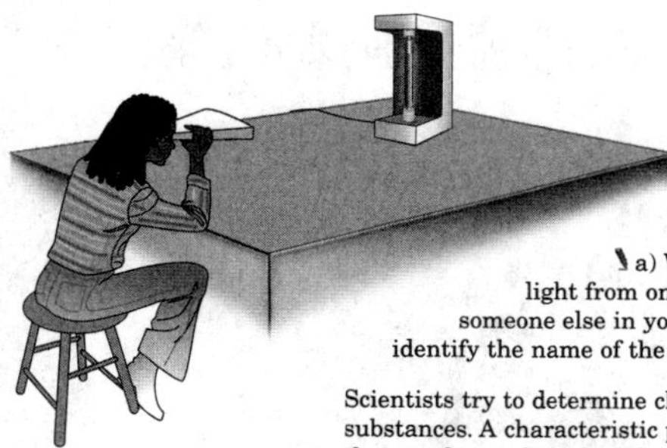

2. Each of the three spectrum tubes has a distinct pattern of colors that comprise it. These color bands correspond to specific wavelengths of light. Each gas has its own distinct set of wavelengths.

   a) Write down three wavelengths of light from one gas tube. Pass this list onto someone else in your lab group. Have that person identify the name of the gas. Were you all successful?

Only your teacher will handle the power supply and the tubes of gas. The power supply uses high voltage, which can be dangerous. The tubes of gas are glass and can be broken, making sharp edges. Do not look into bright lights or the Sun with the spectrometer.

Scientists try to determine characteristic properties of substances. A characteristic property is a unique attribute that can be used to identify that substance and distinguish it from other substances. Fingerprints or DNA patterns for humans are characteristic properties. No two people have been found who share the same fingerprint or the same DNA. The spectrum of light from a gas is a characteristic property of that gas.

3. When viewing the Sun, a set of observed wavelengths had the following values: 434 nm, 471 nm, 486 nm, 588 nm, 656 nm, and 668 nm.

   a) Which gas on Earth emits three of these wavelengths of light?

   b) Which gas on Earth emits the other three wavelengths?

   c) From these observations, what can you conclude about the gases that comprise the Sun?

4. In 1885, Johann Balmer, a Swiss high school physics teacher, created a simple equation to calculate the wavelengths of the hydrogen lines (the color bands you observed).

$$\lambda_n = 3646 \frac{n^2}{n^2 - 4} \times 10^{-10} \text{ m}$$

All wavelengths known to Balmer can be calculated by using this equation and substituting $n = 3, 4, 5, 6$.

© It's About Time

Activity 4 Hydrogen Spectra and Bohr's Model

> Example:
> Calculate the wavelength λ corresponding to $n = 3$.
>
> $$\lambda_3 = 3646 \frac{3^2}{3^2 - 4} \times 10^{-10}\ \text{m}$$
> $$= 6562 \times 10^{-10}\ \text{m}$$
> $$= 656\ \text{nm}$$

a) Calculate the wavelengths corresponding to $n = 4$ and $n = 5$.

5. Niels Bohr explained the significance of Balmer's equation. In the Bohr model of an atom of hydrogen, the proton is the nucleus and the single electron orbits the proton. The electron is able to move in a circle about the proton because of the attractive Coulomb force.

a) Using Coulomb's equation:

$$F = \frac{kq_1q_2}{d^2}$$

calculate the force between the proton and electron (each has a charge of 1.6×10^{-19} C). The distance between them is 5×10^{-10} m.

Niels Bohr received the Nobel Prize in Physics in 1922.

6. Bohr was able to calculate the radii of the orbits and their corresponding energies. The energy levels derived from the Coulomb force equation can be found using this *monster* of an equation:

$$E = -\frac{2\pi^2k^2mq^4}{h^2}\frac{1}{n^2}$$
$$= -13.6\frac{1}{n^2}\ \text{eV}$$

(You don't have to memorize this equation. However, there are some features that you can understand.)

a) Write down which of the symbols in the equation you recognize and what they represent. You should be able to provide the meanings for π, q, m, and k. The n has the values of 1, 2, 3... and corresponds to the energy level. The h is Planck's constant and is a number always equal to 6.63×10^{-34} Js. This is a very, very tiny number. It has an exponent of – 34! It is one of the most important numbers in physics and is the key to all of quantum physics.

ANSWERS

For You To Do
(continued)

4.

n	Wavelength
3	656.2 nm
4	486.1 nm
5	434.0 nm
6	410.1 nm

5. $F = kq_1q_2/d^2$
$= (9 \times 10^9\ \text{Nm}^2/\text{kg}^2)(1.6 \times 10^{-19}\ \text{C})(1.6 \times 10^{-19}\ \text{C})/(5 \times 10^{-10}\ \text{m})^2$
$F = 9.2 \times 10^{-10}$ N

6. a) $\pi = 3.14$
q = charge on the electron $= 1.6 \times 10^{-19}$ C
m = mass of the electron $= 9.1 \times 10^{-31}$ kg
k = Coulomb's constant $= 9 \times 10^9\ \text{Nm}^2/\text{kg}^2$

© It's About Time

ANSWERS

For You To Do

(continued)

6. b)

$\pi = 3.14$

$q = 1.6 \times 10^{-16}$ Coulombs

$m = 9.1 \times 10^{-31}$ kg

$k = 9.0 \times 10^{9}$ N m^2/C^2

6. c)

n	Energy (eV)
1	– 13.6
2	– 3.4
3	– 1.51
4	– 0.85
5	– 0.54
6	– 0.34

7. a) Since its energy is – 3.4 eV, it needs that much energy added to it to free it. Total energy must be 0 to be free.

b) E_3: 1.51 eV
E_4: 0.85 eV
E_5: 0.54 eV
E_6: 0.34 eV

8. a) n = 3: 1.51 eV
n = 4: 0.85 eV
n = 5: 0.54 eV
n = 6: 0.34 eV

b) Since all of these numbers are constants, you can do the calculation and find that together they equal 13.6 eV. (The eV is the symbol for an electron volt. It is the energy an electron gains from a one-volt battery. It is much smaller than one joule and is very convenient to use when discussing these small energies. 1 eV = 1.6×10^{-19} J.) What are the values of the constants π, q, m, and k?

c) Calculate the energies of each Bohr orbit using the equation:

$$E = -13.6\frac{1}{n^2}\text{ eV}$$

for n = 1, 2, 3, 4, 5, 6.

7. In physics, when a particle is bound to another particle, it is defined to have "negative" energy. In hydrogen, an electron in orbit about the proton has an energy of – 13.6 eV. An electron in E_1 would have to be given 13.6 eV to free it.

a) Explain why an electron in E_2 would have to be given 3.4 eV to free it.

b) How much energy would have to be given to an electron in E_3, E_4, E_5 and E_6 to free it?

8. The energy required to free an electron from a nucleus is called its ionization energy. The ionization energy of an electron in the E_1 or ground state is 13.6 eV. The ionization energy of an electron in the E_2 state is 3.4 eV.

a) What are the ionization energies of the electron in the n = 3, 4, 5, 6 states?

9. Niels Bohr explained why light was given off by the hydrogen atom. The electron begins in one orbit and then jumps to a lower orbit. During that jump it loses energy. Energy is conserved. In an isolated system, if one object loses energy, another object must gain energy so that the total energy remains the same. The electron lost energy. The light was created with this energy.

© It's About Time

Example:
Calculate the energy of the light when the electron jumps from E_3 to E_2.
The energy of $E_3 = -1.51$ eV
The energy of $E_2 = -3.40$ eV
An electron jumping from E_3 to E_2 would have a change of energy ΔE.

$$\Delta E = E_{final} \text{ to } E_{initial}$$
$$= E_2 \text{ to } E_3$$
$$= E_2 - E_3$$
$$= -3.40 \text{ eV} - (-1.51 \text{ eV})$$
$$= -3.40 \text{ eV} + 1.51 \text{ eV}$$
$$= -1.89 \text{ eV}$$

The electron lost 1.89 eV of energy. Light was created with exactly this energy.

a) Calculate the change of energy ΔE when an electron jumps from E_4 to E_2, E_5 to E_2, E_6 to E_2.

10. Albert Einstein postulated the relationship between the energy of light and its frequency in 1905.

$$E = hf$$

where E is the energy of light and
f is the frequency of light.

Wavelengths of light (λ) can be found from the wave equation

$$c = f\lambda$$

where c is the speed of light (3.0×10^8 m/s).

It is possible to combine Bohr's calculation of the energy given to the light when the electron jumps from E_3 to E_2 and Einstein's calculation of the corresponding frequency (or wavelength). You must also convert from the energy unit of electron volts to joules so that the wavelength will be in meters.

$$\lambda = \frac{hc}{|\Delta E|}$$
$$= \frac{1.24 \times 10^6 \text{ (m)(eV)}}{|\Delta E|}$$

ΔE is the energy change in electron volts (eV).

ANSWERS

For You To Do
(continued)

9. a) $E_4 - E_2$: 2.55 eV
$E_5 - E_2$: 2.856 eV
$E_6 - E_2$: 3.06 eV

© It's About Time

ANSWERS

For You To Do
(continued)

10.a) $E_4 - E_2$: 4.86×10^{-7} m

$E_5 - E_2$: 4.34×10^{-7} m

$E_6 - E_2$: 4.05×10^{-7} m

b) They are almost identical to the wavelengths observed.

Example:
When the electron jumps from E_3 to E_2, the change in energy is 1.89 eV. Calculate the corresponding wavelength of light.

$$\lambda = \frac{hc}{|\Delta E|}$$

$$= \frac{1.24 \times 10^{6}(\text{m})(\text{eV})}{|\Delta E|}$$

$$= \frac{1.24 \times 10^{6}(\text{m})(\text{eV})}{1.89 \text{ eV}}$$

$$= 654 \text{ nm}$$

This corresponds to the measured wavelength of one of the lines of hydrogen. It also gives the same value as Balmer's earlier equation. (See **Step 4**. Notice that there is a slight difference between the two numbers because some of the numbers in the calculation were rounded.)

a) Calculate the wavelengths of light when the electron jumps from E_4 to E_2, E_5 to E_2, E_6 to E_2.

b) How do these values compare with the ones you found in your observations of the hydrogen spectra?

FOR YOU TO READ
Discovery of Helium

When viewing the Sun, the following set of wavelengths can be observed: 434 nm, 471 nm, 486 nm, 588 nm, 656 nm, and 668 nm. From your observations in this activity, you concluded that these values match the wavelengths emitted by hydrogen and helium. This led you to the conclusion that the Sun is comprised of hydrogen and helium.

The history of these values, however, is a bit more interesting. The set of values corresponding to hydrogen were known from the lab. Nobody had ever observed the second set of wavelengths in a lab. In 1868, Pierre Jenson observed one line from the Sun during a total eclipse in India. J. Norman Lockyear interpreted this yellow line as being evidence of a new element. This set of wavelengths did not correspond to any gas on the Earth and so this unknown gas of the Sun was named "helium" after Helios, the Greek god of the Sun. Years later, in 1895, helium gas was discovered on Earth by William Ramsey of Scotland and independently by Per Cleve of Sweden.

© It's About Time

Activity 4 Hydrogen Spectra and Bohr's Model

Bohr's Model of an Atom

In this activity you observed the spectral lines of several gases. How are the spectral lines created? Are the lines telling you something about the structure of hydrogen, helium, neon, and the other elements? Are the lines a means by which nature is revealing its secrets?

Johann Balmer created a simple equation to calculate the wavelengths of the hydrogen lines.

$$\lambda_n = 3646 \frac{n^2}{n^2 - 4} \times 10^{-10} \text{ m}$$

Balmer's equation worked well. It illustrated the pattern in the wavelengths of light from hydrogen, but it opened up new mysteries. Where did this "magic" number 3646 come from? Why is n^2 used? These questions would be answered by Niels Bohr in the early 20th century.

Niels Bohr took the Rutherford model of the atom, which had a tiny nucleus in the center and orbiting electrons. From this, he created a new model for the emission of atomic spectra. Hydrogen has only one electron and one proton. In the Bohr model, the proton is the nucleus and the single electron orbits the proton in the same way that the planets orbit the Sun. The electron is able to move in a circle about the proton because of the attractive Coulomb force. This force keeps the electron moving in a circle of that radius about the proton. Using certain assumptions, Bohr was able to derive the following equation that gives the energy levels of the various orbits:

$$E = -\frac{2\pi^2 k^2 m q^4}{h^2} \frac{1}{n^2}$$

However, there were still two puzzles you need to understand:

- the – sign in the equation and
- the relationship between the energy levels and the wavelengths of light emitted in the spectra

You were told that when a particle is bound to another particle, in physics, it is defined to have "negative" energy. For instance, the planets orbiting the Sun are bound to the Sun and their total energy would be negative. All objects on the Earth are bound to the Earth and their total energy would also be negative. If you wanted to take an object and free it from the Earth, you would have to add energy to it. That is why incredibly large rockets are needed to lift spaceships off the Earth. When you throw a ball in the air, you also add energy, but the total energy is still negative and it comes back to Earth.

In hydrogen, an electron in orbit in the E_1 or ground state about the proton has an energy of – 13.6 eV. To free the electron, it must receive 13.6 eV of energy. This would bring its total energy to zero and it would be free. If it receives less than – 13.6 eV, it may move away from the proton nucleus but it will return to the nucleus. If it receives more than 13.6 eV, it is freed from the nucleus and has extra energy to move about.

Through calculations that you did in this activity you found that Balmer's equation worked as well as Bohr's equation. What then is gained from Bohr's work? Bohr provided a new theory, an explanation, and a mechanism by which the light is created. Balmer cleverly found a shortcut equation. Balmer invented a number to use to make the wavelengths turn out right. Bohr's theory explained that the number comes from combining Coulomb's Law with Einstein's relationship between the energy and frequency of light and the conservation of energy. Bohr's theory provides an insight into the structure of the atom. Electrons surround the nucleus. The electrons are in fixed orbits. →

© It's About Time

When an electron drops from one orbit to another, light is given off.

Ionization energy is the energy required to free an electron. Energy can be provided to free an electron by having the electron absorb light. When the electron absorbs light, it jumps up to a higher energy state. If the electron absorbs enough light, it can be freed from the nucleus. This is the inverse of the emission process Bohr described to explain why light is emitted by an atom.

The only light that can be absorbed (or emitted) by the electron is the light that has just the right energy and just the right wavelength to get the electron to another fixed orbit. If the light has a little less energy or a bit more energy, it goes right past the electron and has no effect.

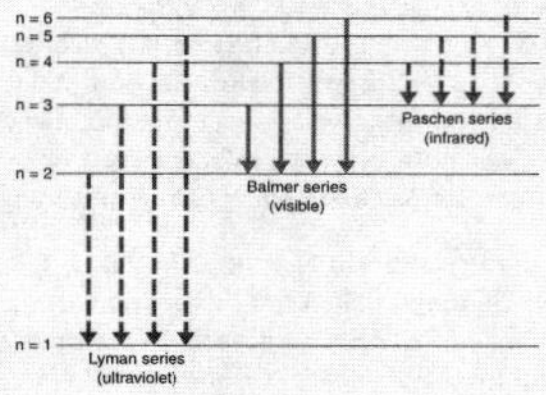

The energy transitions of the light emitted when the electron drops from E_3 to E_2, E_4 to E_2, and E_5 to E_2 can be calculated. You viewed these spectral lines using a spectrometer. You might expect light to be given off when the electron also jumps from E_2 to E_1, E_3 to E_1, and E_4 to E_1. These energy transitions can be calculated, but you were not able to see them. The light with those energies is not visible to the human eye. Visible light is only one part of the electromagnetic spectrum. If you use an ultraviolet light detector, you find that these additional wavelengths of light also exist. The visible light rays are called the Balmer series. The ultraviolet series is called the Lyman series.

Light should also be emitted when the electron drops from E_4 to E_3, E_5 to E_3, and E_6 to E_3. These energies can also be calculated. The light with those energies is also not visible to the human eye. If you use an infrared light detector, you find that these additional wavelengths of light also exist. They are called the Paschen series.

A good new theory should be able to explain what the old theory could explain. It should also be able to explain something that the old theory could not explain. Finally, it should be able to make a prediction of something that nobody had thought of previously. If that prediction turns out to be true, then we sense the power of the theory. The Paschen series had been observed before Bohr's theory. The Lyman series had never been observed. Bohr predicted the wavelengths of the Lyman series and then they were observed with the predicted wavelengths. Other spectra corresponding to transitions to the $n = 4$ state were also found later.

Reflecting on the Activity and the Challenge

The Bohr atom has electrons in fixed orbits surrounding the nucleus. Light is emitted when electrons jump from a higher energy orbit to a lower orbit. An electron that absorbs light can

© It's About Time

Activity 4 Hydrogen Spectra and Bohr's Model

jump from a lower orbit to a higher energy orbit. The wavelengths of light can be calculated, observed, and measured. The values from Bohr's theory and your observations from hydrogen are almost exactly equal. In your museum exhibit, you may try to illustrate the Bohr model of the atom and the emission of light as electrons jump from one energy level to another. You may also wish to show how an atom becomes ionized when the electron absorbs enough energy to free it. Finally, you may choose to show that invisible light in the ultraviolet and infrared regions are also emitted. Electron jumps and emitted light can be an interactive museum display. Will your display create an immediate interest? Provide a means by which the museum visitor will want to stop and see what is going on with hydrogen.

Physics Word

ionization energy: the energy required to free an electron from an atom.

Physics To Go

1. Light of greater energy has a higher frequency. In the hydrogen spectrum, which visible line has the greatest energy? Which transition does it correspond to?
2. Compare the energy of the light emitted from the electron jump from $n = 3$ to $n = 2$ to the light emitted from $n = 5$ to $n = 2$.
3. Given that the speed of light equals 3.0×10^8 m/s and the wavelength of light is 389 nm (389×10^{-9} m), calculate the frequency of the light.
4. Calculate the energies of each Bohr orbit using the equation $E = -13.6 \text{ eV} \left(\frac{1}{n^2}\right)$ for $n = 1, 2, 3, 4, 5$.
5. Make a diagram showing the energies of each Bohr orbit as a vertical number line which goes from – 013.6 eV to 0 eV.
6. How could the electron transitions be creatively displayed in a museum exhibit? Describe a way in which the display could be interactive.
7. The hydrogen spectra is said to be a "fingerprint" for hydrogen. Explain why this is a useful metaphor (comparison).
8. Why can't light of 500 nm be given off from hydrogen?
9. Electrons can jump from the $n = 4$ state directly to the $n = 3$ or $n = 2$ or $n = 1$ states. Similarly, there are multiple jumps from $n = 3$. How many different wavelengths of light can be given off from electrons that begin in the $n = 4$ orbit?

ANSWERS

Physics To Go

1. Light corresponding to the transition from $n = 6$ to $n = 2$.
2. Transition from $n = 3$ to $n = 2$: energy = 3.4 eV

   Transition from $n = 5$ to $n = 2$: energy = 0.85 eV
3. $c = f\lambda$

   $f = c/\lambda$

   $= (3 \times 10^8 \text{ m/s})/(389 \times 10^{-9} \text{ m})$

   $= 7.7 \times 10^{14}$ Hz
4.

n	Energy (eV)
1	−13.6
2	−3.4
3	−1.51
4	−0.85
5	−0.54

5. ----------- 0 eV

   ----------- −0.54 eV

   ----------- −0.85 eV

   ----------- −1.51 eV

   ----------- −3.4 eV

   ----------- −13.6 eV
6. Answers will vary.
7. Each person has a unique fingerprint. Each gas emits a unique set of wavelengths of light.
8. 500 nm does not correspond to one of the transitions in the orbits of the electron in hydrogen.
9. $n = 4$ to $n = 3$

   $n = 4$ to $n = 2$

   $n = 4$ to $n = 1$

   $n = 3$ to $n = 2$

   $n = 3$ to $n = 1$

   $n = 2$ to $n = 1$

   Six distinct wavelengths of light can be given off.

© It's About Time

Blackline Master – Lyman, Balmer, and Paschen Series

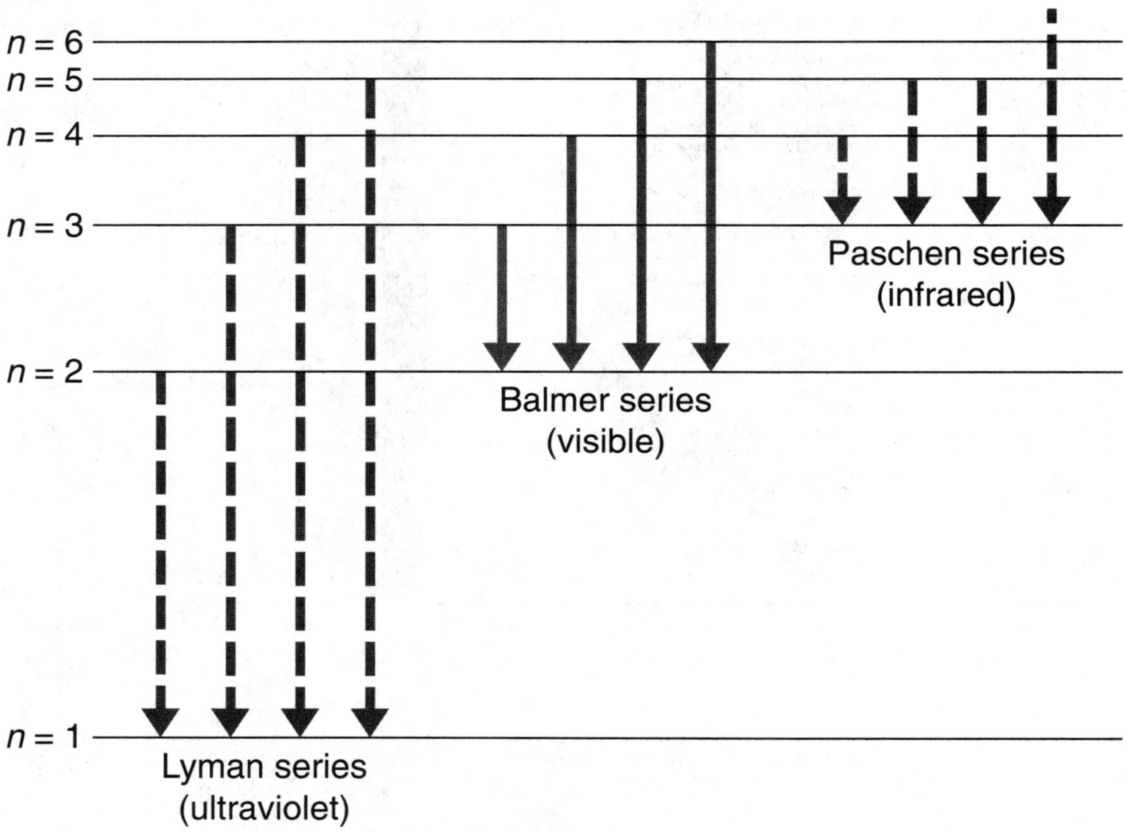

© It's About Time

For use with *Atoms on Display*, Chapter 9, Activity 4: Hydrogen Spectra and Bohr's Model

ACTIVITY 5
Extending and Amending the Model

Background Information

It is suggested that you read the **For You To Read** section of the student text for **Activity 5** before proceeding in this section.

This activity brings together physics principles that have been discussed earlier in the course. Light is shown to diffract (spread out) as it emerges from a small opening. The smaller the aperture, the more the light diffracts. In addition, light is shown to create an interference pattern similar to the interference pattern of sound waves and water waves. This leads us to the conclusion that light is probably a wave.

The photoelectric effect occurs when light (photo) frees an electron (electric) from a metal. By observing the conditions under which light is able to free the electrons, Einstein developed the theory of the photoelectric effect that won him the Nobel Prize. In Einstein's explanation, light must behave as a particle.

Light exhibits both wave and particle characteristics at different times. This is sometimes referred to as the wave/particle duality. Similarly, electrons can exhibit particle characteristics or wave characteristics. The wave/particle duality is not limited to light or to electrons. All matter can have both wave and particle characteristics at the atomic scale. This is another surprising, but valued, conclusions of the quantum theory.

The wave nature of electrons can explain why the electron is restricted to certain orbits in Bohr's model or certain orbitals in Schrodinger's model. The behavior of electrons in atoms is crucial to understanding atomic structure and is the basis for our understanding of chemical behavior.

Planning for the Activity

Time Requirements

Two class periods are required to complete this activity.

Materials Needed

For the class:

- 2 pieces of cardboard, different size slits
- piece of paper

For each group:

- basic calculator
- diffraction grating, 5 cm square
- laser pointer
- ruler, cm
- tuning fork

Advance Preparation and Setup

No advance preparation is required for this activity.

Teaching Notes

Although students will be able to move through the activity without much trouble, it will be up to you to help tie all the parts together in the activity review and the **For You To Read** section. The major point of the activity is for students to realize that the electron, which we usually think of as a particle, also has wave characteristics. These wave characteristics restrict the electron to certain locations. The electrons can move about the nucleus, but the best that we can know is the probability of where to find the electrons. We can never be certain of their exact location.

The electron in the one-dimensional box continues the introduction to the strange world of quantum mechanics. For an electron of a certain energy, the electron is often found in the left side of the box and often found on the right side of the box. The wave function indicates that it will never be found in the middle of the box. The idea that the electron moves from one side of the box to the other makes no sense. We can only talk about where the electron may be, not how it got there.

If you help students develop a good understanding of the wave function of the electron in the one-dimensional box, you can then ask them to imagine the electron being restricted to a three-dimensional space about the nucleus. The Schrodinger equation helps to define this space and the electron's position in that space.

© It's About Time

Activity Overview

The wave and particle nature of light is explored by investigations of two-slit interference and the photoelectric effect. Electrons are restricted to a one-dimensional box so that their wave characteristics can be studied. A new interpretation of the Bohr orbits as standing waves of electrons is introduced, with a non-mathematical introduction of the Schrodinger equation.

Student Objectives

Students will:

- Observe the diffraction of light waves.
- Observe the interference of sound waves.
- Observe the interference of light waves.
- Create a model of wave interference.
- Measure the wavelength of light using a Young's double-slit apparatus.
- Solve simple problems of the photoelectric effect.
- Describe the wave-particle duality of light.
- Describe the wave-particle duality of electrons
- Represent an electron confined to a one-dimensional box and note the probability of where the electron will be located.

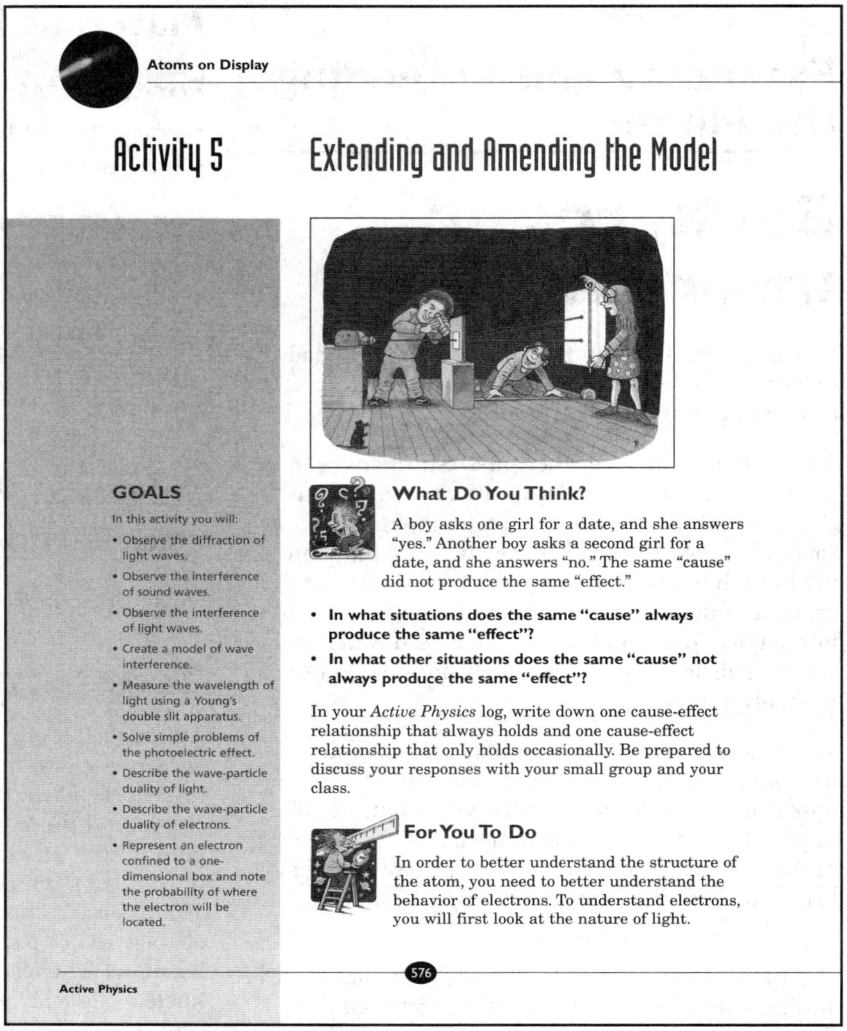

Atoms on Display

Activity 5 Extending and Amending the Model

GOALS

In this activity you will:

- Observe the diffraction of light waves.
- Observe the interference of sound waves.
- Observe the interference of light waves.
- Create a model of wave interference.
- Measure the wavelength of light using a Young's double slit apparatus.
- Solve simple problems of the photoelectric effect.
- Describe the wave-particle duality of light.
- Describe the wave-particle duality of electrons.
- Represent an electron confined to a one-dimensional box and note the probability of where the electron will be located.

What Do You Think?

A boy asks one girl for a date, and she answers "yes." Another boy asks a second girl for a date, and she answers "no." The same "cause" did not produce the same "effect."

- **In what situations does the same "cause" always produce the same "effect"?**
- **In what other situations does the same "cause" not always produce the same "effect"?**

In your *Active Physics* log, write down one cause-effect relationship that always holds and one cause-effect relationship that only holds occasionally. Be prepared to discuss your responses with your small group and your class.

For You To Do

In order to better understand the structure of the atom, you need to better understand the behavior of electrons. To understand electrons, you will first look at the nature of light.

Active Physics 576

Answers for the Teacher Only

What Do You Think?

Students can come up with lots of examples where behavior does not always produce the same result. In physical systems, nature seems to oblige with a strong cause-effect relationship. For instance, if I drop a ball it will fall to the ground. If I push on a cart, it will accelerate.

© It's About Time

1. Shine a laser beam against a wall. Trace the beam onto a piece of paper and measure its thickness.
   a) Record your measurement.
   b) Place a single slit in front of the beam. Measure and record the thickness of the beam again.
   c) Place a thinner slit in front of the beam. Measure and record the thickness a final time.
   d) What happens to the width of the laser beam as it passes through a smaller and smaller opening?

Never look directly at a laser beam or shine a laser beam into someone's eyes. Always work above the plane of the beam and beware of reflections from shiny surfaces.

The process of light bending as it squeezes through a small opening is called diffraction. Light bending around an edge is also referred to as diffraction. Diffraction is one of the properties that water waves and sound waves exhibit. This leads us to the conclusion that light is a wave.

2. In the diagram shown, water waves have traveled through two openings. The solid lines represent the crests of the waves. The dotted lines represent the troughs of the waves.
   a) Sketch the diagram in your log.
   b) When waves meet, they interfere with one another. Constructive interference occurs when the crests of the first wave meet the crests of the second wave. Constructive interference is where you would see large disturbances in the water. On the sketch, mark with an X the regions of constructive interference where the crests intersect.
   c) Destructive interference occurs when the troughs of one wave meet the crests of the other wave. Destructive interference is where you would see undisturbed water. Mark with an O the regions of destructive interference where the crests of one wave interact with the troughs of another.
   d) There are lines of constructive interference which can be seen from the pattern of Xs. Add these lines. There are lines of destructive interference formed by the pattern of Os. Add these lines as well.

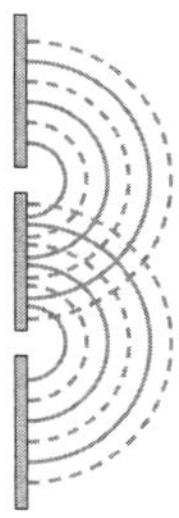

As you can see, there are positions of constructive interference (lots of movement of the water) and positions nearby of destructive interference (where the water is still).

3. Hit a tuning fork gently against a rubber stand. Place the tuning fork near your ear. Slowly rotate the tuning fork so that one prong gets closer to the ear while the other gets further away. Listen carefully to the volume of the sound.

Do not touch the vibrating tuning fork to your skin, especially near your ear.

ANSWERS

For You To Do

1. a) Measurements will vary.
   b) The beam may be thicker than the slit.
   c) The thinner slit will produce a wider beam!
   d) As the size of the opening decreases, the width of the beam increases.
2. a) – d) Students should see a series of lines radiating from between the slits.

© It's About Time

ANSWERS

For You To Do
(continued)

3. a) The volume of the sound increases and decreases as the angle changes.
   b) Students should be given every opportunity to observe the phenomenon.
4. a) The laser will form a series of dots on the far wall.
   b) The dots are an interference pattern. If you compare this to the lines in the diagram in **Step 2**, you can imagine that lines are radiating out from the slits and when they hit the screen, you see dots of light.
   c) If the light behaves in the same way that water waves and sound waves behave, then we can conjecture that light is a wave.

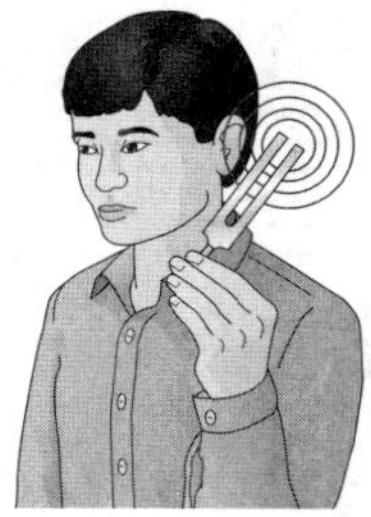

a) Record your observations in your log.

b) A tuning fork produces identical sounds from each prong. At one orientation from the prongs to your eardrum, you heard a loud sound. This was constructive interference. When the orientation from the prongs to your ear was different, the sound diminished. The interference of waves is another property of wave behavior. If you are unsure that you observed the interference of sound, listen to the tuning fork again.

4. Shine a laser beam through two slits or a diffraction grating made up of many slits. Direct the beam at a distant wall. Observe the pattern of the light on the wall.

a) Record your observations in your log.

b) From your observations, what can you conclude about the ability of light to interfere? Explain.

On the distant wall you will see places where the light from neighboring slits interfere constructively (there is maximum light) and places where the light interferes destructively (there is little or no light). Evidence of the interference of light would seem to prove that light behaves like a wave.

Never look directly at a laser beam or shine a laser beam into someone's eyes. Always work above the plane of the beam and beware of reflections from shiny surfaces.

c) Water and sound travel in waves. From your observations, what could you conclude about how light travels? Explain.

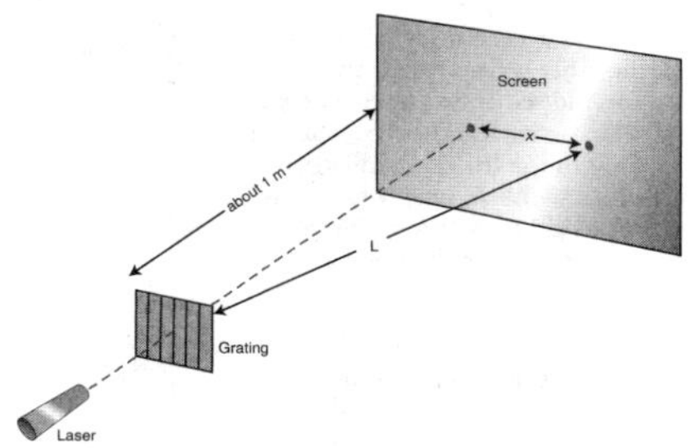

5. You can use the interference of light to measure the wavelength of light. Mount a diffraction grating in the path of a laser beam. Mount a screen several meters away from the grating as shown. Observe the pattern of spots on the screen.

Diffraction of Light

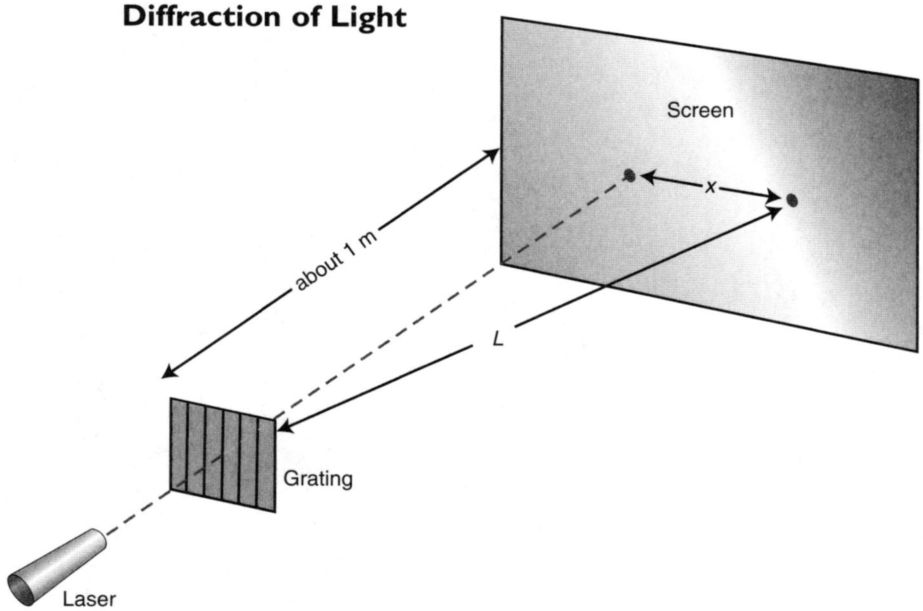

© It's About Time

a) Measure and record the separation between one spot and the next, x.

b) Measure and record the distance from the grating to the screen, L.

c) Measure and record the spacing between the lines of the grating, d. Alternatively, you can use the spacing given by the manufacturer.

6. From your measurements, find the wavelength of the light. You will use the following equation:

$$\lambda = \frac{dx}{L}$$

where λ is the wavelength of laser light,
L is the distance from the grating to the spot on the screen,
x is the separation between the spots,
d is the spacing between lines in the grating.

a) Show your calculations in your log.

The Photoelectric Effect

Now that you have found that light behaves like a wave and measured its wavelength, you can turn to the puzzle of the 20th century that Einstein solved. The puzzle had to do with the behavior of light when it freed an electron from a metal. In the photoelectric effect, beams of light shine on metals. Some wavelengths (and frequencies) of light will always free electrons, while other wavelengths (and frequencies) will never free electrons.

7. Why do some wavelengths (and frequencies) of light always free electrons, while other wavelengths (and frequencies) never free electrons? To understand the photoelectric effect, think of using a vending machine.
A vending machine sells potato chips. Suppose a bag of potato chips costs 10¢. The machine is not able to add coins.
- If you place 10¢ in the machine, a bag of potato chips comes out.
- If you place 5¢ in the machine, your 5¢ is returned.
- If you place 25¢ in the machine, a bag of potato chips comes out and you get 15¢ in change.
- If you place two nickels in the machine, the two nickels are returned (the machine cannot add coins).

ANSWERS

For You To Do
(continued)

5. a) – c) Measurements will vary.

6. a) The values for the wavelength of light will be dependent on the laser that is being used. The values of different lab groups should be approximately equal, given that they have the same types of lasers.

© It's About Time

ANSWERS

For You To Do
(continued)

7. a) 50-cent piece – 1 bag of potato chips and 40 cents in change.
   b) 20 pennies – 20 pennies out.
   c) $1.00 coin in – 1 bag of potato chips and 90 cents in change.
   d) Yes, the example works.

a) What would happen if you placed a 50-cent piece in the machine?

b) What would happen if you placed 20 pennies in the machine?

c) What would happen if you placed a $1.00 coin in the machine?

d) An equation that could describe the behavior of the machine would be:
returned money = money inserted – cost of chips.
Does this equation work for the examples above?

In the photoelectric effect:

- Some frequencies of light will free electrons. The brighter the light, the more electrons freed. This is similar to having lots of dimes to place in the chip machine.
- Some frequencies of light will never free electrons, regardless of how much light there is. This is similar to having lots of pennies or nickels to place in the chip machine.
- Some frequencies of light will free electrons and give them lots of kinetic energy. This is similar to quarters being placed in the chip machine. Each quarter gets a bag of chips and some change.

An equation that describes the photoelectric effect is:
kinetic energy of freed electron =
energy of light – energy required to free an electron (work function).

$$KE_{electron} = E_{light} - w_o$$

The energy of the light E_{light} is equal to hf where h is Planck's constant (6.63×10^{-34} Js) and f is the frequency of light.

The photoelectric effect explanation makes sense only if light were behaving as a particle and there were a collision between the photon and the electron.

8. Light and electrons both exhibit wave and particle behavior. This wave-particle duality is the hallmark finding of the 20th century conception of matter. The electron's particle characteristics include its mass, its charge, and its ability to hit a small target.

   If an electron had wave properties, then it would have a wavelength. Bohr's electrons are in fixed orbits because only specific circumference orbits would be equal in length to multiples of the wavelength of the electrons.

© It's About Time

a) Draw a circle about a nucleus. Try to draw a wave so that it meets perfectly. See the diagram for one such example.

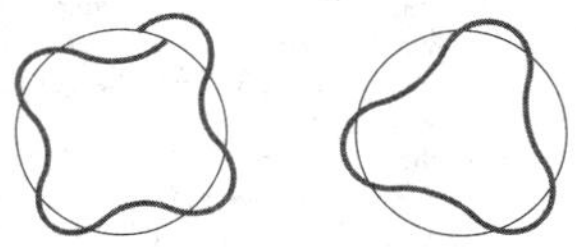

The wave does not fit. This is not a permissible Bohr orbit. The wave fits. This is a permissible Bohr orbit.

9. Imagine placing an electron in a one-dimensional box. A one-dimensional box is a line segment. The electron can move back and forth along the line segment. The electron that can exist in the box must have a specific wavelength and corresponding energy. This wavelength is dependent on the dimensions of the box. A few possibilities are shown.

a) Draw two additional waves that could fit in the box.

10. The probability of finding the electron at any location is equal to the square of the amplitude of the wave that fits in the box.
- In the places where the square of the amplitude is large, there is a high probability of finding the electron.
- In the places where the square of the amplitude is small, there is a small probability of finding the electron.
- In the places where the square of the amplitude is zero, there is no probability of finding the electron.

The amplitudes of three waves.

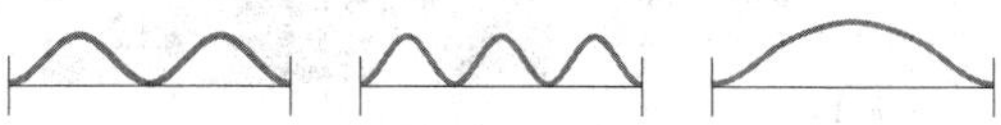

The square of the amplitudes of three waves.

a) Copy these diagrams in your log and describe where you would expect to find the electrons and where you would not expect to find the electrons.

Physics Words

diffraction: the ability of a wave to spread out as it emerges from an opening or moves beyond an obstruction.

constructive interference: the result of superimposing different waves so that two or more waves overlap to produce a wave with a greater amplitude.

destructive interference: the result of superimposing different waves so that two or more waves overlap to produce a wave with a decreased amplitude.

photoelectric effect: the emission of electrons from certain metals when light (electromagnetic radiation) of certain frequencies shines on the metals.

photon: a particle of electromagnetic radiation; a quantum of light energy.

model: a representation of a process, system, or object.

Answers

For You To Do

(continued)

8. a) Students can draw similar diagrams to the one in the text with larger or smaller wavelengths.

9. a) Students can draw waves that have 2 wavelengths, 2.5 wavelengths, 3 wavelengths, etc. in the box.

10. a) You would expect to find the electrons at the positions that correspond to peaks in the graphs. This is where the probability is largest. You would not expect to find the electrons where the graph touches the axis. A zero probability implies that the electron is never there.

© It's About Time

FOR YOU TO READ

Developing the Schrödinger Model

Light is everywhere. You can see because of light, but it is unusual to see light beams. When you do see light rays piercing through the clouds or a laser beam intensely moving forward, you recognize that light travels in straight lines. In this activity you found out that light may also travel a curved path as it squeezes through a small opening.

The process of light bending as it squeezes through a small opening is called **diffraction**. Light bending around an edge is also referred to as diffraction. Diffraction is one of the properties that water waves and sound waves exhibit. This leads toward the conclusion that light might also behave as a wave.

You then explored another property of water and sound waves. When water waves meet, they interfere with one another. **Constructive interference** occurs when the crests of a wave meet the crests of the second wave and there is lots of movement of the water. **Destructive interference** occurs when the troughs of one wave meet the crests of the other wave and the water remains still.

You used a tuning fork to investigate if sound waves also interfere. A tuning fork produces identical sounds from each prong. At one orientation from the prongs to your eardrum, you heard a loud sound. This was constructive interference. When the orientation from the prongs to your ear was different, the sound diminished. The sound from one source can meet with the sound from another source and produce silence! The interference of waves is another property of wave behavior.

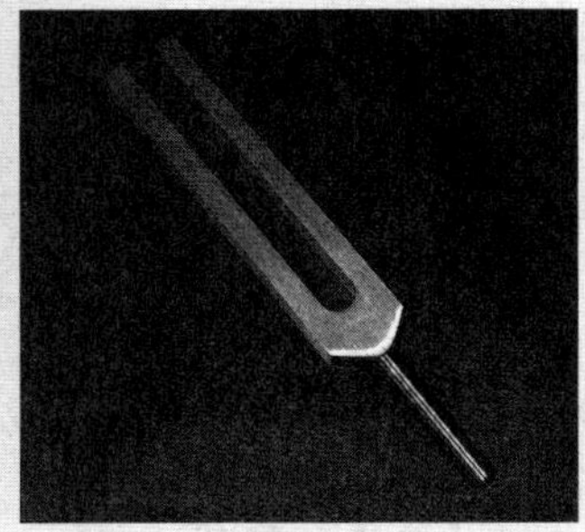

If light behaves like a wave, then it, too, must show interference. When you shone a laser

© It's About Time

beam through two slits onto a distant wall, you saw places where the light from neighboring slits interfered constructively (there was maximum light) and places where the light interfered destructively (there was little or no light). Evidence of the interference of light would seem to prove that light behaves like a wave. The colors you see when light reflects off a CD are interference effects. You were able to use the interference of light to measure its wavelength.

After you found that light behaves like a wave and measured its wavelength, you turned to the puzzle of the 20th century that Einstein solved. The puzzle had to do with the behavior of light when it freed an electron from a metal. In the **photoelectric effect**, beams of light shine on metals. Some wavelengths (and frequencies) of light will always free electrons, while other wavelengths (and frequencies) will never free electrons. Light of high frequency (ultraviolet light) frees many electrons and gives them kinetic energy. Light of low frequency (red light) is not able to free any electrons.

Einstein derived the following equation for the photoelectric effect in 1905:

$$KE_{electron} = E_{light} - W_o$$

The energy of the light E_{light} is equal to hf where h is Planck's constant (6.63×10^{-34} Js) and f is the frequency of light. It was this equation (not his Theory of Relativity) for which Einstein won the Nobel Prize in physics.

The photoelectric effect explanation makes sense only if light were behaving as a particle and there were a collision between the **photon** and the electron. The extraordinary conclusion is that light sometimes behaves as a particle (the photoelectric effect) and sometimes behaves like a wave (diffraction and interference effects).

Now consider the behavior of electrons. Electrons hit the front of your TV screen and make a temporary mark. Electrons have a mass (9.1×10^{-31} kg) and a charge (1.6×10^{-19} C). Electrons behave like particles. Electrons can also interfere. Clinton Davisson and Lester Germer demonstrated this by shooting electrons through a metal crystal foil. They observed some locations on a distant screen where many electrons landed and other locations where no electrons landed. This was explained as an interference effect of electrons. Electrons, like light, can sometimes behave like a particle and sometimes behave like a wave.

If electrons have wave characteristics, then you can expect some unusual behavior. If you place an electron in a one-dimensional box it can

→

© It's About Time

move back and forth along the line segment. When you try to locate the position of the electron, you never find it at the very edge of the box. This is not surprising since you would expect the electron to rebound off the walls at the end of the line segment. In the activity you found that there are times, surprisingly, where you also never find the electron in the middle of the box. This is quite peculiar.

You may be wondering how the electron gets from one side of the box to the other if it is never found in the middle of the box. The explanation is that the electron does not behave like any particle or billiard ball that you are familiar with. You need both the wave and particle languages to fully explain the electron. The electron is neither a wave nor a particle—it's an electron. This need for both models is called the wave-particle duality. A key to understanding wave-particle duality is to recognize that waves and particles are **models**. They are conceptual representations of real things. Wave and particle models are based on experience with water and billiard balls. The "real" nature of an electron is well beyond the range of immediate experience.

Electrons move about the nucleus and in particle language they move in a restricted orbit. However, the model for the structure of the atom also includes a wave model. The electron may move about the nucleus, but it is not restricted to fixed orbits. The electron has wave properties. The best one can do is to describe the probability of where the electron can be found and where it cannot be found. Bohr's orbits correspond to locations where the electron in three dimensions is most likely to be found. Each orbit corresponds to the most probable location for an electron of specific energy. The equation that describes the electron wave and its corresponding probabilities is the Schrödinger equation. It is out with the Bohr model and in with Schrödinger's model. In the Schrödinger model, light is given off or absorbed by the electron when the electron wave pattern changes.

Austrian physicist Erwin Schrödinger (1887–1961). He shared a 1933 Nobel Prize for new formulations of the atomic theory.

The quantum world is not the like the everyday world you experience. The classical view of the world makes sense, but does not provide a complete picture. The quantum world appears to contradict common sense, but it gives correct answers. All the numerical results from predictions based on the Schrödinger model are accurate. It is a dilemma. Do you go with the theory that gives all the right predictions but runs counter to common sense OR go with common sense that does not give accurate experimental results? What do you think?

k© It's About Time

Reflecting on the Activity and the Challenge

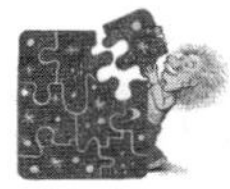

In this activity, you found out that light behaves like a particle in the photoelectric effect and like a wave in diffraction and interference effects. Similarly, electrons behave like particles when they hit a screen and like waves when they move about the nucleus. The atomic model has gotten more complex. You can no longer predict where the electron can be, but only where it is likely to be and where it will never be. Depicting the probability of electron location in an atom will be quite an undertaking for your museum exhibit. Light is given off when the wave pattern (and the probability locations) changes. Your museum exhibit may require you to help visitors explain what an interference effect is and how it is evidence of wave behavior. You may choose to focus on where the electron is found in a 1-dimensional box. Creativity and your imagination will be required to make this part of your exhibit interactive and scientifically correct.

Physics To Go

1. Describe two differences between particles and waves.
2. Someone decides that a laser beam is not thin enough. They decide to pass the beam through a very thin slit to slim it down. Will this work? Explain.
3. What was the principal understanding that emerged from Einstein's explanation of the photoelectric effect?
4. A baseball is bouncing back and forth between two walls. An electron is trapped in a similar "box." How does the ball's behavior differ from the electron's behavior?
5. Suppose the probability function (wave) for an electron in a box is represented below.

a) In which locations would you likely find the electron?
b) Where would you never expect to see the electron?
c) If the electron were to lose energy, how would the probability function (wave) change?

Answers

Physics To Go

1. Particles can hit an object at a specific location while waves are spread out.

   When shot toward two slits, particles go through one slit or the other, while waves go through both slits.

   Waves can diffract and interfere, while particles do not.
2. No, this will not work. As the slit gets thinner, there is greater diffraction and the beam will spread out.
3. Einstein's explanation of the photoelectric effect showed that light can behave as a particle.
4. The baseball has a defined path. It moves left and it moves right. From its past behavior, you can have a good chance of knowing where it will be.

   The electron in a box is restricted in certain ways. This restricts its energy and the probability of where it will be found.
5. a) You have a greater chance of observing the electron at the places where the square of that function has a large amplitude.

   b) You will never observe the electrons where the wave touches the axis in the diagram.

   c) If the electron were to lose energy, it would result in an increased wavelength. This would alter where you expect to find the electron.

© It's About Time

ANSWERS

Physics To Go
(continued)

6. The electrons do not travel in fixed locations. The size of the nucleus is much too large in this depiction.
7. The energy of the emitted electron will be 5.8 eV.
8. $KE_{electron} = E_{light} - w_o$

   The *KE* term represents the kinetic energy of the freed electron

   The E_{light} represents the energy of the incoming light

   The w_o represents the work function, the least energy required to free the electron.
9. Answers will vary.
10. Answers will vary.
11. The electrons cannot travel in ways that we are accustomed to. We can only talk about the probability of locating the electron. The electron can sometimes be found on the left side of a one-dimensional box and sometimes be found on the right side of that box, but will never be found in the middle of the box.
12. Answers will vary and will reflect the students' opinions. Scientists have decided that correct predictions are the way to go.

6. Describe three things wrong with this familiar diagram of an atom.

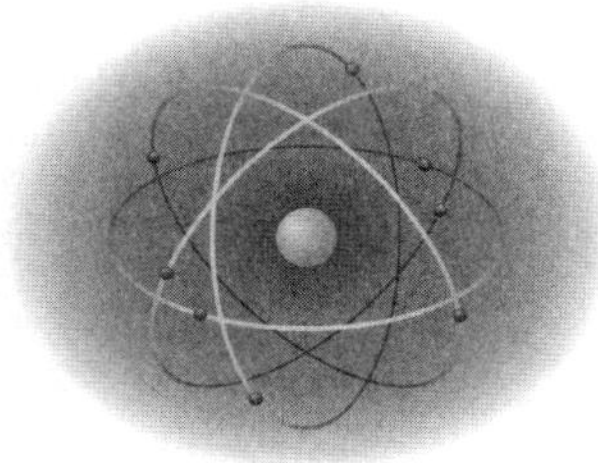

7. In the photoelectric effect, a 10 eV photon of light frees an electron from a metal with a work function of 4.2 eV. What is the energy of the emitted electron?
8. The equation for the photoelectric effect is:

   $KE_{electron} = E_{light} - w_o$

   Explain what each of the terms in the equation represent.
9. In designing your museum exhibit, what might be a creative way to show the unusual behavior of an electron in an atom?
10. For your museum exhibit or perhaps a product for the museum store, can you invent a photoelectric-effect bank? How would it work?
11. The great physicist Niels Bohr once suggested that if the behavior of the electron makes sense to you, then you *don't* understand it at all. Describe some strange behavior of the electron.
12. In the **For You To Read** section you were asked what you would do: go with a theory that gives all the right predictions but runs counter to your common sense OR go with your common sense that does not yield accurate experimental results. Explain your answer to this question and the reasoning you used. (Consider that common sense is based on experience.)

© It's About Time

NOTES

© It's About Time

ACTIVITY 6
Inside the Nucleus

Background Information

It is suggested that you read the **For You To Read** section of the student text for **Activity 6** before proceeding in this section.

Given that the atom is composed of electrons and protons, it was first thought that a nucleus had protons and electrons. The protons accounted for the mass of the nucleus but the embedded electrons could help account for its positive charge. For instance, carbon 12 has a mass of 12 but only has a +6 charge in the nucleus. With the discovery of the nucleus by Chadwick in 1932, an alternative explanation emerged. The carbon nucleus had a mass of 12 (6 protons + 6 neutrons) but only had a charge of +6 (6 protons). With this model, the electrons all reside outside the nucleus.

Electrons residing outside the nucleus is a requirement of quantum mechanics. By the Heisenberg uncertainty principle, if the position (and wavelength) of the electron were confined to the size of the nucleus, it would have enormous momentum and could not reside in the nucleus.

Throughout the discussion of forces of nature, we have always been content to describe the forces. Newton in describing the force of gravitation stated, "I feign no hypothesis." He was not describing why the force exists or how it was transmitted, but rather he described the force in terms of mass and distance. The force at a distance (the Sun attracts the Earth) has always seemed a bit of mystery. How is the force transmitted? The Feynman model (a part of quantum electrodynamics) seeks to explain the transmission of force as an exchange of virtual particles. The diagrams pictorially represent these exchanges but can also be used to calculate the strength of forces. Feynman shared the Nobel Prize with Schwinger and Tomonaga for these insights. The theory of quantum electrodynamics (qed) is probably the most precise theory we have in physics.

Planning for the Activity

Time Requirements

One class period is required to complete this activity.

Materials Needed

For each group:

- basic calculator

Advance Preparation and Setup

No advance preparation or setup is required.

Teaching Notes

This activity requires students to calculate the number of protons, neutrons, and electrons in the atom, to calculate forces between protons, and then to recognize the need for a nuclear force. It concludes with students creating some Feynman diagrams.

The calculations do require an ability to use exponential notation. You may wish to help students understand how to use exponents on their calculators. As an alternative, you can review the rules of exponents. (Add the powers of 10 in the numerator, add the powers of 10 in the denominator and subtract the value in the denominator from the numerator.) By having the students add and subtract the exponents, a simple calculator can be used.

The history of atomic structure in this activity introduces students to additional Nobel Prize winners. Chadwick won the prize for his discovery of the nucleus. Feynman shared the Nobel Prize for insights into the exchange of virtual particles.

© It's About Time

Activity 6 Inside the Nucleus

Activity 6 Inside the Nucleus

GOALS

In this activity you will:

- Determine the number of protons, electrons and neutrons in a neutral atom given the atomic number and atomic mass.
- Calculate the Coulomb force of repulsion between protons in a nucleus.
- Recognize that the nuclear force must be much greater than the Coulomb force.
- Draw Feynman diagrams for the interaction between charges.

What Do You Think?

The alchemist's dream has always been to turn worthless lead into valuable gold.

- **What is the difference between lead and gold?**
- **How can you distinguish one from the other?**

Record your ideas about these questions in your *Active Physics* log. Be prepared to discuss your responses with your small group and your class.

587 Active Physical Science

Activity Overview

With a huge Coulomb repulsion pushing protons apart, the need for a strong attractive force in the nucleus is investigated. Students are then introduced to Feynman diagrams as a means of understanding how forces are transmitted.

Student Objectives

Students will:

- Determine the number of protons, electrons, and neutrons in a neutral atom given the atomic number and atomic mass.
- Calculate the Coulomb force of repulsion between protons in a nucleus.
- Recognize that the nuclear force must be much greater than the Coulomb force.
- Draw Feynman diagrams for the interaction between charges.

ANSWERS FOR THE TEACHER ONLY

What Do You Think?

Lead and gold have different numbers of protons, neutrons, and electrons. They are also different in their atomic weights and their availability on Earth. They have different properties and are very, very different in value.

We are able to distinguish between them on both the microscopic and macroscopic levels. Macroscopically, they have different densities, hardness, luster, and melting points. Microscopically, they differ in the numbers of electrons, protons, and neutrons.

© It's About Time

ANSWERS

For You To Do

1. a) The ratio of the proton mass to the electron mass = $(1.7 \times 10^{-27}$ kg$)/(9.1 \times 10^{-31}$ kg$) = 1868$.

   b) Yes, the proton with a mass 1868 times as large as the electron convinces us that most of the mass is in the nucleus.

2. a) 12 protons equals a mass of 12.

   b) 12 protons = + 12,

   6 electrons = – 6

   Total charge within nucleus + 6.

   The 6 orbiting electrons would then make the atom neutral.

3. a) 6 protons + 6 neutrons = 12

   b) 6 protons + 6 electrons = 0 charge

4. a) 35 – 17 = 18 neutrons

   b) 79 protons, 79 electrons, 118 neutrons

   c) 19 protons, 19 electrons, 20 neutrons

For You To Do

1. The nucleus has all the positive charge and almost all the mass of the atom. Hydrogen is the simplest atom. Hydrogen has one proton and one electron. The proton is the hydrogen nucleus. The mass of a proton is 1.7×10^{-27} kg and the mass of an electron is 9.1×10^{-31} kg.
   a) Calculate the ratio of the proton mass to the electron mass.
   b) Does the value you calculated convince you that most of the mass is in the nucleus? Explain your answer.

2. Carbon is more complicated than hydrogen. Carbon has a mass of 12, but has only 6 electrons surrounding the nucleus. One possibility is that there are 12 protons and 6 electrons within the nucleus and 6 electrons orbiting the nucleus.
   a) Show how this would account for a mass of 12.
   b) Show how this would account for the atom having a net charge of zero.

3. The neutron has a mass almost identical to the proton, but has no electrical charge. It was determined that the carbon atom has 6 electrons, 6 protons, and 6 neutrons.
   a) Show how this would account for a mass of 12.
   b) Show how this would account for the atom having a net charge of zero.

4. The neutral atom of chlorine has 17 electrons and 17 protons. It can be written in the notation $^{35}_{17}Cl$ where the lower number corresponds to the atomic number *and/or* the number of protons in the nucleus *and/or* the number of electrons surrounding the nucleus. The upper number corresponds to the atomic mass *and/or* the sum of the number of protons and neutrons.
   a) Calculate how many neutrons are found in the nucleus of $^{35}_{17}Cl$.
   b) Determine the number of protons, electrons, and neutrons in gold: $^{197}_{79}Au$.
   c) Determine the number of protons, electrons, and neutrons in potassium: $^{39}_{19}K$.

© It's About Time

5. Protons repel other protons. The force can be calculated using Coulomb's Law:

$$F = \frac{kq_1q_2}{d^2}$$

a) Calculate the force between two protons in the nucleus. The charge on each proton is 1.6×10^{-19} C. The distance between them on the average is the size of the nucleus 1.0×10^{-15} m. Coulomb's constant k equals 9×10^9 Nm²/C².

b) Calculate the acceleration of a proton from that force using Newton's Second Law $F = ma$. The mass of a proton is 1.7×10^{-27} kg.

6. This strong repulsive force should push the protons apart. There must be a force holding the nucleus together. This nuclear force is also called the "strong force." It has the following properties:

- strong at short distances (the size of the nucleus 10^{-15} m)
- weak at long distances (greater than 10^{-15} m)
- attraction of protons to protons, neutrons to neutrons, and protons to neutrons
- no effect on electrons

a) Why is it important that the force is strong at short distances and weak at long distances?

b) Draw a graph of the strong nuclear force versus the distance between protons. Note the size of the nucleus on the x-axis of your graph.

7. Neutrons and protons belong to a family of particles called "baryons." Baryons can "feel" the strong nuclear force. Electrons belong to a family called leptons. Leptons do not "feel" the strong nuclear force.

a) List the combination of particles between which the nuclear force can exist.

b) The Coulomb force is only between proton and proton in the nucleus. Why do you think that there is no Coulomb force between neutrons?

ANSWERS

For You To Do
(continued)

5. a) $F = kq_1q_2/d^2$

$F = (9 \times 10^9\ \text{Nm}^2/\text{C}^2)(1.6 \times 10^{-19}\ \text{C})(1.6 \times 10^{-19}\ \text{C})/(1 \times 10^{-15}\ \text{m})^2$

$F = 230$ N

b) $F = ma$

$a = F/m$

$= 230\ \text{N}/ 9.1 \times 10^{-31}$ kg

$a = 2.50 \times 10^{31}\ \text{m/s}^2$

6. a) The protons must be held within the nucleus. However, neighboring nuclei should not be pulled together.

b)

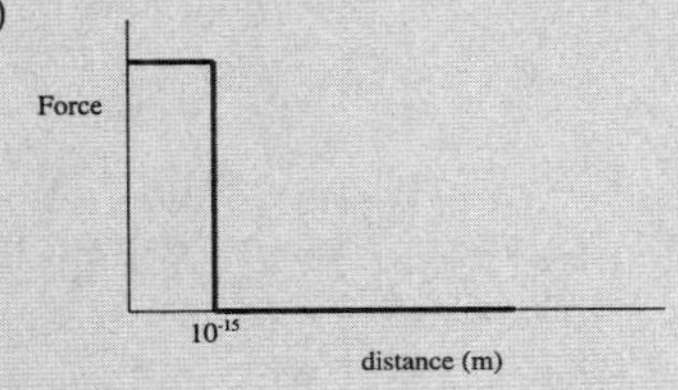

7. a) The nuclear force can exist between protons and protons, neutrons and protons, and neutrons and neutrons

b) Neutrons have no charge.

© It's About Time

Chapter 9

ANSWERS

For You To Do
(continued)

8. a) Student answers will vary.
   b) Student answers will vary.
9. a) An electron at rest is a vertical line on the Feynman diagram.
   b) The Feynman diagram will be identical to that of the electrons in the student text.
   c)

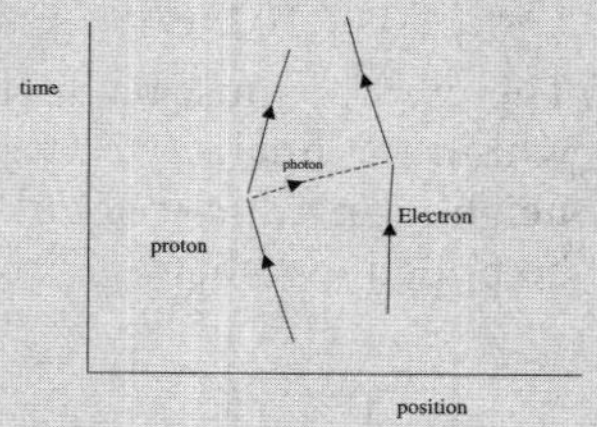

10. a) The diagram will be identical to the Feynman diagram for the attraction of protons in the student textbook.
    b) The diagram will be identical to the Feynman diagram for the attraction of protons.

8. By what means do forces such as gravitational and Coulomb force act at a distance?
   a) You hold out your hand and drop a penny. How does the penny "know" which way is down? How is the gravitational force between the penny and the Earth communicated? Write down your thoughts on this profound question.
   b) You charge up a rod and attract a small Styrofoam ball. How does the ball know the directions to move? How is this electrostatic force different than the gravitational force? Write down your thoughts.

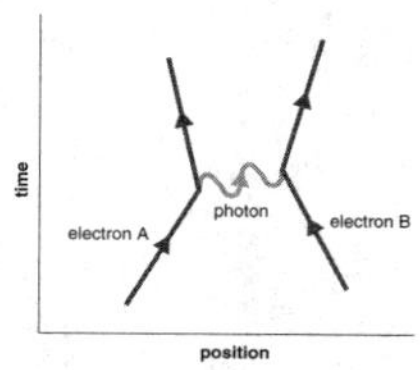

9. Forces act through an exchange of "virtual particles." This can be shown in a Feynman diagram of the repulsion of two electrons.

   Time is on the *y*-axis. Electron A is at first moving to the right as time progresses. Electron B at first is moving to the left. Electron A emits a virtual particle of light (a photon) and recoils to the left. Some time later, electron B absorbs the virtual photon of light and recoils to the right. The effect is that the electrons have repelled each other.
   a) Draw a Feynman diagram of an electron at rest.
   b) Draw a Feynman diagram of a proton and proton repelling.
   c) Draw a Feynman diagram of the attraction between a proton and an electron.

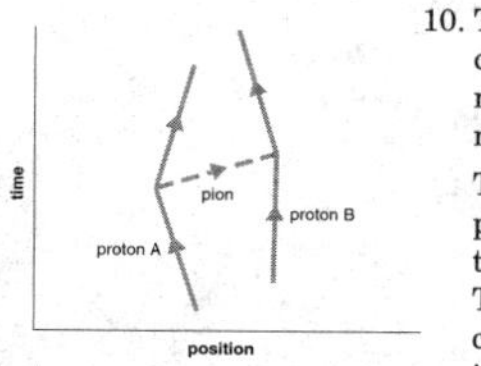

10. The nuclear force is due to the exchange of virtual particles called "mesons." The protons and neutrons exchange mesons; the protons and protons exchange mesons; the neutrons and neutrons exchange mesons.

    The diagram looks very similar to the exchange of virtual photons in the repulsion of two electrons. In this case there is the *attraction* of two protons. The pion is a type of meson. There are pions with a positive (+) charge, a negative (–) charge, and no (0) charge. It is the exchange that is most important in the diagram, not the direction of motion of the particles. The attractive strong nuclear force depicted here is mediated with a virtual pion and is shown with a straight line.
    a) Draw the Feynman diagram for the attractive strong nuclear force between two neutrons.
    b) Draw the Feynman diagram for the attractive strong nuclear force between a proton and a neutron.

© It's About Time

11. In both of the previous Feynman diagrams, the mediating particle was uncharged. The photon is always neutral. The pion comes in three varieties – the neutral, the positive, and the negative pions. The attractive strong nuclear force between a proton and a neutron can also be mediated by a positive pion as shown in the diagram to the right.

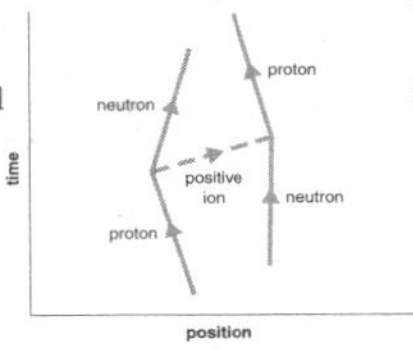

Notice that the proton emits a virtual positive pion. The pion carries the positive charge and the proton becomes a neutron. The neutron absorbs the virtual positive pion and becomes a proton. The strong nuclear force is still attractive. At every vertex of a Feynman diagram, the charge must be conserved.

a) Draw a Feynman diagram where a neutron emits a pion and becomes a proton. Note the charge of the pion on your diagram.

FOR YOU TO READ

Holding the Nucleus Together

Rutherford's scattering experiment provided evidence for a dense nucleus at the center of the gold atom. Early models of the nucleus had included the **proton**. However, when you examined a carbon atom you realized that the proton alone could not account for the mass of the nucleus. Rutherford addressed this problem when he suggested that another particle was present in the nucleus. He proposed a particle with about the same mass as the proton, but with no electric charge. He named this particle the **neutron**. In 1932 James Chadwick, a British physicist, actually discovered the neutron. This discovery added a great deal to the understanding of the nucleus of the atom. However, the model of the nucleus still posed a huge puzzle.

Sir James Chadwick, a British physicist, was winner of the Nobel Prize in 1935. His discovery of the neutron led eventually to the development of the nuclear bomb.

How can all of those protons and neutrons fit into a small space? Protons repel protons. You calculated this force of repulsion in the **For You To Do** section. Also, you calculated the acceleration due to this force. With these kinds of forces and corresponding accelerations, you would expect the nucleus to explode. There must be another force holding the nucleus together. This force must be very large. It must also be limited to very small distances. If the force were long-range, then all the nuclei would clump together. The force must be strong enough to hold all of the protons together but short-range so that one nucleus →

ANSWERS

For You To Do *(continued)*

11.

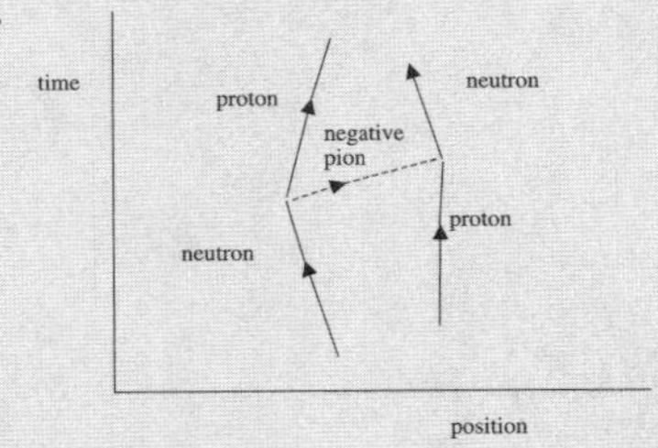

© It's About Time

Atoms on Display

does not affect the neighboring nucleus. There is such a force holding the nucleus together. It is called the **nuclear force** or the strong force. The strong nuclear force exists between neutrons and protons, protons and protons, and neutrons and neutrons. These particles belong to a family of particles called **baryons**. Baryons can "feel" the strong nuclear force. The Coulomb force is only between protons and protons in the nucleus. Neutrons have no charge and do not "feel" the Coulomb force. Electrons belong to a family called **leptons**. Leptons do not "feel" the strong nuclear force.

In physics, you have studied the effect of forces as explained by Newton's Second Law ($F = ma$). Net forces produce accelerations. You have calculated the strength of forces such as the gravitational force $F = \frac{Gm_1m_2}{d^2}$,

weight $F = mg$,

and the Coulomb's electrostatic force $F = \frac{kq_1q_2}{d^2}$.

In this activity, you were asked to consider by what means these forces act at a distance.

When you push on something, where does the force come from? Consider this question at the atomic level. There are electrons in your hand and there are electrons in the table. As the electrons get close, there is repulsion. This repulsion is the force you feel. This repulsion moves the table forward. The electrons in your hand are trying to get closer to the electrons in the table and the table moves because of the electrostatic repulsion as calculated by Coulomb's Law. The force can be huge because there are so many electrons.

How does one electron know the other electron is there? The 20th century breakthrough answer is that forces act through an exchange of "virtual particles." The electrons continuously exchange virtual photons of light so that the repulsion continues. The virtual photons of light are exchanged in a time too short to observe directly. You drew a Feynman diagram to illustrate this exchange.

You also drew Feynman diagrams to illustrate the nuclear force. In this case, the virtual particles that are exchanged are **mesons**. Specifically, they are pions that can have positive, negative, or no charge. The Feynman diagrams you drew depicted the forces and their exchange of particles. These diagrams can provide even more information. They can be used to calculate the strength of the forces. Richard Feynman showed that equations and calculations that at one time required days to complete could be completed in minutes using the diagrams he invented. Feynman was awarded the Nobel Prize for these achievements.

© It's About Time

Reflecting on the Activity and the Challenge

The nucleus is a crowded place. It contains all of the protons and neutrons of the atom. These protons and neutrons are held together by the strong nuclear force. The force is short-range and very attractive. The strong nuclear force overwhelms the Coulomb repulsive force between protons. The force is mediated by the exchange of charged and uncharged pions. Your museum exhibit on the structure of the atom may focus on the nucleus and its teeming micro-world of protons, neutrons and virtual pions. Communicating the size of the nucleus and how it is held together will be quite a creative challenge.

Physics To Go

1. In the following notation for a carbon nucleus $^{13}_{6}C$, what do each of the numbers represent?
2. A proton is represented as $^{1}_{1}p$ and a neutron is represented as $^{1}_{0}n$.
   a) Why do they both have superscripts of 1?
   b) Why do they have different subscripts?
3. Isotopes are elements that have identical atomic numbers but different atomic masses. For the following sets of isotopes, list the number of protons, neutrons, and electrons:
   a) $^{12}_{6}C$, $^{13}_{6}C$, $^{14}_{6}C_6$
   b) $^{40}_{20}Ca$, $^{41}_{20}Ca$, $^{42}_{40}Ca$
   c) $^{232}_{92}U$, $^{235}_{92}U$
4. Calculate the electrostatic force between two protons that are separated by 6×10^{-14} m within the nucleus.
5. Calculate the electrostatic force between an electron and a proton that are separated by 8×10^{-9} m in an atom.
6. Sketch a graph showing how the nuclear force between two protons varies over distance.
7. Sketch a graph showing how the electrostatic force between two protons varies over distance.

Physics Words

proton: a subatomic particle that is part of the structure of the atomic nucleus; a proton is positively charged.

neutron: a subatomic particle that is part of the structure of the atomic nucleus; a neutron is electrically neutral.

nuclear (strong) force: a strong force that hold neutrons and protons together in the nucleus of an atom; the force operates only over very short distances.

baryon: a group of elementary particles that are affected by the nuclear force; neutrons and protons belong to this group.

lepton: a group of elementary particles that are not affected by the nuclear force; electrons belong to this group.

meson: a virtual particle that mediates the strong, nuclear force of an atom; the protons and neutrons exchange mesons; the protons and protons exchange mesons; the neutrons and neutrons exchange mesons.

ANSWERS

Physics To Go

1. The 6 represents 6 protons and 6 electrons; the 13 represents the sum of the protons and neutrons. Since there are 6 protons, there must be 13 – 6 = 7 neutrons.

2. a) Both have a mass of 1.

   b) The proton has a charge of 1; the neutron has no charge.

3. a) All have 6 protons; the number of neutrons are 6, 7, 8.

   b) All have 20 protons; the number of neutrons are 20, 21, 22.

   c) Both have 92 protons; the number of neutrons are 140 and 143.

4. $F = kq_1q_2/d^2$

   $F = (9 \times 10^9 \text{ Nm}^2/\text{C}^2)(1.6 \times 10^{-19}\text{ C})(1.6 \times 10^{-19}\text{ C})/(6 \times 10^{-14}\text{ m})^2$

   $F = 0.064$ N

5. a) $F = kq_1q_2/d^2$

   $F = (9 \times 10^9 \text{ Nm}^2/\text{C}^2)(1.6 \times 10^{-19}\text{ C})(1.6 \times 10^{-19}\text{ C})/(8 \times 10^{-9}\text{ m})^2$

   $F = 3.6 \times 10^{-12}$ N

6 & 7:

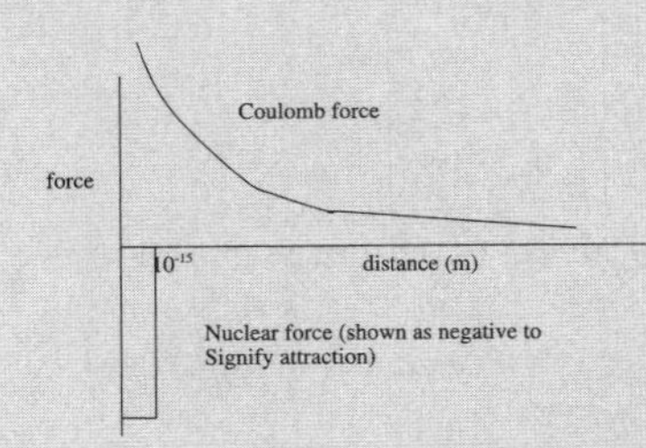

© It's About Time

Chapter 9

ANSWERS

Physics To Go

(continued)

8. (see below)
9. a) $F = Gm_1m_2/d^2$

   $F = (6.67 \times 10^{-11}\ \mathrm{Nm^2/kg^2})(1.7 \times 10^{-27}\ \mathrm{kg})(1.7 \times 10^{-27}\ \mathrm{kg})/(1 \times 10^{-15}\ \mathrm{m})^2$

   $F = 1.9 \times 10^{-34}\ \mathrm{N}$

   b) $F = kq_1q_2/d^2$

   $F = (9 \times 10^9\ \mathrm{Nm^2/C^2})(1.6 \times 10^{-19}\ \mathrm{C})(1.6 \times 10^{-19}\ \mathrm{C})/(1 \times 10^{-15}\ \mathrm{m})^2$

   $F = 230\ \mathrm{N}$

   c) The gravitational force cannot be responsible for holding the nucleus together. The gravitational force is attractive but much too weak to overcome the electrostatic force.
10. The diagram of the proton would be a vertical line.
11. At the first vertex, the proton (+) gives off a positive pion (+); the neutron has no charge.

    At the second vertex, the positive pion (+) is absorbed by the neutron and becomes a proton (+).
12.

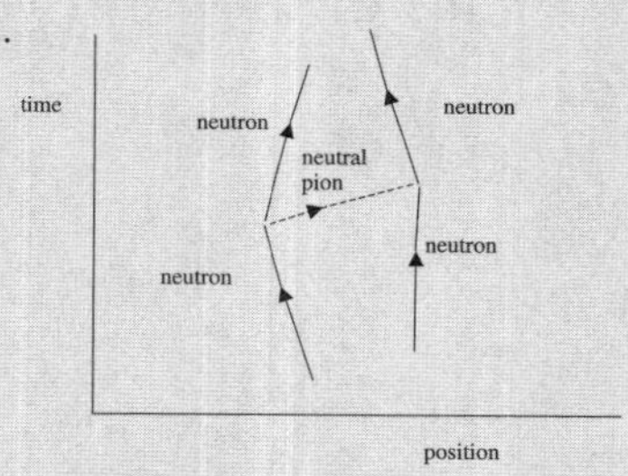

13. The electrons in one fingertip repel the electrons in the other fingertip and the skin gets depressed.
14. Students' answers will vary

Atoms on Display

8. Complete a chart indicating if the following pairs of particles interact by the electrostatic force and/or the nuclear force:

   proton-proton; proton-electron; proton-neutron; neutron-electron; electron-electron; neutron-neutron.
9. Two protons are separated by 1×10^{-15} m in a nucleus.
   a) Calculate the gravitational force between them.
   b) Calculate the electrostatic force between them.
   c) Can the gravitational force be responsible for holding the nucleus together? Explain.
10. Draw a Feynman diagram of a single proton at rest.
11. In the Feynman diagram on the left, show that the charge is conserved at each vertex (i.e., the charge entering the vertex is equal to the charge exiting the vertex).
12. Sketch a Feynman diagram for the interaction between two neutrons.
13. When your fingertips touch, your skin deforms. Explain why this happens and why the deformation increases when you press harder.
14. Could a museum exhibit help visitors understand how the nuclear force is able to hold the nucleus together but not pull neighboring nuclei together? Could it be interactive? Could it capture people's attention within 30 seconds? Describe such an exhibit.

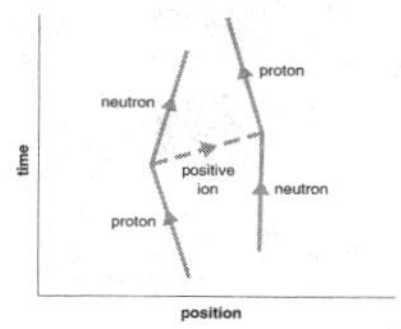

Stretching Exercise

Prepare a research report and/or a simulation of the experiment that Chadwick performed to discover the neutron in 1932.

Active Physics 594

8.

Particles Interacting	Electrostatic Force Interaction	Nuclear Force Interaction
Proton-proton	Yes	Yes
Proton-electron	Yes	No
Proton-neutron	No	Yes
Neutron-electron	No	No
Electron-electron	Yes	No
Neutron-neutron	No	Yes

© It's About Time

NOTES

© It's About Time

ACTIVITY 7
Radioactive Decay and the Nucleus

Background Information

It is suggested that you read the **For You To Read** section of the student text for **Activity 7** before proceeding in this section.

One element can become another element through radioactive decay. In radioactive decay, the parent nucleus "spits out" a particle. The remaining daughter nucleus is now a different element. The particle that is emitted from the nucleus can be an alpha ray (helium nucleus), a beta ray (a high energy electron), or a gamma ray (a photon of light or electromagnetic energy).

Radioactive decay is a probabilistic event. We do not have any way of knowing which nuclei will decay but we have a very good notion of how many will decay during a given time. This is similar to throwing dice or spinning a roulette wheel. We have no indication of what random number will come up during the next throw or spin but we do know that after a given time, the numbers will form a distribution based on the statistics. The more spins, the closer the results will come to the statistical average. In radioactive decay, we are dealing with 10^{23} particles and the statistical average is quite close to what truly happens.

Since radioactivity is not affected by heat or pressure, we can use the radioactivity of elements as a clock to determine the age of artifacts. Some radioactive elements are good for artifacts that are millions of years old while others are good for artifacts that are a few thousand years old. In medicine, we use radioactive elements to aid in diagnosis. Here, the radioactive isotopes used have very short lifetimes.

Planning for the Activity

Time Requirements

One class period is required to complete this activity.

Materials Needed

For each group:

- basic calculator
- felt-tip marker
- graph paper, pad of 50 sheets
- interlocking gram cubes (substituted for sugar cubes)
- plastic beaker, 1000 mL

Advance Preparation and Setup

Be sure that the box allows free rotation of the sugar cubes

Teaching Notes

We have tried to ensure that certain misconceptions about radioactivity are avoided in the use of our simulation. We use sugar cubes rather than coins or M&M's® to overcome the misconception that the half-life is always equal to one toss. With the sugar cubes, the half-life can vary depending on how many faces of the cube get dots. We also require that the decayed sugar cube get replaced with a penny so that students are reminded that after decay there is a daughter nucleus and that mass is conserved.

Another misconception that students adopt after a decay simulation is that the radioactive elements have to do with tosses as do the cubes. In radioactive decay, nothing is tossed. Some of the nuclei decay for a reason that we have yet to understand. Just like the toss, we have no idea which nuclei will decay, but we do know that they follow a statistical pattern and we can predict very accurately how many will decay.

To avoid this misconception, we immediately have the students change their graph from "toss" to time. This allows them to interpolate to times like 2.4 hours while 2.4 tosses would make no sense at all. Try to impress upon students that the decays are taking place continuously–they do not all happen at an instant as they do in our tosses.

An extension would allow for the pennies (the daughter nucleus) to also decay if it landed with heads up. The decayed penny could be replaced with a stable daughter nucleus, like a marble. The mathematics of the changes in the number of the pennies would be an interesting study for interested students.

© It's About Time

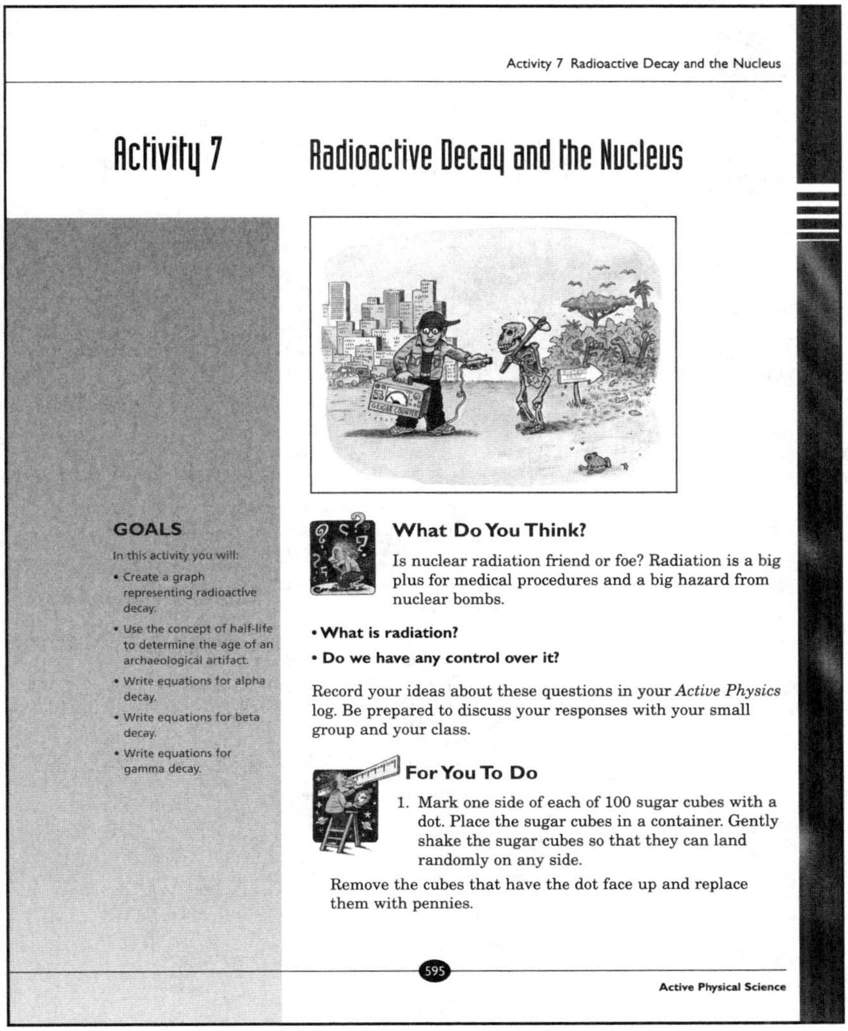

Activity 7 Radioactive Decay and the Nucleus

Activity 7 Radioactive Decay and the Nucleus

GOALS

In this activity you will:

- Create a graph representing radioactive decay.
- Use the concept of half-life to determine the age of an archaeological artifact.
- Write equations for alpha decay.
- Write equations for beta decay.
- Write equations for gamma decay.

What Do You Think?

Is nuclear radiation friend or foe? Radiation is a big plus for medical procedures and a big hazard from nuclear bombs.

- **What is radiation?**
- **Do we have any control over it?**

Record your ideas about these questions in your *Active Physics* log. Be prepared to discuss your responses with your small group and your class.

For You To Do

1. Mark one side of each of 100 sugar cubes with a dot. Place the sugar cubes in a container. Gently shake the sugar cubes so that they can land randomly on any side.

Remove the cubes that have the dot face up and replace them with pennies.

595

Active Physical Science

Activity Overview

Students investigate the statistical properties of randomly tossing sugar cubes. This is then related to the radioactive decay of nuclei and the use of half-life as a clock. Students then complete nuclear equations for alpha, beta and gamma decays.

Student Objectives

Students will:

- Create a graph representing radioactive decay.
- Use the concept of half life to determine the age of an archaeological artifact.
- Write equations for alpha decay.
- Write equations for beta decay.
- Write equations for gamma decay.

ANSWERS FOR THE TEACHER ONLY

What Do You Think?

Radiation is either alpha rays (helium nuclei), beta rays (high speed electrons), or gamma rays (photons of light – electromagnetic radiation). Radiation is emitted from the nuclei of atoms.

We are not able to change the rate of decay. We are able to create radiation by bombarding stable nuclei with other particles to make them unstable. We are also able to shield against radiation so that it cannot penetrate a barrier.

© It's About Time

ANSWERS

For You To Do

1. a) Students' results will vary. The data table is available as a Blackline Master in this Teacher's Edition at the end of the activity.
2. Students' results will vary.

   A sample data table is shown below.

Trial	Number of sugar cubes	Number of pennies	Total number of cubes and pennies
0	100	0	100
1	83	17	100
2	69	31	100
3	58	42	100
4	48	52	100
5	40	60	100
6	33	67	100
7	28	72	100
8	23	77	100
9	19	81	100
10	16	84	100
11	13	87	100
12	11	89	100

Atoms on Display

Do not eat the sugar cubes. As with anything used in a lab activity, they are contaminated and not edible.

a) Record your result in a data table like the one shown.

Trial	Number of sugar cubes	Number of pennies	Total number of cubes and pennies
0	100	0	100
1			
2			
3			
Etc.			

2. Repeat with the remaining cubes. Continue with at least 10 additional trials.
   a) Record your results each time.
   b) Graph the information from the chart. Place the trials on the x-axis and the number of cubes on the y-axis. Draw the best-fit smooth curve through most of the cube data points.
3. Use your graph to answer the following questions:
   a) How many tosses did it take to remove half the sugar cubes?
   b) In this model, let each toss of the box represent a time interval of one hour. Change the "trials" label on your graph to "hours." How many hours did it take to remove half the sugar cubes? This unit of time is the half-life of your sample, the amount of time it takes for half of the sample to change.
4. Compare your results with that of other groups.
   a) Are they similar?
   b) Combine the data from all groups. Graph the class data.
   c) How does the class data compare to your graph?
   d) During each shake of the box, what was the probability that any one sugar cube would land with the dot facing up?
5. Consider what would happen if the sugar cube had a dot on two faces.
   a) How would you expect the data to change?
   b) During each shake of the box, what was the probability that any one sugar cube would land with a dot face up?
   c) Sketch a graph that you might expect to get.
   d) What may be the value of the half-life for this new set of sugar cubes with two dots?

Active Physics 596

b) The graph will be similar in shape to the graph shown below.

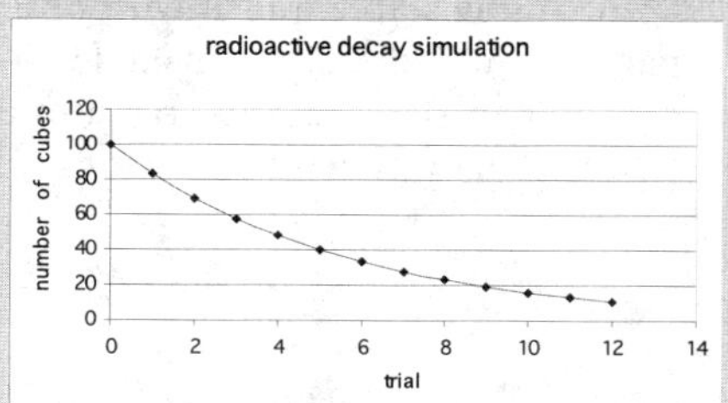

3. a) According to the sample graph, it takes between 3 and 4 tosses.
   b) 3.5 hours
4. a) Groups compare results.
   b) Class data should come very close to sample shown above.
   c) Students comment on the similarity between their graph and the graph of the class data.
   d) The probability of a dot landing up is 1/6 since there is one side with a dot and 6 sides on the cube.
5. a) We would expect the number of cubes to decrease faster because of the higher probability of a dot landing face up.
   b) It was 1/6, now it would be 2/6 or 1/3.
   c)

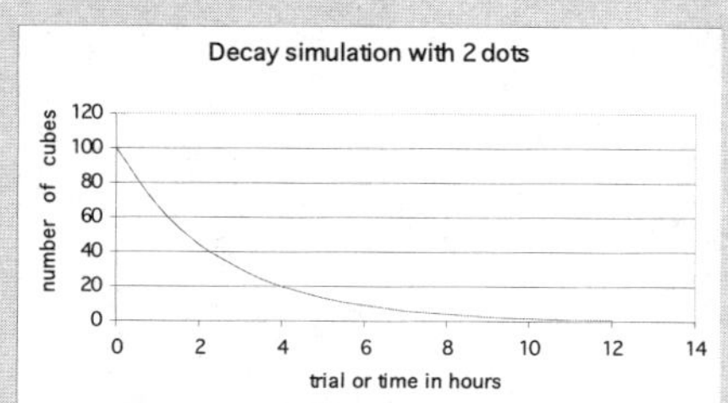

   d) The value of the half-life might be 1.6 hours.
6. a) We would expect the number of cubes to decrease faster because of the higher probability of a dot landing face up.

© It's About Time

6. Consider what would happen if the sugar cube had a dot on three faces.
   a) How would you expect the data to change?
   b) During each shake of the box, what was the probability that any one sugar cube would land with a dot face up?
   c) Sketch a graph that you might expect to get.
   d) What may be the value of the half-life for this new set of sugar cubes with three dots?

7. Consider what would happen if the sugar cube had a dot on five faces.
   a) How would you expect the data to change?
   b) During each shake of the box, what was the probability that any one sugar cube would land with a dot face up?
   c) Sketch a graph that you might expect to get.
   d) What may be the value of the half-life for this new set of sugar cubes with five dots?

8. The half-life of carbon-14 is 5730 years. That is, every 5730 years, half of the carbon-14 decays by emitting an electron and becomes nitrogen-14. By measuring the remaining carbon-14 in a long-dead sample, scientists can find out for how many half-lives the organism has been dead. If half the carbon-14 remains, then they know that 5730 years have passed. If $\frac{1}{16}\left(\frac{1}{2} \text{ of } \frac{1}{2} \text{ of } \frac{1}{2} \text{ of } \frac{1}{2}\right)$ of the carbon-14 remains, then scientists know that four half-lives, or 22,900 years, have passed.
   a) How many years have passed when the sample contains $\frac{1}{8}$ of the carbon-14?

9. Here is a graph showing the radioactive decay of a 500-g sample of C-14.

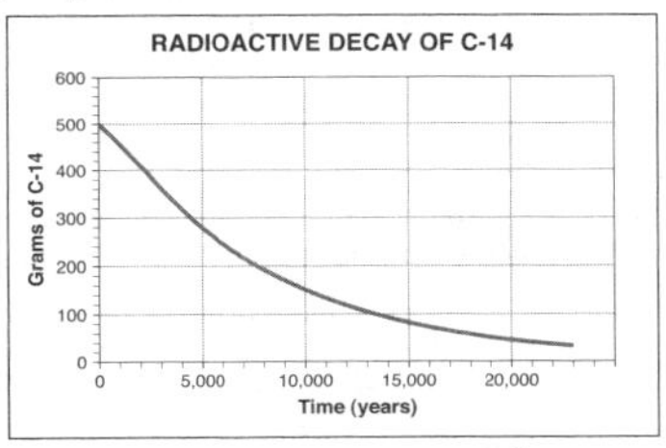

ANSWERS

For You To Do *(continued)*

b) Now it would be 3/6 = 1/2

c)

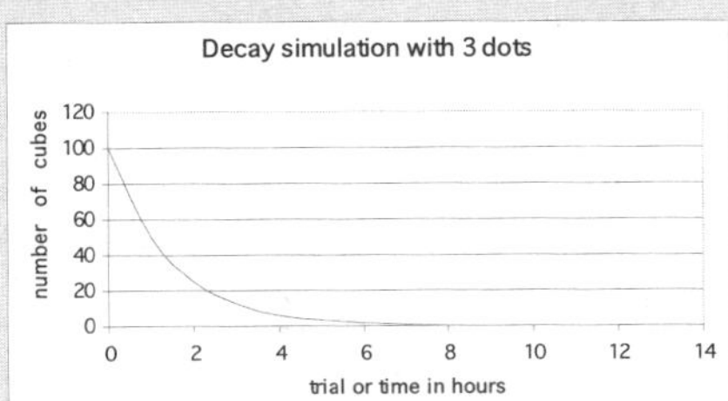

d) The value of the half-life might be 1 hour.

7. a) We would expect the number of cubes to decrease faster because of the higher probability of a dot landing face up.

b) Now it would be 5/6.

c)

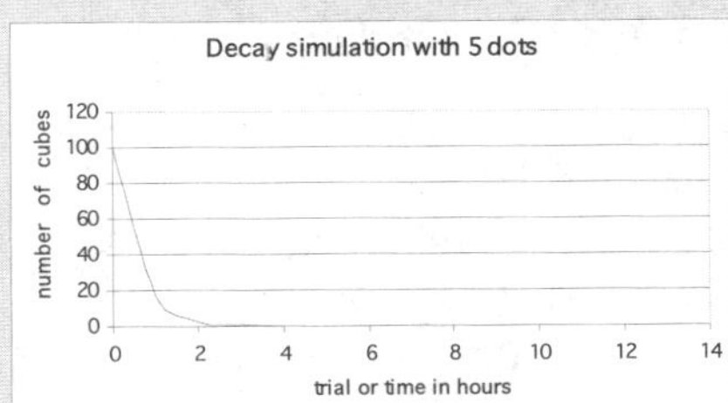

d) The value of the half-life might be 0.6 hours.

8. a) 1/8 is three half-lives

$\left(\frac{1}{2} \text{ of } \frac{1}{2} \text{ of } \frac{1}{2}\right)$

$= 5730 \text{ years} \times 3$

$= 17{,}200 \text{ years}$

© It's About Time

a) What is the half-life of the sample?

b) How many grams have undergone decay after 7000 years?

c) The original sample had 500 g. The sample now has 400 g of C-14. How much time has elapsed?

10. A 1600-g sample of iodine-131 is observed over the course of a few months. A graph of the remaining iodine-131 is shown below.

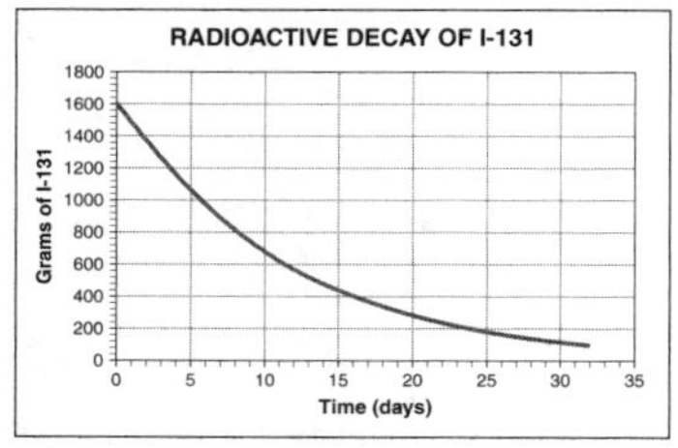

a) What is the half-life of the sample?

b) How many grams have undergone decay after 25 *days*?

c) The original sample had 1600 g. The sample now has 1400 g. How much time has elapsed?

d) The original sample had 1600 g. The sample now has 600 g. How much time has elapsed?

11. Carbon-14 (called the parent nucleus) decayed and became nitrogen-14 (called the daughter nucleus). In the last example, iodine-131 (parent nucleus) decayed and became xenon-131 (daughter nucleus). In the simulation using sugar cubes, the sugar cube (parent) "decayed" and became a penny (daughter). The total number of sugar cubes plus pennies always remained at 100. In radioactive decay, the total number of parent and daughter nuclei must also remain the same.

a) Construct a graph of the growth in the pennies over time from your earlier data.

12. Elements become other elements through radioactive decay. Three different decay processes can take place. The nucleus can emit an alpha particle. The alpha particle is a helium nucleus. It has two protons and two neutrons. It is written

ANSWERS

For You To Do *(continued)*

9. a) From the graph, the half-life of the sample is approximately 6000 years.

   b) There is 210 g remaining. Therefore 500 – 210 = 290 g have decayed.

   c) Approximately 2500 years have elapsed.

10. a) From the graph, the half-life is approximately 8 days.

    b) There is 200 g remaining. Therefore 1600 – 200 = 1400 g have decayed.

    c) Approximately 2 days.

    d) Approximately 24 days.

11. a)

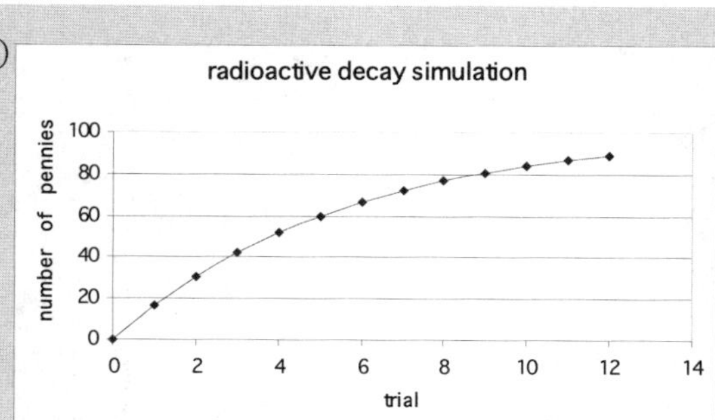

© It's About Time

as ${}^{4}_{2}He$. In all alpha decays, the number of protons must remain constant and the atomic mass (protons + neutrons) must remain constant. If an alpha particle is emitted from the parent nucleus, then the new daughter nucleus will have two fewer protons and two fewer neutrons. Writing the nuclear equation makes this much simpler.

$$ {}^{238}_{92}U \rightarrow {}^{4}_{2}He + {}^{234}_{90}X $$

where X is as yet unknown.

Notice that the numbers of protons (the numbers on the bottom) remain constant.

$$92 = 2 + 90$$

The total mass of protons and neutrons (the top number) also remains constant.

$$238 = 4 + 234$$

The new daughter element has an atomic number of 90. Referring to the periodic table, the element is thorium and its chemical symbol is Th (X = Th).

$$ {}^{238}_{92}U \rightarrow {}^{4}_{2}He + {}^{234}_{90}Th $$

Although the half-life of the element is not indicated in the decay scheme, the half-life of ${}^{238}_{92}U$ is 4.5 billion years.

a) Write the decay scheme for ${}^{235}_{92}U$ emitting an alpha particle (half-life = 700 million years).

b) Write the decay scheme for ${}^{232}_{90}Th$ emitting an alpha particle (half-life = 14 billion years).

c) ${}^{238}_{92}U$ and ${}^{232}_{92}U$ are isotopes of uranium. They both have 92 protons in the nucleus. They have different numbers of neutrons. It is the number of protons that determine the name of the element. The half-lives are quite different. Calculate the number of neutrons in these isotopes of uranium.

13. A second possible nuclear decay occurs with the emission of a beta particle from the nucleus. A beta particle is a high-speed electron or a high-speed positron (identical to the electron but with positive charge). A neutron within the nucleus becomes a proton and emits the beta particle. Potassium-40 undergoes beta emission. The decay process:

$$ {}^{40}_{19}K \rightarrow {}^{0}_{-1}e + {}^{40}_{20}X $$

ANSWERS

For You To Do
(continued)

12.

a ${}_{92}U^{235} \rightarrow {}_{2}He^{4} + {}_{90}Th^{231}$

b) ${}_{90}Th^{232} \rightarrow {}_{2}He^{4} + {}_{88}Ra^{228}$

c) ${}_{92}U^{238}$ has $238 - 92 = 146$ neutrons

${}_{92}U^{235}$ has $235 - 92 = 143$ neutrons

13. a) ${}_{6}C^{14} \rightarrow {}_{-1}e^{0} + {}_{7}N^{14}$

b) ${}_{55}K^{137} \rightarrow {}_{-1}e^{0} + {}_{56}Ba^{137}$

© It's About Time

The beta particle emission causes an increase in the number of protons (the neutron became a proton). There is no change in the atomic mass since the mass of a proton and neutron are almost identical.

The new daughter element has an atomic number of 20. Referring to a periodic table, the element is calcium and its chemical symbol is Ca.

$${}^{40}_{19}\mathrm{K} \rightarrow {}^{0}_{-1}\mathrm{e} + {}^{40}_{20}\mathrm{Ca}$$

a) Write the decay scheme for ${}^{14}_{6}\mathrm{C}$ emitting a beta particle (half-life = 5730 years).

b) Write the decay scheme for ${}^{137}_{55}\mathrm{Cs}$ emitting a beta particle (half-life = 30 years).

During some radioactive decays, additional energy is emitted in the form of gamma rays. Gamma rays are high-energy light. Their symbol in the radioactive decay is ${}^{0}_{0}\gamma$ since they have no mass.

Physics Words

probability: a measure of the likelihood of a given event occurring.

radioactive: a term applied to an atom that has an unstable nucleus and can spontaneously emit a particle and become the nucleus of another atom.

FOR YOU TO READ

Models

Lots of people go to Las Vegas or other gaming centers to gamble. They think that because dice are involved that it is all a matter of luck whether they win or lose. Have you ever looked at the size of the hotels in Las Vegas? From the size and numbers of these hotels, you have to assume that the "house" always wins and that there is less luck at the casinos than some people expect. **Probability** is not the same as luck. In this activity you studied probability as it relates to atomic structure

You modeled an event in the real world. Scientists make many kinds of models. Models are conceptual representations of real things. You are probably familiar with physical models of the planets, moons, and the Sun and their interactions in the solar system. This type of model can help you "see" something too big to see all at once in nature. The model of the atom that you will create for your museum exhibit is another example of a model. The model is based on many observations and experimental results. Many models in science are complex mathematical models.

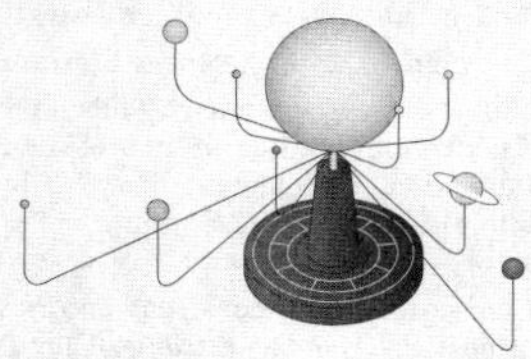

The **radioactive** model in this activity used sugar cubes to represent nuclei. When the sugar cubes were tossed and a "dot" emerged face-up, the

© It's About Time

face-up, the sugar cube "decayed" and became a penny. In radioactive samples like carbon-14 or iodine-131, the nuclei are not being tossed. As time goes on, some parent nuclei decay and become daughter nuclei. You had no idea which of the sugar cubes were going to land with the dot face-up. You also have no idea which of the carbon-14 nuclei will decay. You did get a good sense how many sugar cubes would land with the dot face-up. You have a very, very precise understanding of how many carbon-14 parent nuclei will decay in a given time.

Scientists use radioactive decay as precise clocks. Geologists want to know the age of rocks to provide an understanding of the Earth. Paleontologists want to know the age of fossils to better understand life on Earth. Archaeologists want to know the age of tools to better understand humans. Scientists in each of these fields make use of the known half-lives of radioactive elements for dating purposes.

One type of carbon atom, called carbon-14, is used in radioactive dating. As long as an organism lives, the ratio of carbon-14 to carbon-12 in the body stays the same. When organisms die, no new carbon-14 is added to their tissues. The decay of carbon-14 continues. If you know how much carbon-14 was in the organism when it lived, you can figure out how long ago it died by measuring the carbon-14 that remains.

Entombed in ice for over 5000 years, the iceman was discovered in 1991 by hikers in the Tyrolean Alps in Austria. Carbon-14 dating, along with many other scientific tools, was used to investigate this archaeological find.

Other elements are important tools in radioactive dating. Potassium-40 has a half-life of 1.3 billion years. Uranium-238 has a half-life of 4.47 billion years. Thorium-232 has a half-life of 14 billion years. These radioactive elements, as well as others, are found in rocks. Often, all three of these elements are used to date a specific rock. By comparing the radioactive decays, the age of a rock can be determined with more precision than if only one element was used. Rocks that are several billion years old have been discovered, indicating that the Earth must be at least this old.

Reflecting on the Activity and the Challenge

Some nuclei are stable while some decay. The parent nuclei can decay through alpha, beta or gamma emission. The decay process is random in that you cannot tell which of the individual nuclei will decay. However, the number of nuclei that will decay is quite predictable. The decay rate is defined by the half-life of the element. The half-life is the time required for one-half of the parent nuclei to become daughter nuclei. You can show nuclei decaying as part of your museum exhibit. It may be a bit tricky to show the mechanism of decay and to adequately describe half-life, but these are both worthwhile topics to pursue.

© It's About Time

ANSWERS

Physics To Go

1. a)

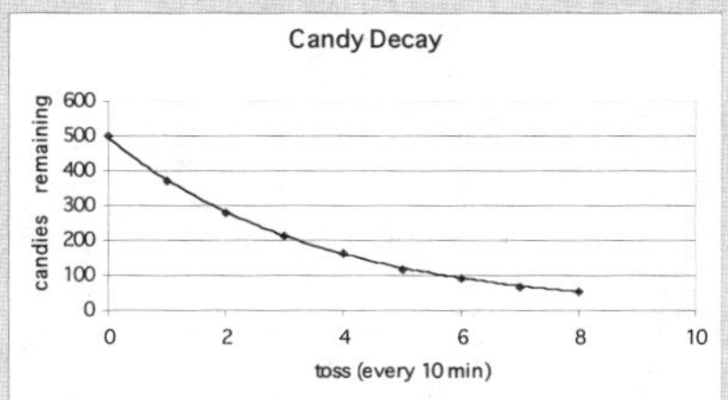

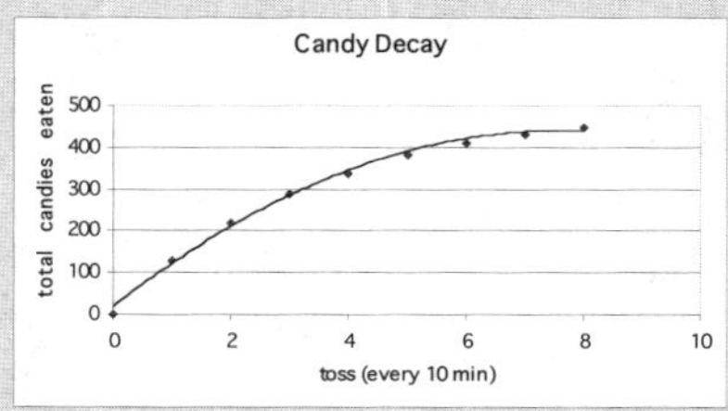

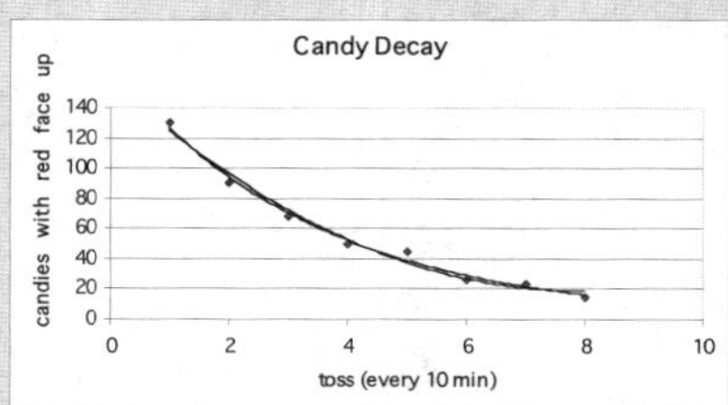

Physics To Go

1. A box of 500 candies was dropped on a table. If the red side of the candy was facing up, the candy was eaten. The remaining candy was shaken and dropped on the table again. A data table was created.

Toss (every 10 minutes)	Candies with red face up and then eaten	Total candies eaten	Candies remaining
0		0	500
1	130	130	370
2	90	220	280
3	68	288	212
4	50	338	162
5	45	383	117
6	26	409	91
7	23	432	68
8	15	447	53

a) Construct three graphs of the data.
b) From the graph, determine the half-life of the candy in this tossing simulation.
c) From the percentage that "decay" each toss, determine the shape of the candy.

2. There are 600 g of radioactive iodine-133. The half-life of iodine-133 is 21 h.

a) How much of this radioactive substance would exist after 21 h?
b) After 42 h?
c) After 126 h?

3. The half-life for the radioactive gas radon-220 is 51.5 s. Suppose a sealed bottle contains 100 g of radon-220. What mass of radon-220 will still be in the bottle after:

a) 51.5 s? b) 103 s? c)257.5 s?

4. Radioactive decay can be measured. One type of measuring tool gives a reading called a counting rate. A sample of radioactive material initially produces 2000 counts per

ANSWERS

Physics To Go *(continued)*

b) From the graph of candies remaining, we can see that the half-life is approximately 2.5 hours

c) Since it appears that $\frac{1}{4}$ of the candies "decay" after each toss, the candy probably is 4-sided. This would make it a tetrahedron.

2. a) After 21 hours, $\frac{1}{2}$ of 600 g = 300 g would exist.

b) After 42 h, $\frac{1}{2}$ of $\frac{1}{2}$ of 600 g = 150 g would exist.

c) After 126 h or 5 half-lives, 1/32 of 600 g = 19 g would exist.

3. a) After 51.5 s, $\frac{1}{2}$ of 100 g = 50 g would exist.

b) After 103 s, $\frac{1}{2}$ of $\frac{1}{2}$ of 100 g = 25 g would exist.

c) After 257.5 or 5 half-lives, 1/32 of 100 g = 3.13 g would exist.

© It's About Time

second. Four hours later, the same sample produces 500 counts per second. What is the half-life of the sample?

5. Carbon-14 has a half-life of 5730 years. What fraction of the original amount of carbon-14 will remain after 23,000 years?

6. Suppose a sample of carbon is extracted from a partially burned log found in a prehistoric fire site. The sample is $\frac{1}{8}$ as radioactive as carbon extracted from a modern log. How old is the sample taken from the ancient log?

7. For what age materials is uranium dating more useful than carbon dating? Explain your reasoning.

8. Radioactive uranium-238 decays by alpha emission. The daughter product, thorium-234 then decays by beta emission. A chart of the decay series is shown. The first line shows the decay of uranium-238. The thorium-234 is then placed on the next line. It then produces Pa-234 by beta emission. The Pa-234 is then placed on the next line and its decay is written.

   Complete the chart. (The half-lives are given for interest and are not needed to complete the chart.) The periodic table will help you identify the element symbols from their atomic number. If the periodic table is not available, you can complete the chart with the atomic number and mass and leave a blank for the symbol.

Element	Half-life	Emission	Reaction
$^{238}_{92}U$	4.9×10^9 years	alpha	$^{238}_{92}U \rightarrow\, ^{4}_{2}He +\, ^{234}_{90}Th$
$^{234}_{90}Th$	24 days	beta, gamma	$^{234}_{90}Th \rightarrow\, ^{0}_{-1}e +\, ^{0}_{0}\gamma +\, ^{234}_{91}Pa$
$^{234}_{91}Pa$	1.14 minutes	beta, gamma	$^{234}_{91}Pa \rightarrow\, ^{0}_{-1}e +\, ^{0}_{0}\gamma +\, ^{234}_{92}U$
$^{234}_{92}U$	3×10^5 years	alpha	$^{234}_{92}U \rightarrow\, ^{4}_{2}He +\, ^{230}_{90}Th$
$^{230}_{90}Th$	83,000 years	alpha, gamma	
	1600 years	alpha, gamma	
	3.8 days	alpha	
	3.0 minutes	alpha	
	26.8 minutes	beta, gamma	
	19.7 minutes	beta, gamma	
	10^{-5} seconds	alpha, gamma	
	22 years	beta	
	5 days	beta, gamma	
	140 days	alpha	
	Stable		

Answers

Physics To Go

(continued)

4. Since the decay rate has changed from 2000 counts/s to 500 counts/s, 2 half-lives have transpired in the 4 hours. The half-life is 2 hours.

5. 23,000/5730 = 4 half-lives. After 4 half-lives, 1/16 will remain.

6. If the sample is 1/8 as radioactive, then 3 half-lives have transpired. Each half-life is 5730 years. The elapsed time equals 3 × 5730 years = 17,200 years.

7. Uranium dating is more useful when the artifact is billions of years old (U-238) or tens of thousands of years old (U–234).

8.

Element	Half-life	Emission	Reaction
$_{92}U^{238}$	4.9×10^9 years	alpha	$_{92}U^{238} \rightarrow\, _2He^4 +\, _{90}Th^{234}$
$_{90}Th^{234}$	24 days	beta, gamma	$_{90}Th^{234} \rightarrow\, _{-1}e^0 +\, _0\gamma^0 +\, _{91}Pa^{234}$
$_{91}Pa^{234}$	1.14 minutes	beta, gamma	$_{91}Pa2^{34} \rightarrow\, _{-1}e^0 +\, _0\gamma^0 +\, _{92}U^{234}$
$_{92}U^{234}$	3×10^5 years	alpha	$_{92}U^{234} \rightarrow\, _2He^4 +\, _{90}Th^{230}$
$_{90}Th^{230}$	83,000 years	alpha, gamma	$_{90}Th^{230} \rightarrow\, _2He^4 +\, _{88}Ra^{226}$
$_{88}Ra2^{26}$	1600 years	alpha, gamma	$_{88}Ra^{226} \rightarrow\, _2He^4 +\, _{86}Rn^{222}$
$_{86}Rn^{222}$	3.8 days	alpha	$_{86}Rn^{222} \rightarrow\, _2He^4 +\, _{84}Po^{218}$
$_{84}Po^{218}$	3.0 minutes	alpha	$_{84}Po^{218} \rightarrow\, _2He^4 +\, _{82}Pb^{214}$
$_{82}Pb^{214}$	26.8 minutes	beta, gamma	$_{82}Pb^{214} \rightarrow\, _{-1}e^0 +\, _0\gamma^0 +\, _{83}Bi^{214}$
$_{83}Bi^{214}$	19.7 minutes	beta, gamma	$_{83}Bi^{214} \rightarrow\, _{-1}e^0 +\, _0\gamma^0 +\, _{84}Po^{214}$
$_{84}Po^{214}$	10^{-5} seconds	alpha, gamma	$_{84}Po^{214} \rightarrow\, _2He^4 +\, _{82}Pb^{210}$
$_{82}Pb^{210}$	22 years	beta	$_{82}Pb_{210} \rightarrow\, _{-1}e^0 +\, _0\gamma^0 +\, _{83}Bi^{210}$
$_{83}Bi^{210}$	5 days	beta, gamma	$_{83}Bi^{210} \rightarrow\, _{-1}e^0 +\, _0\gamma^0 +\, _{84}Po^{210}$
$_{84}Po^{210}$	140 days	alpha	$_{84}Po_{210} \rightarrow\, _2He^4 +\, _{82}Pb^{206}$
$_{82}Pb^{206}$	Stable		

© It's About Time

Stretching Exercises

1. Iodine-133 is a common by-product of nuclear fission in electric power plants. You are given a sample of material contaminated by iodine-133. Find out how long you have to wait before you can dump the iodine into the environment. How do you know it is "safe?" Report your findings to the class.
2. Find out about radon, a radioactive element found in your homes. Why is it considered a hazard? Report your findings to the class.
3. Medical doctors use radioactive materials in many ways. Research this and make a list of at least five uses for radiation in medicine. Create a poster of your findings.

© It's About Time

Blackline Master – Data Table

Trial	Number of sugar cubes	Number of pennies	Total number of cubes and pennies
0			
1			
2			
3			
4			
5			
6			
7			
8			
9			
10			

© It's About Time

For use with *Atoms on Display*, Chapter 9, Activity 7: Radioactive Decay and the Nucleus

ACTIVITY 8
Holding the Nucleus Together

Background Information

It is suggested that you read the **Physics Talk** and the **For You To Read** sections of the student text for **Activity 8** before proceeding in this section.

Albert Einstein, in the year 1905, recognized that mass and energy are related and described that relationship in his now famous equation, $E = mc^2$. Each bit of mass has a tremendous amount of energy. We have daily reminders of this as many cities receive their entire electrical energy from the conversion of mass to energy in nuclear power plants. We also live at a time where we have nuclear weapons capable of destroying entire cities.

When protons and neutrons come together to form a nucleus, some of the mass is released as energy. This binding energy holds the nucleus together. It is necessary to add energy in order to remove a proton or a neutron from the nucleus. By adding up the masses of the constituent parts of a nucleus and comparing this total mass with the mass of the formed nucleus, we can calculate the mass difference and then compute the binding energy of the nucleus. We find that the binding energy of a proton to a nucleus is equivalent to millions of electron volts. This is millions of times larger than the binding energy of an electron to an atom that was found earlier to be of the order of a few electron volts.

Creating nuclear energy will be the topic for the next activity.

Planning for the Activity

Time Requirements

One class period is required to complete this activity.

Materials Needed

For each group:

- basic calculator

Advance Preparation and Setup

You will need to decide how you will acquire the items needed for groups who wish to pursue alternate designs.

Teaching Notes

As Einstein once stated, "The equivalence of mass and energy is a difficult concept for the average mind." Students are guided through some calculations of the energy equivalent of specific masses. You may want to model the calculation to show how to square the speed of light in the equation $E = mc^2$.

To get a sense of the enormity of the energy in a small mass, students are then asked to find equivalent energies in forms that they have studied previously: kinetic energy and gravitational potential energy.

Students will need some encouragement to correctly calculate the binding energies. The masses of protons, neutrons, and the nuclei are listed to the millionth of the atomic mass unit. This accuracy is required to get a correct value for the mass defect.

You should take the time to review the **For You To Read**. In that section, four distinct descriptions of Einstein's equation, $E = mc^2$ are provided. Students will certainly be interested in the idea of particle/antiparticle annihilation and the creation of energy. One of the **Physics To Go** questions focuses on the idea of the matter/antimatter generator in *Star Trek*.

© It's About Time

Activity 8 Holding the Nucleus Together

GOALS

In this activity you will:

- Explain the meaning of Einstein's equation: $E = mc^2$.
- Calculate the energy equivalent of different masses.
- Determine the binding energy of protons and neutrons in the nucleus.
- Compare the binding energy of protons in the nucleus with the binding energy of electrons in the atom.

What Do You Think?

Nuclear bombs, nuclear radiation, and nuclear medicine conjure up very different images than chemical explosions, chemical reactions, and chemical medicines.

- **How are nuclear reactions different from chemical reactions?**

Record your ideas about this question in your *Active Physics* log. Be prepared to discuss your responses with your small group and your class.

For You To Do

1. Many people can recite Einstein's famous equation:

$E = mc^2$

where E represents energy in joules (J), and m represents mass in kilograms (kg).

Write down the approximate mass of the following objects:

a) an eight-year-old child
b) a bowling ball
c) a box of spaghetti
d) a compact car

ANSWERS

For You To Do

1. There will be some variations in the students' answers.

a) child = 20 kg

b) bowling ball = 5 kg

c) a box of spaghetti = 0.4 kg

d) a compact car = 500 kg

Activity Overview

Students calculate the mass defect and binding energies of nuclei by using Einstein's famous equation $E = mc^2$.

Student Objectives

Students will:

- Explain the meaning of Einstein's equation: $E = mc^2$.
- Calculate the energy equivalent of different masses.
- Determine the binding energy of protons and neutrons in the nucleus.
- Compare the binding energy of protons in the nucleus with the binding energy of electrons in the atom.

ANSWERS FOR THE TEACHER ONLY

What Do You Think?

Nuclear reactions are interactions among and within nuclei while chemical reactions are interactions among and within atoms. Chemical reactions are much weaker than nuclear reactions. A chemical reaction may have the energy of a few electron volts. Nuclear reactions have the energy of millions of electron volts. A violent chemical reaction is a stick of dynamite that can blast a hole in a rock face. A violent nuclear reaction is an atomic bomb that can level an entire city.

© It's About Time

ANSWERS

For You To Do
(continued)

2. a) $t = d/v$

$t = 100 \text{ m}/(3 \times 10^8 \text{ m/s})$

$= 3.3 \times 10^{-7} \text{ s}$

b) $t = d/v$

$t = 10^6 \text{ m}/ (3 \times 10^8 \text{ m/s})$

$= 3.3 \times 10^{-3} \text{ s}$

c) $t = d/v$

$t = 3.8 \times 10^8 \text{ m}/(3 \times 10^8 \text{ m/s})$

$= 1.3 \text{ s}$

d) $t = d/v$

$t = 1.5 \times 10^{11} \text{ m}/(3 \times 10^8 \text{ m/s}$

$= 5.0 \times 10^2 \text{ s}$

3. $E = mc^2$

$E = (1 \times 10^{-3} \text{ kg})(3 \times 10^8 \text{ m/s})^2$

$= 9 \times 10^{13} \text{ J}$

4. a) $KE = \frac{1}{2} mv^2$

$KE = \frac{1}{2} (50 \text{ kg})(8 \text{ m/s})^2$

$= 1600 \text{ J}$

b) The energy equivalent of a pea (problem 3) $= 9 \times 10^{13}$ J.

$9 \times 10^{13} \text{ J} = \frac{1}{2} (50 \text{ kg})v^2$

$v = 1.9 \times 10^6 \text{ m/s}$

5. $GPE = mgh$

$9 \times 10^{13} \text{ J} = (6 \text{ kg})(9.8 \text{ m/s}^2)(h)$

$h = 1.5 \times 10^{12} \text{ m}$

2. The c in the equation $E = mc^2$ represents the speed of light, which is equal to 3×10^8 m/s. Use the equation:

$d = vt$

to calculate how long it takes light to travel from:

a) one side of a football field to the other (approximately 100 m)

b) from Canada to Texas (approximately 1000 km or 10^6 m)

c) from the Moon to Earth (approximately 380,000 km or 3.8×10^8 m)

d) from the Sun to Earth (approximately 150,000,000 km or 1.5×10^{11} m)

3. Consider the energy equivalent of a small mass. The mass of a pea may be 1 g or 1×10^{-3} kg.

a) Calculate the energy equivalent in joules of this mass using $E = mc^2$. (Assume 100% conversion.)

4. There are many forms of energy that can be calculated. Kinetic energy, associated with moving cars, people, planets, and subatomic particles can be calculated using the equation:

$KE = \frac{1}{2} mv^2$

where KE is kinetic energy in joules (J)
m is the mass in kilograms (kg), and
v is the velocity in meters per second (m/s)

a) If a sprinter with a mass of 50 kg is running at 8 m/s, calculate the kinetic energy in joules.

b) How fast would a sprinter have to be running to have the energy equivalent of one pea?

5. Gravitational potential energy is energy associated with position. It can be calculated using the equation:

$GPE = mgh$

where GPE is gravitational potential energy in joules (J)
m is the mass in kilograms (kg)
g is the acceleration due to gravity equal to about 9.8 m/s², and
h is the elevation in meters (m)

a) How high would a bowling ball (mass = 6 kg) have to be elevated to have the same gravitational potential energy as the energy equivalent of a pea?

© It's About Time

6. In **Activity 3**, you studied electrons in the atom. The binding energy of an electron in hydrogen is the energy that holds the electron to the nucleus. In the $n = 1$ (ground) state of hydrogen, the binding energy was – 13.6 eV. The negative sign reminded you that the electron is bound to the nucleus. That electron requires +13.6 eV in order for it to be free of the nucleus.

   Neutrons and protons are also bound to the nucleus. A nitrogen-15 $\left({}^{15}_{7}N\right)$ nucleus has 7 protons and 8 neutrons. To free a nucleon (proton or neutron) requires an addition of energy equal to the binding energy of the nucleon. You can calculate that binding energy.

Example:

- *Calculate the mass of all the protons and neutrons.* The "atomic mass unit" is used for masses of atoms. It is defined in terms of the mass of carbon-12. Each atomic mass unit (u) is approximately equal to 1.7×10^{-27} kg. (You can see from this value why people prefer to use atomic mass units instead of kilograms.)

Mass of 7 protons	= 7 (1.007825 u)	= 7.054775 u
Mass of 8 neutrons	= 8 (1.008665 u)	= 8.069320 u
Total nucleon mass (protons + neutrons)		= 15.124095 u

- *Compare this to the mass of the nitrogen-15 nucleus.* The mass of the nitrogen-15 nucleus (15.000108 u) can be found from a chart of the nuclides in a reference book. The nitrogen-15 mass will be less than the total mass of the protons and neutrons. Energy must be supplied to remove a neutron. That implies that energy was given off when the nitrogen nucleus was first formed.

Total nucleon mass	= 15.124095 u
Mass of nucleus of ${}^{15}_{7}N$	= 15.000108 u
Mass difference	= 0.123987 u

ANSWERS

For You To Do
(continued)

6. a)

Chlorine 37

mass of 17 protons	17.133025	u
mass of 20 neutrons	20.1733	u
Total nucleon mass	37.306325	u
Mass of chlorine 37	36.965898	u
Mass difference	0.340427	u
Total binding energy	317.1	MeV
Average binding energy per nucleon	8.6	MeV

b) 15.000108 u – (14.003074 u + 1.008665 u) = 0.011631 u

0.011631 u × 931.5 MeV/u = 10.8 MeV

c) The 10.8 MeV to remove the first nucleon required much more energy than the average binding energy per nucleon (derived in the text as 7.7 MeV/nucleon).

d) Two isotopes are identical elements (the same number protons), but have different numbers of neutrons.

© It's About Time

- *Calculate the total binding energy.*
  (1 u = 931.5 MeV) The explanation for 931.5 MeV is given in the **For You To Read** section.
  Total binding energy = 0.123987 × 931.5 MeV
  = 115.5 MeV

- *Calculate the average binding energy per nucleon.*
  Total number of nucleons = 15 (7 protons and 8 neutrons)
  Average binding energy $= \frac{115.5 \text{ MeV}}{15} = 7.7$ MeV

On average, it would require 7.7 MeV or 7.7 million eV to remove a proton or neutron from the nucleus of ${}^{15}_{7}N$. (The removal of the first nucleon actually requires more than this energy—this is an average.) Recall that removing an electron from hydrogen required only 13.6 eV.

a) Calculate the average binding energy per nucleon for ${}^{37}_{17}Cl$. The mass of chlorine-37 is equal to 36.965898 u. Use the masses of protons and neutrons given in the sample problem.

b) A neutron can be removed from ${}^{15}_{7}N$ to create ${}^{14}_{7}N$. Compare the mass of ${}^{15}_{7}N$ to the masses of ${}^{14}_{7}N$ + 1 neutron to determine the actual energy required to free the neutron. The mass of nitrogen-14 equals 14.003074 u. The other values are given above.

c) Compare this binding energy to remove the neutron with the average binding energy calculated in the sample problem.

d) ${}^{15}_{7}N$ and ${}^{14}_{7}N$ are isotopes. What defines two elements as isotopes?

© It's About Time

Activity 8 Holding the Nucleus Together

PHYSICS TALK

Converting Mass Units to Energy Units

The conversion of **atomic mass units** (u) to energy units in electron volts (eV) required the use of the equation $E = mc^2$ to find the corresponding energy from the mass defect.

This would require certain conversions:

- The mass difference between the nucleus and constituent parts is called the mass defect. It is given in atomic mass units (u).
- The mass units would have to be converted to kilograms ($1 \text{ u} = 1.7 \times 10^{-27}$ kg).
- The energy would then be calculated in joules ($E = mc^2$).
- The joules could then be converted to electron volts ($1 \text{ eV} = 1.6 \times 10^{-19}$ J).

All of these steps can be combined into one step:

$1 \text{ u} = 931.5$ MeV.

Physics Words

atomic mass unit: a standard unit of atomic mass based on the mass of the carbon atom, which is assigned the value of 12.

FOR YOU TO READ

Energy and Matter

The speed of light is the "speed limit" of the universe. No material object can travel faster than this speed. In this activity you examined Einstein's equation:

$$E = mc^2$$

that equates energy and mass with the speed of light.

Einstein's equation can be interpreted in a number of different ways:

- It provides the energy equivalent of a piece of mass. (In the activity, you found the equivalent energy of a pea.)
- It tells you that energy and mass are equivalent entities but one is given in joules and the other is given in kilograms.
- It tells you that the conversion factor between energy and mass is the square of the speed of light (c^2). To change kilograms to joules, you have to multiply by c^2, equal to 9×10^{16} m^2/s^2.
- It explains how an electron and a positron (identical to the electron but with a positive charge) can annihilate each other and create light energy. This is called particle-antiparticle annihilation.

In **Activity 3**, you investigated the binding energy of electrons in an atom. In the $n = 1$ (ground) state of hydrogen, the binding energy was −13.6 eV. The binding energy of the

→

© It's About Time

ANSWERS

Physics To Go

1. a) i) $E = mc^2 = (9.1 \times 10^{-31}$ kg$)(3 \times 10^8$ m/s$)^2 = 8.2 \times 10^{-14}$ J

ii) $E = mc^2 = (1.0 \times 10^{-3}$ kg$)(3 \times 10^8$ m/s$)^2 = 9.0 \times 10^{13}$ J

iii) $E = mc^2 = (50$ kg$)(3 \times 10^8$ m/s$)^2 = 4.5 \times 10^{18}$ J

b) i) none

ii) 2.3×10^{12}

iii) 1.1×10^{17}

electron in the $n = 2$ state is – 3.4 eV. You would have to give that electron +3.4 eV to free it. Similarly, if the nucleus of hydrogen captures an electron to make a hydrogen atom and the electron drops into the $n = 1$ state, 13.6 eV is given off as a photon of light.

The conservation of energy requires that the energy before must equal the energy after the event. An electron absorbs a photon of light of +13.6 eV bringing its total energy to 0 and it is free. Or, a free electron of 0 energy becomes bound to the proton nucleus with an energy of −13.6 eV and gives off a light photon with energy equal to +13.6 eV.

In this activity, you calculated the binding energy on a **nucleon**. You found that to remove a proton or neutron from the nucleus of nitrogen-15 $\left({}^{15}_{7}N\right)$ requires 7.7 MeV or 7.7 million eV. Compare that to the binding energy of an electron in hydrogen, which is 13.6 eV, and you can begin to appreciate the difference between chemical and nuclear reactions. Chemical processes have to do with the exchange of electrons. Nuclear processes have to do with the exchange of nucleons. You can get a sense of the strength of these processes by comparing the binding energies. The nuclear binding energies are millions of times larger than the electron's binding energies. This is one reason why nuclear bombs are so much more devastating than dynamite.

Reflecting on the Activity and the Challenge

The nucleons (neutrons and protons) are held tightly in the nucleus by the strong nuclear force. You are now able to use the equation $E = mc^2$ to calculate the energy required to free a nucleon. The energy is of the order of a million electron volts while removing an electron requires only a few electron volts of energy. You might try to find a way to add numerical calculations to your atomic structure museum exhibit. You may want to compare the binding energy of electrons to the binding energy of nucleons. The nucleus is a wonderful place to introduce Einstein's famous $E = mc^2$ equation and its multiple interpretations.

Physics Words

nucleon: the building block of the nucleus of an atom; either a neutron or a proton.

Physics To Go

1. a) Calculate the energy equivalent of the following objects:

i) an electron

ii) a pea

iii) a 50-kg student

© It's About Time

b) Lifting a 4-kg shovel full of snow 1 m requires 40 J of energy. How many shovels full of snow could be lifted with the energies calculated on the preceding page?

2. a) A direct observation of the equivalence of mass and energy occurs when an electron and a positron (same mass as an electron, but opposite charge) annihilate each other and create light. Calculate the total energy of the light produced. (The mass of an electron is $9.1 \times x\ 10^{-31}$ kg.)

   b) Calculate the total energy of the light produced when a proton and an antiproton (same mass as a proton, but opposite charge) annihilate each other.

3. Lots of quantities can be measured using different units. Length can be measured in meters, centimeters, and miles. Volume can be measured in liters, milliliters, gallons, meters cubed, and cubic feet. In a similar way, mass and energy are related where mass is measured in kilograms and energy is measured in joules. How could this be depicted in a museum exhibit to make clear the notion that energy and mass are identical?

4. a) Calculate the total binding energy of phosphorus–31 with an atomic number of 15. The atomic mass is 30.973765 u.

   b) Calculate the binding energy per nucleon.

5. a) Describe binding energy in a way that a child visiting a science museum may understand. Is there some way that you can make the explanation visually appealing?

   b) Compare and contrast the binding energy of an electron to a nucleus with the binding energy of a proton with the nucleus.

6. In *Star Trek*, the matter-antimatter energy source provides all the energy the spaceship needs to operate. How might a matter-antimatter energy source work?

7. Two energy units are commonly used, the joule (J) and the electron-volt (eV). 1 eV is equivalent to 1.6×10^{-19} J. If both of these units are energy, why are two different units useful?

8. Sketch a graph showing the relationship between energy and mass. What would be the slope of the graph?

9. The mass difference between the nucleus and its constituent part (also called the mass defect) is greater for nucleus A than nucleus B. Compare their binding energies.

ANSWERS

Physics To Go *(continued)*

8.

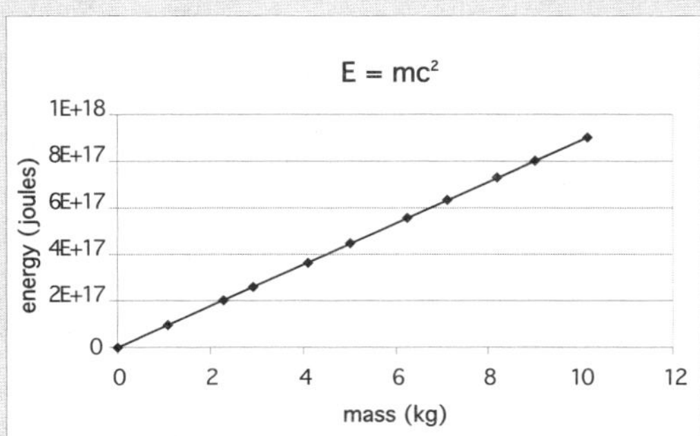

b) The slope will be the velocity of light squared (from the equation: $E=mc^2$).

9. Larger mass defects correspond to larger binding energies.

ANSWERS

Physics To Go *(continued)*

2. a)
   $E = mc^2$
   $= (2 \times 9.1 \times 10^{-31}\text{ kg})(3 \times 10^8\text{ m/s})^2$
   $= 1.6 \times 10^{-13}$ J

   b)
   $E = mc^2$
   $= (2 \times 1.67 \times 10^{-27}\text{ kg})(3 \times 10^8\text{ m/s})^2$
   $= 3.0 \times 10^{-10}$ J

3. Answers will vary.

4. a) – b)

Phosphorus – 31		
mass of 15 protons	15.117375	u
mass of 16 neutrons	16.13864	u
Total nucleon mass	31.256015	U
Mass of chlorine 31	30.973768	U
Mass difference	0.2822475	U
Total binding energy	262.9	MeV
Average binding energy per nucleon	8.5	MeV

5. a) Answers will vary

   b) The binding energy of a proton to the nucleus is millions of times larger than the binding energy of an electron to an atom.

6. The matter and antimatter will annihilate each other with the mass becoming energy.

7. When dealing with small energies, it is often easier to discuss the energy in eV rather than joules. This is similar to discussing room dimensions in meters rather than miles.

© It's About Time

Chapter 9

NOTES

© It's About Time

ACTIVITY 9
Breaking Up Is Hard to Do

Background Information

It is suggested that you read the **For You To Read** section of the student text for **Activity 9** before proceeding in this section.

If the average binding energy of nucleons were to increase, then additional energy would be released. This is the basis for nuclear energy releases. The increase in average binding energy of nucleons can occur when small nuclei fuse or when large nuclei break into two pieces (fission.)

The binding energy curve of all of the elements has a peak at Fe−57. This element has the greatest average binding energy per nucleon and is therefore the most stable element. Removing a nucleon (proton or neutron) or absorbing a nucleon would both result in a decrease of the average binding energy per nucleon. This would mean that energy must be supplied for this to occur.

Uranium−235 becomes unstable after it absorbs a neutron. When the uranium undergoes fission and becomes two elements, there is a release of additional neutrons. These additional neutrons can then be absorbed by other uranium and the process can quickly become a chain reaction. This is what physicists realized could be the basis for an uncontrolled chain reaction – the first atomic bomb. By controlling the number of neutrons, the chain reaction can be controlled. In this way, we can have nuclear power plants.

Planning for the Activity

Time Requirements

One class period is required to complete this activity.

Materials Needed

- graph paper, pad of 50 sheets

Advance Preparation and Setup

No advance preparation and setup is required for this activity.

Teaching Notes

Students can use practice calculating the average binding energy per nucleon. However, it would be both tedious and unproductive to have each student repeat the calculation for all of the elements. Each student or pair of students should complete a calculation. The results can be pooled and the binding energy graph can then be constructed.

It doesn't seem to make sense that both the fusion and the fission of nuclei can both create energy. The paradox is removed when it becomes clear that only small elements can fuse and create energy. Those elements must absorb energy if they were to break apart. Similarly, the large elements like uranium and undergo fission and release energy. Large elements cannot fuse without the addition of energy. By constructing the binding energy curve and ensuring that students can explain how the average binding energy per nucleon changes for the different elements affords students the opportunity to best explain both fission and fusion.

© It's About Time

Activity Overview

After graphing the binding energy per nucleon curve for all of the elements, students explore nuclear fission and fusion. Chain reactions in fission reactions are also studied.

Student Objectives

Students will:

- Calculate the binding energy of various nuclei.
- Create a graph of average binding energy per nucleon versus the number of nucleons.
- Explain how a fusion reaction can release energy.
- Explain how a fission reaction can release energy.
- Describe a mousetrap model for a fission chain reaction.

ANSWERS FOR THE TEACHER ONLY

What Do You Think?

Nuclear energy is created when small nuclei fuse to make larger nuclei (fusion) and when very large nuclei break apart to make smaller nuclei (fission).

The energy in nuclear power plants is produced by fission. The energy in the Sun is produced by fusion. We have both fission and fusion nuclear weapons.

Activity 9 Breaking Up Is Hard to Do

GOALS

In this activity you will:

- Calculate the binding energy of various nuclei.
- Create a graph of average binding energy per nucleon versus the number of nucleons.
- Explain how a fusion reaction can release energy.
- Explain how a fission reaction can release energy.
- Describe a mousetrap model for a fission chain reaction.

What Do You Think?

Nuclear energy is used for nuclear power plants, for creating energy in the Sun, and for atomic weapons.

- **How is nuclear energy created?**
- **Is all nuclear energy produced the same way?**

Record your ideas about these questions in your *Active Physics* log. Be prepared to discuss your responses with your small group and your class.

For You To Do

In the previous activity you learned to calculate binding energy. The binding energy is the energy required to disassemble a nucleus into its separate protons and neutrons. The higher the binding energy, the more energy is required to free the nucleons. The higher the binding energy, the more stable the nucleus.

1. To see if there is a pattern to binding energies and the related stability of nuclei, you can calculate the binding energy per nucleon for all known elements and make a graph. In the spirit of learning the physics and making the most efficient use of your time, limit your calculations to some specific elements.

© It's About Time

You know the mass of the proton and neutron. The only additional information that is needed to calculate the binding energy for a given nucleus is the mass of that nucleus. The masses of a number of elements are given below.

Element	Atomic number	Mass number	Atomic mass (u)	Element	Atomic number	Mass number	Atomic mass (u)
neutron	0	1	1.008665	Co	27	59	58.933190
proton	1	1	1.007825	Zn	30	66	65.926052
H	1	1	1.007825	Br	35	79	78.918330
He	2	4	4.002603	Zr	40	91	90.905642
Li	3	7	7.016004	Rh	45	103	102.905511
Be	4	9	9.012186	Sn	50	119	118.903314
B	5	11	11.009305	Cs	55	133	132.905355
C	6	12	12.000000	Nd	60	145	144.912538
N	7	14	14.003074	Yb	70	173	172.938060
O	8	16	15.994915	Hg	80	200	199.968327
F	9	19	18.998405	TI	81	205	204.974442
P	15	31	30.973765	Pb	82	207	206.975903
Ca	20	40	39.962589	Bi	83	209	208.981082
Mn	25	55	54.938051	Th	90	232	232.038124
Fe	26	57	56.935398	U	92	235	235.043915

Example:

Calculate the binding energy per nucleon for phosphorus-31 $\left({}^{31}_{15}P\right)$. (You will use the same method as you used in the previous activity.)

The atomic number of phosphorus is 15.

Number of protons = atomic number = 15
Number of neutrons = atomic mass – number of protons = 16

Mass of 15 protons = 15 (1.007825 u)	= 15.117375 u
Mass of 16 neutrons = 16 (1.008665 u)	= 16.138640 u
Total nucleon mass	= 31.256015 u
Mass of nucleus of ${}^{31}_{15}P$	= 30.973765 u
Mass difference	= 0.282250 u

ANSWERS

For You To Do

(continued on next page)

© It's About Time

ANSWERS

For You To Do

1.

Element	Atomic number	Mass number	Number of protons	Number of neutrons	Atomic mass
neutron					1.008665
proton					1.007825
H	1	1	1	0	1.007825
He	2	4	2	2	4.002603
Li	3	7	3	4	7.016004
Be	4	9	4	5	9.012186
Be	5	11	5	6	11.009305
C	6	12	6	6	12.000000
N	7	14	7	7	14.003074
O	8	16	8	8	15.994915
F	9	19	9	10	18.998405
P	15	31	15	16	30.973765
Ca	20	40	20	20	39.962589
Mn	25	55	25	30	54.938051
Fe	26	57	26	31	56.935398
Co	27	59	27	32	58.933190
Zn	30	66	30	36	65.926052
Br	35	79	35	44	78.918330
Zr	40	91	40	51	90.905642
Rh	45	103	45	58	102.905511
Sn	50	119	50	69	118.903314
Cs	55	133	55	78	132.905355
Nd	60	145	60	85	144.912538
Yb	70	173	70	103	172.938060
Hg	80	200	80	120	199.968327
Tl	81	205	81	124	204.974442
Pb	82	207	82	125	206.975903
Bi	83	209	83	126	208.981082
Th	90	232	90	142	232.038124
U	92	235	92	143	235.043915

1. a)

Element	Total mass of protons	Total mass of neutrons	Total mass	Mass defect	Binding energy	Binding energy per nucleon
neutron						
proton						
H	1.007825	0	1.007825	0.000000	0	0
He	2.01565	2.01733	4.03298	0.030377	28.29618	7.074044
Li	3.023475	4.03466	7.058135	0.042131	39.24503	5.606432
Be	4.0313	5.043325	9.074625	0.062439	58.16193	6.462436
Be	5.039125	6.05199	11.09112	0.081810	76.20602	6.92782
C	6.04695	6.05199	12.09894	0.098940	92.16261	7.680217
N	7.054775	7.060655	14.11543	0.112356	104.6596	7.475687
O	8.0626	8.06932	16.13192	0.137005	127.6202	7.97626
F	9.070425	10.08665	19.15708	0.158670	147.8011	7.779006
P	15.11738	16.13864	31.25602	0.282250	262.9159	8.481157
Ca	20.1565	20.1733	40.3298	0.367211	342.057	8.551426
Mn	25.19563	30.25995	55.45558	0.517524	482.0736	8.764975
Fe	26.20345	31.26862	57.47207	0.536667	499.9053	8.770269
Co	27.21128	32.27728	59.48856	0.555365	517.3225	8.768178
Zn	30.23475	36.31194	66.54669	0.620638	578.1243	8.759459
Br	35.27388	44.38126	79.65514	0.736805	686.3339	8.68777
Zr	40.313	51.44192	91.75492	0.849273	791.0978	8.693382
Rh	45.35213	58.50257	103.8547	0.949184	884.1649	8.584125
Sn	50.39125	69.59789	119.9891	1.085821	1011.442	8.499515
Cs	55.43038	78.67587	134.1062	1.200890	1118.629	8.410745
Nd	60.4695	85.73653	146.206	1.293487	1204.883	8.309539
Yb	70.54775	103.8925	174.4402	1.502185	1399.285	8.088354
Hg	80.626	121.0398	201.6658	1.697473	1581.196	7.90598
Tl	81.63383	125.0745	206.7083	1.733843	1615.075	7.878413
Pb	82.64165	126.0831	208.7248	1.748872	1629.074	7.869924
Bi	83.64948	127.0918	210.7413	1.760183	1639.61	7.845026
Th	90.70425	143.2304	233.9347	1.896556	1766.642	7.614836
U	92.7199	144.2391	236.959	1.915080	1783.897	7.591051

ANSWERS

For You To Do *(continued)*

b) The graph is shown in text.

2. a) Student sketch the graph in notebook.

b) Features of the graph include:

- its general trend is a sharp increase in average binding energy for small numbers of nucleons
- its general trend is a gradual decrease in average binding energy for large numbers of nucleons
- the graph a maximum value of binding energy corresponding to Fe (iron) with 57 nucleons.

3. a) When two small nuclei with average binding energies of 7 MeV per nucleon combine, they form a larger nucleus. This nucleus, according to the graph, may have an average binding energy of 8 MeV per nucleon. This would mean that each nucleon has a larger binding energy and has released energy in forming the larger nucleus.

© It's About Time

NOTES

© It's About Time

> Total binding energy = 0.282250 u × 931.5 MeV/u
> = 262.9 MeV
> Total number of nucleons = 31 (15 protons + 16 neutrons)
> Average binding energy per nucleon = $\frac{262.9 \text{ MeV}}{31}$ = 8.48 MeV
> On average, it would require 8.48 MeV or 8.48 million eV to remove a proton or neutron from the nucleus of phosphorus–31.

a) Calculate the binding energies for the elements assigned to you by your teacher.

b) Use combined class data to plot a graph of the binding energy per nucleon versus the number of nucleons for each element in the periodic table.

2. If you made a graph of all the elements, it would look like this:

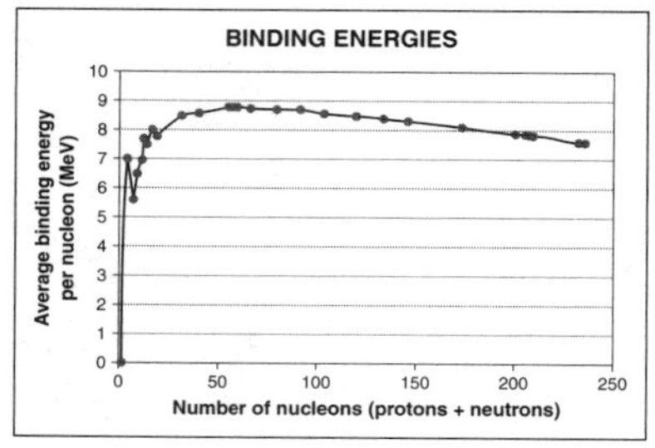

a) Sketch this graph in your log.

b) Describe three features of the graph that you took into account when you made your sketch.

3. Small nuclei can be put together to form larger nuclei.

a) Show evidence from your graph how the binding energy per nucleon increases if two small nuclei are put together to form one larger nucleus.

If the nucleon becomes more tightly bound to the nucleus, additional energy is given off. It takes additional energy to free this nucleon.

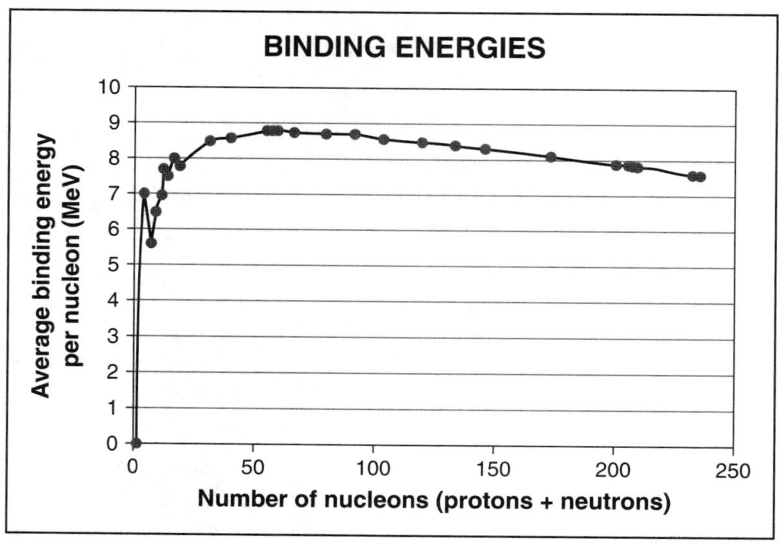

© It's About Time

Larger binding energy means:
- more energy is given off
- the nucleon is more tightly bound
- more energy is needed to free the nucleon

b) Use the graph of nuclear binding energy versus atomic number to answer the following questions:
   i) Which element has the most tightly bound nucleons?
   ii) Which element is most stable?
   iii) Which element requires the most energy to free a nucleon?

c) Suppose two atoms of a lighter element (e.g., atomic number 5) are fused together to create a larger nucleus (e.g., atomic number 10). This process is called nuclear fusion. How does the binding energy of the nucleons in the larger nuclei compare with that of the smaller nuclei? Will energy be emitted or absorbed during this transition?

d) Not all elements can fuse and provide energy. In order to produce energy by nuclear fusion, the created nucleus must have a larger binding energy per nucleon. Draw a sketch of the binding energy graph and indicate the portion of the graph where smaller nuclei can fuse and produce a larger nucleus with an increase in binding energy per nucleon. Label this portion—fusion.

4. There is another means by which to create nuclei with larger binding energy. If a very heavy element like uranium were to break apart, the two daughter nuclei would have a greater binding energy per nucleon.

a) What is the average binding energy per nucleon for uranium-235?

b) From the graph, estimate the average binding energy per nucleon of barium-137.

c) From the graph, estimate the average binding energy per nucleon of krypton-84.

Since the daughter products of barium-137 and krypton-84 have more binding energy per nucleon than the parent uranium-235, then energy is given off. This energy release is due to nuclear fission. It happens when a large nucleus splits into smaller nuclei.

5. Uranium-235 will become unstable and undergo fission when it absorbs a neutron. The reaction can be written as:

$${}^{235}_{92}U + {}^{1}_{0}n \rightarrow {}^{144}_{56}Ba + {}^{89}_{36}Kr + 3\,{}^{1}_{0}n$$

Answers

For You To Do
(continued)

3. b) i) Fe (57 nucleons) has the largest average binding energy per nucleon.

   ii) Fe is the most stable.

   iii) Fe requires the most energy to free each nucleon.

   c) The larger nucleus has more average binding energy for each nucleon. Some energy is emitted when these nuclei fuse.

   d) The binding energy graph should have a vertical line drawn at Fe (number of nucleons = 57) and "fusion" should be noted for the portion to the left of this line.

4. a) 7.59 MeV

   b) 8.35 MeV

   c) 8.7 MeV

© It's About Time

ANSWERS

For You To Do
(continued)

b) Each of the 5 mousetraps that are hit will close. This will release 5 more marbles into the air and another 5 mousetraps that are hit will close. This will continue until all the mousetraps have released their marbles.

c) Each of the 5 mousetraps that are hit will close. This will release 10 more marbles into the air and another 10 mousetraps that are hit will close. 10 marbles will hit 10 mousetraps and 20 marbles will be released. This will continue until all the mousetraps have released their marbles.

d)

Cycle	Marbles released
1	5
2	10
3	20
4	40
5	80
6	160
7	320
8	640
9	1280
10	2560

e)

10	2560
11	5120
12	10,240
13	20,480
14	40,960
15	81,920
16	163,840
17	327,680
18	655,360
19	1,310,720
20	2,621,440

a) Show that the number of protons on both sides of the reaction is equivalent.

b) Show that the total number of nucleons (the sum of protons and neutrons) is conserved.

6. Only the heaviest elements can undergo fission and provide energy. In order to produce energy by nuclear fission, the created nuclei must have a larger binding energy per nucleon. Refer to your sketch of the binding energy graph.

a) Indicate the portion of the graph where a larger nucleus can undergo fission and produce nuclei with an increase in binding energy per nucleon. Label this portion—fission.

7. Notice that the fission of uranium-235 with the absorption of a neutron yields three additional neutrons. Those three neutrons are responsible for our ability to create a chain reaction. The chain reaction permits two technologies – nuclear power and nuclear bombs.

In order to understand how the chain reaction takes place, imagine a set of 100,000 mousetraps. When a small marble is dropped onto any mousetrap, the mousetrap closes with a large snapping sound.

a) What will happen if 5 marbles are dropped?

b) Imagine that each mousetrap now has a marble balanced on it. When that mousetrap closes, its marble jumps in the air and can land on another mousetrap. What will happen if 5 marbles are now dropped?

c) Imagine now that each mousetrap has 2 marbles balanced on it. When that mousetrap closes, the 2 marbles jump in the air and can land on other mousetraps. What will happen if 5 marbles are now dropped?

d) The mousetraps get out of control very rapidly. The 5 marbles release 10 marbles that then release 20 marbles. Create a chart to show how many marbles are released in the first 10 cycles.

e) Assume that you have an unlimited supply of mousetraps. Continue your chart to show how many marbles are released in the first 20 cycles.

This enormous release of mousetraps is similar to the enormous release of energy in a nuclear chain reaction.

Physics Words

nuclear fusion: a nuclear reaction in which nuclei combine to form more massive nuclei with the release of a large amount of energy.

nuclear fission: a nuclear reaction in which a massive, unstable nucleus splits into two or more smaller nuclei with a release of a large amount of energy.

ANSWERS

For You To Do *(continued)*

5. a) Number of protons are the bottom numbers: $92 + 0 = 56 + 36 + 0$

b) The number of nucleons are the top numbers: $235 + 1 = 144 + 89 + 3$

(The 3 comes from the fact that there are 3 neutrons emitted.)

6. a) The binding-energy graph now has a vertical line drawn at Fe (number of nucleons = 57) and "fission" should be noted for the portion to the right of this line.

7. a) Each of the 5 mousetraps that are hit will close.

© It's About Time

FOR YOU TO READ

Nuclear Fusion and Nuclear Fission

The structure of the atom has a nucleus made up of protons and neutrons. These nucleons are held tightly together with a specific binding energy per nucleon. To free a nucleon would require the addition of energy equal to this binding energy per nucleon. In this activity you calculated the average binding energy per nucleon for each nucleus. The resulting calculations were plotted on a graph of binding energy versus atomic number. The graph revealed that iron has the highest average binding energy per nucleon and is therefore the most stable nucleus.

Nuclei of smaller mass than iron have less binding energy. This makes **nuclear fusion** possible. In nuclear fusion two lighter nuclei fuse together to produce a larger nucleus. The larger nucleus has a greater average binding energy per nucleon. Energy is therefore released in the creation of the larger nucleus. This nuclear fusion is responsible for the Sun's energy. The Sun has provided the Earth's energy for over five billion years. It is expected to continue to produce energy through the fusion of hydrogen into helium for another five billion years.

Nuclei of larger mass than iron also have less binding energy. This makes **nuclear fission** possible. In nuclear fission a heavy nucleus can break apart into two smaller nuclei. The smaller nuclei have a greater average binding energy per nucleon and energy is therefore released during their creation.

You saw that the fission of uranium-235 with the absorption of a neutron yields three additional neutrons.

$$^{235}_{92}U + ^{1}_{0}n \rightarrow ^{144}_{56}Ba + ^{89}_{36}Kr + 3^{1}_{0}n$$

Those three neutrons have changed our politics, our culture, and our lives. Those three neutrons are responsible for our ability to create a chain reaction. The chain reaction permits two technologies—nuclear power and nuclear bombs.

You compared the release of mousetraps to the enormous release of energy in a nuclear chain reaction. The uranium-235 nucleus absorbs one neutron, but gives off three neutrons. Each of those three neutrons can be absorbed and more uranium-235 can undergo fission. With each fission reaction, more energy is released. In a matter of moments, a huge fission explosion can take place. The fission explosion can be a nuclear bomb. However, the fission chain reaction can also be controlled. By removing neutrons before the other uranium absorbs them, a controlled reaction takes place. In a nuclear power plant, the control rods remove these neutrons, so that the uncontrolled chain reaction does not take place.

Reflecting on the Activity and the Challenge

Both nuclear fusion and nuclear fission provide energy. Your museum exhibit on atomic structure can certainly include information about binding energy and nuclear fission and fusion. It will require some special insight to find ways to help

© It's About Time

museum visitors understand how breaking up a nucleus can provide energy while fusing together other nuclei can also produce energy. You may decide to draw people into your exhibit with something concerning fission or fusion, sunlight or bombs, endless energy sources, or political drama.

Physics To Go

1. a) Calculate and compare the binding energies per nucleon for these three elements:
   i) oxygen – 16
   ii) lithium – 7
   iii) calcium – 40
   b) Which nucleus in part (a) is most stable?
2. a) Sketch the general shape of the binding energy curve where the number of nucleons is on the x-axis and the average binding energy per nucleon is on the y-axis.
   b) Label the peak of the graph as the most stable element. Identify that element.
   c) Label the part of the graph where fusion could occur.
   d) Label the part of the graph where fission could occur.
3. Should a single museum exhibit attempt to discuss both fusion and fission or should it focus on one? Discuss your reasons.
4. A simulation of a chain reaction can be constructed using dominoes.
   a) How would such a simulation be set up?
   b) Is there a way to set up such a simulation at a museum exhibit? You probably don't want everybody taking the time to set up the dominoes by hand. Is there a mechanical way of easing the setup time?
5. The Sun creates energy by the fusion of hydrogen into helium. Describe in detail (using equations and numbers) how this energy can be created. How long do you think that the Sun could continue to produce energy in this way?
6. In a nuclear reactor, control rods absorb excess neutrons. Why would the absorption of neutrons slow the reaction?
7. Nuclear medicine, nuclear energy, nuclear bombs… some people have very strong opinions about these applications of our knowledge of the nucleus. Could a museum exhibit poll

© It's About Time

ANSWERS

Physics To Go

1.

Element	Atomic number	Mass number	Number of protons	Number of neutrons	Atomic mass
neutron					1.008665
proton					1.007825
Li	3	7	3	4	7.016004
O	8	16	8	8	15.994915
Ca	20	40	20	20	39.962589

Element	Total mass of protons	Total mass of neutrons	Total mass	Mass defect	Binding energy	Binding energy per nucleon
neutron						
proton						
Li	3.023475	4.03466	7.058135	0.042131	39.24503	5.606432
O	8.0626	8.06932	16.13192	0.137005	127.6202	7.97626
Ca	20.1565	20.1733	40.3298	0.367211	342.057	8.551426

2. 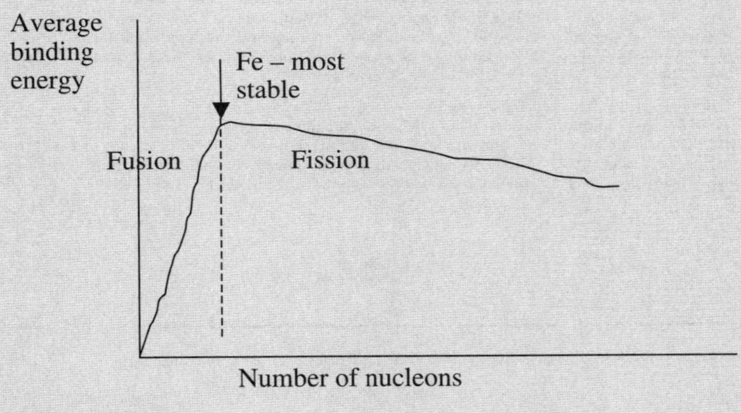

3. Answers will vary.

4. a) One falling domino would hit two dominoes. These two dominoes would hit four that would then hit eight and that would then hit sixteen dominoes, etc.

   b) Answers will vary. A tool can probably be set up where the dominoes are hinged at the bottom and they can all be set standing with a rotation of the base.

5.

Element	Total mass of protons	Total mass of neutrons	Total mass	Mass defect	Binding energy	Binding energy per nucleon
neutron						
proton						
H	1.007825	0	1.007825	0.000000	0	0
He	2.01565	2.01733	4.03298	0.030377	28.29618	7.074044

© It's About Time

ANSWERS

Physics To Go
(continued)

To create helium (which has 4 nucleons) from hydrogen (which is 1 proton), would require 4 hydrogen joining. (The protons of 2 of the hydrogen would have to become neutrons by losing their charge. How this occurs is physics beyond this activity. The fusion process can be investigated by students as a **Stretching Exercise**.) The mass defect of 2 neutrons and 2 protons is shown above to be 7.1 MeV per nucleon.

[numbers should both be on left-hand side of the element symbol and the positron.]

The equation may be ${}_1H^2 + {}_1H^2 + {}_1H^2 + {}_1H^2 \text{ --> } {}_2He^4 + {}_1\text{positron}^0 + {}_1\text{positron}^0$

Estimates are that the Sun can continue to supply energy through fusion for a lifetime of 5 billion years.

6. If the extra neutrons are absorbed, then fewer neutrons can be absorbed by the uranium and less uranium will undergo fission.

7. Answers will vary.

PHYSICS AT WORK

Dr. Allen Friedman

MAKES SCIENCE FUN

As director of the New York Science and Technology Center—an interactive museum in Queens, New York—it is Dr. Friedman's job to introduce complex scientific principles to people of all ages and levels of knowledge. "I have to communicate science to the general public and that is an extremely interesting intellectual puzzle."

Born in Brooklyn, Dr. Friedman earned his bachelor degree from the Georgia Institute of Technology and a Ph.D. in physics from Florida State University. He had every intention of becoming an experimental scientist. That all changed the day he walked into the Lawrence Hall of Science, an interactive museum at the University of California at Berkeley. As he walked from exhibit to exhibit he noticed something extraordinary: People were actually having fun.

Inspired by the museum's energy, he applied for a fellowship, received it, and has never looked back. "One of the very first exhibits we ever created was a three-dimensional illustration of the structure of an atom," he explains. "By projecting images on a moving screen we were able to simulate three dimensions. Quantum theory—which is how we understand atoms today—basically says that the traditional, rigid, mechanical model isn't nearly sophisticated enough to explain what we see in the physics of the atom. And that same principle can be transferred to the teaching of science."

"There are really three crucial aspects of every exhibit," he explains. "First, it has to attract an audience—and that means everyone from a small child with virtually no scientific knowledge to a high school science student and even practicing scientists. Then, you have to find ways to hold their attention long enough to communicate the information. And lastly, every good museum exhibit must be entertaining—the audience has to like it."

© It's About Time

Chapter 9 Assessment

The time has come for you to complete your **Chapter Challenge**. It's time for your *Active Physics* group to put the final touches on your science museum exhibit on atoms. With what you learned in this chapter, you should be ready to meet the challenge.

You have probably already designed the item that will be for sale at the museum gift shop. However, before you finalize your exhibit, carefully review all the features that you were asked to include. You do not want to receive a poor grade just because you forgot to include something.

Your exhibit must:

- **include distinct features of the structure of the atom**
- **communicate the size and scale of the parts of the atom**
- **provide information on how the atom is held together**
- **capture the visitor's attention within 30 seconds**
- **include written matter that will further illuminate the concepts**
- **have a model, a T-shirt, a poster, a booklet, or a toy that can be sold at the museum store**
- **have a building plan and cost estimate for the exhibit**
- **include safety features**
- **be interactive—visitors should not merely read**

Before you began this chapter, you and your classmates decided on the criteria by which your **Chapter Challenge** will be evaluated. Revisit those criteria now.

- **What weight will be placed on the 30-second criterion?**
- **How many points will be allocated to the building plan and cost?**
- **How much should the physics content be worth?**

You may wish to revise your grading system in light of what you learned in this chapter. The better you understand how you will be graded, the better the likelihood of receiving a good grade.

© It's About Time

Physics You Learned

Coulomb's Law

Conservation of charge

Millikan's oil-drop experiment

Size of nucleus

Neutron, proton, electron

Rutherford scattering experiment

Light spectra

Models of the atom

Conservation of energy

Diffraction of light

Interference of light

Photoelectric effect

Electron wavelength

Isotopes

Nuclear forces

Feynman diagrams

Radioactive decay

Binding energy

Nuclear fission

Nuclear fusion

© It's About Time

NOTES

© It's About Time

Alternative Chapter Assessment

1. Which phenomenon is *best* explained by the particle nature of light?

   a) the Doppler effect

   b) polarization

   c) interference

   d) the photoelectric effect

2. According to the quantum theory of light, the energy of light is carried in discrete units called

   a) photons

   b) photoelectrons

   c) alpha particles

   d) protons

3. The energy of a photon varies directly with its

   a) rest mass

   b) wavelength

   c) frequency

   d) speed

4. What is the energy of a photon with a frequency of 5.0×10^{15} Hz?

   a) 3.3×10^{-18} J

   b) 1.5×10^{24} J

   c) 2.0×10^{-16} J

   d) 7.5×10^{48} J

5. In the photoelectric effect, the speed of emitted electrons may be increased by

   a) decreasing the intensity of illumination

   b) increasing the intensity of illumination

   c) decreasing the frequency of the light

   d) increasing the frequency of the light

6. The work function of a metal is 4.2 eV. If photons with an energy of 5.0 eV strike metal, the maximum kinetic energy of the emitted photoelectrons will be

   a) 9.2 eV

   b) 0 eV

   c) 3.8 eV

   d) 0.80 eV

k© It's About Time

7. Which is conserved when a photon collides with an electron?
   a) momentum and energy
   b) momentum only
   c) velocity
   d) energy only
8. In Rutherford's model of the atom, the positive charge
   a) revolves about the nucleus in specific orbits
   b) is concentrated at the center of the atom
   c) occupies most of the space in the atom
   d) is distributed throughout the atom's volume
9. According to the Rutherford model of the atom, an atomic nucleus contains
   a) all of the atom's electric charge, but none of the atom's mass
   b) all of the atom's mass, but none of the atom's electric charge
   c) most of the atom's mass and all of the atom's positive charge
   d) most of the atom's mass and all of the atom's negative charge
10. The Rutherford scattering experiments suggested that the mass of the atom is composed mostly of
   a) electrons
   b) alpha particles
   c) positrons
   d) nucleons
11. Which statement most accurately describes the interaction that binds a nucleus together?
   a) long-range and weak
   b) long-range and strong
   c) short-range and weak
   d) short-range and strong
12. Which type of force causes the hyperbolic trajectory of alpha particles in Rutherford's scattering experiment?
   a) gravitational
   b) electrostatic
   c) magnetic
   d) nuclear

© It's About Time

13. Rutherford observed that most of the alpha particles directed at a metallic foil appear to pass through unhindered, with only a few deflected at large angles. What did he conclude?

a) Atoms have most of their mass distributed loosely in a electron cloud.

b) Atoms can easily absorb and reemit alpha particles.

c) Atoms consist mainly of empty space and have small, dense nuclei.

d) Alpha particles behave like waves when they interact with an atom.

14. When an electron changes from a higher energy state to a lower energy state within an atom, a quantum of energy is

a) emitted

b) fused

c) fissioned

d) absorbed

15. In returning to the ground state from the $n = 2$ state, the maximum number of photons a hydrogen atom can emit is

a) 1

b) 2

c) 3

d) 4

16. In the Balmer series, during which of the following energy-state changes of the hydrogen atom is the photon with the most energy emitted?

a) $n = 4$ directly to $n = 2$

b) $n = 2$ directly to $n = 5$

c) $n = 2$ directly to $n = 4$

d) $n = 5$ directly to $n = 2$

17. Which electron transition in the hydrogen atom results in the emission of a photon of greatest energy?

a) $n = 4$ to $n = 2$

b) $n = 5$ to $n = 3$

c) $n = 2$ to $n = 4$

d) $n = 3$ to $n = 2$

18. Which part of an atom is most likely to be transferred as a body acquires a static electric charge?

a) electron

b) proton

c) neutron

d) positron

© It's About Time

19. A neutral atom must contain equal numbers of

a) protons and electrons only

b) protons, neutrons, and electrons

c) protons and neutrons only

d) electrons and neutrons only

20. A wool cloth becomes positively charged as it

a) loses protons

b) gains protons

c) loses electrons

d) gains electrons

21. One of two identical metal spheres has a charge of + q. The spheres are brought together and then separated. Compared to the total charge on the two spheres before contact, the total charge on the two spheres after contact is

a) less

b) the same

c) greater

22. Two identical conducting spheres carry charges of + 3 C and – 1 C, respectively. If the spheres are brought into contact and separated, the final charge on each will be

a) + 1 C

b) + 2 C

c) – 2 C

d) – 1 C

23. The Millikan oil-drop experiment showed the existence of

a) an elementary charge

b) an atom

c) a photon

d) an atomic nucleus

24. The coulomb is a unit of electrical

a) charge

b) potential

c) current

d) resistance

© It's About Time

25. The electrical force of attraction between two point charges if F. The charge on one of the objects is quadrupled and the charge on the other object is doubled. The new force is

a) 6 F

b) 1/2 F

c) 8 F

d) 2 F

26. Which procedure will double the force between two point charges?

a) doubling the distance between the charges

b) halving the distance between the charges

c) doubling the magnitude of one charge

d) halving the magnitude of one charge

27. As the distance between two point charges is tripled, the electrostatic force between the charge will become

a) 9 times as great

b) 3 times as great

c) 1/3 as great

d) 1/9 as great

28. Which particle has *no* charge?

a) alpha particle

b) proton

c) neutron

d) electron

[both numbers (super- and subscript in 31 – 33 should be on the left side of the chemical symbol]

29. The total number of nucleons in an atom of $^{8}B_5$ is

a) 5

b) 13

c) 8

d) 3

30. What is the total number of neutrons in the nucleus of an atom of $^{234}Pa_{91}$?

a) 91

b) 143

c) 234

d) 325

© It's About Time

31. In the equation below, what is particle Y?

$^{14}N_7 + {^4He_2} = {^{17}O_8} + Y$

a) a proton

b) a neutron

c) an alpha particle

d) an electron

32. Compared to electrostatic forces, nuclear forces are

a) weaker and of shorter range

b) stronger and of a longer range

c) stronger and of shorter range

d) weaker and of longer range

33. As two nuclei are moved closer together, the electrostatic force of repulsion between them

a) decreases

b) remains the same

c) increases

34. Atoms of different isotopes of the same element contain the same number of

a) neutrons, but a different number of protons

b) neutrons, but a different number of electrons

c) protons, but a different number of neutrons

d) electrons, but a different number of protons

[both numbers (super- and subscript in 37 – 38 should be on the left side of the chemical symbol]

35. Which pair correctly represents isotopes of the same element?

a) $^{210}Pb_{82}$ and $^{214}Pb_{82}$

b) $^{210}Pb_{82}$ and $^{210}Po_{84}$

c) $^{210}Pb_{84}$ and $^{210}Po_{84}$

d) $^{210}Pb_{82}$ and $^{210}Pb_{84}$

36. The following represents a nuclear equation.

$^{30}P_{15} \quad {^{210}Pb_{82}} + {^0X_{+1}}$

If $^{30}P_{15}$ has a half-life of 2.5 min, how long will it take for 16 g of the element to decay to 2.0 g?

a) 5.0 min

b) 7.5 min

c) 10.0 min

d) 2.5 min

© It's About Time

37. The half-life of an isotope is 14 days. How many days will it take 8 g of this isotope to decay to 1 g?

    a) 42

    b) 28

    c) 21

    d) 14

38. The splitting apart of a heavier nucleus to form lighter nuclei is called

    a) fission

    b) positron emission

    c) fusion

    d) beta emission

39. During nuclear fission, energy is released as a result of the

    a) splitting of light nuclei

    b) combining of light nuclei

    c) splitting of heavy nuclei

    d) combining of heavy nuclei

Alternative Chapter Assessment Answers

1. D
2. A
3. C
4. A
5. D
6. D
7. A
8. B
9. C
10. D
11. D
12. B
13. C
14. A
15. A
16. D
17. C
18. A
19. A
20. C
21. B
22. A
23. A
24. A
25. C
26. C
27. D
28. C
29. C
30. B
31. A
32. C
33. C
34. C
35. A
36. B
37. A
38. A
39. B

k© It's About Time

NOTES

© It's About Time

NOTES

© It's About Time

NOTES

© It's About Time

NOTES

© It's About Time

NOTES

© It's About Time